'THE GREAT TRIAL'

A SWALEDALE LEAD MINING DISPUTE IN THE COURT OF EXCHEQUER
1705–1708

The officers of the Yorkshire Archaeological Society would like to thank
the Trustees of the Elisabeth Exwood Memorial Trust for
a grant towards this publication and
their continuing support for the Record Series

THE YORKSHIRE
ARCHAEOLOGICAL SOCIETY
FOUNDED 1863 *INCORPORATED* 1893

RECORD SERIES

VOLUME CLXII

FOR THE YEAR 2011–12

'THE GREAT TRIAL'

A SWALEDALE LEAD MINING DISPUTE IN THE COURT OF EXCHEQUER 1705–1708

EDITED BY

TIM GATES

YORKSHIRE ARCHAEOLOGICAL SOCIETY

THE BOYDELL PRESS

2012

First published 2012

A publication of the Yorkshire Archaeological Society
in association with The Boydell Press
an imprint of Boydell & Brewer Ltd
PO Box 9, Woodbridge, Suffolk IP12 3DF, UK
and of Boydell & Brewer Inc.
668 Mt Hope Avenue, Rochester, NY 14620–2731, USA
website: www.boydellandbrewer.com

ISBN 978 1 903564 56 1

A CIP catalogue record for this book is available from the British Library

Papers used by Boydell & Brewer Ltd are natural, recyclable products made from wood grown in sustainable forests

MIX
Paper from responsible sources
FSC® C013604

Printed in Great Britain by
CPI Group (UK) Ltd, Croydon, CR0 4YY

To the people of Swaledale, past and present

Contents

Illustrations

Figures

Location Map

Acknowledgements

Many people have helped in the course of preparing this book and I am grateful to them all. My first and greatest debt is to my editor, Chris Webb of the Borthwick Institute, University of York, and to the committee of the Yorkshire Archaeological Society Record Series, for encouraging me to publish this work in book form. At all stages in this project I have received encouragement and timely advice from Mike Gill, Ian Spensley and Les Tyson of the Northern Mine Research Society, all of whom know a great deal more about the history of lead mining in Swaledale than I do. In the United States, I wish to thank Professor Hamilton Bryson, of the University of Richmond, Virginia, and Professor Henry Horwitz, of the University of Iowa, who patiently answered questions about the Court of Exchequer, its history and procedures. Thanks are also due to Professor Hamilton Bryson, Bernard Jennings and Chris Webb who read and made helpful comments on draft versions of the introduction, while Chris Webb also commented on my transcriptions of court documents. Chris Webb and Keith Walls gave invaluable assistance in checking and revising my translations of Latin documents. For their help and unfailing courtesy, I thank the staff of the following record offices and archives: Alnwick Castle Archives; the British Library; the Brotherton Library, University of Leeds; the Codrington Library, All Souls, Oxford; Durham Record Office; Gray's Inn Library and Archives; Lincoln's Inn Law Library; the London Metropolitan Archives; the National Archives at Kew; North Yorkshire County Record Office; and York Minster Library. Trevor Pearson kindly made a digital version of the location map. Finally, for friendship and hospitality on successive visits to Kew, I owe a special debt of thanks to Alan, Alice and Sally Ann Morris.

Preface

My interest in the dispute between Reginald Marriott and Thomas, Lord Wharton, over the Grinton lead mines in Swaledale dates back to 2007 when I bought a manuscript map bearing the title 'A Plan of the Manor of Grinton' (Fig. 1). The map in question was neither signed nor dated but, as enquiries soon showed, was made by order of the Court of Exchequer in 1708 in the context of the Marriott v. Wharton court case.[1] Showing the whole of the dale from Grinton upstream as far as Muker, it is therefore one of the earliest, if not the earliest, map of upper Swaledale in existence.

Investigations into the origin of this map inevitably led to the National Archives (TNA) where the official records of the Court of Exchequer are preserved. There, it was the witness depositions which first attracted my attention, not least because of the unexpectedly large number that was involved. For, taking the witnesses on both sides, there were more than three hundred depositions in total, occupying no fewer than thirty one large parchment membranes, each closely written on both sides. The range of material presented by these witnesses went far beyond lead mining and covered many other aspects of life in Swaledale as it was lived at the very beginning of the eighteenth century, and indeed in earlier years also. Although this valuable resource has occasionally been mined by historians of Swaledale, until now no attempt has been made to put the whole of it into the public domain. The possibility of doing this was first raised by Chris Webb of the Borthwick Institute at York University and his proposal was then enthusiastically endorsed by Dr Amanda Bevan, Principal Records Specialist, Legal Records, at TNA. Daunting though this task seemed, it has been undertaken in the hope that the result will prove useful to more than one potential readership.

As well as its intrinsic value as a source for the study of Swaledale's history, it is hoped that the Marriott v. Wharton case will also prove useful as an example of a lawsuit litigated in the Court of Exchequer. Indeed this may well be the first time that an attempt has been made to draw together all the records for a single case and to publish them in their entirety. In doing this I have received encouragement from Professors W.H. Bryson and Henry Horwitz whose books on the Court of Exchequer and its records have proved invaluable as guides to the procedures of the court and the whereabouts of specific classes of document in TNA.[2] Although almost 100,000 equity suits were litigated in the Court of Exchequer between its founding in the mid-sixteenth century and its final merging with Chancery in 1841, the records of the court have, with some exceptions, been relatively little used by historians.

1 For the sake of simplicity I refer here to the case as Marriott v. Wharton. However, as the original bill of complaint was in the name of the Attorney General, representing the interest of the Crown, the technically correct formula is Attorney General *ex relatione* Reginald Marriott v. Wharton.

2 Bryson 1975; Bryson 2001; Horwitz 2001.

References and Abbreviations

ACA	Alnwick Castle archives
Alumni Cantabrigiensis	*Alumni Cantabrigiensis*, 1922 Cambridge
Alumni Oxoniensis	*Alumni Oxoniensis*, 1891 London
Bailey 2001	J. Bailey, 2001, 'Voices in Court: lawyers' or litigants'?', *Historical Research*, 74, 392–408
Baker 1984	J.H. Baker, 1984, *The Order of Serjeants at Law*, Selden Society, Supplementary Series, vol. 5, London
Berry 1998	E.K. Berry, 1998, *Swaledale Wills and Inventories*, YAS, Record Series, vol. 152 (for 1995–6)
BL	British Library
Blackburn 1994	Alan Blackburn, 1994, 'Mining without laws: origins and practices of the Weardale moormasters', *Bulletin of the Peak District Mines Historical Society*, vol. 12, no. 3, Summer 1994, 69–75
BnL	Brotherton Library, University of Leeds
Bryson 1975	W.H. Bryson, 1975, *The Equity Side of the Exchequer*, Cambridge (reprinted 2008)
Bryson 2001	W.H. Bryson, 2001, *Cases Concerning Equity and the Courts of Equity 1550–1660*, 2 vols., Selden Society, vols. 117–18, London
Bryson 2007	W.H. Bryson, 2007, *Equity cases in the Court of Exchequer, 1660 to 1714*, Medieval and Renaissance Texts and Studies, vol. 313, Tempe, Arizona
Chambers 1996	C. Chambers, 1996, '"The most unconvincing testimony": the genesis and historical usefulness of the country depositions in Chancery', *The Seventeenth Century*, 11, 209–227
Chancery Case	1898 Chancery Case, Lyell and Vyner v. Broderick and Close, 1899. Evidence privately printed as *Rights of Shooting over Crackpot and Satron pastures: copies, translations and extracts from ancient documents*, vol. 1 (1899). NYCRO, ZYL, mic. 2688
Cooke and Horwitz 2000	J. Cooke and H. Horwitz, 2000, *Samples of Exchequer Equity Pleadings and Suits: 1685–86, 1734–35, 1784–85 and 1818–19*, List and Index Society, vol. 278
Cressey 1980	D. Cressey, 1980, *Literacy and the Social Order: Reading and Writing in Tudor and Stuart England*, Cambridge
CTB	*Calendar of Treasury Books … preserved in the Public Record Office*, prepared by W.A. Shaw, 32 vols., 1904–62, HMSO, London
DNB	*Dictionary of National Biography*, 2004 Oxford
EDD	*English Dialect Dictionary*, 1961, Oxford
Fieldhouse and Jennings 1978	S. Fieldhouse and B. Jennings, 1978, *A History of Richmond and Swaledale*, London
Fleming 1998	A. Fleming, 1998, *Swaledale: valley of the wild river*, Edinburgh

Gill 2004	M. Gill, 2004, *Swaledale: its mines and smelt mills*, Ashbourne, Derbyshire
Harrison 1879	G.H. de S.N.P. Harrison, 1879, *The History of Yorkshire, vol. I: The Wapentake of Gilling West*, London
Hartley and Ingilby 1986	M. Hartley and J. Ingilby, 1986, *Dales Memories*, Dalesman Books, Lancaster
History of Parliament 1983	B.D. Henning, 1983, *The History of Parliament: The House of Commons, 1660–1690*, Cambridge
History of Parliament 2002	E. Cruickshanks, S. Handley and D.W. Hayton, 2002, *The History of Parliament: The House of Commons, 1690–1715*, Cambridge
Horwitz 1999	H. Horwitz, 1999, 'Chancery's "Younger Sister": the Court of Exchequer and its Equity Jurisdiction, 1649–1841', *Historical Research*, 72, 160–82
Horwitz 2001	H. Horwitz, 2001, *Exchequer Equity Records and Proceedings, 1649–1841*, Public Record Office Handbook No. 32, Richmond, Surrey
LCL	Leeds City Library
LI	Lincoln's Inn Library
LMA	London Metropolitan Archives
Luttrell 1857	N. Luttrell, 1857, *Brief Historical Relation*, London
Milhous and Hume 1997	J. Milhous and R.D. Hume, 1997, 'Eighteenth-century Equity Lawsuits in the Court of Exchequer as a source of historical research', *Historical Research*, 70, 231–46
NYCRO	North Yorkshire County Record Office
ODD	*Oxford Dialect Dictionary*, Oxford
OED	*Oxford English Dictionary*, Oxford
Sainty 1983	J.C. Sainty, 1983, *A List of English Law Officers, King's Counsel and Holders of Patents of Precedence*, List and Index Society, Special Series, vol. 18
Spearman 1765	R. Spearman, 1765, *A Supplement to the Life of J. Hutchinson*, London
TNA	The National Archives
Tyson 1989	L.O. Tyson, 1989, *A History of the Manor and Lead Mines of Marrick, Swaledale*, British Mining no. 38, Northern Mine Research Society, Sheffield
Tyson 1995	L.O. Tyson, 1995, *The Arkengarthdale Mines*, British Mining no. 53, Northern Mine Research Society, Keighley
Tyson, Spensley and White 1995	L.O. Tyson, I.M. Spensley and R.F. White, 1995, *The Grinton Mines*, British Mining no. 51, Northern Mine Research Society, Keighley
Valor Ecclesiasticus	J. Caley and J. Hunter, eds., 1810–34, *Valor Ecclesiasticus*, London
VCH	*Victoria History of the Counties of England: Yorkshire, North Riding*, 1914, vol. 1, edited by W. Page, London
Wood 1999	A. Wood, 1999, *The Politics of Social Conflict: The Peak Country, 1520–1770*, Cambridge
YAS	Yorkshire Archaeological Society
YML	York Minster Library

INTRODUCTION

At stake in the contest between Wharton and Marriott was the right to mine for lead within the bounds of the manor of Grinton, 'the parish town of Swaledale'.[1] At the time of the dispute Reginald Marriott, the plaintiff, was employed as a deputy auditor in the Exchequer.[2] He claimed an exclusive right to mine for lead over 'all the territories and fields of Grinton Whitaside and Harkerside' by virtue of a lease granted to him by the Crown in 1696 in the name of a nominee, George Tushingham.[3] Marriott's adversary was Thomas, first Marquess of Wharton, a leading Whig politician and landowner who possessed large estates in Swaledale and neighbouring parts of Westmorland. These included the ancient manor of Healaugh, a small village situated two miles upstream of Grinton on the opposite (north) side of the river Swale (see the location map at the end of the book). Lord Wharton himself claimed the right to mine for lead on the moors and wastes of Harkerside and West Grinton, two hamlets of scattered houses on the south side of the Swale, which, he contended, belonged to Healaugh manor.[4] Lord Wharton's claim was disputed by Marriott on the grounds that the manor of Grinton referred to in his Crown lease encompassed all the moorlands south of the Swale from Grinton westwards as far as Haverdale Beck, including those of Harkerside and West Grinton.

The dispute between Marriott and Lord Wharton came to a head in September 1705 when Lord Wharton ordered his bailiff, Cornelius Fryer, to ride the customary boundaries of Harkerside and West Grinton. The riding, which was clearly intended as a public demonstration of Lord Wharton's claim of lordship over the disputed territory, took place on 26 September and was attended by at least sixty local inhabitants, both mounted and on foot. Two or three days later, workmen employed by Lord Wharton started to sink three new shafts near the summit of Grinton Greets at a place called the How Groves, less than one hundred yards from mines already being worked by Marriott. There can be little doubt that Lord Wharton's actions were intended both to forestall any attempt by Marriott to extend his mining operations westwards across the traditional boundary separating the moors of Harkerside and West Grinton from those of Grinton itself, and to assert his own ownership of the disputed territory.

The timing of Lord Wharton's intervention on Grinton How can most probably be explained by the dramatic rise in the value of lead raised by Marriott from the Grinton mines in the period 1702–4. For whereas the mines made a net loss

1 Grinton is described as 'the parish town of Swaledale' on both the Alnwick and NYCRO copies of the Grinton map.

2 For a brief summary of Marriott's career, see his profile in Appendix 4.

3 An English translation of the Latin lease is given in the Appendix 6(a). Some writers have been misled into thinking that Marriott obtained an earlier lease in 1660. That this was not the case is explicitly stated in a contemporary Treasury minute (*CTB*, vol. 11, 1933, p. 224). Before Marriott, the previous Crown lessee was Humphrey Wharton of Gillingwood who was granted a lease in May 1628 for a period of twenty-one years. Humphrey Wharton died in January 1635 and no new lease was granted until Marriott heard of the vacancy and took the opportunity to obtain one for himself in the name of a nominee, George Tushingham. It was the lack of a Crown lessee in the period 1650 to 1696 which enabled two local families, the Swales and the Hillarys, to exploit the mines unchallenged by the Crown. Marriott himself first took possession of the mines on 1 July 1697 (TNA LRRO 3/85).

4 The place-name 'West Grinton' was used in contradistinction to 'East Grinton', or simply 'Grinton', which refers to the 'town' proper.

of £173 in the year 1702, Marriott's accounts show that in the following year they returned a modest profit which rose to the startlingly high figure of £4093 in 1704 and thereafter remained high for the next five years.[5] From 1710 onwards, however, profits declined steeply and by 1715 the mines were again being run at a loss. This may help to explain Lord Wharton's apparent lack of interest in pursuing further legal action after the verdict went against him at the trial held in May 1708, even though his lawyers immediately gave notice of their intention to move for a retrial and/or an appeal to the House of Lords.

Nor was Lord Wharton the only person who showed an interest in the Grinton mines at this time. For, early in 1704, the Treasury received a petition from Ambrose Pudsey Esq. who offered the Crown £30 a year plus a one third share of what the mines would produce over the £30 if he were allowed to take over Marriott's lease. Pudsey's petition was immediately followed by a counter proposal from Marriott's father-in-law, Thomas Pulleine, who offered the vastly increased annual rent of £300, no doubt as a spoiling action intended to keep the lease in Marriott's hands. The Treasury responded by asking the Surveyor General of the Crown lands in Yorkshire to investigate and report on the value of the mines. The result was that Marriott retained the lease, though the Auditor was instructed to make certain that in future the Crown received the full value of its one tenth of the net profit.[6]

News of Marriott's windfall would doubtless soon have reached the ears of Lord Wharton and it was most probably this which spurred him on to try his luck on land immediately adjacent to Marriott's mines on Grinton How whilst taking care to remain on the right side of the boundary of Harkerside moor which, he argued, lay outside the frame of Marriott's lease. He may even have calculated that, even were Marriott to take legal action in defence of his lease, the ensuing court proceedings would be likely to drag on so long as to make the venture worth his while even if the court eventually found in Marriott's favour.[7]

It is also worth mentioning that Ambrose Pudsey's petition to the Treasury had been made on behalf of a third party, namely John Blackburne of Fryers Head, near Skipton, who was himself a partner in lead mines at Marrick, near Grinton, and therefore a potential rival to Marriott.[8] Four years later, in 1708, when the Marriott v. Wharton case finally came to trial, Blackburne was chosen to serve on the jury. By this time his interest in taking over the Grinton mines on his own account had evaporated and, as we shall see, he may well have been bribed in an attempt to influence his fellow jurors in Marriott's favour.

From Marriott's point of view, the action undertaken by Lord Wharton in September 1705 was both provocative and confrontational and his immediate response

5 Net annual profits for Marriott's Grinton mines in the years 1703–9 were as follows: £323 8s 2¼d (1703); £4093 11s 0d (1704); £3531 8s 4d (1705); £4904 11s 9d (1706); £4749 7s 9½d (1707); £4553 11s 9¼d (1708); and £3873 7s 6d (1709) – see TNA LRRO 3/85. The tonnage of smelted lead sold in the years 1700 to 1705 can be estimated from Marriott's accounts as follows: 1701 – 70 tons; 1702 – 26 tons; 1703 – 30 tons; 1704 – 805 tons; 1705 – 791 tons; 1706 – 1032 tons. Marriott's accounting year ended at Michaelmas (29 September).

6 For the involvement of Ambrose Pudsey Esq. see BnL, Special Collections, Marrick MSS, 5/4/90–99. For the Treasury's response see *CTB*, vol. 19, pp. 26, 127, 180, 316, 334, 447, 553. I am grateful to Les Tyson for drawing my attention to Ambrose Pudsey's petition.

7 At Grinton lead ore occurs mainly in the form of 'flots' or 'floats' – sheets of ore spreading out horizontally on either side of a vein. While these might be rich, they were quickly exhausted (Gill 2004, p. 120). Miners would certainly have known these facts and been aware that Marriott's recent good fortune at Grinton How was likely to be short-lived.

8 See Tyson 1989. For Blackburne's involvement with Ambrose Pudsey see BnL, Special Collections, Marrick MSS, 5/4/98.

was to file an affidavit asking the Court of Exchequer at Westminster for an injunction restraining Lord Wharton from any action that might disturb his, Marriott's, own mining operations. Granting Marriott's request, the court issued an injunction on 24 October 1705 confirming Marriott in such possession of the mines as he had at the time of the injunction and for three years beforehand. The day before the injunction was granted Marriott's lawyers exhibited a bill of complaint in the same court naming Lord Wharton and others, known and unknown, as defendants in the lawsuit.

As will already be apparent, the dispute revolved around a question of manorial boundaries and specifically whether the moors and wastes of Harkerside and West Grinton, where the disputed mines lay, belonged to the manor of Grinton or to the manor of Healaugh. This was not a matter that was easily resolved, as the post-Conquest history of the two manors was neither well documented nor understood. Also confusing was the fact that for different administrative purposes Harkerside and West Grinton on the one hand and Grinton on the other belonged to different and sometimes overlapping territorial units, namely the wapentakes of Gilling West and Hang West and the constableries of Reeth and Grinton. Lord Wharton also refers somewhat vaguely to a less well defined entity, the 'lordship of Swaledale',[9] which is mentioned in medieval documents and which he sometimes equates with Healaugh manor itself.

At the same time there was a degree of uncertainty as to the validity of Marriott's Crown lease as it applied to the territory in dispute. For the Crown's title to the lead mines of Grinton ultimately derived from rights formerly belonging to Bridlington priory which passed to the Crown at the Dissolution in 1537.[10] As these rights had been acquired piecemeal, as a result of successive grants of land made by Count Alan and his descendants in the eleventh and twelfth centuries, the precise extent of the priory's ownership was difficult to establish from documentary evidence. To complicate matters further, in 1599 the manor of Grinton had been sold by the Crown to two citizens of London, Richard Wiseman and Francis Fitch. Though the lead mines were reserved to the Crown, the wastes and moors on which the mines were situated were included in the sale, thus leaving unclear the status of any mines which might be opened there in the future. Subsequent piecemeal sales of land to the tenants, some of whom thereby became freeholders, coupled with errors perpetrated by lawyers in making out the title documents, confused the issue still further and left the ownership of the mineral rights of the manor of Grinton in some doubt.

At least three copies of the Crown lease granted to Marriott on 18 August 1696 have been preserved in TNA. One of these appears to be an early draft prepared by officials in the Exchequer while the other two are copies of the enrolled version, dated 16 December 1696.[11] As the lease makes clear, the lessee was expected to assert and defend the Crown's right to the mines, if necessary by taking legal action at his own expense. Perhaps as an incentive he was also empowered to recover, and

9 The phrase 'lordship of Swaledale' occurs five times in Lord Wharton's bill and seven times in the interrogatories put to his witnesses (e.g. numbers 2, 4, 6, 7, 15 and 18). At one point the bill refers to unspecified 'writings and ancient evidences' in which 'the lordship of Swaledale is sometimes called by the name of the lordship or mannor of Helaugh & at other times called by the name of the lordship or mannor of Reeth'. The clear intention here is to suggest that the terms 'lordship of Swaledale', 'mannor of Helaugh' and 'mannor of Reeth' are synonymous.

10 The *Valor Ecclesiasticus* valued the annual 'profits of the lands and tenements' (*exitus terrae et tenementum*) of Grinton at £20 (*Valor Ecclesiasticus*, vol. 5, p. 120, 1831). Interestingly, Grinton is not referred to here as a manor.

11 TNA E367/3645, LRRO 3/85 and LR 1/207. For an English translation of the lease see Appendix 6(a).

keep, the arrears and mesne profits arising from any lead mining that had been undertaken without authorisation of the Crown in the period since the expiry of Humphrey Wharton's lease on 25 March 1649. This explains why Marriott, acting in the name of Tushingham, his trustee, took legal action against one of the usurpers, Sir Solomon Swale, only a matter of months after he obtained his lease, though there is no evidence to show that any money was recovered from Sir Solomon as a result.[12]

As Horwitz has demonstrated, only a small percentage of equity cases litigated in the Exchequer progressed beyond the pleadings stage to the issue of a decree and even more rarely to a trial.[13] In both these respects the Marriott v. Wharton case is therefore somewhat atypical but at the same time more instructive as a case study. Another attractive aspect of the case is the very large number of witnesses who made depositions – 175 for Marriott and 142 for Lord Wharton. This already large body of testimony is further increased by the survival of no fewer than 34 affidavits in TNA, including several that were made by miners and other individuals who did not go on to make formal witness depositions. Unlike the other court documents, these affidavits include some compelling narrative accounts of the confrontations which took place at the mines on Grinton How between Marriott's agents who were attempting to serve the injunction and miners working at Lord Wharton's shafts.

Study of this unusually large body of evidence raises some interesting questions about the truthfulness of witnesses and the extent to which they may have been coached in advance of their examination so as to ensure that certain specific points of evidence were put before the court in the most favourable light. For example, on the fundamental questions of whether or not the river Swale was a boundary separating the manors of Healaugh and Grinton, or to which manor the hamlets of Harkerside and West Grinton belonged, the two opposing sets of witnesses gave consistent but mutually contradictory answers, thereby raising questions about the extent to which their testimony can be accepted as either disinterested or truthful. On this point it should be borne in mind that the great majority of Lord Wharton's witnesses were either tenants or freeholders on his Swaledale estate, and/or were employed in his mines, be it on a full time or part time basis. By contrast, most of Marriott's witnesses came either from Fremington or other parts of Swaledale outside Wharton's immediate sphere of influence, or else from Wensleydale, all areas where Marriott was a major employer of labour in his lead mines. In these circumstances, allegiance to a powerful landowner or mineral lord, born out of a condition of economic dependency, was likely to carry more weight with witnesses called upon to give testimony in a court case in whose outcome they had a clear economic interest than any moral obligation to tell 'objective' truth. Indeed it could hardly be otherwise, especially in this part of upper Swaledale where farm holdings were often too small to yield an income large enough to support a family, and where mining made a vital contribu-

12 For further details of this and other related court actions involving Sir Solomon Swale and others, see Tyson, Spensley and White 1995. That Marriott never succeeded in recovering any of the profits accruing from the period of vacancy is made clear by the following note appended by an Exchequer official to Marriott's accounts ending at Michaelmas 1706, viz. Marriott 'is not charged with the receipt of the mesne profits of the mines or premises from the year 1660 for that the said accomptant hath alleadged upon oath that the same hath [hitherto] been illeviable', TNA LRRO 3/85. What is not clear is why the writer has chosen the date 1660 as the start of the vacancy, rather than 1649 which was the expiry date of Humphrey Wharton's lease. For a brief period, in *c.*1672, the Duke of Bolton took a lease of the Grinton mines from the Crown in the mistaken belief that it was a fee farm rent. But on discovering that it was in fact an old expired lease rent, and illeviable, he surrendered it to the Crown. TNA LRRO 3/85.

13 Horwitz 2001, p. 31. Of a sample of 150 cases filed in 1685, a decree was issued in only fifteen instances.

tion to the local economy, if only on a seasonal basis. No less important, mining also provided welcome opportunities for the employment of both women and children who commonly worked on the surface washing and sorting the ore or operating the windlass, or else underground carrying ore from the 'forefield' or working face to the bottom of the shaft. Unlike in the Peak District, where a tradition of 'free mining' survived well into the eighteenth century, the right to mine for lead in Swaledale was the exclusive preserve either of the Crown or the great mineral lords and their lessees.[14] Consequently, a miner's ability to work depended entirely on the good will of powerful men who, more often than not, were also their feudal lords.

Court Proceedings and Documents

Pleadings

The Marriott v. Wharton case progressed through three distinct stages as described by Bryson and Horwitz.[15] The initial pleadings stage was followed by the gathering of proofs in the form of witness depositions and finally by a series of court hearings aimed at the determination of the suit. Hearings were conducted before one or more of the barons of the court, as the four judges in the Exchequer were styled. At the time of the Marriott v. Wharton case, the chief baron was Sir Edward Ward. He was assisted by three 'puisne' barons – Robert Price, John Smith and Thomas Bury. When, after a lengthy series of hearings, the suit had still not been resolved, the court ordered a trial at common law to be held at the bar of the court before a jury of twelve Yorkshire freeholders.

As mentioned above, the pleading stage of Marriott's suit was initiated on 25 October 1705 when Marriott and the Attorney General (representing the Crown's interest) exhibited their bill of complaint naming Thomas, Lord Wharton and thirteen others as defendants.[16] Because of the Crown's involvement in the case, the bill was technically known as an 'information' with both the Attorney General, Sir Edward Northey, and Marriott named as plaintiffs. Because the bill was exhibited by the Attorney General at the relation (*ex relatione*) of Marriott, he is referred to as the 'relator'.

In TNA, Marriott's bill and its accompanying answers are filed together in portfolio E112/915. The bill is endorsed with the date and regnal year (4Anne) in which it was filed, together with the county of origin of the suit and the number of the bill for that particular county and reign (EBOR 80), indicating that it was the eightieth bill received for Yorkshire in the year 1705.[17] The bill itself follows a standard format. It opens with a formal address to the Lord High Treasurer and Chancellor, who were the two most senior office holders in the Exchequer, and to the chief baron, Sir Edward Ward. Having established Marriott's status as lessee of the Grinton mines, and

14 Wood 1999.

15 Bryson 1975, pp. 93–159; Horwitz 2001, pp. 19–30.

16 On 23 October 1705 Marriott swore an affidavit alleging that Edward Hartley, Anthony Wharton and others unknown had disturbed him and his servants in their legitimate operation of the mines (E207/111/1). At a hearing the next day, the court issued an injunction granting Marriott peaceful possession of the mines. In 1705 the Michaelmas term began on 23 October, so Marriott's first court hearing took place at the earliest available opportunity. Order, E127/24, f. 178r; Bill, E112/915 80.

17 Bryson 1975, p. 104.

therefore a debtor of the Crown, there follows the stating part of the bill in which the facts of the case, as understood by the complainants, are presented. Here Marriott outlines his claim to all the mines and veins of lead in Grinton under his Crown lease, including those situated in the wastes and moors of Harkerside and Whitaside, and refers to the successful defence of his title in lawsuits previously undertaken against Sir Solomon Swale. There then follows his allegation against Lord Wharton and the other 'confederates' for their illegal entry onto Grinton How and disturbance of him in his rightful occupation of the mines. One purpose of the charge of confederacy 'with persons unknown' was to leave open the possibility of adding the names of other defendants to the bill at a later date, as indeed did happen. The bill ends with a request to the court for aid in settling the dispute by clarifying the respective boundaries of the manors of Grinton and Healaugh. To this end the court is asked to compel the defendants to produce any maps or written documents which might substantiate their claim to the mines. Meanwhile the court is asked for an injunction giving Marriott peaceful possession and working of his own mines for the time being. In order to give the court personal jurisdiction over the defendants so that the court can enter an order against them, the Attorney General asks that writs of subpoena be sent to the defendants requiring them to appear and answer the charges in the Court of Exchequer.

The court responded to Marriott's bill by issuing an injunction restraining the defendants and their agents from disturbing 'the relator' (Marriott) and confirming him 'in such quiet possession of the said mines as he had at the time of exhibiting the said bill and for three years before'.[18]

Twice, on 8 December 1705 and 9 May 1706, the court permitted the Attorney General and Marriott to amend their bill. On the second occasion the name of Philip Wharton, Lord Wharton's six year old son, was added as a defendant. As a result of these amendments interlineations were inserted into the text of the bill, one of which is the additional charge that Lord Wharton had offered to indemnify Marriott's workmen provided that they change sides and work for him.

The second stage of pleading consists of the answers made by the defendants to the complainants' bill. In this case answers were returned by three of the defendants – Lord Wharton, Philip Wharton (Lord Wharton's young son) and Sir Solomon Swale.

Lord Wharton's answer was received by the court on 4 May 1706. In it he states that his father, 'old' Lord Philip Wharton (d.1696), was for sixty years seised of the 'dominion lordshipp and mannour of Swaledale and Helaugh alias Helaw alias Ryeth' which, he asserts, contain the vills and townships of West Grinton and Harkerside.[19] As lord of divers demesne lands in these places, his father enjoyed certain royalties and

[18] Order, E127/24, f. 178r. A copy of the injunction, written in Latin, survives attached to an affidavit made by Edward Smithson on 2 June 1707 (E207/114/2). The order issued by the court on 24 October 1705 was for an injunction confirming Marriott in such possession of the mines as he had at the time of exhibiting his bill '*and* for three years before'. Yet the wording of the injunction was '*or* for three years before' which was not warranted by the order. This error was the cause of a good deal of concern until Lord Wharton himself waived the defect. Had he not done so the court doubted whether it could proceed to a trial. Another source of difficulty was uncertainty as to what possession Marriott could rightly be said to have in a vein which crossed into a neighbouring manor. On this point the barons judged that the situation in Swaledale was different from that pertaining in either Derbyshire or Cornwall where a tradition of 'free mining' existed, and the court concluded that Marriott could therefore have no possession of a vein where it crossed into an adjacent manor. These were some of the reasons why the injunction ultimately proved ineffective as a means of stopping Lord Wharton's miners from continuing to work (LI MS Misc. 536, Ward MSS, notebooks vol. 6, ff. 95v–96r; Bryson 2007).

[19] Lord Wharton's answer, p 7.

privileges, including rights of common over the moors and wastes of Harkerside and West Grinton. Likewise his father's tenants attended manor courts held at Healaugh where they performed the duty of suit and service. Lord Wharton also informs the court that before his, Thomas's, marriage, in 1673, his father settled the estate on Thomas as tenant for life with the remainder descending to Thomas's first and other sons in tail male and to their sons in turn. It was for this reason that it became necessary to add the name of Philip Wharton, Thomas Wharton's first and only son, to the Attorney General and Marriott's bill, as any decision made by the court affecting the estate would necessarily affect Philip's inheritance. The rest of Lord Wharton's answer consists of a point by point denial of the various assertions made by the Attorney General and Marriott in their bill. Amongst other things, he claims not to know of the existence of any such manor as the manor of Grinton, or East Grinton as it is otherwise referred to, or that King William III had a right to any mines in Whitaside, Harkerside or on Grinton How. With regard to the riding of the bounds of Harkerside and West Grinton in September 1705, he says that he only ordered this to be done after he had received information that Marriott was about to extend his mining operations on Grinton How beyond the boundaries of East Grinton and into the territories of Harkerside and West Grinton where he, Lord Wharton, claims the mining and other rights. He further asserts that all the land taken in at the riding of the bounds of Harkside and West Grinton belongs to his manor of Healaugh which, he says, is synonymous with his 'lordship of Swaledale'. Finally, he denies all unlawful combination or confederacy.

Annexed to Lord Wharton's answer is a schedule describing the boundaries of Harkerside and West Grinton as they were ridden on 26 September 1705. The schedule is endorsed with the names of sixty-four of those who were present, including Matthew Smales of Gilling, who was Lord Wharton's solicitor and local agent, and Cornelius Fryer, his steward or bailiff, who had given prior notice of the riding in Reeth market place and Grinton churchyard. One point to note is that no mention is made in the schedule of any *manorial* boundaries nor is the question addressed of to which, if any, manor Harkerside and West Grinton belonged.

The Attorney General, Sir Edward Northey, responded to Lord Wharton's answer by means of a replication, a formulaic third pleading whose purpose was to close off any further involvement by him, Sir Edward, in the proceedings.[20]

In his answer, dated 19 June 1706, Sir Solomon Swale questions whether the manor of Grinton as alleged in the bill is in fact a real manor at all and attempts to document the history of Bridlington priory's possessions by reference to various ancient royal grants and a lawsuit litigated in the Court of Exchequer in the time of James I, the records of which he alleges have been removed from their files and concealed or destroyed and so cannot be produced.[21] As for mines on the moors or wastes of Harkerside or West Grinton, he claims them as his own in his right as lord of a manor of West Grinton whose existence, he says, is proved by an eleventh-century grant (a document which had in fact already been shown to be a forgery). Admitting that Marriott had previously obtained verdicts against him in two lawsuits at York assizes, he says that Marriott only won because he had removed documents from the Court of Exchequer so denying him, Sir Solomon, the opportunity to make use of them in his defence. Now some of these same documents have again been concealed

20 See Bryson 1975, p. 124.

21 Such 'losses', whether caused by bureaucratic inefficiency or otherwise, were apparently not uncommon. See, for example, Horwitz 1999, p. 167.

or lost. He also denies that Marriott ever had quiet possession of the mines at the How Groves, he himself having brought four actions of ejectment against Marriott for possession of the same mines. Indeed, on one occasion, in Trinity term 1704, he succeeded in obtaining an order in Chancery for Marriott's ejectment but was prevented from appearing in court because, he alleges, Marriott had him kidnapped and imprisoned in the Fleet on a fraudulent writ. Unfortunately no more is said about this intriguing episode, nor can we be certain what, if any, truth there was in Sir Solomon's assertion that he was kidnapped at Marriott's instigation. Accordingly we should be cautious before accepting his account at face value.

As noted above, the name of Philip Wharton, Lord Thomas Wharton's son and heir, was only belatedly added to his father's bill of complaint, in May 1706, following Lord Wharton's disclosure that his estate was subject to an entail. When, a year later, no answer from Philip Wharton to Marriott's bill had been received, the court issued a commission of rebellion ordering his arrest for contempt.[22] This situation may have arisen either because Lord Wharton had not realised that a formal answer from his young son would be required once his name had been added to Marriott's bill as a defendant, or because process had not in fact been served on him. In any event, after the Serjeant at Arms reported that Philip Wharton could not be found, a commission was granted to take Philip Wharton's answers at his home in the country.[23]

As Philip Wharton was a minor, an answer to Marriott's bill was made on his behalf by his father who was also his legal guardian. The answer states Philip's interest as heir apparent to an estate in tail and begs the court to preserve his inheritance without loss. When Philip Wharton's answer was received by the court, Sir Simon Harcourt, who succeeded Sir Edward Northey as Attorney General in April 1707, replied with his replication, thereby bringing the pleading stage of these proceedings to a close.[24]

In Easter term 1706, at almost the same time as he submitted his answer to the Attorney General and Marriott's bill, Lord Wharton filed a cross bill in which Marriott, George Tushingham (Marriott's trustee), Sir Solomon Swale, and Toby West, vicar of Grinton, are, with others, named as defendants.[25] The effect of this was to make Lord Wharton the complainant in a separate but inter-linked suit. Thereafter the two suits proceeded together side by side in tandem.

Lord Wharton's bill is a lengthy document engrossed on four large sheets of parchment. One point to bear in mind is that the defendant (i.e. Marriott) was obliged to answer *seriatim* all the points made in the bill of complaint. Seen in this light, one objective of Lord Wharton's bill was to force Marriott to make a full declaration of the grounds on which his claim to the disputed mines was based. As Sir Edward Ward noted, it was a bill of discovery as well as for relief.[26]

In his bill, Lord Wharton sets out to substantiate his claim to a historically attested 'lordship of Swaledale', of which his manor of Healaugh formed a part. According to

22 Order, E127/26. f. 100v.

23 E112/915 80. The commission is dated 5 May 1707. Philip Wharton's answer was given at Winchendon where Lord Wharton had his main residence. It was witnessed by Laurence Kingford (vicar of Winchendon, 1682–1714) and Robert Meade. Philip Wharton's answer was delivered to the court by Robert Meade on 26 May 1707.

24 See note 20 above. The wording of Sir Simon Harcourt's replication is identical with that of his predecessor and has not been transcribed.

25 Tushingham's name is variously spelt 'Tussingham' or 'Tuchingham'. He signs his name 'Tushingham' on original orders in TNA class E131.

26 The bill and the answers which followed are filed in portfolio E112/915 EBOR 100. For Sir Edward Ward's comment see LI MS Misc. 536, Ward MSS, notebooks vol. 6, f. 97v; and Bryson 2007, p. 538.

Lord Wharton, the boundaries of this 'lordship' extended south of the Swale taking in the moors and wastes of Harkerside and West Grinton where the disputed mines lay. The bill also attempts to discredit Marriott's title to the lead mines of the manor of Grinton, or East Grinton as it is sometimes called, arguing that Grinton was described as a vill, not a manor, in medieval grants to Bridlington priory. As previously mentioned, the right of the Crown to the mines, and of Marriott as the Crown's lessee, derived ultimately from grants made by Count Alan and the Gant family to Bridlington priory in the twelfth and thirteenth centuries. Wharton challenges Marriott to produce these grants and other documents claiming that, were they to be produced, they would show that the moors and wastes of Harkerside and West Grinton were never conveyed to the priory and so could not have passed to the Crown at the Dissolution. Rather, Lord Wharton and his ancestors, as lords of Swaledale and the manor of Healaugh, and their tenants under them, have had continuous and uninterrupted use of the moors and wastes of Harkerside and West Grinton since time beyond the memory of man. The bill also repeatedly asserts that Marriott has colluded with others to conceal documents relevant to the case, including records of the Court of Exchequer to which, as an official of the Exchequer, he, Marriott, had privileged access. The missing documents include the bills and answers and witness depositions in an equity case litigated in the court in the early seventeenth century. That case was initiated by the incumbent vicar of Grinton, Henry Simpson, who tried, but failed, in his attempt to claim lordship over a manor of Grinton with the privilege of holding manor courts and of enclosing pieces of land taken in from its moors.[27] Wharton's bill further seeks to discredit Sir Solomon Swale's claim to be lord of a 'manor of West Grinton' alleging that it was based on forged documents which 'are much different in style and form from the deeds of that age in which they are pretended to be made' (an interesting example of palaeography and diplomatic in the service of a lawsuit). It also accuses the current vicar of Grinton, Toby West, of colluding with Marriott to conceal the boundaries of the various vills and townships in Grinton parish in return for gifts and in the expectation of financial reward should a future court case establish his right to tithes on any lead mined in the disputed territory.[28] The assistance of the court is requested in forcing Marriott to produce documentary evidence supporting his and the Crown's claim to the mines. Finally, the bill closes with a request that the court grant a commission to preserve the testimony of witnesses and issue an injunction to grant him, Lord Wharton, possession of the disputed mines.

Marriott's answer was received by the court on 24 May 1706. It consists of a point by point denial of the allegations contained in Lord Wharton's bill. He denies knowledge of any lordship or manor of Swaledale and instead says he intends to prove that West Grinton and Harkerside belong to the manor of Grinton which he himself claims. Nor does he know what, if any, grants of land in Grinton manor were made by the Gants to Bridlington priory. At the same time he insists that the whole manor, and the wastes and commons belonging to it, had belonged to the priory rather than just scattered parcels of enclosed land as alleged by Lord Wharton. Admitting that it was when he was an auditor in the Court of Exchequer that he discovered that the Grinton mines belonged to the Crown, he intends to use the records of the court to show that the Grinton mines had been subject to a succession of Crown leases from the time of Elizabeth I until the Civil War. Having also found that the Crown's

27 For the witness depositions in this case, dated 1618, see TNA E134/16Jas1/Hil12.

28 Elsewhere in the country, as for example in the parish of Stanhope in Weardale in the later seventeenth century, tithes on lead ore were payable to the vicar (Blackburn 1994).

ownership of the mines had subsequently been 'concealed', he took advantage of the opportunity to obtain a lease on his own account in the name of a trustee, George Tushingham.[29] Marriott denies knowledge of any lawsuit between Henry Simpson, vicar of Grinton, and the tenants of Grinton, or of any decree issued by the Court of Exchequer prohibiting Simpson from holding manor courts or enclosing parts of Grinton moor. Marriott concludes his answer by saying that he purchased the title to the manor of Grinton from the Hillarys and reiterating that he holds the Grinton mines by lease from the Crown.

The answer of Toby West, vicar of Grinton, to Lord Wharton's bill was given in Wensleydale on 6 November 1706. He states that he well knows the manor of Grinton, having lived there for thirty-three years, and says that both Marriott's and Lord Wharton's mines are situated within its bounds. He denies acting in partnership with Marriott or that he has any interest in the mines except in so far as he may be entitled to tithes on the lead produced. West's answer is accompanied by two schedules. The first describes the bounds of Grinton manor, taking in both Harkerside and Whitaside, while the second states that Marriott's mines lie on land which forms part of Grinton moor. Each schedule is endorsed by ten signatories, five of whom – Ralph Wensley, James Hird, Giles Metcalfe, Miles Close and Anthony Close – were present when the bounds of Harkerside and West Grinton were ridden in September 1705.[30]

George Tushingham's answer simply states that he was Marriott's clerk in the Court of Exchequer and acted only as his trustee in the matter of the lease which he afterwards assigned to Marriott by a deed dated 19 August 1696.[31]

Presumably because it was thought that he had not sufficiently answered some points contained in Lord Wharton's bill, Marriott made a second answer in June 1707 saying that he did not know what lands may have been granted to Bridlington priory by the Earl of Richmond or Walter de Gant. While admitting that he had indeed produced some records of the Court of Exchequer at York assizes during the trial of Sir Solomon Swale, he denies having had custody of them and states that they were afterwards returned safely to the court.[32]

Marriott's second answer concludes the pleading stage of Lord Wharton's suit.

The Collection of Evidence

Once the pleadings in both suits had been completed, the scene was set for the next stage of proceedings, namely the collection of evidence in the form of testimony given by witnesses on both sides. As all the witnesses lived at a considerable distance from London, the court appointed locally based commissioners to oversee the taking

29 Since before the time when Marriott obtained the lease, Tushingham was himself employed as a clerk in the court of Exchequer. In this capacity he acted as Marriott's sworn clerk during the Marriott v. Wharton case and it is for this reason that his signature appears on several orders issued by the court – see TNA E131/13.

30 For the schedules attached to Toby West's answer to Lord Wharton's bill, see p. 219. For the names of those who rode the boundaries for Lord Wharton in September 1705 see the schedule attached to Lord Wharton's answer to Marriott's bill, p. 13.

31 For George Tushingham's answer, see p. 221.

32 A draft affidavit made by the auditor in the Exchequer for the county of York, Henry Shales, in 1701, states that if Marriott had indeed removed records from the Exchequer for use at York assizes in 1696 he did so without authorisation and that no such records had been returned (TNA LRRO 3/85). This carefully worded affidavit, which was intended to be used by Sir Solomon Swale in a case against Marriott, stops short of an outright accusation against Marriott that he had actually removed documents without authorisation, whilst at the same time leaving open the possibility that he might have done so.

of their depositions. In this instance each of the contesting parties put forward a list of nominations from which four commissioners were chosen, two from each side. The commissioners appointed were Thomas Thomson, Thomas Gyll, Thomas Waite and Robert Hilton.[33] Power to administer the oath and take depositions from witnesses was conveyed to the commissioners by writs of *dedimus potestatem*. Written in Latin and following the standard format described by Bryson, two writs, one for each set of witness statements, were issued by the court on 2 July 1707. Both writs were to be returned with their respective depositions within one month of the feast of St Michael (29 September) following. According to the *liberatur*, or certificate of delivery, written on the *verso* of each writ, both sets of depositions were delivered to the court at Westminster on 25 October 1707.[34]

Before witnesses could be examined, counsel for each side prepared a set list of questions known as interrogatories which, by the rules of the court, had to be based on matter that had already been introduced in the bills and answers. At the same time, a set of counter-interrogatories was drawn up, some or all of which were to be answered by witnesses appearing for the opposite side. Each set of interrogatories and counter-interrogatories was sealed and delivered to the commissioners by the party's solicitor when he presented his witnesses at the time of their examination. The examinations were conducted in private, one witness at a time, with only the commissioners and a clerk or scrivener present. The commissioners then read out the relevant interrogatories to the witness whose answers were written down by the clerk, presumably in the form of notes or in shorthand. When all the witnesses had been examined, their answers were engrossed on parchment and delivered to the court at Westminster. In the Marriott v. Wharton case no single witness answered all the pre-prepared interrogatories nor, given the number of interrogatories involved, does it seem possible that all of them can have been read out to each and every witness. Most probably the solicitor representing the party whose witnesses were being examined gave the commissioners a list of the interrogatories that were to be answered by each individual witness.[35]

As mentioned above, one of the striking features of the Marriott v. Wharton case is the unusually large number of witnesses who gave testimony – 317 in all. The resulting answers number more than 1500 in total.[36] In view of this, the process of examination must have been a long drawn out affair and the time taken to administer the oath to each witness, read out the relevant interrogatories and record their answers must have been considerable. Although the steward or solicitor representing the party whose witnesses were being examined was not allowed to be present during the examination itself, he would very likely have been in attendance nearby, marshalling the witnesses and determining the order in which they were to appear. Prior to the examination there would have been ample opportunities to coach witnesses in their testimony and the temptation to do so must have been strong.[37]

Some idea of the importance of a particular interrogatory to the argument being advanced by one or other of the contesting parties can be gauged by the number

33 As Thomas Thomson and Thomas Gyll represented Lord Wharton at the view in April 1708 (see below), it is reasonable to suppose that they were also chosen by him or his solicitor, Mathew Smales, to act as commissioners. If so, Thomas Waite and Robert Hilton must have been chosen by Marriott.

34 Bryson 1975, p. 119, gives an example of the formula.

35 The procedure for taking depositions by commission is described in Bryson 1975, pp. 138f.

36 The depositions of witnesses appearing for Marriott are referenced in TNA as E134/6Anne/Mich 44 and those for Lord Wharton as E134/6Anne/Mich 38.

37 For the coaching of witnesses, see Chambers 1996, pp. 209–27.

of witnesses who responded to it.[38] For example, 169 (97%) of Marriott's witnesses gave answers to the second interrogatory asking to which manor Harkerside and West Grinton belonged and whether Lord Wharton or any of his ancestors had ever mined there for lead. Similarly, 125 (88%) of Lord Wharton's witnesses answered the fourth interrogatory drafted by his lawyers which asked whether the river Swale had ever been reputed to be the boundary of the manor of Healaugh.

Where an interrogatory was answered by just one or two witnesses, it is safe to assume that it had been especially drafted with those particular persons in mind in order to bring some particular piece of evidence before the court. Thus, for example, of the twenty-two interrogatories submitted by Wharton's lawyers, numbers nine and nineteen were each answered by one witness only, namely Cornelius Fryer in the first instance and Leonard Brackenbury in the second. Fryer, who was Lord Wharton's steward or bailiff, took the opportunity to put on record specific evidence about the yearly rents paid by Lord Wharton's tenants in Swaledale that he was best qualified to supply, while Brackenbury testified as to the accuracy of a map that he had made of the disputed territory.[39]

Marriott's witnesses were examined at Christopher Bowes' inn in Wensley, on Tuesday, 26 August 1707. While it is impossible to do full justice to the resulting mass of testimony in the context of this short summary, or mention more than a few of the incidents rehearsed by witnesses, some themes can usefully be picked out. For example, many of Marriott's witnesses described what they knew about the history of lead mining in the vicinity of Grinton and some recalled that as far back as 1583 the Grinton mines had been leased by the Crown to Henry, Lord Scrope, and Arthur Philips.[40] Others testified that in the early seventeenth century the mines were leased to Humphrey Wharton of Gilling and, in 1696, to Reginald Marriott.[41] After the Civil War, lead and coal were mined without permission of the Crown by two local families, the Swales of Swale Hall, and the Hillarys of Grinton. Three generations of Swales – 'Old' Sir Solomon Swale (d.1678), Sir Henry Swale (d.1682) and Sir Solomon Swale (d.1733) – and their numerous lessees, worked for lead on Harka Hill and Grinton How. By contrast the Hillarys, one of whose ancestors, Robert Hillary (d.1623), had purchased large swathes of land in Grinton manor from Henry Simpson, in 1622, worked mainly for coal on Grinton moor though they also re-worked the lead wastes from 'Old' Sir Solomon Swale's mines on Harka Hill.

The first of Marriott's witnesses to be examined was Benjamin Purchase of Middleham and it seems very likely that his placing at the head of the list was a deliberate tactic designed to emphasise the importance of his testimony and create an impression of authority. Nor, in this regard, is it likely to be a coincidence that Purchase was a wealthy individual of high standing in the local community with considerable experience of the management of lead mines in general and the Grinton mines in particular. Thus, for example, Purchase had acted as agent for John Bathhurst's mines in Arkengarthdale in the 1660s and '70s, and had also leased the lead mines on Grinton How and Harka Hill from 'Old' Sir Solomon Swale in partnership with four others. One of his partners in this latter enterprise was Philip Swale

38 Alphabetical lists of witnesses and the interrogatories answered by each are presented in Appendix 2, tables 1 and 2.

39 For the depositions of Cornelius Fryer and Leonard Brackenbury see pp. 344 and 355.

40 The history of successive Crown leases is more fully described in Tyson, Spensley and White 1995.

41 Humphrey Wharton (d.1635) was distantly related to the seventeenth-century ancestors of Thomas, Lord Wharton, the two branches of the family having split apart in the late fifteenth or early sixteenth centuries.

of Hartforth who at the time also happened to be Philip, Lord Wharton's, steward in which capacity he managed the Wharton lead mines in Swaledale. Purchase made the telling point that if Philip Swale had known of any legitimate claim that his master might have had to the mines of Grinton or Harkerside he, Philip Swale, would never have become involved in taking a lease of the mines on Grinton How on his own account.[42] Benjamin Purchase's testimony on this and other points was evidently considered particularly telling for he was one of the witnesses later called to London by Marriott to give evidence in person at the trial.[43]

Marriott's witnesses sought to establish that neither Thomas, Lord Wharton, nor any of his ancestors, had ever been in possession of lead mines on the moors of Grinton or Harkerside. Nor indeed had they laid claim to the mineral rights there before the boundaries of Harkerside and West Grinton were ridden in September 1705. By naming all the past Crown lessees, and describing the mining operations undertaken by the Swales and the Hillarys without official sanction, they hoped to show that the mines had been continuously worked by a succession of speculators since the mid-sixteenth century, and by Marriott himself from 1696 to the present, and that throughout this period no claim had ever been made by the Whartons. In this way they hoped to convince the court that Lord Wharton's claim was of no more than very recent origin and no doubt fuelled by rumours of the rich returns that Marriott had recently been getting from his mines on Grinton How.

Marriott's witnesses also tried to undermine Lord Wharton's claim that the straggling hamlets of Harkerside and West Grinton, with the moors and commons that belonged to them, formed part of the manor of Healaugh rather than Marriott's manor of Grinton. To this end some witnesses deposed that Lord Wharton's freehold and copyhold lands in Harkerside and West Grinton were of only minor value compared with other lands that belonged to the Crown or were formerly held by Bridlington priory. Also, they say, Lord Wharton owned only two of the two dozen or so 'fire houses' in West Grinton and Harkerside.[44] And while Lord Wharton's tenants did indeed attend his manor courts at Healaugh, where they performed the duty of suit and service, tenants of the Crown lands and lands formerly belonging to Bridlington priory appeared at the sheriff's turn held periodically at either Middleham or Askrigg in Wensleydale. If, as this evidence implies, Lord Wharton was no more than a minority landholder in Harkerside and West Grinton, then how could his claim to lordship of a manor which included these places be sustained?

The validity of Lord Wharton's claim to be lord of a manor that included Harkerside and West Grinton is likewise called into question by witnesses who rehearsed their knowledge of the manorial rights and customs that applied over the disputed territory. Thus witnesses deposed that for many years past waif and stray animals were taken up off the moors of Harkerside and West Grinton by bailiffs employed not by Lord Wharton but by the Hillarys and the Swales as the supposed lords of the manor of Grinton or, in the case of Sir Solomon Swale, of the (spurious) manor of 'West Grinton'. Similarly, witnesses assert that cattle were allowed to move freely between the commons of Harkerside and West Grinton on the one hand, and those of Grinton on the other, by means of a recognised cattle 'rake', again implying that Harkerside and West Grinton must belong to the manor of Grinton rather than Healaugh. Again,

42 Philip Swale's involvement in the lease was also mentioned by John Simpson, p. 108.

43 See below.

44 The term 'firehouse' also occurs in two sixteenth-century Swaledale wills (Berry 1998, nos. 16 and 126). It is presumably intended to distinguish an occupied dwellinghouse from a barn or other uninhabited building.

they say, the boundaries ridden by order of Lord Wharton in September 1705 were not manorial boundaries at all but existed only for the purpose of distinguishing those parts of the moors where the tenants of West Grinton and Harkerside had a customary right to take peat, turf, ling and stone for their own use.

Marriott's witnesses further insist that the river Swale had *always* been regarded as a boundary dividing the townships and settlements on the north side of the river from those on the south, not only in upper Swaledale but as far downstream as Richmond as well. To demonstrate this they rehearse a host of small incidents involving the recognised right of the inhabitants on either bank to fish only to the mid-stream of the river; disputes between landowners over the ownership of certain islands; the right to take stones from the river bed; and even the obligation devolving on the inhabitants of a particular township to pay the funeral charges when, as occasionally occurred, the body of a drowned person chanced to be washed up on their side of the river.

The examination of witnesses appearing for Lord Wharton took place on 29 August 1707 in the same inn at Wensley. Twenty-two interrogatories were submitted by Lord Wharton's attorneys who hoped to elicit testimony showing that Harkerside and West Grinton, and their moors and commons, belonged to his, Lord Wharton's, manor of Healaugh and not to the manor of Grinton, as claimed by the other side. In this they were not disappointed. Witness after witness asserted that the river Swale had *never* been reputed a boundary separating Harkerside and West Grinton on the south side of the river from Healaugh on the north. Also, they say, many tenants and freeholders of Harkerside and West Grinton attended the manor courts held by Lord Wharton at Healaugh where they performed suit and service. To demonstrate this, William Waller, one of Lord Wharton's stewards, placed the manor court records for Healaugh, dating back to the mid-sixteenth century, before the commissioners.[45] Numerous witnesses described the tenurial arrangements by which they held land in Harkerside and West Grinton. Whereas most tenants in Grinton held their property from the Crown on twenty-one-year leases, the majority of Lord Wharton's tenants in Harkerside and West Grinton paid copyhold or customary rents, consisting of a fixed annual or 'ancient' rent of eight pence per acre, plus various entry fines, or gressoms, paid on alienation or death of the tenant. The level of these entry fines varied according to whether it was 'old lands' or 'new lands' that were in question, the 'new lands' constituting that part of Healaugh manor which had been purchased by Sir Philip Wharton from the Molineux and Vachell families in 1635.[46] As part of his evidence, Cornelius Fryer, Lord Wharton's bailiff, presented a table showing the yearly rents paid by different categories of tenant within the manor of Healaugh.[47] This was intended to show that the rents paid by the inhabitants of Harkerside were comparable to those paid by Lord Wharton's tenants in other townships within Healaugh manor where his lordship was not in dispute. This being the case, the rents paid by the tenants of Harkerside and West Grinton could not simply be passed off as 'petty rents', as alleged in Marriott's bill, and so did not negate Lord Wharton's claim that Harkerside and West Grinton did indeed belong to his manor of Healaugh.

Lord Wharton's witnesses also describe some of the traditional practices which, they say, link Harkerside and West Grinton to Healaugh manor but set these hamlets

45 At the trial, Lord Wharton's counsel attempted to have several of these ancient manor rolls and books put before the jury, but the court refused to accept them as evidence – see the report by Baron Price in Bryson 2007, p. 541. Price does not explain why this evidence was refused.

46 See *VCH*, p. 241; Fieldhouse and Jennings 1978, chap. 6.

47 Deposition of Cornelius Frier, p. 344.

apart from Grinton. For example, the custom of 'gavelkind', or partible inheritance with one third reserved to the widow during her lifetime, applied in Harkerside and West Grinton but not in Grinton. Likewise, the tenants of Harkerside and West Grinton had the right to take peat, turf, ling and stones from their own moors and commons but not from the moors and commons belonging to Grinton, the boundary between the two being the same as that which was ridden on the orders of Lord Wharton in September 1705.

Witnesses also deposed that Lord Wharton, like his ancestors before him, had since time out of mind claimed and exercised in Harkerside and West Grinton those rights and privileges which might be expected to be enjoyed by them as lords of the manor of Healaugh. Hence their bailiffs took up waif and stray animals ('goods') on their behalf, releasing them only on payment of a fine. Again, no land could be taken in and enclosed from Harkerside moor without the Whartons' express permission, and several witnesses recall how a certain Mr Hutchinson of Gilling had been prevented by Lord Wharton's bailiff from making a new enclosure even though he had the prior agreement of the tenants to do so.[48] Another privilege claimed by the Whartons was that of free chase over the whole of Swaledale, including Harkerside and West Grinton. Consequently any red deer which happened to wander into the dale were reserved exclusively for the Whartons' table.[49] To deter poachers and 'preserve' game of all kinds, the Whartons employed gamekeepers and 'rangers', whose duties included the issuing of licences to anyone wanting to fish in the Swale. Those who fished without a licence were liable to be presented at Healaugh manor court where they could be fined and have their fishing gear confiscated. The Whartons also reserved the right to take wood from within the enclosed grounds of their copyhold tenants who might themselves only do so on payment of a fine known as a 'greenhew'.[50]

Like Healaugh, Harkerside and West Grinton were said to belong to the constablery of Reeth and not to the constablery of Grinton, and it was for this reason that stray animals were impounded in the pinfold at Reeth rather than the one at Grinton. By the same token, the service of watch and ward was performed in Reeth, and a yearly modus in lieu of tithe was paid there too, as well as other taxes. Witnesses also state that in former times there was a cattle 'rake', or path, leading between Reeth and Harkerside, and that the tenants of Reeth and Healaugh on the north side of the Swale enjoyed reciprocal grazing rights with the inhabitants of Harkerside and West Grinton on the south. At the same time they deny the existence of any such arrangement with respect to Grinton. Also, they say, the men of Reeth were entitled to take peat, turf and ling on a part of Harkerside moor known as 'Reeth peat moor'. In the opinion of Lord Wharton's witnesses, these were all reasons why Harkerside and West Grinton must be considered as belonging to the manor Healaugh and not to the manor of Grinton.

One striking aspect of the testimony of witnesses appearing for both Marriott and Lord Wharton is the long reach of their collective memory, even for incidents which at first sight might seem to be of quite trivial importance. For example, many of Lord Wharton's witnesses recalled an occasion some twenty or thirty years previously, when one Richard Harker, an inhabitant of East Grinton, pulled ling on Harkerside

48 See, for example, the testimony of Edward Chirry, p 316.

49 By the early eighteenth century red deer were becoming increasingly scarce in Swaledale and had disappeared altogether by mid-century, if not before (Hartley and Ingilby 1986, chap. 1).

50 Otherwise referred to as 'green hows', 'greeheugh', or 'greenhugh'. See the depositions of Richard Close, p. 232, William Waller, p. 245 and Mathew Alderson, p. 281.

moor without permission of the Harkerside tenants who, when they heard of it, took the ling away from him and gave it John Hird, a poor man of Harkerside, to repair his roof. Or again, James Galloway of Healaugh tells how his aunt, Elizabeth Forster, who had died sixteen years previously at the age of ninety-five, said to him that when she was a young girl her father used to drive his cattle across the Swale from Healaugh to graze on Harkerside moor where she herself also went to milk his ewes.[51] In this instance the events referred to must have taken place around a century earlier, in the first decade of the seventeenth century.[52]

These often vivid witness statements echo Professor Bryson's observation that 'each set [of depositions] gives an incident or vignette of life in England; ... Here is a fertile field from which can grow many things besides stubble and thorns.'[53]

Injunction and Affidavits

The progress of the case can most conveniently be followed by reference to the minutes of court hearings recorded in the common minute books and to the orders of the court as transcribed in the entry books of orders.[54] Although these formal records give a comprehensive account of proceedings in court, they afford no more than occasional glimpses of events as they unfolded on the ground at Grinton. Fortunately, however, this deficiency is to a great extent made good by the numerous affidavits which have been preserved in the 'bille', a miscellaneous class of document which includes not only affidavits but much other material relating to both the equity and revenue sides of the Exchequer.[55]

The affidavits in this case fall into two main groups. The first comprises statements of purported fact offered by individuals many of whom later went on to make witness depositions for one party or the other. As the contents of these affidavits are to a great extent duplicated in the witness depositions, we do not need to concern ourselves further with them here. The affidavits in the second group are, by contrast, more illuminating as they consist of complaints made to the court by Marriott's agents who were given the task of enforcing the injunction issued by the court in October 1705. To this end several men, most if not all of whom were employed by Marriott in his own mining enterprises, made repeated visits to the mines on Grinton How from the spring of 1706 onwards in order to serve the injunction on Lord Wharton's miners. This brought them into direct, and sometimes violent, confrontation with Wharton's men who stubbornly resisted all attempts to stop them working and at the same time made plain their contempt for those who served the injunction.

As we have seen, the injunction ordered by the court on 24 October 1705 was to confirm Marriott in such possession of the mines as he had at the time of the exhibiting of his bill and for three years before.[56] As Marriott himself made no claim to have been in possession of any mines within the bounds of Harkerside or West Grinton in the preceding three years, Lord Wharton's lawyers argued that he could therefore have no possession of any of the shafts newly opened by Lord Wharton within the previous few weeks. Consequently the injunction did not forbid Lord

[51] See the deposition of James Galloway, p. 305.

[52] Attention has also been drawn to the length of the collective memory by Wood 1999, p. 131.

[53] Bryson 1975, p. 173.

[54] In TNA the common minute books are classed as E161 and the entry books of orders as E127. The original orders are in class E131. These and other documents are calendared in Appendix 1.

[55] TNA E207. See Horwitz 2001, p. 65.

[56] See footnote 18 above.

Wharton's miners from continuing to work these shafts. In any case, as previously noted, the court had already decided that Marriott could have no possession of the unworked part of a vein where it extended into the territory of an adjacent manor. Therefore, until the status of Harkerside and West Grinton as part of either Grinton or Healaugh manors could be clarified, the court was of the opinion that Marriott had no claim on that part of the vein being worked by Lord Wharton. Lord Wharton's miners were therefore free to continue working provided that they refrained from any action which might adversely affect Marriott's own operations.[57]

Whether intentionally or otherwise, these important qualifications seem not to have been made clear by Marriott's representatives when they arrived at Grinton How ready to serve the injunction in April 1706. On the contrary, the miners seem to have been given the impression that the injunction was to stop them from working altogether. In any case, even those miners who were literate could hardly be expected to read or understand it for themselves since the injunction was written in Latin. On the other hand even if they were not equipped to debate its precise meaning, the miners may well have been better informed as to the limitations of the injunction than those who were attempting to serve it had imagined.[58] Either way Marriott's representatives met with a uniformly hostile reception when they appeared at the mines and any expectations that the injunction would prove effective in preventing Lord Wharton's miners from continuing to work were soon dispelled.

The injunctions were served by three men, all of whom were probably Marriott's own employees, namely Matthew Sisson of Ripon, William Bodinson of Middleham, and Jacob Baynes of Grewelthorpe.[59] They in turn were assisted by two local men, Henry Place, a miner and a resident of Grinton, and Thomas Pratt, who is described as a yeoman of Grinton. Place and Pratt appear to have acted as informers, identifying by name those who continued to work in apparent defiance of the injunction. According to affidavits made by all five of Marriott's agents, they personally handed out copies of the injunction to more than three dozen of Lord Wharton's miners. The recipients included James Clerkson, who is described as Lord Wharton's steward or overman, and Robert Robinson, a mason, who was in the process of building a smith's shop close to the newly opened shafts. According to Henry Place, Clerkson responded to the injunction by saying that 'it was the old thing over agayne' and that 'they did not vallue it' and by telling the other workmen to take no notice of the injunction and to continue working. For his part, James Baynes reported Robinson as saying that 'he valued not their papers and if Lord Wharton had another house to build there, he would doe it, or begin on it tomorrow', while Clerkson had said 'they cared not a farthing for the injunction'. More ominously, both Place and Sisson stated that one of the miners, John Sheldon, had fired a 'gunn or foweling piece' in their direction as they rode their horses back down the hill to Harkerside and that he then ran after them, pursuing them all the way to the village where 'he said he would

57 LI MS Misc. 536, Ward MSS, notebooks vol. 6, f. 96v; Bryson 2007, p. 537.

58 See footnote 18 above.

59 It should not be assumed that any of these men were neutral parties in the dispute as they all probably had a vested interest in the outcome from an early stage. At any rate when Marriott finally won control of the mines on Grinton How, William Bodinson was employed as the purser and Roger Baynes, who was presumably a relative of Jacob Baynes, was appointed to the post of barmaster – see Tyson, Spensley and White 1995, p. 20. Roger Baynes was Marriott's solicitor, acting for him not only in his case against Lord Wharton but also in an earlier case against Sir Solomon Swale, in June 1697. It is for this reason that his name appears on writs of *dedimus potestatem* for the taking of witness depositions in both actions. TNA E134/9Wm3/Trin 18 and E134/6Anne/Mich 38.

either fight or else doe any exercise with them or any man or servantt Mr Marriott had'.[60] At this, Sisson and Place, fearing for their safety, turned and rode away.

Bodinson, Sisson, Baynes, Place and Pratt also alleged that the miners had not only shown disrespect to the injunction, and by extension to the court, but that they had also disturbed Marriott in the legitimate operation of his mines by stopping up and diverting the channels or leats which carried water to his works. Moreover, by cutting 'the greate springs of water within [the] ground and haveing noe draine to carry the said water [they] will unavoidably drowne the said works'.[61] If true, this would surely be a clear breach of the injunction which explicitly forbade any interference with Marriott's working of his own mines.[62] Accepting this version of events, the court reacted by ordering the arrest of nineteen named miners, including Clerkson, Robinson and Sheldon, for breach of the injunction, noting that some had also spoken 'slightly and disrespectfully' of it.[63]

Before any arrests could be made, however, the court received affidavits from several of the miners flatly contradicting the version of events given by Bodinson and his associates. The miners informed the court that they had not treated the injunction with disrespect and gave assurances that they fully intended to comply with it in future. For his part, John Sheldon denied that it was he who had fired the 'small fowling peice', adding that in any case there was 'onely a little powder' in it and that Sisson and Place were over two hundred yards away when it was discharged by another man. He also denied having threatened anyone. On the contrary, he alleged that Sisson 'did in his expressions boast himselfe of severall lewd acts done by himselfe and challenged [him] to such beastly actions as were not fitt for any Christian to name'.[64]

Accepting these assurances of future compliance, the court set aside the order for the miners' arrest. This was not the last time, however, that Lord Wharton's miners appeared to defy orders of the court with seeming impunity or that Marriott's agents found themselves impotent in the face of the miners' determination to continue working. In practice, the claim made by several of the miners that Lord Wharton would 'render them harmless', i.e. keep them safe from prosecution, proved to be well founded.

Injunction or no injunction, Lord Wharton's miners were still working on Grinton How in May 1707 when Edward Smithson of Richmond, attended by William Bodinson, returned to the mines and again served injunctions on eleven miners, eight of whom had previously been served in the spring of 1706 and who, under threat of arrest, had promised obedience to the court. According to Smithson, this time the miners ridiculed the injunctions, pretending they were ballads, and said 'they would get new tunes for them, and sing them'. Bodinson added that several of the miners 'made a jest thereof calling the copies of the said injuncions deliver'd to

60 Affidavits of Henry Place, Jacob Baynes and Matthew Sisson, 18 May 1706, E207/112/1.

61 Affidavit of William Bodinson, 18 May 1706, E207/112/1.

62 Counsel for Lord Wharton argued that as Marriott had had no legal possession of the disputed territory in the three years before the injunction was granted, he could have no legal possession of any leats that had been dug there during this period. Therefore Lord Wharton's miners could not be held to be in breach of the injunction if they filled them in. LI MS Misc. 536, Ward, notebooks vol. 6, f. 96r.

63 Order, E127/25, f. 156r. In his notes dated 28 May 1706, Sir Edward Ward recorded the names of three miners – Robinson, Williamson and Sheldon – underlining the words 'said valued not the injunction' after each name. Evidently he regarded the implied disrespect shown to the court as a particularly serious matter and possibly more serious than their other alleged actions. LI MS Misc. 536, Ward MSS, notebooks vol. 6, f. 95v.

64 Affidavit of John Sheldon, 1 June 1706, E207/112/2.

them ballads and the old bable over again and particularly the said Thomas Harrison to make himselfe & the rest sport with tooke the copy thereof which was delivered to him & sung parte of an old song upon it & said if he had but had his other eyes (meaning his spectacles as this deponent apprehended) he wo'd have sung them it all'.[65] Given that at least half the miners were probably illiterate, and so would have been unable to read the injunction even had it been written in English, their reaction to these proceedings seems all too predictable.[66] Particularly as on this occasion neither Smithson nor Bodinson alleged that any interference with Marriott's own mines was taking place, the miners were probably aware that they were not infringing the injunction simply by continuing to work.

Be that as it may, when the court read Smithson and Bodinson's account of these proceedings, on 16 July, it ordered the arrest of those who were alleged to have broken the injunction and used 'foul and disrespectfull language' against it, unless cause could be shown to the contrary.[67] No arrests were actually made, however, and on 21 November the order was stayed until a further court hearing could be arranged. Meanwhile work at the mines went on as before.

The ineffectiveness of the injunction as a means of stopping Lord Wharton's men from working, and the contradictory affidavits made on both sides, illustrate some of the difficulties faced by a court located far from the scene of the dispute. Equally, a close reading of the evidence shows that the court went out of its way to act even-handedly, without favouring either one side or the other. At the same time it is abundantly clear that Sir Edward Ward was in no doubt as to the importance of the issue at stake, noting 'It is the right of the Crown & to a very great value that is in question in the cause & possibly the greatest the court hath at present'.[68]

Decree

The witness depositions on both sides were formally published in November 1707, thereby allowing the lawyers for each party to read the statements made by their opponents' witnesses for the first time.[69] Then, after further hearings had failed to bring about a resolution of the dispute, the court issued a decree on 25 February 1708.[70] After summarising the substantive points made by both sides in their pleadings, the decree ordered that a trial be held at common law before a jury of Yorkshire freeholders at the bar of the court in Westminster. The date of the trial was set for 7 May. Meanwhile, the court ordered maps of the disputed territory to be made by two surveyors, one nominated by each of the contesting parties. The completed maps were to be shown to members of the jury at a view of the disputed ground scheduled to take place three weeks before the trial, on 22 April. At the view, the jury was to be shown the mines by four members of the local gentry, Sir Hugh Smithson and Thomas Dickonson representing Marriott, and Thomas Gyll and Thomas Thomson

65 Affidavits of Edward Smithson and William Bodinson, E207/114/2. Unfortunately we are not told what the 'old song' was.

66 The tentative suggestion that up to 50% of the miners may have been illiterate is based on a sample of forty individuals who signed affidavits, of whom twenty did not sign their name fully, using either a letter or a simple mark instead of a signature. It is admitted, however, that such evidence cannot be conclusive. For further discussion on this topic, see Cressey 1980.

67 Order, E127/26, f. 242r.

68 LI MS Misc. 536, Ward MSS, notebooks vol. 6, f. 120r.

69 Order, E127/26, f. 309v. For the 'publication' of witness depositions see Bryson 1975, pp. 141–2.

70 Decree, E126/19, ff. 13v–14r.

representing Lord Wharton.[71] The purpose of the view was to familiarise the jury with the location of the mines in relation to established boundary marks and no evidence was to be presented to them at this time.

Maps and Plans

The decree issued by the court ordered that 'John Hutchinson a surveyor named by the relator and Leonard Brackenbury another surveyor named by the defendant the Earl of Wharton doe make a mapp of the whole estate in question with all the known places or markes but without setting forth the claimes of either party'.[72] John Hutchinson, who was born in Spennithorne, near Middleham, in North Yorkshire, later became a well known writer on theology and natural philosophy. But the early part of his career was spent working as a steward on a number of northern estates, including those of Charles Bathurst and the Duke of Somerset. It was in the course of this work that he acquired skill as a surveyor and also developed a keen interest in geology, which led him to make original observations on lead mining and the formation of ore deposits.[73] By contrast, little is known about Brackenbury's career as a surveyor and map maker, though a letter written by Lord Wharton to his solicitor, Matthew Smales, in July 1703, makes it clear that Brackenbury had previously been employed by Lord Wharton to make a map and a three dimensional model in the context of at least one other lead mining dispute.[74]

The respective roles played by Hutchinson and Brackenbury in making the maps ordered by the court have yet to be clarified. Nor do we know for certain what map was shown to Leonard Brackenbury when he gave his evidence before the commissioners in August 1707, six months before the court issued its decree. All we can be sure of is that Brackenbury had already made a map by this time, as he explicitly says so in his deposition, adding that he had taken great care to settle the boundaries of the moors and enclosed fields according to the best information given to him by local inhabitants.[75] Since Brackenbury made no mention of John Hutchinson at the time of his examination, it seems that he cannot have been involved in making whatever map it was that was shown to Brackenbury then and which he claimed as his own work.

As mentioned at the beginning of this introduction, a manuscript map entitled 'A Plan of the Manor of Grinton' was offered for sale in 2007 and bought by the

[71] Decree, E126/19, f. 13v. Sir Hugh Smithson of Stanwick (1657–1733) was the grandfather of the first Duke of Northumberland and his involvement presumably explains how a copy of the Grinton map comes to be in the Alnwick Castle archives. Thomas Dickonson (not Dickenson, as his name is spelled in the decree) was an alderman of Richmond, who had earlier acted as an intermediary between Marriott and John Blackburne over the purchase of the Marrick priory lands. Thomas Gyll, was one of the commissioners who took the depositions of witnesses in August 1707. He can be identified as Thomas Gyll of Barton, near Darlington, whose signature on the depositions matches that on various deeds and other documents dated between 1712 and 1725 (NYCRO, Kiplin Hall papers, ZBL/1/1/363–368, 373, 388). The identity of Thomas Thomson remains uncertain.

[72] Decree, E126/19, f. 14r.

[73] *DNB*; Spearman 1765.

[74] Letter from Lord Wharton to Mathew Smales – YML, Hailstone Collection, Box 4.32. The dispute in question was over the boundaries of Wetshaw, at the head of Bleaberry Gill in Arkengarthdale, which had been running since at least 1675 – see Tyson 1995, pp. 18–26. Original sketch surveys of the disputed territory at Wetshaw survive, one in the NYCRO, dated 1710 (ZX/9820), and another in the Northumberland Record Office (3410/Plans/1/32). Though neither is signed, either or both could be the work of Leonard Brackenbury.

[75] For Leonard Brackenbury's deposition see p. 355.

writer.[76] As the map bore no date or signature, there was no immediate means of identifying the mapmaker or establishing the purpose for which it had been made. In the hope of answering these questions, and providing a context for the map, a search was made of the Manorial Documents Register on TNA website at the suggestion of Chris Webb. By a stroke of good luck this immediately revealed the existence of a map with precisely the same title in the archives at Alnwick Castle.[77] When, soon afterwards, the two maps were compared side by side, it was found that they bore exactly the same legend and were likewise identical in scale and content. On the other hand, they were quite dissimilar in appearance, as different colour schemes and means of portraying relief had been employed in each case. While it was clear that the two maps must be derived from one and the same survey, they had evidently been drawn and coloured by different hands. Most important of all, the Alnwick map bore the additional inscription 'Surveyd by John Hutchinson April 1708' (Figs. 2 and 3).

The two copies of Hutchinson's map are reproduced here as Figures 1 and 2. The towns of Grinton and Reeth and all the smaller settlements in Swaledale, from Cogden upstream to Muker, are carefully depicted together with the boundaries of their respective moors and commons. The bounder marks of Grinton manor, as rehearsed by the vicar, Toby West, and many other of Marriott's witnesses, are individually identified by name. Starting in the east, the boundary runs uphill from the river Swale to Stallerstone Stile, near Cogden, and then to Snowden Man, before turning westwards along the watershed separating Swaledale and Wensleydale which forms the southern limit of the map.[78] At Summer Lodge Tarn the boundary turns north down Bloody Vale to Haverdale Beck and so back to the Swale.

Within the territory covered by the map, 184 parcels of enclosed land are colour-coded to indicate different categories of ownership. Thus, freehold land, customary or copyhold land, land belonging to the Crown and land formerly held by Bridlington priory are respectively coloured brown, green, red and yellow. 'Abbey Land', coloured orange and lying to the west of Oxnop Gill, represents territory that was once owned by Rievaulx abbey. Presumably the map was originally accompanied by a field book or terrier listing the acreage and the name of the occupier for each numbered enclosure but, unfortunately, no such document has yet come to light.

One reason why it was considered important to show these different categories of land was because Lord Wharton claimed to be the owner of a significant acreage of enclosed land in Harkerside, some of which, referred to by witnesses as 'my lord's own land', was said to be demesne land belonging to his manor of Healaugh. The fact that the Crown's title to the lead mining rights of Grinton derived, at least in part, from land confiscated from Bridlington priory at the Dissolution, explains why it was thought important to show the extent of 'Priory land'. Other land, shown as 'Crown land', may also have formed part of the priory estate before it became progressively more fragmented after the Dissolution.

Lead mines or shafts are represented conventionally on the map by hollow squares. Four groups of shafts are depicted – at Green Hills, near Whitaside; on Harkay (now Harka) Hill, at the head of Grove Beck; on Grinton How; and at Devis Hole, beside Cogden Gill. 'Hushes' are also shown in two places, at Green Hills and at the western end of Harkay Hill (Fig. 3). The depiction of hushes is especially interesting for hushing, a process by which water is channelled down a vein in order to strip away

76 The map is now in the NYCRO collection, reference M98.

77 ACA, ref. X. II. 6 Box 34.

78 Instead of the term 'watershed', the evocative phrase 'as heaven water deales' was in common use at this time.

Fig. 1. The North Yorkshire County Record Office copy of the Grinton map. NYCRO, M 98 (© North Yorkshire County Record Office).

Fig. 2. The Alnwick copy of the Grinton map inscribed 'Surveyd by John Hutchinson April 1708'. ACA, X.II.6. box 34 (© Collection of the Duke of Northumberland, Alnwick Castle).

Fig. 3. Detail of Fig. 2 showing the disputed lead mines and the township boundaries. Note the depiction of 'Hushes' on 'Harkay Hill'.

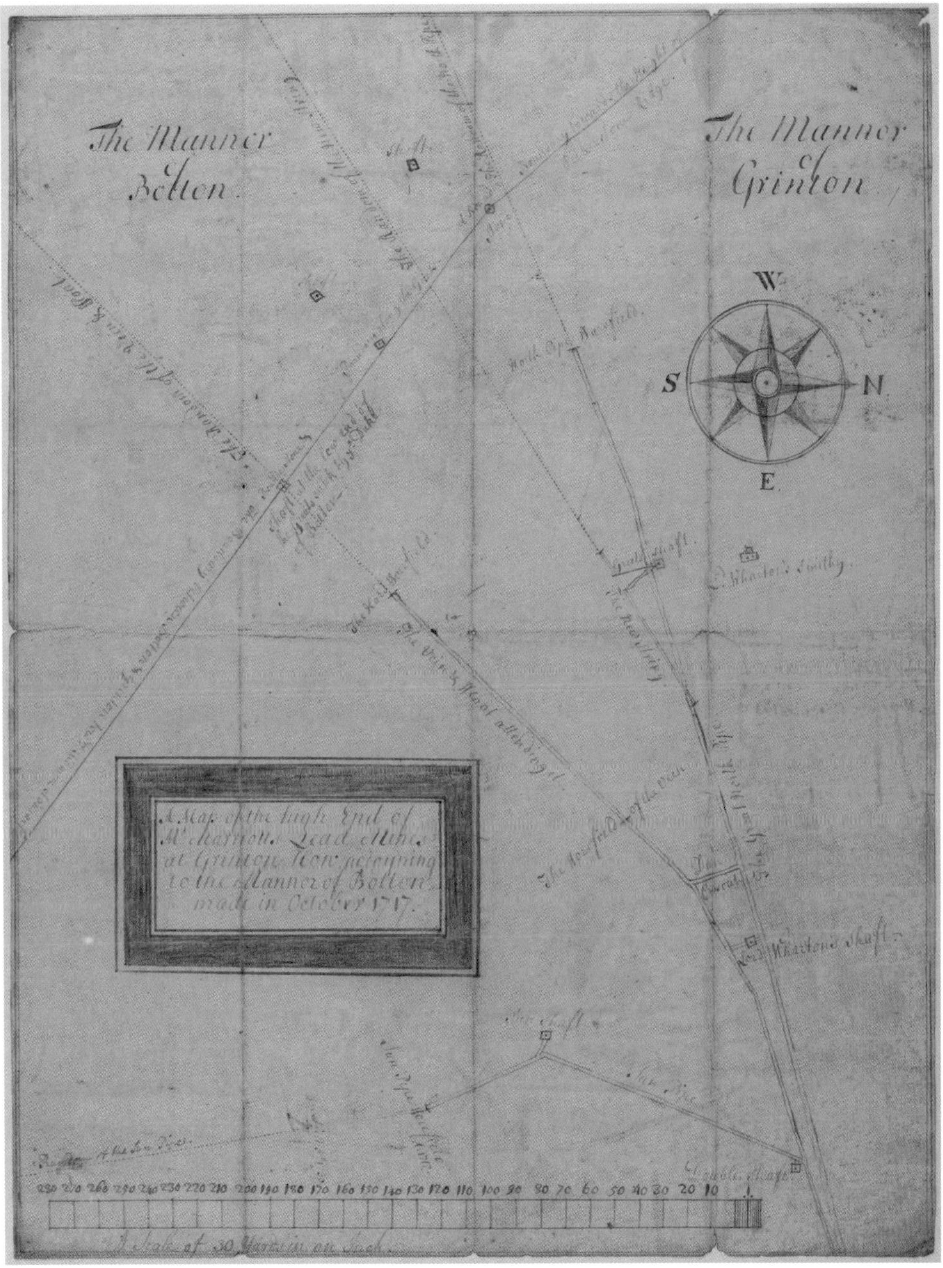

Fig. 4. A plan by an unknown surveyor showing the disposition of the shafts on Grinton How in October 1717. 'Ld. Wharton's smithy' is shown just below the compass rose and 'Lord Wharton's shaft' at lower right. NYCRO, ZBO L 1/2 (© North Yorkshire County Record Office).

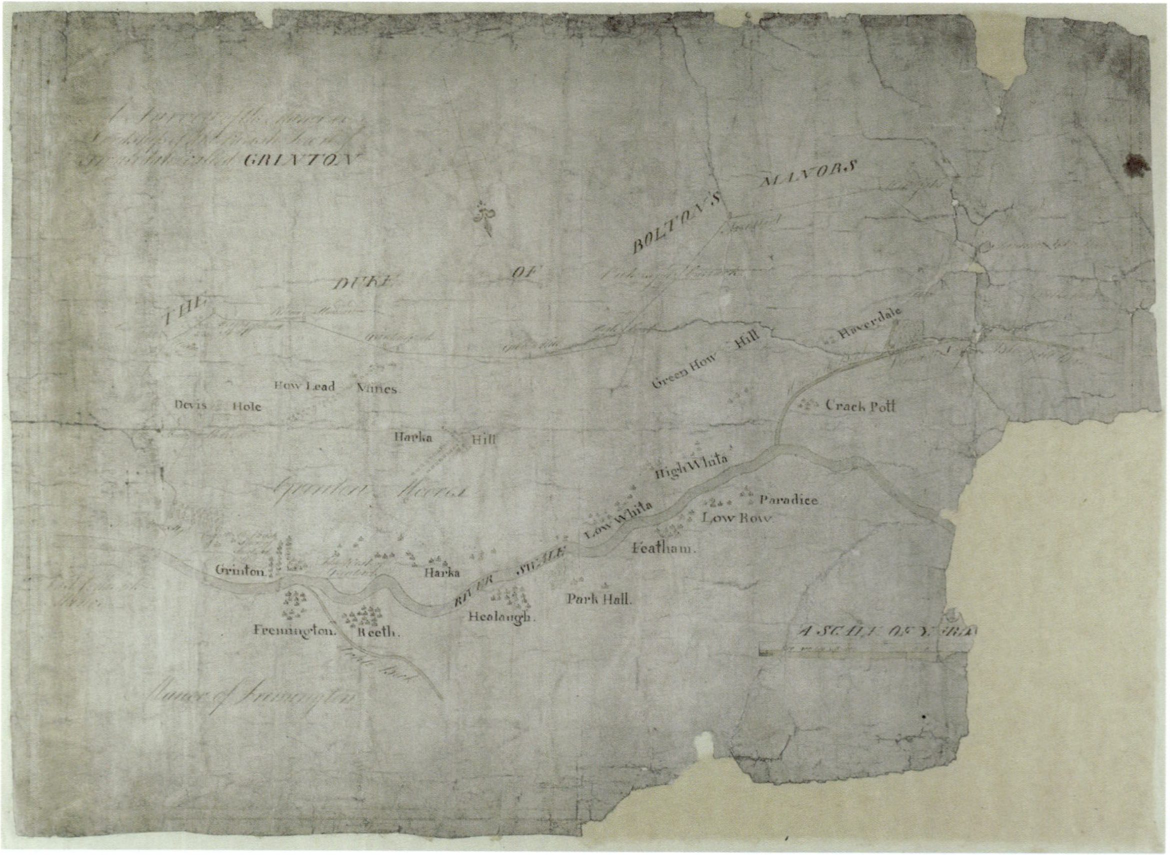

Fig. 5. *An undated plan of Grinton manor showing the lead mines on Grinton How, which may well be related to the dispute.* NYCRO, ZBO L 1/1 (© North Yorkshire County Record Office).

Fig. 6. Detail of Fig. 5. Note the buildings shown near the lead mines on Grinton How and Harka Hill. Curracks (piles or pillars of loose stones) mark significant points on the parish boundary.

soil and other debris in order to expose the ore-bearing rock, has not previously been documented in Swaledale at such an early date.[79]

Dotted lines are used to portray the traditional boundaries separating the moors of all the various townships in upper Swaledale. The division between the moors of Grinton and those of Harkerside and West Grinton starts from the south-east corner of field number 163, which can only be the 'Ox Intack' referred to by numerous witnesses in their depositions.[80] From here, the boundary runs in a direct line across the open moor to the bounder mark known as 'the Grey Stone on the How' where it changes course by a few degrees to the west before continuing uphill to the point where a currack formerly stood on Grinton Greets, on the watershed between Swaledale and Wensleydale.[81]

The Grey Stone on the How remains in situ on the crest of the How about sixty yards east of the point where the modern post and wire fence makes a slight change of direction to the west as it runs uphill.[82] From this position there is a clear view uphill to the fence line on the Greets where a nineteenth-century boundary stone now lies face down amidst the debris of an old shooting butt.[83] Although no longer upright, there is good reason to believe that this stone occupies a position close to where there once stood a currack on the Grinton parish boundary marking its point of intersection with the traditional bounds of Harkerside and West Grinton. This assumption is strengthened by the fact that it is only here that a slight break in the slope allows a clear line of sight from the watershed down to the Grey Stone on the How. By contrast, the Grey Stone on the How cannot be seen from the highest point on the Greets which lies well back from the edge of the broad summit plateau. It was no doubt precisely for this reason that there were *two* curracks situated close together on the Greets, as shown on the early eighteenth-century map illustrated here as Figure 6. One stood at the point where the division separating the moors of East Grinton on the one hand, and those of Harkerside and West Grinton on the other, ran up against the parish boundary, while the other was situated a little further west, on the summit itself.

As depicted on Hutchinson's map, the boundary separating the moors of Harkerside and West Grinton from those of Grinton cuts through a string of mines on the How, leaving three open squares (i.e. shafts) on the west side and three dozen or so to the east. As this was the site of the disputed mines, there is every reason to suppose that the map accurately represents the position on the ground in April 1708. Indeed this is confirmed by the testimony of witnesses appearing for both Marriott and Lord Wharton who describe three shafts sunk by Lord Wharton's miners close to the Harkerside and West Grinton boundary and only sixty to one hundred yards distant from the nearest of Marriott's shafts on the Grinton side to the east. This is precisely the situation as it is portrayed on both contemporary copies of the 'Plan of Grinton manor'.

79 Information kindly supplied by Mike Gill. Earlier references to hushes do, however, occur elsewhere in the country. For example, they are referred to in witness depositions in an Exchequer case concerning the parish of Stanhope, in Weardale, county Durham, in 1680 (TNA E134/32Chas2/Mich18). According to the *OED*, the earliest use of the word in print dates to 1750.

80 On the 1841 tithe map for Grinton, the same field is simply named 'Intake' (NYCRO DN3).

81 A 'currack', or 'currick', is a north country word meaning a cairn or pile of stones often used as a waymark or boundary marker (*EDD*).

82 The stone is in fact a weathered block of unshaped limestone, measuring *c.* 0.8m by 0.6m by 0.5m high. It is located at NGR SE 0378 9662.

83 At NGR SE 0297 9561. I am grateful to Les Tyson and Ian Spensley for drawing my attention to this stone. One face of the stone is inscribed with the letter 'B' for 'Bolton'. The other face is blank.

The disposition of the mine shafts on Grinton How is also recorded by another, larger scale, survey dated 1717 (Fig. 4).[84] Entitled 'A map of the high end of Mr Marriott's lead mines at Grinton How adjoyning to the manner of Bolton made in October 1717', this manuscript plan identifies a number of mine shafts and other features by name including 'Lord Wharton's shaft' and 'Lord Wharton's smithy'. Another shaft, labelled the 'Greets shaft', may also have been one of those sunk by Lord Wharton's miners though it is not specifically identified as such.

While the inscription on the Alnwick map positively identifies John Hutchinson as the surveyor, it is not certain whether the signature is autograph or not nor can we be sure if this particular copy was drawn by him personally. On the other hand both the Alnwick and NYCRO copies can plausibly be attributed to the early eighteenth century on stylistic grounds and there is therefore every reason to suppose them to be contemporary with the Marriott v. Wharton court case.

What is certain, however, is that two *different* surveys of Grinton manor were undertaken, one by Hutchinson and the other by Brackenbury. The evidence for this is a note made by Sir Edward Ward who states that the jury had a 'view of the mapps made on both sides'.[85] Were it not for the fortunate survival of this note there might perhaps be some question as to whether two different maps had in fact ever been produced. However, Sir Edward Ward's careful phrasing leaves little room for doubt on this point, so we must conclude that Leonard Brackenbury also made a map which was shown to the jury at the time of the view and again during the trial.

Unfortunately, only two maps bearing Brackenbury's name are extant and one of these is a joint production with another surveyor. As neither map exhibits any distinctive stylistic traits they are of no use in helping us to recognise him as the author of any hitherto unattributed surveys.[86] Therefore until such time as a second map of Grinton manor surfaces which can positively be identified as Brackenbury's, we remain in the dark as to the nature of the map which we know he made by order of the Court of Exchequer in 1708. Meanwhile, however, it is legitimate to draw attention to two unsigned plans in the NYCRO collection which are clearly related to early eighteenth century disputes involving Lord Wharton's lead mines in Swaledale and which on that account might plausibly be considered as potentially the work of Leonard Brackenbury.

The larger of these maps depicts the whole boundary of Grinton parish at a scale of 5 furlongs to the inch, or roughly 1:39,600.[87] Its scale is thus a little less than half that of Hutchinson's map of 1708 which is drawn at a scale of about 1:16,900. For present purposes what is interesting is that 'Mr Marriott's mines in the How' are identified by name as well as other mines at 'Pearl Groves [the] works in question' on the opposite (north) side of the river Swale near the head of Blaeberry Gill. The reference to 'works in question' at Pearl Groves strongly suggests that this map was

84 NYCRO ZBO (L) 1/2. Another version of the same plan, with some later additions, is in the LCL, Backhouse collection, SRQ 622.344 B127, f. 47.

85 LI MS Misc. 536, Ward MSS, notebooks vol. 6, f. 119r.

86 The two maps in question are an outline plan of Gateley Moor near Middleton Tyas, North Yorkshire, dated 1709–10 (NYCRO ZKU/VII); and a survey of the lordship of Winston, Co. Durham, jointly surveyed with Edward Rawsthorn in 1700 (Durham Record Office, D/HH/8/1/125).

87 NYCRO ZKU X 5. Unfortunately the original plan is too large and too faintly drawn to be successfully reproduced here but a re-drawn and slightly simplified version of this plan is reproduced as Fig. 2.4 in Fleming 1998. The suggested date of 1738 for this map is misleading as it does not in fact relate to the date of its production but rather to the date when it was borrowed from Thomas Swales, probably in connection with a sale of land in Helaugh manor which took place in that year. A date for this map in the first decade of the eighteenth century is a distinct possibility.

made in the context of a long running lawsuit between Lord Wharton on the one part, and Charles Bathurst and his partners on the other, over the ownership of lead mines in Blaeberry Gill, a dispute which began in 1675 and was still continuing in 1710. A letter dated July 1703 from Lord Wharton to his solicitor, Matthew Smales, at Gilling, contains instructions for Leonard Brackenbury to make a map – 'by his owne notes without goeing upon the ground againe' – and a three-dimensional model both of which were to be shown to the jury in this case. It therefore seems probable that Brackenbury did make a map for Lord Wharton in the context of this dispute and that he used some pre-existing notes or sketches to do this. For our purposes it is a matter of interest that the three shafts dug by Lord Wharton on Grinton How, on the west side of the Harkerside/Grinton boundary, are carefully depicted on this sketch as well as 'Mr Marriott's works' on the east. This would seem to indicate that the survey was made some time after October 1705 but before *c.* 1710 or, alternatively, that details of the mines on Grinton How were added to a pre-existing survey at the time of the Marriott v. Wharton lawsuit. Either way, a case can be made for seeing this map as the work of a surveyor, possibly but not necessarily Leonard Brackenbury, who was acting on instructions from Lord Wharton.

An even stronger case can be made for seeing a second map, also in the NYCRO, as being connected with the Marriott v. Wharton dispute which, as it does not seem to be related to Hutchinson's survey, may potentially be the work of Leonard Brackenbury (Fig. 5).[88] Drawn at a scale of approximately 1:22,500, this map shows not only the Grinton parish boundary but also the division between the moors of Harkerside and West Grinton on the one hand and those of Grinton on the other in considerably greater detail than is the case with any of the other maps described thus far. As shown in the extract from this map (Fig. 6), the lead mines on Grinton How and on Harka Hill, including Lord Wharton's three shafts on the west side of the Harkerside/Grinton boundary, are all carefully depicted. Other details include a line of stone-built curracks at various significant points on the parish boundary as well as three buildings near the mines on the How and on Harka Hill. The purpose of these buildings is not specified though they may well be smith's shops or else shelters for the miners. Some of the field or enclosure boundaries in the vicinity of Grinton are also shown as faint dotted lines, though in less detail than they are on Hutchinson's map. While the evidence again suggests a connection of some sort with the Marriott v. Wharton dispute, the identity of the surveyor must remain conjectural.

Given that two apparently contemporary copies of Hutchinson's map have now been identified, it seems likely that more than one copy of Brackenbury's map may also have been in circulation at the time of the court case. What we do know is that additional copies of Hutchinson's map were produced from time to time in later years, most probably in the context of other lawsuits.[89] Thus, for example, an uncoloured copy of Hutchinson's map, dated September 1898, exists which is inscribed with a note in the bottom margin stating that it is a fourth generation copy ultimately derived from Hutchinson's original survey of 1708.[90]

88 NYCRO ZBO (L) 1/1.

89 For these later lawsuits see Tyson, Spensley and White 1995.

90 In a private collection. According to this inscription other copies were made in November 1712, February 1811, and November 1894.

The trial and its aftermath

For an equity suit litigated in the Court of Exchequer, one would not normally expect any record of the outcome other than the final decree or, where a trial at common law took place, the verdict given by the jury. It is therefore exceptionally fortunate that we have Sir Edward Ward's own report which covers not only all the substantive hearings in this case but also includes a lengthy and detailed account of the trial itself.[91] Additionally, a handful of other manuscript sources and at least one contemporary newspaper report survive, all of which help to build up a picture of what took place before, during and after the trial.

According to Sir Edward Ward's report, some discussion took place between the parties as to what mechanism should be employed to bring about a trial, whether by means of an action of ejectment or a trial at common law with issues directed by the court.[92] After some initial disagreement, it was eventually decided to hold a trial at common law at the bar of the Court of Exchequer using the device of a 'feigned issue', a collusive action involving the fiction of an unpaid debt.[93]

Two issues were put before the jury, as set out in the decree of 25 February 1708. These were, first, 'whether the place or places where the shafts lately sunck or wrought by the said Earle of Wharton in the wasts or moores of Harkerside or West Grinton or any and what part thereof are or is parcell of the mannor of Grinton or the mannor of Healaugh; second whether the wasts or moores of Harkerside or West Grinton or any and what part thereof are or is parcell of the mannor of Grinton or the mannor of Healaugh'.[94]

The common minute book of the court records that on 28 February 1708, three days after the trial date had been set, a report was read from the Deputy Remembrancer whose responsibility it was to prepare a strictly factual account of the case and the issues to be decided, though without taking sides. Unfortunately no trace of this document has come to light, though we can guess at its contents as it is likely that the decree was based on it.[95]

Before the trial the parties had an opportunity to prove exhibits – that is to introduce any documents or other material helpful to their case – and have them accepted as evidence by the court. Both sides took advantage of this arrangement and submitted long lists of documents to the auditor of the Exchequer which give some insight into the nature of the exhibits that were produced.[96] Thus most of the documents asked for by Marriott were extracts from books of account or records of leases kept either by ministers or receivers of the Crown rents. From the references

91 Sir Edward Ward's contemporary record of the trial occupies twenty-four consecutive pages of his notes, LI MS Misc. 536, Ward MSS, notebooks vol. 6, ff. 106v–118r.

92 LI MS Misc. 536, Ward MSS, notebooks vol. 6, ff. 105r–105v.

93 See Horwitz 1999, p. 176. For a definition of the 'feigned issue' see, for example, *The Law Dictionary*, Sir Thomas Edlyne Tomkins, 3 edn, 1820. The precise nature of the feigned issue in this case is rehearsed in the verdict which was later recorded in the Queen's Remembrancer's memorandum roll – E159/552 (see Appendix 6(b)).

94 Decree, E126/19, ff. 13v–14r. This same wording appears on a slip of paper headed 'Issue to be tryed betwixt my Lord and Mr Marriott', now in the Durham Record Office (D/HH/6/4/139). A second, almost identical slip, has been preserved amongst the papers of John Blackburne who we know served on the jury (BnL, Special Collections, Marrick MSS, 5/4/61). Such slips may perhaps have been given to all members of the jury as *aide memoirs*.

95 E161/66, f. 102r. In February 1708 the Deputy Remembrancer was Henry Stevens who had been appointed on 6 November 1707 (Bryson 1975, p. 189). The report is not present amongst Henry Stevens' surviving papers (E219/719); or in E195 (miscellaneous reports).

96 TNA LRRO 3/85.

provided it is evident that these items were intended to document the history of the Crown's title to the Grinton mines, from the dissolution of Bridlington priory in 1537 to the granting of Humphrey Wharton's lease in 1628, and to prove that the appropriate rents and royalties had actually been paid during this period. As many of the same documents were also requested by Lord Wharton it is evident that both sides intended to subject the Crown's title to the Grinton mines to very close scrutiny. For this same reason at least two copies of the 1696 Crown lease to George Tushingham were copied out for use in court.[97]

As mentioned above, the court had decreed that the jury should have a view of the disputed ground on 22 April, two weeks before the trial. As set out in the decree, the jury consisted of twenty-four Yorkshire freeholders chosen from the 'freeholders book' brought to London by the sheriff.[98] The Deputy Remembrancer himself made an initial selection of forty-eight names from which each of the contesting parties had the right to strike out twelve. The names of the remaining twenty-four jurymen are recorded in two lists preserved amongst the papers of John Blackburne of Fryers Head, Winterburn, near Skipton, who was himself one of those chosen to serve on the jury.[99] Jury members invited to attend the view were asked to present themselves at Mr Raper's inn, the Queen's Arms, in Richmond, on the previous evening where they were to be 'moderately entertained' at the joint expense of the plaintiff and the defendant. On the following day, the jury were shown over the ground in the manner prescribed by the court.

After the jury had been selected a period of about six weeks intervened before the view, followed by a further interval of two weeks before the trial. This gave both parties ample opportunity to try to influence members of the jury and there is evidence to show that such tampering did indeed take place. For example, it was later alleged by John Blackburne who, as we have seen, was selected to serve on the jury, that he was approached at the view by one Reuben Orton with an offer to purchase some land of Blackburne's at Marrick for 'a very good price'.[100] This offer by Orton was said to have been made with the authority of Thomas Pulleine, Marriott's father-in-law, on condition that Blackburne use his influence with the jury and see to it that at the view the boundaries of Harkerside were ridden according to Marriott's wishes. As Blackburne was known to be in financial trouble at this time, an attempt at bribery

97 TNA LRRO 3/85. For a translation of the lease see Appendix 6(a).

98 Henry Iveson Esq. of Black Bank was High Sheriff of Yorkshire in 1707–8 – see *The Lord Lieutenants and High Sheriffs of Yorkshire, 1066–2000*, ed. W. Mark Ormrod, 2000 Barnsley.

99 BnL, Special Collections, Marrick MSS, 5/4/59–60. The names of those finally selected are as follows: Griffith Boynton of Burton Agnes, Bart.; John Legard of Ganton, Bart.; Richard Osbaldiston of Hunmanby, Kt.; Robert Constable of Catfoss, Kt.; Mathew Pierson of Lowthrop, Kt.; Roger Talbot of Thornton le Street, Gent.; William Dawson of Farlington, Gent.; Henry Darley of Buttercrambe, Gent.; James Hill of Thornton Dale, Gent.; Thomas Langley of North Grimston, Gent.; Hugh Bethell of Rise Park, Gent.; Walter Calverley of Calverley, Gent.; Thomas Micklethwaite of Swine, Gent.; John Blackburn of Flasby and Winterburn, Gent.; John Bourchier of Newton, Gent.; Thomas Robinson of Pickering, Gent.; Jacob Heblethwaite of Norton, Gent.; Boynton Boynton of Rawcliffe, Gent.; Josias Morley of Beamsley, Gent.; Thomas Fountaine of High Melton, Gent.; Isaac Knight of Letwell, Gent.; Westby Gill of Greasebrooke, Gent.; Thomas Stillington of Kelfield, Gent.; and William Swaine of Bradford, Gent. On 3 May, three days before the trial, Hugh Bethell sent an affidavit saying that he was unable to attend the trial due to infirmity (E207/117/1). The first list of names is dated 4 March, so it is clear that the final selection of jury members was made no more than a week after the issuing of the decree. This left ample time for jury tampering before the trial.

100 BnL, Special Collections, Marrick MSS, 5/4/74. Undated notes for an affidavit. Since at least 1697 Reuben Orton had been overseer of Lord Pawlett's lead smelting mills at Marrick and Preston where all the lead from Marriott's Grinton mines was smelted (TNA LRRO 3/85). In these circumstances it is entirely credible that he was acting as Marriott's mouthpiece.

is certainly plausible. Nor was this all, for there is also evidence to show that before the trial Thomas Dickonson, one of the two gentlemen chosen to represent Marriott at the view, signed articles of agreement with Reuben Orton for the purchase of a lead smelting mill at Marrick, known as the Cupulo, for the considerable sum of £400.[101] At the time Reuben Orton was a partner in the mill with John Blackburne. Dickonson's offer may therefore be seen as a way of bringing indirect influence to bear on Blackburne, especially as Dickonson made it clear that he was acting as agent for Reginald Marriott in this business and that the prospective purchase was conditional on Marriott obtaining a favourable verdict at the forthcoming trial. In these circumstances it is impossible to see the offer to buy the mill as anything less than another attempt at bribery. In later years, Blackburne further alleged that Orton repeated his offer to buy Blackburne's Marrick estate, and the Cupulo mill, when they met in a Piccadilly coffee house four days before the trial.[102] As before, Thomas Pulleine and Thomas Dickonson were said to have guaranteed these purchases on behalf of Reginald Marriott, provided that the verdict went Marriott's way. Whether or not Blackburne did in fact attempt to influence the jury on Marriott's behalf cannot finally be proved, nor does it seem that any money was actually paid to him after the trial. Certainly at no time, then or later, did Blackburne ever admit that he had accepted such bribes, though both he and his son pursued Marriott for what they said were these unpaid debts for the next twenty years, all the time threatening to expose him for this attempted corruption.[103] Although Blackburne himself is careful to say that he was not corrupted, his denials have a hollow ring in view of his dogged pursuit of Marriott for money that he said was owed to him by Marriott for the Marrick priory lands and his share of the Cupulo mill. If in the end he had not done Marriott's bidding, then what claim could he have on this money?

Without corroborating evidence these charges of corruption levelled against Marriott remain unsubstantiated, though his cautious use of intermediaries and avoidance of written communications is certainly deeply suspicious. Nor do we know what degree of influence Blackburne may actually have been capable of exerting over other members of the jury. On the other hand it would not be surprising if Marriott had acted in the way alleged given that he had a record of corrupt practice dating from well before the trial. In 1697, for example, when he was a deputy auditor of the Crown revenues, he was implicated with John Knight MP in a scandal involving the fraudulent conversion of Exchequer bills to the value of at least £7000 and was lucky to escape a prison sentence, or worse. Again, in May 1701, Marriott was cited in articles of impeachment against Lord Somers who had made use of inside information supplied by Marriott to obtain lucrative royal grants of fee farm rents for which Marriott himself was rewarded with grants to the value of £500 per annum.[104]

If Marriott did attempt, or even succeed in, swaying the jury by such methods he may not have been the only guilty party for, in what appears to be a statement of his case for council's opinion written some years after the events described, Blackburn's son made the additional allegation that Lord Wharton asked his father's advice about

101 BnL, Special Collections, Marrick MSS, 5/4/63.

102 We know that Reuben Orton was in London for the trial because he appeared in court as a witness for Marriott. Sir Edward Ward's note of his testimony records that 'for 27 years smelted for Swale above 5000 pigs of lead about 14 years ago that came from Harkahill near place where defendants works are'. LI MS Misc. 536, Ward MSS, notebooks vol. 6, f. 113v.

103 For affidavits and letters written by Blackburne and his son accusing Marriott of attempted corruption, see BnL, Special Collections, Marrick MSS, 5/4/74–80.

104 *History of Parliament* 2002, vol. 4, pp. 760–1.

the composition of the jury and 'told him he would gett him concerned in the Custom house at Leverpoole and other kind offers which ended in nothing'.[105]

As mentioned above, we have a very full and detailed contemporary note of the trial made by Sir Edward Ward. According to this record, the trial began early on the morning of Friday, 7 May 1708, and went on continuously for thirty-four hours until 7.00pm on Saturday evening when the jury withdrew, returning three hours later to give their private verdict.[106] For those who had travelled up to London by coach from Yorkshire, including not only the twelve members of the jury but over one hundred witnesses as well, the trial must have been a considerable ordeal. This at any rate was the experience of Sir Walter Calverley who recorded in his memorandum book:

> 3° May. Set forth towards London upon the account of that jury in Wakefield stage coach; Mr Swayne of Bradford going up at same time, and gott hither on Thursday night 6 Maij exhausted And on Friday morning (the day following) the cause came on about 8 of clock & continued all that whole day & night & all the day following & till Saturday night about 10 of clock, what time the jury withdrew & having considered of it gave in their verdict for Mr Marriott.[107]

The trial was held in a courtroom at the north-west end of Westminster Hall and was presided over by the four barons of the court.[108] The case for the Attorney General and Reginald Marriott was presented first. Many documents were produced and sixty-one witnesses (not all of whom had previously made depositions) gave oral testimony and in some cases were also cross-examined. A further three depositions made by witnesses who were either dead or too old to travel were read out. Lord Wharton's counsel then put his case which involved a further detailed examination of the Crown's title to Grinton manor and the mines, and oral testimony from another forty-two witnesses.[109]

As we have seen, the jury withdrew at 7.00pm on Saturday evening and took three hours to reach a verdict, which at this stage was supposed to remain strictly private (though it soon became known that they had in fact decided in favour of Marriott and the Crown). On the Monday following, the jury returned to court and were asked by an official if they remembered their verdict and abided by it, to which questions the foreman replied that they did. But before the verdict could formally be announced in open court counsel for Lord Wharton attempted to introduce a bill of exceptions which, had it been accepted by the court, would have had the effect of overturning the verdict and bringing about a retrial. This last minute manoeuvre by Lord Wharton's lawyers caused the court a great deal of difficulty and the question of whether the bill should be accepted by the court or not was the subject of many hours of debate between the barons and Lord Wharton's lawyers. In the end it was not until 23 June that the court finally decided that the bill could not be accepted for reasons that are explored at length in reports by Sir Edward Ward and Baron Robert

105 BnL, Special Collections, Marrick MSS, 5/7/85, f. 15v.

106 LI MS Misc. 536, Ward MSS, notebooks vol. 6, f. 162r.

107 BL Additional MSS 27418, ff. 103v–104r. The published version of Sir Walter Calverley's memorandum book silently omits this passage. For further relevant entries see *Memorandum Book of Sir Walter Calverley, Bart.*, ed. S. Margerison, Yorkshire Diaries and Autobiographies in the Seventeenth and Eighteenth Centuries, Surtees Society, 77, 1886, pp. 118–19.

108 The building that housed the courtroom no longer exists. For a depiction of the court in operation see the aquatint by J. Sadler (1808) reproduced on the cover of Horwitz 2001. Sadler's print was based on a drawing by Pugin and Rowlandson.

109 LI MS Misc. 536, Ward MSS, notebooks vol. 6, ff. 106v–118r.

Price. As both of these reports have been published by Professor Bryson we need not dwell further on the issue here except to note that had the bill been presented earlier, and in the proper manner, the outcome of the trial could well have been different.[110]

Meanwhile the trial did not escape the notice of the London public. Thus the memorialist Narcissus Luttrell, writing in his diary on Saturday, 8 May, recorded that

> Yesterday at 10 in the morning came on a tryal before the barons of the exchequer, between Mr Marriot, plaintiff, and the earl of Wharton, defendant, about the boundaries of a mannor in Yorkshire, wherein there are very rich lead mines; the court satt all night, and like to doe so till 12 this night, 15 council and above 70 witnesses being on each side.[111]

While it is known that Luttrell gathered much of his information from the newspapers this may not have been the case here as the first newspaper report of the trial seems not to have appeared until Wednesday, 12 May, when *The Daily Courant* reported:

> On Friday last at Exchequer-Barr about 10 a clock in the morning, came on the great trial between the Earl of Wharton and Marriot, concerning lead-mines in York-shire. It lasted till about 7 a clock in the evening on Saturday; at which time the jury withdrew, and in about an hour's time gave a private verdict for Mr. Marriot. Abundance of records were produc'd on both sides, and a great number of witnesses; and on Monday morning the jury confirm'd their verdict in court.[112]

Further entries in Luttrell's diary followed the progress of the court's deliberations on Lord Wharton's bill of exceptions and noted the possibility of an appeal to the House of Lords.[113]

After the trial, the validity of Marriott's, and the Crown's, title to the Grinton mines continued to be the subject of intense scrutiny and the issue was debated in a series of meetings which took place in the Treasury between November 1708 and March 1709.[114] According to Treasury minutes, these discussions involved Treasury ministers, the Attorney General, the Solicitor General and Lord Wharton. Unfortunately no record of the substance of these meetings has come to light so it is not known precisely what aspects of the Crown's title or Marriott's lease were being called into question. What is clear, however, is that the Solicitor General, Sir James Montague, and the Attorney General, Sir Robert Eyre, took contrary views of the matter. Meanwhile the verdict was enrolled in the Queen's Remembrancer's memo-

110 Bryson 2007, pp. 536–45. Additional information concerning the bill of exceptions is contained in a four-page note now in York Minster Library, Hailstone MSS, Box 4.7. Although the author of the note is not identified it probably derives from the papers of Mathew Smales of Gilling who was Lord Wharton's attorney in Yorkshire. The note lists counsel who appeared on both sides at hearings which followed the trial and lists a number of cases cited by Lord Wharton's lawyers most of which are concerned with defective leases.

111 Luttrell 1857. The relevant passages in the published version of Luttrell's Diary have been checked against the original manuscript, now in All Souls Library, Oxford (MS 165), and found to be accurate with nothing omitted. I am grateful to the librarian of the Codrington Library, Dr Norma Aubertin-Potter, for carrying out these checks.

112 BL, Burney Collection of newspapers, consulted in digitised format.

113 Lutrell 1857, entries for 11 May, 15 and 24 June. I have found no evidence that such an appeal ever materialised.

114 Meetings took place on 1 November 1708, and on 23 and 24 March 1709. See *CTB*, vol. 22, 1952 and vol. 23, pt. 2, 1949.

randum roll.[115] This was a relatively unusual procedure involving extra costs and was almost certainly done at the insistence of either Marriott or the Crown, the intention being to make the verdict more difficult, and expensive, to challenge.[116]

Already in October 1708, before any of these meetings in the Treasury took place, Lord Wharton's counsel had asked the permission of the court to move for a new trial.[117] Permission was granted but although the matter was raised in court on three subsequent occasions in November, and three times adjourned to a later hearing, no further court proceedings ensued.[118] Again, the court records themselves are silent as to the reason why court action fizzled out at this point though it would not be surprising if it was because the production of ore from the Grinton mines was already starting to decline.[119]

Meanwhile, on Grinton How Marriott's favourable verdict seems to have made little immediate difference, for Edward Smithson and Henry Place swore fresh affidavits on 2 October 1708 alleging that Lord Wharton's men were continuing to mine lead ore from shafts originally dug by Marriott as well as from those which had been newly opened by Lord Wharton's miners in September 1705.[120] Served yet again with injunctions, the miners persisted in their refusal to stop working. How long this situation persisted is uncertain, though conflict evidently continued into the autumn of 1708. For, on 25 October, three of Marriott's miners swore fresh affidavits stating that they had gone up to Harka Hill to salvage timber from a 'house' built by Marriott 'ten or eleven years ago' which had already been partly dismantled and some of the wood stolen by Lord Wharton's men.[121] While Marriott's men were engaged in this work, a group of Lord Wharton's miners arrived on the scene and, having themselves pulled down the rest of the building, carried away all the usable timber by force.

Even if skirmishing of this kind continued into the winter of 1708/9, by the spring of 1709 Marriott seems finally to have gained complete control of all the mines on Grinton How and Harkerside. In any event there is no evidence that Lord Wharton continued to maintain an interest in the Grinton mines after the last recorded discussions about the validity of Marriott's Crown lease took place in March of that year. By this time Lord Wharton's political career had in any case taken a new turn with his appointment as lord lieutenant of Ireland.[122]

In 1715, the anonymous author of a memoir on the life of Lord Wharton, tentatively identified as John Oldmixon, included a short account of the Marriott v. Wharton contest in his narrative. Nakedly partisan in Lord Wharton's favour, the writer accuses the Tories of seeking to make a party political issue of the case while at the same time endeavouring to portray Lord Wharton himself as an indefatigable and energetic champion in his own cause, acting 'as his own sollicitor, following his

115 TNA E159/552.

116 Horwitz 2001, p. 69. The enrolled verdict is written in Latin. An English translation is given in Appendix 6(b).

117 Order, 23 October 1708 – E127/27, f. 216v.

118 Decrees of 4, 11 and 15 November 1708 – E126/19, f. 74.

119 Marriott's accounts show that receipts from the sale of smelted lead remained buoyant in the accounting year ending at Michaelmas 1709 but fell sharply thereafter (see footnote 5 above). Depending on the length of time it took to process and smelt the ore prior to sale, and allowing for fluctuations in the market price of lead, it may be that the production of ore at the mines was already declining by early 1709. In any case, in the year ending at Michaelmas (29 September) 1709 Marriott's clear profit was given as £3486 0s 9d. One year later it stood at £787 7s 3¾d and by Michaelmas 1711 it had fallen to only £52 0s 1¾d. TNA LRRO 3/85.

120 E207/117/2.

121 E207/118/1.

122 Lord Wharton was appointed lord lieutenant of Ireland on 4 December 1708 – *DNB*.

lawyers, and attending courts in person, and that with so much industry and perseverance, that his antagonist found it would be more for his advantage to comply with him than continue a suit that was so vigorously prosecuted'.[123] This portrayal of Lord Wharton's part in the court case is to be regarded as a piece of overt political propaganda rather than an attempt at historically accurate description.

Afterword

In several respects the Grinton mines dispute of 1705–08 stands out as an exceptional, and exceptionally interesting, example of an equity case litigated in the Court of Exchequer. In the first place, apart from the Deputy Remembrancer's report, the documentation in TNA appears to be virtually complete, thus enabling the progress of the case to be followed stage by stage, from the exhibiting of the Attorney General and Marriott's bill of complaint in Michaelmas term 1705, through to the issue of the final decree and the trial at common law in Easter term 1708. Thereafter the attempt by Lord Wharton's lawyers to introduce a bill of exceptions and move for a retrial extended proceedings into the following Michaelmas term. The entire case therefore ran for three years, or twelve legal terms, which is slightly longer than the average for Court of Exchequer suits in this period that progressed beyond the pleadings stage.[124]

As previously noted, Horwitz has studied a sample 150 equity bills filed in the year 1685 and found that only 10% progressed as far as the issue of a decree.[125] The fact that a decree was issued in the Marriott v. Wharton case is therefore relatively unusual. It is unusual too that the case features a cross bill filed by the defendant, Lord Wharton, who thereby became the plaintiff in a separate, parallel, suit. In Horwitz's combined sample of 300 suits from the years 1685 and 1816, only four cross bills were found.[126]

The documentation in the Grinton mines dispute is more than usually extensive and the bills and answers, witness depositions, orders, and affidavits in the two interlinked suits, together with the final decree, are all preserved in readable condition in TNA. This already substantial archive is further supplemented by other relevant material in archives and collections elsewhere, including several contemporary maps and plans of Grinton manor and the disputed mines, a list of the jurors who attended the trial, letters, entries in memorandum books, and a newspaper report, all of which help to deepen our understanding of the case. More important still, we have Sir Edward Ward's personal record of the trial, while other notes on Lord Wharton's bill of exceptions made by him and by Baron Price have been published in full by Professor Bryson.[127] Thanks to these additional sources, we have a much better understanding of the case and its context than we otherwise would.

As stressed above, a particularly notable feature of this case is the very large number of witness depositions made on both sides – more than three hundred in all; a fact which clearly struck contemporary observers as noteworthy, including Luttrell. As can be imagined, the decision to transcribe and publish these depositions in their

123 Anon. [J. Oldmixon?], *Memoirs of the life of the most noble Thomas late Marquess of Wharton*, 1715 London, pp. 43–4.
124 Horwitz 2001, p. 44, Table XIIIb.
125 Horwitz 2001, p. 31, Table I.
126 Horwitz 2001, p. 22, n. 52.
127 LI MS Misc. 536, Ward MSS, notebooks vol. 6; Bryson 2007.

entirety was not taken lightly. One justification for doing so is that it helps to bring out certain important aspects of this 'evidence' that would not otherwise be so clearly apparent. For example, as even a superficial reading makes clear, depositions made by successive batches of witnesses tend to be repetitive, rehearsing the same 'facts' and incidents, and often employing similar, if not identical, words and phrasing. The most likely explanation for this is that witnesses were coached in advance of their appearance before the commissioners.[128] Again, the sheer bulk of the depositions raises some potentially awkward questions about the procedures that were involved in the examination of witnesses and the recording of their testimony. For example, given the number and length of the interrogatories prepared in advance by the lawyers – twenty-two on Lord Wharton's side and seventeen on Marriott's – considerations of time make it very unlikely that all were read out to each and every witness in turn. Rather, the parties' solicitors probably gave lists to the commissioners on the day of the examination indicating which interrogatories were to be addressed to each individual witness. If witnesses were coached in advance, as we suspect, they too are likely to have had advance knowledge of the questions that were to be put to them, even though this would be contrary to the rules of the court. The presentation of witnesses in a predetermined sequence by a solicitor in attendance might also help to explain why witnesses who gave similar or near identical testimony often appear one after another in batches. Alternatively, this bunching together of repetitive testimony, which is unlikely to have come about by chance, could be explained if the original order in which witnesses appeared was changed when their statements were engrossed so as to bring together depositions whose content was broadly similar. Either way, the intention may have been to use repetition as a way of giving emphasis to particular aspects of their evidence which would not otherwise come across so strongly. Repetition of witness testimony was also a marked feature of the trial. Time and again, Sir Edward Ward simply writes '*ut supra*' or 'the like' against the names of successive witnesses who gave similar testimony. One can only imagine the soporific effect this is likely to have had on the jury and barons alike, especially as the trial wore on through the night and into a second day.

Another problematic aspect of the taking of testimony arises from a consideration of the dates of the examinations as given on the parchment engrossments. For, taken at face value, these suggest that the depositions of Marriott's 175 witnesses were taken in just three days.[129] If this was indeed the case it is difficult to see how such a large number of witnesses could have been sworn and examined in so short a time and one struggles to imagine the interrogatories being read out and the statements of so many witnesses taken down by a clerk (even if in the form of notes or some kind of shorthand) in, say, three twelve or fourteen hour days. Given that the depositions of Marriott's witnesses occupy no fewer than seventeen parchment membranes, all closely written on both sides, their testimony would have to have been recorded at the rate of between two and three thousand words an hour, without a break, for the whole of the three day period if everything that appears in the final engrossment was spoken and written down *verbatim* at the time of the examination. By some means the process of examination must surely have been made shorter though, in the absence of any draft notes made at the time, it is not immediately apparent how this may have been achieved.

[128] For the coaching of witnesses generally, see Chambers 1996.

[129] The examination of Marriott's witnesses is stated as having begun on 26 August and that of Lord Wharton's on 29 August 1707.

One possible scenario might involve some testimony being submitted in writing at the time of the examination or even after the formal examination had taken place. Certainly where financial accounts or other complex material is concerned, it is highly likely that it was presented in writing at the time of the oral examination or afterwards.[130] Again, in the absence of any notes made during the examination itself, it is impossible to know exactly what took place during the examination of witnesses, or how the rules of the court were interpreted in practice. Unconventional methods may have been tolerated in the interest of speeding up what would otherwise be an almost intolerably long procedure, especially if it suited both parties to do so. Since the commissioners were chosen by the parties themselves, and were not necessarily thought of as impartial, they may have arranged matters to suit themselves even if in so doing they went beyond the accepted rules which were supposed to govern the gathering of evidence. While suggestions of this nature can be no more than tentative, it is interesting to note that some of the affidavits sworn by Lord Wharton's miners are similar, and in places identical, in both content and phraseology to depositions that were made twelve months or more later, either by these same individuals or by others on the same side. This observation lends some weight to the suggestion that at least some of the evidence given by witnesses appearing for Lord Wharton was prepared long in advance by his solicitor, Matthew Smales of Gilling, whose signature appears on many of these same affidavits in his capacity as a commissioner for oaths. As a local man, Smales would anyhow have been best placed to choose and rehearse Lord Wharton's witnesses in advance of their formal examination. He can also confidently be assumed to have had a leading role in drafting the interrogatories that were to be addressed to these same witnesses and which were supposed to remain secret until formally unsealed by the commissioners on the day of the examination. In the context of this discussion it is also worth remembering that the final engrossment of the witness statements on parchment was done locally by scriveners employed by the parties' own solicitors and under their direct supervision. In these circumstances there would have been ample opportunities to tidy up or rationalise testimony that was originally given orally, and perhaps less coherently, at this later stage. Nor, given the exceptional volume of testimony given in this case, is it easy to imagine how the commissioners would have been capable of checking the accuracy of the statements as finally written down before they endorsed the final parchment membrane with their signatures. This realisation may again have increased the temptation to make such adjustments during the process of engrossment as might seem desirable to the party concerned.

How far witness testimony, in its final, written form, can be accepted as truthful or as accurately representing statements made orally at the time of the examination, are questions which come into sharp focus, especially in a case like this where a large number of witnesses on one side make almost identical statements on crucial issues of fact which are then directly contradicted by an almost equally large number of statements made by witnesses on the other. Indeed one justification for publishing in their entirety all the depositions in this case is to bring out precisely this point.

The distance between the court at Westminster and the scene of the dispute in Swaledale is more than two hundred miles, a fact which may have caused the court some difficulty in its attempts to arrive at a balanced understanding of events as they unfolded on the ground. Time and time again allegations made in affidavits by agents

130 See, for example, the tables and lists incorporated in the testimony of William Waller and Cornelius Frier, pp. 245 and 347.

attempting to enforce the injunction on behalf of the plaintiffs were flatly denied in further affidavits made by those who were accused of breaching it. As a consequence, penalties imposed at one court sitting were set aside at the next. In these circumstances the task of determining where truth lay was fraught with difficulty, not least because some or all of those who were charged with serving the injunction were Marriott's own employees and therefore seen as biased in the eyes of the miners. Again, the fact that the injunction in this instance was recognised by the court as technically defective added further to the problems of enforcement. The commissioning of two detailed maps of the disputed territory as a means of helping the court in its deliberations is an interesting development and may be seen as an attempt by the court to remain strictly objective. Whereas references to 'maps' in lead mining disputes of this period sometimes turn out to mean no more than a written list, here, where the issues that were to be tried involved determining a precise relationship between the mines and their topographical setting, an actual survey of the territory was required.[131] In the case of Hutchinson's map, colour-coded to show different kinds of land tenure, the intention may have been that it should be used in conjunction with written documents, including the long series of Crown leases, *inquisitions post mortem* and so forth, to reconstruct the holdings of Bridlington priory as they were at the time of dissolution, since it was from the priory that the Crown claimed to derive its title to the mines. How successful this attempt at what we would now call regression analysis may have been in practice is impossible to judge, and it could well be that in the end the jury decided to place more weight on the *de facto* existence of a long succession of Crown leases of the Grinton mines dating back to the mid-sixteenth century.

Despite Marriott's favourable verdict and Sir Edward Ward's observation that the Crown's interest in the mines is 'soe very considerable & (as it seemed throughout the whole cause) soe clear a title', the right of the Crown to the Grinton mines continued to be called into question right up until the end of the nineteenth century but seemingly without any final resolution of the underlying legal problems involved.[132] Thus, for example, when asked by the Commissioners of Crown lands to enquire into the Crown's title to the lead mines in 1870, one lawyer concluded that 'It is a matter of some doubt whether [Bridlington] Priory ever possessed a manor of Grinton and the point appears to be of considerable importance to the present question.'[133] In doubt were not only the rights enjoyed by the priory and the territory over which these rights applied, but there were also difficulties in determining precisely what rights had been granted by the Crown to a succession of tenants and lessees after the Dissolution. Bureaucratic errors in the drafting of leases and uncertainty about the precise meaning of such phrases as 'the territories and fields' of Grinton, all created difficulties in defining the Crown's right as it might apply to the unenclosed moors and wastes separate from the enclosed fields. Forged documents, documents that could not be traced, and the ambiguous wording of certain key passages in grants and leases, are recurring features of these legal wrangles. The resolution of the resulting problems lies well beyond the scope of this enquiry which is limited to a single court case and to the evidence as it was presented and understood at the time. To reconstruct in detail

131 In a dispute over lead mines in Weardale, Co. Durham, in the early 1680s, a straightforward list of over one hundred mines headed a 'schedule lyst or mapp' was offered in evidence (TNA E134/36Chas2/Mich 18 and Mich 47). In this instance these three terms appear to be being used interchangeably.

132 LI MS Misc. 536, Ward MSS, notebooks vol. 6, f. 119v.

133 TNA CRES 34/211.

the history of the title to the Grinton mines, with its many twists and turns, would be an eminently worthwhile task but not one to be undertaken lightly.

As stated at the beginning of this introduction, this enquiry began as an investigation into a map whose purpose, date and authorship were all in doubt. Now, at the conclusion of our researches, not only have answers been found to these questions but, more importantly, the map has been restored to its original historical context, something that could never have been achieved simply by an analysis of its overt, visual content. As with many other maps, its true meaning proved to lie beneath the surface.

Editorial Conventions

Capital letters: The use of capital letters has been modernised throughout. However, where a significant new clause begins, the capital letter of the introductory word has been retained (usually And, That, But, Whereas) or, less frequently, raised from lower to upper case. This most often occurs in the bills and answers which otherwise consist of long, or very long, passages of unpunctuated and unbroken text with little or nothing to indicate to the reader where a fresh point is being introduced.

Spelling: The original spelling has been retained throughout. The thorn has been transliterated as th. The long 's' has been rendered as 's'.

Punctuation: Most documents have little or no punctuation and no new punctuation marks have been introduced.

Abbreviations: Abbreviations have been silently extended where their meaning is clear.

Deletions: Deleted text that is legible is enclosed in braces {text}. Deleted text that is not legible is indicated [x x x] accompanied by a footnote indicating the approximate length of the text that has been deleted.

Illegible text: This is indicated [. . .]

Insertions and interlineations: Where single words or lines of text have been inserted into previously written text, either by way of correction or addition, the relevant words or passages are contained in angled brackets <text>

Restored text: Due to the repetitive nature of the documents, it is often possible to guess words or phrases that are illegible. Where a suggested reading has been inserted by the editor it is contained in square brackets [text]

Damage to the documents: Where holes have been made in the parchment after it was written on, this is indicated by a footnote. Wear to the parchment resulting in text that has become unreadable is treated as illegible text.

Blanks: Blanks in the text are indicated by empty square brackets [ ]

Endorsements: Endorsements that indicate the date that a document was brought into court, or which relate to amendments ordered by the court, are given in a footnote. Other endorsements, such as the *fiat*, where this is not accompanied by a date, are silently omitted.

Signatures: Holograph signatures are prefixed */s/*

Line endings: Line endings in the longer texts are indicated by a forward slash /

Exchequer documents: In TNA Exchequer documents are prefixed with a capital E. In the text all documents with E prefixes are in TNA.

TEXTS

Attorney General *ex rel.* Reginald Marriott v. Thomas Lord Wharton

Pleadings[1]

E112/915 EBOR 80 – document 1[2]

[membrane i *recto*][3]

To the right honourable Sidney Lord Godolphin[4] lord high treasurer of England Henry Boyle[5] esquire chancellor and under treasurer of her majesty's Court of Exchequer at Westminster Sir Edward / Ward[6] knight lord chief baron of the said court and to the rest of the barons there /

Informing sheweth unto your lordships Sir Edward Northey[7] knight her majestie's Attorney General as well on the behalf of her majesty as also for and on the behalf and at the relacion of Reginald Marriott esquire her majesty's lesee farmer and accomptant as here<in>after mencioned / That the mannor of Grinton in the county of York and the territories and fields thereunto belonging are and immemorially have been a large tract or piece of ground conteining several demesnes inclosures lands grounds wastes and moores of a very great extent all lying [or being] together / and butted and bounded by severall other mannors lands wastes and other boundarys in the said county of York [x x x][8] / That the said mannor of Grinton is [but]ted and bounded northwards by the river Swale and by severall mannors northwards from the said river of Swale And the said Attorney General at the relation aforesaid further sheweth that the townes or places called Whitaside & Harkaside and the / hills places moores and wastes thereunto belonging and also the moor or waste called Grinton how are part and parcell of the said mannor and within the territories and fields of Grinton aforesaid which said mannor of Grinton and the moores and wastes of Grinton Harkaside [. . .] / Whitaside and other the premises were formerly parcell of the possessions of the late dissolved monastery of Bridlington in the said county of York And the said Attorney General at the relacion aforesaid further sheweth that his late majesty King William the third being seised and [. . .] by / right of the crown

1 In TNA, the pleadings are contained in portfolio E112/915 EBOR 80.

2 Bill of complaint.

3 Parchment, 830mm x 545mm. The right-hand margin is badly decayed and rubbed. The bill is endorsed in the top left-hand corner with the legal term, Michaelmas 4 Anne, and the date, 25 October 1705.

4 Sidney, Lord Godolphin, first Earl of Godolphin, 1645–1712, politician, was Lord High Treasurer 1700–1701 and 1702–1710.

5 Henry Boyle, 1669–1725, first Baron Carleton, politician, was Chancellor of the Exchequer 1701–1708.

6 Sir Edward Ward, 1638–1714, lawyer, was Attorney General from March 1693 until June 1695 in which year he was appointed chief baron of the Court of Exchequer.

7 Sir Edward Northey, 1652–1743, lawyer and politician, was Attorney General twice, from 29 June 1701 to 26 April 1707; and again from 14 October 1710 to 18 March 1718.

8 More than half a line of text has been erased.

of England of and in all those mines and veines of lead as well open as to be found arising happening or being within the said mannor of Grinton and the territories fields wastes and moors thereof and particularly arising or being in Whitaside Harkaside [Grinton] / -How and the moors and wastes thereof or thereunto belonging did by his letters patents[9] lease or grant under the seale of this court made and bearing date on or about the eighteenth day of August in the eighth year of his reigne for the consideracions therein mencion'd / grant and to farm let to George Tussingham[10] gentleman (among other things) All and every those mines and veines of lead as well open as to be found arising and being within the territories and fields of Grinton Whitaside and Harkaside aforesaid in the county of York [parcell of] / the lordship or mannor of Grinton aforesaid and sometime parcell of the possessions of the said late monastery of Bridlington in the said county of York and free liberty to dig for the same together with all edificies lands commons waters watercourses comodities and emoluments [what] / -soever any ways relating or belonging to the premises or together with them theretofore demised occupied reputed or enjoy'd (and which premises were amongst other things formerly demised to Humphrey Wharton[11] esquire by letters patents of the late King Charles the First bearing date [the] / first day of May in the fourth year of his reigne for the term of twenty one years as therein mencion'd)[12] To have and to hold all and singular the mines and premises aforesaid with their appurtenances unto the said George Tussingham his executors administrators and assignes from [the] / making of the said letters patents lease or grant to the said George Tussingham to and for the end and term of one and thirty years from thence next ensueing yielding and paying yearly to his then majesty his heires and successors the rent of twenty shillings and one full tenth / part of the yearly profits from time to time proceeding growing hap'ning or arising out of and for the said mines in mony as in and by the said letters patents or grant remaining upon record in this court and to which the said Attorney General referreth more at large it doth / and may appear And the said Attorney General further informeth your lordships that there formerly being (as yet there are in fact) severall mines and veines of lead in and under the said mannor wastes and moors of Grinton Harkaside and Whitaside one Sir Solomon Swale pretending some right / or title to the said mines And having got into possession of the same the said relator Marriott in the name and upon the demise of the said Tussingham to assert the right of the crown and his own right did bring an ejectment about eight years ago and upon a full and fair trial recovered / the possession of the same and in a short time after there was another trial (and which was at the bar of this court) concerning the title of the same where again a verdict was given in affirmance of the title of the crown and of the relator or his trustee both which trials and proceedings [were] / carry'd on by the relator at his very great charge and expence for whom the said Tussingham was only a trustee and who hath since duly granted and assign'd the said lease or grant so to him made unto the said relator Marriott his executors administrators and assignes And the said Attorney General / informeth your lordships that the title of the crown and the relator to the said mines

9 The phrase 'letters patents' is used instead of the more modern 'letters patent'.

10 George Tushingham was a sworn clerk in the Exchequer and also acted as Marriott's trustee in the matter of the Crown lease. Marriott leased the Grinton mines from 18 August 1696 for a term of thirty-one years.

11 Humphrey Wharton Esq. of Gillingwood, Yorkshire (d. 1 Jan. 1635) was in possession of a twenty-one year lease of the Grinton mines from the Crown at the time of his death. The lease was due to expire on 1 May 1649.

12 This bracket is opened after 'enjoy'd'.

being thus settled and affirmed by two solemn tryals at law and the relator having such grant and assignment of the said crown lease from the said Tussingham his trustee And there being severall / mines or veines of lead in the wastes and moores of Grinton Harkaside and Whitaside he the said relator did proceed to dig and work the said mines and particulerly the mine or vein at Harka hill upon Harkaside moor and the mine or vein at Grinton how being part of Grinton moor both within / the said mannor of Grinton and demised by the crown to the said Tussingham as aforesaid and hath ever since been in the quiet possession thereof and digged and worked the same and hath constantly answer'd and paid the reserved rent to the crown and yearly answerd accounted for and paid to the crown the clear [tenth] / part of the produce in mony amounting to very great summes according to the said lease or grant from the crown as appears by the said account remaining upon record in this court by means whereof the said Attorney General on the behalf of her majesty as well as the relator well hoped that [. . .] / title of the crown would have been in quiet and no further pretences made to the said mines by any person whatsoever But now so it is may it please your lordships that the right honourable Thomas Lord Wharton <the honourable Philip Wharton esquire his son & heir apparent Robert Robinson William Arundel Ralph Wensley Miles Close Christopher Harker Leonard Galloway Ralph Addison John Hird> and Mathew Smale [blank] Hartley[13] [blank] Hetherington[14] Sir Solomon Swales [blank] Gibson[15] [. . .] / <and divers other persons> at present unknown to the said Attorney General whom when discovered the said Attorney General prays liberty to make parties defendants and with apt & proper words & clauses to charge 'em having confederated them <selves for that purpose they the said Smale Hartly Hetherington Anthony Wharton and other persons aforesaid or some of them> have lately enterd upon part of the said mannor and moors of Grinton and particularly on Harka Moor and H[. . .] / on Grinton how and on several places within the territories and fields of Grinton in the said county of York and laid some claim to the said mines and disturbed the crown and relator therein and when the relator apply'd in all respectfull manner to know the meaning thereof or what right / or title they pretended to the same they the said Thomas Lord Wharton Smale Hartley Hetherington <Anthony Wharton> and Sir Solomon Swale <and the rest of the confederates> pretended and insisted upon it that they or some of them had some right or title to the same and sometimes pretended that the said Lord Wharton <or the said Philip Wharton his son> was seised in fee of the mannor / or reputed mannor of Helaugh in the said county of York or of some other mannor lands or tenements and that the moors and wastes in Grinton did belong to the same and particulerly that the places called Harkaside and Harka hill or moor thereunto belonging and the said place [called] / Grinton how and the moor of Grinton and all the mines and veines therein were parcell of the said mannor of Helaugh and did belong to the same and were noe part of the said mannor of Grinton or the moors or wastes thereunto belonging or any part of or within the territories or fields of

13 The Hartley referred to could be the Mr Hartley who is described as an agent or steward for Lord Wharton (see pp. 265 and 279). An Edward Hartley was named by Marriott in his affidavit dated 23 October 1705 (E207/111/1). This is presumably the same Edward Hartley, a miner, who was served with an injunction by Mathew Sisson on 16 April 1706 (E207/112/1).

14 This Hetherington is probably to be identified with Joseph Etherington, one of the 'agents, servants or workmen' working at Lord Wharton's mines, who was served with an injunction by Mathew Sisson on 17 April 1706 (E207/112/1). Joseph Etherington, a gentleman of Hartford [Hartforth] near Gilling, aged 39, made a deposition on behalf of Lord Wharton in August 1707 (see p. 355).

15 Richard Gibson, a husbandman of Reeth, aged 70, made a deposition on behalf of Lord Wharton in August 1707 (see p 285).

Grinton / and under such pretences they the said Lord Wharton <& Philip Wharton> or the said Smale Hartly Hetherington <Anthony Wharton> Sir Solomon Swale and several others <aforesaid> by his or their order direction comand or appointment pretended within these two months last past or some other short space of time to set out and ride the boun[daries] / of the said mannor of Helaugh and under such pretences did actually ride over and take in <a> great part of the said mannor of Grinton and the wastes and moors of the same and particulerly the said places calld Harkaside Whitaside Grinton how and the said wastes called Harka hill Grinton how [. . .] with / the mines aforesaid and several other parts and places of the wastes and moors within the territories and fields of Grinton aforesaid and then and there laid claim to or pretended some right or title to the said mines and veines so demised to the said Tussingham by the crown and particulerly to the mines on / Whitamoor Harka hill and Grinton how and forbid the relator's agents and servants who were working the same to proceed any further therein or further or any more to work the same <and the said Lord Wharton's agents have since offer'd to indempnify the relator's workmen provided they will turn and work the mines for him the said lord> and altho in their riding their pretended boundaries they did not take in the mouth of the said mines or veine there [. . .] / working by the relator at Grinton how yet they came very near the same and have under such their pretences set down <and sunk> one or more shaft or shafts upon the vein of the said mine and before and near adjoining to the relator's shafts there <and have so far advanced & carried on their works that they give out [. . .] / that they can sink into the said veins before Martinmass[16] next and [. . .] / quantitys of ore as will not only defray the charge & expense of their labour [. . .] to support their> / <possession there & defend them in any suits at law that may be comenced by or against them or any of the confederates for their actions & doings in the premises and hereupon the said confederates pretend to bring actions & prosecute suits at law against the relator his servants agents or workmen touching the said mines & the relator's right & title thereunto & to disturb him in the possession of the same> ------- And the better to carry on the said design and to compass their ends they the said Mathew Smale Hartley Hetherington and other the said confederates did write or agree to write some paper / as mencioning to contain the bounders of the said mannor of Helaugh and give out in speeches that they have sign'd and got several hands for the same whenas if any such paper or writing there be the same was got made and obteind by fraud and surprize and does not contein the true bounders of the said mannor / but is false and mistaken in itself and was offer'd and tendered to severall knowing persons to have signed and attested the same who refused so to do and declared the said paper was not true and did comprehend much more lands and grounds than the mannor of Helaugh or estate of the said Lord Wharton / <or Philip Wharton> And yet the confederates now conceal the same and refuse to let the relator have a copy thereof or to discover who have signed and who refused to sign the same and in truth the said Attorney General does believe that the said Lord Wharton did know nothing thereof or was privy or consenting to the same and that / the same was carried on and contrived by the other persons and confederates on their own head and without the order or privity of the said Lord Wharton All which doings and pretences are done and carried on with design to create some evidence of a right hereafter and to disinherit the crown / and dispossess the relator of the said mines and veines and the profits and produce of the same and to defeat the title of the crown and the relator thereunto when the

16 The feast of Martinmas was held on 11 November.

aged witnesses that could disprove and contradict the same are dead until when they pretend not to try the same / Whereas the truth was and is and so the Attorney General on the behalf of her majesty and the relator hereby chargeth the same to be that the said Lord Wharton <or Philip Wharton his son> hath no right or title to the said wastes moors or mines {there} <of> or in Grinton aforesaid or any of them and that the same are no part / of the said mannor or pretended mannor of Helaugh or within the boundarys of the same or any other part of the said Lord Wharton's <or Philip Wharton's estate> estate[17] but are part of and within the mannor of Grinton or within the territories and fields of Grinton Harkaside and Whitaside aforesaid And the <said> mannor / or pretended mannor of Helaugh lyeth on the northside of the river Swale and is bounded by the said river and by other known and notorious markes and boundaryes from the said mannor moores and wastes of Grinton Harkaside and Whitaside and does no where lye within or break into the said mannor of Grinton All which does plainly [. . .] / in and by the antient deeds mapps surveys terriers and writings in the possession power or custody of the said defendants in case they would read over peruse search or examine the same or cause the same to be so carefully done by some other person or persons on their behalfe And the said mannor of Grinton and the moores / and wastes of the same doe extend and alwayes have been reputed to extend from the mannor or lordship of Ellerton in the east to Haverdale beck in the west and from the river Swale in the north to and beyond the high Carle in the south adjoyning to his grace the duke of Boltons[18] mannor there and [. . .] / part of the mannor of Helaugh does extend or lye within the territories and fields of Grinton aforesaid, but the Crowne and relator are duly intitled to all the mines lying within the boundaries aforesaid And that [from][19] time immemorial Harkaside and Whitaside and Harka Harka hill or moore Whita Somer Lodge and [Grinton] / –how are all lying within and parcel of Grinton mannor and within the territories and fields of Grinton aforesaid and are part of the wastes and moores of the same And the said Lord Wharton <& Philip Wharton> under colour or pretence of haveing some houses or freehold <or copyhold> estate in or neere Harkaside or [. . .] / [x x x][20] pretends and gives out in speeches that the mannor of Helaugh extends to and takes in Harkaside and the moores and wastes there [. . .] / layes claimes to the mines there and endeavours to conceale and confound the boundaries of the said mannors of Grinton and Helaugh in prejudice to the Crowne and the relator and to the disherison of the Crowne of the said mines and veines Whenas neither the said Lord Wharton <Philip Wharton> [nor any of] / his <or theer> ancestors ever pretended or layde clayme to any such boundaries right or title or to have seniory or lordship within the mannor of Grinton or upon the moores and wastes thereof or any right or title to any mines open or not open in or upon the same or in Harkaside [. . .] / Whita or Grinton how untill now lately within this month or two by rideing or pretending to ride the boundaries aforesaid. Now for that the title and inheritance of the Crowne and rents reserved and profits made and to be made of and in the said mines are and stand in [this] / court and are under the care and inspection of the same and ought not to be brought in question elsewhere than before your lordshipps, the records and evidences concerning the same and the right and title of the Crowne thereunto being all in this court [. . .] /

17 The word 'estate' is repeated in the original.

18 Charles Paulet (1661–1722), politician and nobleman, succeeded to the titles of Marquis of Winchester and Earl of Wiltshire in 1675. In 1699 he was created second Duke of Bolton.

19 The word 'from' is omitted in the original.

20 Half a line of text has been erased.

proper offices before your lordshipps And for that her majestie and the relator ought to be quieted in the possession of the said mines and the workeing of the same by the aid and assistance of this court and for that the Crowne and relator are so very much concerned [. . .] / of the said mannor of Grinton and territoryes and fields of Grinton aforesaid and particularly in the moores and wasts of Harkaside Whitaside and Grinton how (clayming and having a possession of and in and right unto all the mines open or not open ariseing upon [. . .] / the same) And for that the said Attorny General on the behalfe of the Crowne and the relator cannot discover the matters and things aforesaid or the pretences of the said Lord Wharton <& Philip Wharton> and confederates or the meets or bounds of the said mannors of Grinton and Helaugh [. . .] / the bounds or boundaries of the same or preserve the testimony of the witnesses who are old and unable to travel in this behalf but by the answers of the said defendants and the aid of the honourable court To the end therefore that the said Lord Wharton <Philip Wharton> and the other confederates may true and perfect answers [. . .] / and singuler the premisses and may severally answer and set forth whether they claim any and what right or title estate or interest in all or any the mines in Harkaside Whitaside Grinton how or the wastes or moores of the same or within the mannor territories and fields of Grinton [aforesaid and] / the moors or wastes of the same And how and by what ways or means and under and from whom and by what right or title whether he the said Lord Wharton did not direct an order and whether the rest of the defendants or some and which of them did not ride the pretended boundaries aforesaid and [. . .] / the same the mines and moores aforesaid or any or which of them or how near to the same And that the said confederates may answer and set forth whether they or some and which of them did not write such paper and sign or subscribe the same as above charged or some other or like paper [. . .] / to some such effect And who it was that subscribed the said paper and to whom the same was offerd or tendred that refused to sign the same and why they refused to do so And that they may set forth a true copy of the said paper and the contents thereof And what bounders the same doth contain [And that] / Lord Wharton <& Philip Wharton> may set forth whether he <or they> be seised of the said mannor of Helaugh and what are the particuler metes and bounds thereof And whether any part thereof do extend into the said mannor of Grinton or the moors and wastes of the same And that the said Lord Wharton [. . . con-] / -federates may set forth and discover what they respectively know believe have heard or been informed are the boundarys of the said mannors of Grinton and Helaugh respectively and what particulerly appears relating thereunto in and by the respective records or copies of records deeds / maps surveys papers or writings in their or any of their custody or power And that they may severally and respectively peruse and inspect the same for that purpose And that the said Lord Wharton may set forth whether he or any of his predecessors did ever before ride [. . .] boundary / or took in or laid claim to the respective places and mines aforesaid And who when and upon what occasion and whether the said Lord Wharton or any of his ancestors and who and when were ever in possession of any and which of the said moors wastes or mines or wrought [. . .] / and when And to the end that the boundarys of the said mannors and wastes of Grinton and Helaugh and the extend of the mines and veins aforesaid so leased to the relator may be set out and ascertained by the aid of this court And that a comission for that purpose [. . .] / awarded And that the crown and relator may be quieted in the possession and working of the said mines by that aid and injunction of this court And that the ancient witnesses for the crown and the relator concerning the premises be examined and [. . .] / evidence preserved to be made use of as occasion shall require And that the said Attorny General on the behalf of the crown and relator may be relieved in all and singuler the premises according to equity and good conscience And that all proceedings [at law be] / stay'd May it [please][21] your lordships to grant unto the said Attorny General your lordship's letter directed to the right honourable Thomas Lord Wharton and her majesties most gracious writ or writs of subpaena directed to the said <Philip Wharton> Mathew Smale [blank] Hartley [blank] Hetherinton Sir Solomon Swale Robert Robinson [. . .] / William Arundel Ralph Wensley Miles Close Christopher Harker Leonard Galloway Ralph Addison John Hird Anthony Hird [blank] Gibson Anthony Wharton and John Milner thereby desiring the said Lord Wharton and requiring and commanding the rest of the defendants and every of them on a certain day and [under a] / certain pain personally to be and appear before your lordships in this honourable court then and there to answer all and singuler the premisses herein before alleged and further to stand to and abide such order and decree of this honourable court as to your lordships shall seem meet / And the Attorny General shall ever pray

/s/ Sam Dodd[22]

E112/915 EBOR 80 – document 2[23]

[membrane i][24]

The severall answers of the right honourable Thomas Lord Wharton in the county of Westminster one of the defendants / to the informacion of Sir Edward Northey knight her majesties Attorney Generall as well on the behalf of her majesty as on the behalf and / the relacion of Reginald Marriott esquire complainants /

This defendant saveing to himself now and at all times hereafter all and all manner of benefitt and advantage of excepcion to the manifold incertainties and insufficiences in the said informacion contained for a full and perfect answere thereunto or unto soe much thereof as this defendant / is advised is any way materiall for him this defendant to make answere unto hee this defendant answereth and sayeth That Philip late Lord Wharton this defendants father as this defendant doubts not to prove was in his life time for sixty yeares and upwards and / to the time of his death seized of and in the dominion lordshipp and mannour of Swaledale and Helaugh alias Helaw alias Ryeth in Swaledale in the county of York within the antient boundaryes whereof there are as this defendant is informed beleives / & doubts not to prove diverse vills towneshipps or hamletts with diverse moors wasts and commons thereunto belonging and divided or distinguished from each other and in particular the vills of East Grinton West Grinton and Harkaside And this defendant / saith that hee is crediblely informed and doubts not to prove that this defendants said late father dureing all his time aforesaid did quietly and peaceabley hold and enjoy diverse demesne lands rents services royalties and priviledges and their proffitts in / Harkaside and West Grinton aforesaid <&> the moors wasts and commons thereunto belonging and that the tenants thereof did dureing the said time pay their rents and performe their accustomed suite and services at the respective courts held for the said mannour / and that the said copyhold tenants who hold their respective <copyhold> tenements by copy

[21] Word omitted in the original.

[22] Sir Samuel Dodd (1652–1716), lawyer, was made Lord Chief Baron of the Exchequer in November 1714 (*DNB*).

[23] Lord Wharton's answer.

[24] Parchment, 780mm x 615mm. Lord Wharton's answer is endorsed in the left-hand margin with the *jurat*, or note of swearing, dated 4 May 1706, signed by baron R[obert] O. Price as witness.

of court roll of the mannour of Helaugh in Swaledale or Rieth aforesaid as alsoe the said freehold and leasehold tenants paying their said rents and doeing their / accustomed suite and service did all along under the right of this defendants said father enjoy a right of common and other profitts upon the moors and wasts of Herkeside and West Grinton separate and distinct from the moors wasts and commons of East / Grinton aforesaid And this defendant further saith that the said Philip late Lord Wharton this defendants said father in or before the marriage of this defendant in the yeare one thousand six hundred and seaventy three made a settlement of / all and singular the premises to the use of himself for life remainder to this defendant and his first and other sons in taile with diverse remainders over as by the same when produced may appeare And this defendant sayeth that Philip Wharton this / defendants son is now liveing and as this defendant is advised is intituled to the remainder of the premises after this defendants death by and under the said settlement and afterwards about tenn yeares since dyeing seized as aforesaid And hee this defendant / being eldest son and heir at law to him did thereupon some short time afterwards enter into all and singular the said premises whereof his father dyed soe seized and hath hitherto enjoyed and claimes the same by and under the said deeds of settlement / and not by reason or colour of the particular tenements or estate that this defendant and his ancestors have enjoyed in Herkaside and West Grinton And this defendant further answereth and saith that hee does not know of any such / mannour as the mannour of Grinton as <in> the bill alledged nor if the same be a mannour doth this defendant know or can sett forth what are the feilds and teretorries thereto belonging nor how the same if any such mannour there be doe lye / nor how the said pretended mannour or the territories thereof or any of them are bounded This defendant has heard that the vill of East Grinton aforesaid has been sometimes by the owners thereof pretended to be a mannour but whether the same / be a mannour or not this defendant knows not it haveing been as this defendant has heard frequently in dispute whether the same be a mannour or no but this defendant denyes that to his knowledge or beleife the townes and places called West Grinton and / Harkaside and the moores and wastes thereto belonging and particularly the places where the defendant as is after sett forth hath caused a shaft to be dugg or any part thereof is or are a parcell of the mannour of Grinton or East Grinton (if the same be a mannour) to this defendants knowledge or beleife and is informed beleives and doubts not to prove that the same are and have been enjoyed by this defendant and his ancestors and their tennants distinct from the said pretended mannour of / Grinton or East Grinton (if any such mannour there be as aforesaid) and beleives <that> the same are parcell of the mannour of Helaugh but whether the same are or not this defendant insists and hopes to prove that hee this defendant is well intituled / thereto as part of his estate And this defendant further sayeth that he doth not know that the premises all any or which of them in the informacion for that purpose mencioned were part of the possessions of the monastery of Bridlington But this / defendant hath heard and hath reason to beleive that the said monastery was seized or possessed of the vill of East Grinton in this defendants answere before mencioned or of some particular tenements with their appurtenances lying and being within / the said vill and alsoe of some particular tenements and rights of common within the places in the informacion called Whitaside and Harkaside and Grinton how all or some of them but this defendant cannot sett forth the particulars or certainties / thereof And this defendant further sayeth that he doth not know that his late majesty King William the third was seized of and in the mines and veines of lead in the severall places in the informacion menciond or sett forth or any of them But this defendant / hath heard and doth beleive that the late King William did under the seale of his exchequer grant a lease to George Tushingham in the informacion mencioned of diverse lead mines But this defendant never haveing seen the said lease or grant or any / copy or inrolement thereof to the best of his knowledge cannot sett forth the same but refers to the said lease when produced to this honourable court nor does this defendant know for whom the said lease was made in trust nor to whom the said lease / was assigned Hee has heard and does beleive that sometime after the granting of such lease there were some suites and tryalls between the relator Marriott or some one in trust for him and Sir Solomon Swale (one other of the defendants / in the said informacion) upon which said tryals this defendant hath heard that there were verdicts obtained against the said Sir Solomon Swale But this defendant is advised and doth insist that hee being noe partie to the said suites / or any of them nor his right in question therein that the same is noe way asserted thereby nor can the same give <the> relator any colour or pretences to be quieted in his possession against this defendant or any claimeing under him / And this defendant further sayeth that he has been informed and does beleive that some time after the said tryalls the said relator did proceed to digg & work the mine or veine at Harka hill upon Harkaside Moore and the meine or veine / at Grinton How and continued the same work at Harkahill for some short time and afterwards desisted But this defendant doth not know or beleive or hath heard that the said relator hath ever since had the quiet possession of the said mines or either / of them And this defendant sayeth that he doth not know whether any and what rents the said relator has answered to the Crown for the said mines or which of them now claimed by the <said> relator nor whether any or what profitt has been answered by / the said relator or any account made by him to the Crowne for any produce of the said mines or any of them But this defendant confesseth it to be true that there being diverse mines and veines of lead lyeing and being within the / moores and wasts of East Grinton aforesaid and other the places to which the said relator Marriott pretends title And hee the said Marriott haveing wrought the said mines called by the name of How Groves in and through / the wastes of East Grinton or that part thereof commonly called by the name of the How and very near to the utmost extremity or extent thereof where the boundaryes thereof divide the same from the moores and wastes of Harkaside / and West Grinton unto which this defendant is intituled And hee this defendant receiveing informacion by his agents that the relator or his agents were about to digg within the boundarys of West Grinton and Harkaside aforesaid and / particularly in that part of the moors wasts and commons now wrought by this defendant hee this defendant did in affirmance and assertion of his right thereto give direcions to his servants and agents to disturb and forbid the relator / from working therein and did in further affirmance and assertion of his right aforesaid on or about the eleventh day of September last (and before the said relator Marriott had carried his works beyond the said boundary) send an / authority to severall of his agents (who are named defendants in the said informacion) thereby authorizeing & requireing them or any three or more of them to ride this defendants boundaries of Herkaside and West Grinton takeing to their / assistance such of the most antient and knowing inhabitants as could give them proper informacion and giveing due notice to the neighbouring lords of the time and certain places of meeting in order to ride the same And this defendant sayeth / that the said boundary was accordingly ridd and attested and subscribed by above sixty persons then present thereat as this defendant is informed and beleives and as by a copy of the said boundary soe ridd attested and subscribed / which this defendant hath annext to this his answere and prays the same may be taken and accepted as part thereof and to which this defendant refers himself for his greater certainty more fully may appeare And / this defendant confesseth it to be likewise true that the said boundary being soe ascertained this defendants agents did by his order sink one or more shafts pitts or groves very near the mouth of the said mine then in / workeing by the said relator his agents or partners at Grinton how and upon the veine of the same mine or before or near adjoyning to the said relators shafts there as in the informacion is sett forth But this defendant sayeth / that the same was soe sunk by this defendants agents within the said boundaries of this defendants said estate And this defendant does lay claime to the mines within the said boundaries but what pretences this defendants agents made / use of in his behalf when they (pursuant to his directions) forbid the relator further proceeding this defendant knows not nor can sett forth And this defendant further sayeth that upon the intelligence of the relators carrying on his work into the / estate of this defendant hee did give directions for the rideing of the boundaries as before is sett forth and beleives the same were accordingly ridd & that in rideing the same they took in some of the places in the bill mencioned (that is to / say) Harka and Harka hill and Harkaside Moore but this defendant did not nor did any of the other defendants in the informacion mencioned to his knowledge enter upon any or take in any other of the places in the / informacion mencioned nor did they in rideing the aforesaid boundary take in or claime to take in any of the places in the informacion except the lands moors & wasts of Harkaside and West Grinton to which this defendant claimes title nor / does this defendant know that his said agents or any of them insisted upon any right or made any claime to any other places either as parcell of the mannour of Helaugh or any other the estate of this defendant or that they / insisted that the said other places were noe part of the said pretended mannour of Grinton nor did they to the knowledge of this defendant or by his order or direction forbid the relators agents or servants to proceed in working any / of the said mines save those in the moores and wasts of Herkaside and West Grinton aforesaid Nor does this defendant know that the said agents have since offered to imdempnify the relators workmen provided they will turne / and work the said mines for this defendant Nor can this defendant sett forth whether his agents have given out any thing relating to their workeing and proceeding in the said mines as in the informacion is alleged Nor does this defendant / know that his said agents pretended to bring actions or disturb the relator in any of the said mines save those of Harkaside and West Grinton aforesaid or the moores and wasts thereof And this defendant confesseth that a paper of the / boundary aforesaid was wrote a copy whereof is hereunto annext but by whom the same was writt this defendant cannot sett forth And this defendant denies that the said paper or boundary to his knowledge or beleif was made or obtained by / fraud or surprize or is to his knowledge or beleif false or mistaken but if the same should be misinformacion [or][25] in any particular appeare mistaken this defendant humbly insists that the same ought not to conclude this defendants right as the / same shall appeare And this defendant sayeth that he does not know that Harkaside and West Grinton and the moores and wasts thereof are part of the mannour of Grinton (if any such mannour there be) and doubts not to / prove the contrary but if the same should appeare ever to have been so this defendant is advized and insists that hee his ancestors and their tenants haveing enjoyed soe long a possession and received the profitts thereof as aforesaid / has a good title thereto And this defendant does beleive that his agents upon rideing the said boundary did in his this defendants right lay claime to the places last mencioned together with the mines therein and this defendant / knows not nor does he beleive that the paper

[25] Word omitted in the original.

drawn upp of the boundary and hereunto annext did comprehend more lands or grounds than the mannour of Helaugh or the estate of this defendant nor does he know that any / of the defendants concealed the same nor refused to lett the relator have a copy thereof or to discover who signed or who refused to signe the same nor doth this defendant know that any person refused to signe the same or that / any of the doeings or pretences of this defendants agents were with designe to create an evidence of a right hereafter or to disinherritt the Crowne or to dispossess the relator of any of the said mines to which he can make out any / title And this defendant further answereth and denyeth that the river of Swale to the knowledge or beleife of this defendant is a boundary either to the dominion lordship or mannour of Swaledale or to the mannour of Helaw And this / defendant doubts not to prove that diverse of the demesne lands moores wasts and commons with the rights royalties members and appurtenances of and belonging to the said lordshipps have always extended and been reputed to extend as well to the / south as northside of the river of Swale And this defendant saith that he doubts not to prove by antient records that the boundarys of Swaledale in the times of the reign of King Edward the first were claimed by Gilbert de Gant / then lord of Swaledale to extend and that the same were then allowed to extend from the antient boundary mark called Stallrestone alias Stallestone or Stallerstone stile which lyes on the south side of the river of Swale unto another bounder / mark on the west and at or near the head of the river Swale called Helaw Min alias Helaw Miln or Hollow Mill Cross and that the north and south side of the said lordshipp were then boundered and divided from the adjoyning dales of / Wensleydale on the south and Arkingarth Dale on the north as heaven water deales (that is to say) as the rain water which falls on the summitt or ridge of the hills or mountains lying betwixt the aforesaid dales deales or divides it self from them / running part into one dale and part into the other And this defendant sayeth that within the aforesaid antient boundaryes not onely the lands of Herkaside and West Grinton are included but severall other places now claimed by the said relator / And this defendant is advised that hee is well intituled to all such rights royalties and priviledges within the extent of the aforesaid boundaryes as have not yet been granted off by the lords of Swaledale or Helaw aforesaid whose estate this defendant / now has and claimes of and in the said lordshipps and every of them And this defendant further sayeth that hee this defendant cannot discover or sett forth the boundaries of this defendants mannour of Helaw different or distinct / from the aforesaid boundaries of the lordshipp of Swaledale And this defendant sayeth that hee is informed and beleives that the severall places in the informacion mencioned by the names of Harka Harka Hill and Harkaside Moore are part / and parcell of Herkeside and West Grinton or the moores wasts and commons thereunto belonging And this defendant further answereth and sayeth that hee cannot sett forth what claime or claimes this defendants ancestors or any of / them had made to the mines of West Grinton and Herkeside or either of them Nor can hee sett forth how often or when they or any of them have dugg or wrought the mines within the boundaries aforesaid Nor has this defendant since / the estate aforesaid come to him searcht or dugg for any mines in Herkeside or West Grinton but does hope and insist that the same shall noe way prejudice this defendants right thereto hee this defendant and his ancestors haveing / valuable mines constantly or generally open in working in other parts of the mannour of Swaledale or Helawe and haveing by himself or tenants a continued enjoyment of other proffitts within the moores and wasts of Herkeside & / West Grinton Nor <doth he> know whether at any time or when or how often the boundaries of Herkeside and West Grinton or of the moores wasts and commons thereunto belonging have been ridd by this defendants ancestors or any of them Nor does / this defendant know that the mines of Herkeside or West Grinton have been wrought or any proffitt or advantage made thereof till the relator Marriott and Sir Solomon Swale did of late years pretend to work the same And / this defendant denyes that he endeavours to conceale or confound the boundaries of the mannour of Helawe or the said pretended mannour of Grinton or that hee this defendant to his knowledge has in his custody or power any records copys of records / deeds mapps surveyes papers or other writeings which evidence or prove the meets or boundarys of any of the premises now in question to be otherwise than as the same are herein sett forth And this [defendant][26] denyeth all and all manner of unlawfull / combinacion and confederacy wherewith he is charged in and by the said informacion without that that any other matter or thing in the said informacion contained materiall or effectuall for him this defendant to make answer unto & herein & hereby not well & sufficiently answered unto confessed / avoided traversed or denyed are true all which matters & things he this defendant shall & will be ready to averr maintain & prove as this honourable court shall award and prays to be hence dismissed with his reasonable costs & charges in this behalfe most wrongfully had & without any proper cause sustained etc

/s/ N. Lechmere[27]

E112/915 EBOR 80 – document 3[28]

[membrane i][29]

The schedule to which the answer annext referrs

In pursuance of the within written authority and command publique notice was given upon Fryday the one and twentieth day of / September at the markett cross in Reeth about one a clock the said day as alsoe the Sunday following in the parish church yard of Grinton / after divine service and sermon ended per Cornelius Freyer bayliffe of the right honourable the Lord Wharton of the rideing the / bounds within mencioned by his lordshipps agents or commissioners upon Wednesday the twenty sixth day of September next / meeting at tenn a clock of the same day att the elme tree in the lane upon the east side of Swale hall Now wee whose names / are under written to testify and declare our being present at the said place according to the said notice and did then ride claime or / goe the bounds of Herkaside and West Grinton in manner following (vizt) From the midstreame of the river Swale near the east / corner of a close called Holme from thence cross the lane to the east corner of Low Close haveing the elme tree upon the west and / Cuddy Spout on the east up a ridge to Benson cowhouse or Ginney wrenn cowhouse soe up by the east wall of Ox Intaik to the / corner from thence by direct line to the White Stone near unto the west end of the How and soe by direct line from the said stone / to the place where the Gray Stone at the east end of the Greets formerly laid and lately removed from thence towards the south / where an antient standard or hurrock stood and alsoe lately removed where the duke of Bolton adjoynes to Harkaside for / his mannor of Bolton the former places being the bounders between Herkaside and East Grinton and soe westwards as heaven / watter deales to a place called Gibbon Hill and in the like manner westwards to a hurrock or standard called Gibbon Hill standard or /

26 Word omitted in the original.

27 Sir Nicholas Lechmere (1675–1727), first Baron Lechmere, lawyer and politician, was Solicitor General from 1714 to 1715 and Attorney General from 1718 to 1720 (*DNB*).

28 A schedule annexed to Lord Wharton's answer.

29 Parchment, 505mm x 410mm.

hurrock from which place westwards comes in the bounder of Whitaside and from the said standard or hurrock northward downe a / gill called Brownah Gill to a place called Horse Pasture corner and soe downe the said gill to a yate or gate called Scabba Yate / and soe to the river Swale which last mencioned bounder divides Harkaside and Whitaside /

Mathew Smales	James Lonesdale	Mark Whipps	George Butterfeild
Joseph Etherington	Mathew Alderson	Henry Machell	James Close
William Waller	John Metcalfe	Simon Robinson	Hugh Stones
Edward Hartly	Thomas Hall	Anthony Hird	Richard Sidgewick
Thomas Kinnerseley	John Daveson	Anthony Hird Junior	William Clarkson
Joseph Coates	Leonard Raw	Robert Robinson	Mathew Buckton
Anthony Fryer	James Hird	John Simpson	Jerimy Robinson
Ralph Fryer	John Miller	Anthony Close	James Clarkson
Cornelius Fryer	Leonard Brackenbury	John Bland	Michael Cowpland
Richard Close	Francis Bell	Ja: Close sen	Luke Robinson
James Hutchinson	Richard Walker	Ja: Close jun	Robert Kirkland
Ralph Addison	Ja: Gorton	John Hird	James Fryer
Anthony Collier	William Arrundell	Christopher Hall	James Fryer Junior
Miles Close	Giles Metcalfe	William Robinson	Leonard Galloway
William Kearton	John Routh	Ralph Metcalfe	Thomas Garth
Ralph Wenesley	Francis Chard	Anthony Hird	Robert Bradbury

E112/915 EBOR 80 – document 4[30]

[membrane i][31]

The severall answers of Sir Solomon Swale baronet one of the defendants to the informacion of Sir Edward Northey knight her majesties Attorney Generall at the relacion / of Reginald Marriott relator /

This defendant saveing to himselfe now & att all times hereafter all & all manner of benefitt & advantage of exception to the many incertainties & insufficiences in the said informacion contained for a full & present answere thereunto or unto soe much thereof as this defendant is advised doe any way materially cause / him this defendant to make answere unto He this defendant answereth & saith that he does not know that the mannor of Grinton in the informacion named is a reall mannor or that the same as such was at any time enjoyed by the monastery of Bridlington But this defendant saith that it may be true that the estate / of the said monastery in the village of East Grinton might after the dissolucion of the said monastery assume the name or reputacion of a mannor & that the ministers of the Crowne might in some of their accounts after the dissolucion of the said monastery stile the same by the name of a mannor & he beleives the same might / be granted out of the Crowne about the forty first yeare of Queen Elizabeth (with an excepcion of the mines therein) by the name of a mannor as severall other granges & farmes frequently were at the nominacion & request of the grantees therein as this defendant has been informed And this defendant further / saith that the reallity of the said mannor has been frequently in dispute as this defendant verily beleives & perticularly in a suite

30 Sir Solomon Swale's answer.

31 Parchment, 760mm x 510mm. The document is endorsed at the foot with the *jurat*, dated 19 June 1706, and the signature of the chief baron, Sir Edward Ward.

which long depended in this honourable court during the reigne of King James the first wherein Henry Simpson[32] clerke (under whose title the relator gives out that he had lately purchased / the inheritance of the said pretended mannor of Grinton) was plaintiff & Richard Hutchinson & others (the tenants of East Grinton) were the defendants in which said cause it does appeare by the interlocutory order still remaining that the deposicions therein taken were ordered to be published & the cause to be heard / but this defendant saith that he is informed that the bill answeres deposcions & other proceedings in the said cause are now taken off the proper fyles & imbezelled destroyed or otherwise concealed from the knowledge of this defendant And this defendant further saith that he doth not beleive that the said East Grinton / was originally granted by the name of a mannor to the monastry of Bridlington by Walter de Gant (by whose grant the relator Marriott by the former informacion in this cause insisted that the same was originally granted to the said monastry) For this defendant saith that he this defendant had cause to beleive that the said Walter de / Gant in or about the latter end of the reigne of King William Rufus or the beginning of the reigne of King Henry the first being seized of the whole dominion lordshipp & mannor of Swaledale or of Helaugh in Swaledale now claymed by the other defendant the Lord Wharton & of the church of Grinton cum Grinton / otherwise called the village of East Grinton (wherein the said church still stands) as part & parcell of the lordshipp of Swaledale He the said Walter de Gant did thereout grant in free almes to the said monastery the said church & village of East Grinton aforesaid by the name of *Ecclesiam de* Grinton / *cum* Grinton or by such like name And this defendant further saith that the said monastry being so seized of the vill of East Grinton & holding the same of the mannor of Helaugh did afterwards obtaine diverse other subsequent grants from the heirs or descendents of the said Walter de / Gant the then lords of Helaugh & Swaledale aforesaid of diverse other perticular lands tenements & farmes scituate & being in Harkaside Whitaside Cogden & other places in the informacion under which said severall grants this defendant verily beleives that the said monastery did hold such / particular lands & tenements with right of common only in respect thereof upon the moores wasts & commons now in question as by the said severall & respective grants to the said monastrey or antient entyrs or copies thereof still remaining more fully may appeare And this defendant / further saith that in case the said vill of East Grinton was ever enjoyed by the said monastry of Bridlington as a reall mannor or that any of the tennants of the said monastry did for any of their farmes or tenements in Herkaside or other places pay their rents to the said monastery at the court / held for such said mannor he this defendant does beleive that the same was so paid for convenience only & not of right But this defendant saith whether the said East Grinton be a mannor or not or be stiled as the relator pretends itt to be by the generall name of the mannor of Grinton or was / soe held & enjoyed by the monastry of Bridlington (which yet this defendant does not admit the same ever was) he this defendant is advised that the same is noe way materiall to conclude this defendants right & title to the moores in question for this defendant saith that he this defendant / does verily beleive that the said pretended mannor of Grinton did never extend to comprehend or take in the soyle or mines of the moores wasts or commons now in question nor does this defendant know or

32 Henry Simpson was vicar of Grinton at the beginning of the seventeenth century. In 1617 he exhibited a bill of complaint in the Exchequer against the tenants of East Grinton in an attempt to force them to attend manor courts and establish his right to enclose land from the moors. The depositions in this case are in TNA E134/16Jas1/Hil12 [1618]. When his suit failed, Simpson sold his lands to Roger Hillary (*VCH*, p. 239; Fieldhouse and Jennings 1978, pp. 114–15).

beleive that either of the said monastry or the Crowne since the desolucion thereof / did in right of the possessions of the said monastry ever worke the mines or enjoy the soyle of the moores wasts or commons now in question or had any other right upon the same save a tennant right only or right of common with other tennants And this defendant saith that if the said monastry / was intituled to any lead mines that it was only to such as were lying or being within their inclosed grounds feilds or territories of Grinton Herkaside & Whitaside in the relators said lease mencioned as other freehold tennants were to the mines within their respective / freeholds <and> the said lead mines of & in the said inclosures of Grinton Harkaside & Whitaside comeing to the Crowne by the attainder of the prior of Bridlington[33] or by the desolucion of the said monastry this defendant saith that if the Crowne before the relator obtained the said lease did make any former leases / of the said mines <the same were only> at the annuall rent of twenty shillings And this defendant beleives that the lead mines within the moores wasts & commons now in question were not expressed in such said leases but the said moores wasts & commons being of very great & vast extent this defendant conceiving / that the mines therein could never be intended to pass thereby under such rent of twenty shillings nor were the mines of the said moores wasts & commons before set forth to the knowledge or beleife of this defendant ever wrought by the Crowne or under such said lease or leases within / the memory of any man liveing till the relator presumed to worke the same And this defendant saith that he beleives it to be true that the relator Marriott whilst he was deputy auditor of this court pretending to make diverse discoveries of concealed lands for the benefitt of the Crowne did by / ways & means for the benefitt of himself procure grants & leases from the Crowne in the names of others & in particular did obtain the said lease now in question yet this defendant saith that he beleives there are noe mines intended to be granted by the said lease but such as lye within the / inclosed feilds or grounds of Grinton Whitaside & Harkaside as aforesaid & such as were parcell of the mannor of Grinton & of the possession of the late dissolved monastry of Bridlington And this defendant saith the he doth not beleive that the relator Marriott is by the words of his said lease / intituled to all the mines lying & being within the whole & intire mannor or reputed mannor of Grinton aforesaid as by the informacion is insinuated but only within the feilds or inclosed grounds parcell thereof or that any of the mines within the moores wasts or commons now / in dispute are expressly granted or intended to be granted by the said lease & this defendant beleives it to be true that the said lease was not taken in the said Marriotts owne name but in the name of George Tushingham in the informacion named as is by the informacion sett forth And this defendant / further answereth & saith that he this defendant being at the time of granting the relators said lease in actuall possession of severall of the mines in the informacion mencioned & the Crowne haveing no manner of possession of any of the said mines at the time of granting of the said lease nor having / ever wrought any mines within the moores in question within time of memory as aforesaid to the knowledge or beleife of this defendant true it is that he this defendant did clayme the same as parcell of his this defendants mannor called West Grinton under certain antient grants / made by the aforesaid Walter de Gant the then lord of the dominion lordshipp or mannor of Swaledale now the estate of the said Lord Wharton <of whom this defendants said mannor of West Grinton is held & not of the relator nor of his pretended mannor of Grinton> which said grants

33 At the time of the Dissolution, in 1537/8, the prior of Bridlington was William Wood or Wode (*Yorkshire Monasteries – Suppression Papers*, J.W. Clay ed., YAS Record Series, vol. 48, 1912, pp. 41–7).

under which this defendant so claymed were made to one Alured the ancestor of this defendant & nephew to the aforesaid Walter de Gant And this / defendant doth verily beleive that the ancestors of this defendant for many years last past have claymed under the said grants But this defendant saith that if the said grants under which this defendant claymes as aforesaid are not valid (which yet this defendant does not admit) or in case the right & interest / of the said Walter de Gant (the then lord of Helaugh or Swaledale aforesaid[34] was not sufficiently granted or conveyed to this defendants ancestors by the aforesaid grants nor was legally vested in this defendant then he this defendant does verily beleive that in such case the estate & interest of & in the mines / & premises now in question does still remaine ungranted & in the other defendant the Lord Wharton (the now Lord of Swaledale & Helaugh aforesaid) And this defendant further answereth & confesseth it to be true that the relator Marriott obtained two verdicts against this defendant as in the informacion is / set forth But this defendant saith that he hath discovered & found out fresh evidence & severall other matters very materiall & not before given in proofe in either of the said tryalls And this defendant further saith that upon neither of the said tryalls with this defendant he the said Marriott did either / prove or pretend to prove that the monastry of Bridlington before the dissolucion of the same had been ever seized of their said estate in Grinton aforesaid as a mannor save by a pretended copy of an inquisicion pretended to be taken after the attainder of the prior of Bridlington wherein it is / pretended to be named as a mannor but he the said relator Marriott did not produce soe much as the originall commission or inquisicion or any other record proveing the same precedent to the time of the dissolucion of the said monastry And this defendant saith that upon neither of the said / tryalls he the said relator Marriott did either prove or pretend to prove that the said monastry did ever enjoy the soyle or mines of the moores wasts or commons now in question & therefore he this defendant is advised that he ought not to abide by the said verdicts but to endeavour by all lawfull / means to prosecute the recovery of this defendants right And this defendant further saith that some person or persons (as this defendant beleives[35] having before the tryall with this defendant at Yorke assises presumed to take out great numbers of originall records out of some of the / treasurys or record offices of or belonging to this honourable court & which concerned or did relate to the said places now in question & the same being carryed or conveyed into Yorkshire & severall of them afterwards produced in evidence at the said assizes by his the said Marriotts attorney or agents / without any officer attending the same or any order of councel authorizing the relator or agents or any other person whomsoever to take the same off their proper fyles this defendant saith that he this defendant was deprived of the liberty of takeing copies of & inspecting the same in / order to make his defence thereunto at the said tryall & this defendant does apprehend that the said records not being under the care & custody of an officer of this court were in the utmost danger of being vitiated mutilated imbezelled or destroyed And this defendant saith that he doth not / beleive that the records relating to the said proceedings are all now remaining in the proper offices of this honourable court as by the informacion is alledged for this defendant saith that the records relating to the said suite betwixt the aforesaid Henry Simpson (the then pretended lord of Grinton[36] / & the tennants of Grinton are not only concealed or taken off the proper fyles but diverse

34 This bracket is not closed.
35 This bracket is not closed.
36 This bracket is not closed.

other records relating to the premises now in question are not to be found as this defendant has reason to beleive And this defendant denyeth that he the said Marriott hath since the said tryalls or verdicts / aforesaid had the quiet possession of the said premises as by the informacion is falsely suggested by the relator for this defendant saith that he this defendant hath since the aforesaid verdicts entred upon the works called the How Groves now wrought by him the said Marriott & claymed the same & has / brought four severall ejectments for the same & endeavoured by all the ways possible for him this defendant to have his right tryed over againe & he the said Marriott has been soe farr from haveing a quiet possession for these three years last past before the exhibiting of the / said informacion as is therein suggested that he this defendant haveing within the said three years pursuant to an order of the high court of Chancery brought his ejectment vizt in Trinity terme which was in the third yeare of her present majesties reigne for trying / his title to the same And this defendant thereupon goeing into Yorkshire to make preparacion for his said tryall but the relator Marriott being apprehensive of this defendants interest in the neighbourhood of Grinton & fearing the neighbourhood in assertion of / this defendants right to the aforesaid works called the How Groves now wrought as aforesaid by him the said Marriott would in conjunction with this defendant have entred upon him & turned him out or that he this defendant would have proceeded to tryall upon the said ejectment / he the said Marriott to prevent the same did contrive to send or employ one Charles Germaine[37] with some other persons to assist him for seizing or apprehending this defendant whilst he continued in the country under the pretence of some warrant which they some times gave / out was from the Queen & councell & at other times from the Attorney Generall & sometimes from the warden of the Fleet & under colour or pretence of such said pretended warrant or warrants the said persons did meet & attempt to seize this defendant soe that he was forced for his security / to leave the country but he the said Marriott being as this defendant beleives informed of this defendants taking coach for London did sett severall persons to waite this defendants arrivall at London & the said persons when this defendant stept out of the coach did seize & take him into custody and / immediately conveyed him to the prison of the Fleet whence for about two months together this defendant was kept under close confinement & denyed the liberty of the rules of the Fleet till he the said Marriott had obtained the injuncion of this court to stay this defendants proceedings at law / touching the said premises upon the said ejectment But this defendant could never yet discover by what authority or warrant the said persons seized this defendant or continued him in custody haveing immediately upon notice discharged what accounts they then had against this defendant / And this defendant saith that he the said Marriott since his obtaining the said injuncion (which was procured for want of this defendants answere to an informacion contrived by the relator to be exhibited against this defendant during the time of his imprisonment aforesaid in order to be quieted in the relators said / possession upon the pretence of the aforesaid verdicts) did never yet proceed further in the said cause nor has the said Marriott ever yet obtained any decree upon the same to quiet him in the possession of the premises And this defendant hopes he never will be able to obtaine any such And this / defendant further answereth & denyeth that he this defendant was any way concerned in the boundary ridd by the other defendant the Lord Wharton or his agents nor was this defendant privy to or acquainted with his lordshipps intencions therein nor of any of the transaccions in the / informacion mencioned relating

37 I have not been able to find out anything about the identity Charles Germaine.

to the proceedings of the said Lord Wharton or of his agents or of the other defendants in the said informacion before the same were respectively transacted And this defendant denys all & all manner of combinacion & confederacy wherewith he is / charged in & by the said informacions without that that any other matter or thing in the said informacion contained materiall or effectuall for him this defendant to make answere unto & herein & hereby not well & sufficiently answered unto confessed or avoyded traversed or / denyed are true all which matters & things he this defendant is ready to averr & prove as this honourable court shall award & humbly prays to be hence dismissed with his reasonable costs & charges in this behalfe most wrongfully had & sustained:

/s/ Jno: Walker[38]

E112/915 EBOR 80 – document 5[39]

[membrane i][40]

The replication of Sir Edward Northey knight her majesties Attorney / Generall at the relacion of Reginald Marryott esquire complainant to the answer / of the right honourable Thomas Lord Wharton defendant /

The said Attorney <Generall> for replication to the said defendants answer doth say in all & every mater & thing / as in & by his bill of complaint he hath already said & alledged & doth & will averr justify / maintaine prove the said bill of complaint & all & every the matters & things therein contained / to be just & true certain & sufficient in the law to be answered unto by the said defendant and that the answer of the said defendant is untrue uncertaine & insufficient in the law to be replyed / unto by the said Attorney <Generall> for divers manifest errors untruthes incertainties & insufficiencies / therein contained without that that any other matter or thing in the said defendants answer / contained materiall or effectuall in the law for the said Attorney <Generall> to reply unto & not herein / & hereby well & sufficiently replied unto confessed & avoided traversed or denyed is true All / which the said Attorney <Generall is> ready to averr & prove as this honourable court shall award & / pray as in & by the said bill hee hath already prayed. /

/s/ Edw Northey[41]

E112/915 EBOR 80 – document 6[42]

E112/915 EBOR 80 – document 7[43]

[38] John Walker was presumably a lawyer acting for Sir Solomon Swales. I have not been able to find any further information about him.

[39] Replication of Sir Edward Northey, Attorney General 28 June 1701 to 26 April 1707.

[40] Parchment, 265mm x 130mm. The document is endorsed in the top left-hand corner with the term (Michaelmas), the regnal year (5 Anne) and the date, 25 October [1706].

[41] Sir Edward Northey – see p. 1 footnote 7.

[42] Commission of *dedimus potestatem* for taking the answer of Philip Wharton in the country. Parchment, 265mm x 210mm. The commission is written in Latin, following the formula given by Bryson, 1975, p. 134, and is dated 5 May 1707. It is addressed to Robert Meade, John Watts, Lawrence Kingford and William Ridle and signed by the Queen's Remembrancer, [Henry] Ayloffe. On the *verso* is written the memorandum of execution which is signed by Lawrence Kingford and Robert Meade. From 1682 to 1714 Lawrence Kingford was vicar of Winchendon where Philip Wharton was living. This document has not been transcribed.

[43] Answer of Philip Wharton.

[membrane i][44]

The severall answeres of the honourable Philipp Wharton esquire commonly called Lord Winchindon <an> infant under the age / of one and twenty yeares by the right honourable Thomas Earl of Wharton (the father and guardian) one of the defendants to / the informacion of Sir Edward Northey Knight her majesties Attorney Generall at the relacion of Reginald Marriott esquire / complainant. /

This defendant now and at all times hereafter saveing and reserving to himselfe all and all manner of benefitt and advantage of exceptions that may be taken to the manifold errours / untruths insufficiences and imperfections of and in the said complainants said informacion contained for answere thereunto or unto soe much thereof as this defendant is advised is materiall / for him this defendant to make answere unto Hee this <defendant> by his said guardian answereth and saith that he this defendant is now at the time of putting in this his answere of about the age of / eight yeares and six months and noe more And that he this defendant is a meer stranger to all the matters things charges and allegacions in the said complainants said informacion / sett forth and charged But this defendant saith that this defendant is oldest son and heir apparent to the right honourable Thomas Earl of Wharton his said guardian and as / such this defendant now hath or hereafter may have some right title or interest of in or to all or some part of the premises in question either as tenant in taile maile or otherwise / under some deed or deeds settlement or settlements grant or grants duely made and executed by the late right honourable Philipp Lord Wharton deceased the defendants grand father / or by this defendants father or some one of his this defendants ancestors whereby this defendant hath as heire in taile or heir maile of the family or otherwise a right title or interest / in or to the said premises in question or some part thereof in revercion And therefore the defendant humbly hopes that this honourable court will take effectuall care that he this defendant / may and shall have the same fully and amply preserved to him and see as he shall not by proceedings in this cause or otherwise in any wise be deprived thereof or debarred therefrom / and the rather for that he this defendant is an infant as aforesaid and submitts himself to the due consideracion and protection of this honourable court And this defendant traverseth / and saith without that that any other matter or thing in the said informacion materiall effectuall in the law to be answered unto by the defendant and not herein and hereby well and sufficiently / answered unto confessed and avoided traversed or denied is true all which matters and things this defendant is ready to aver justify maintaine and prove as this honourable court shall direct and / award And prays to be hence dismissed with his reasonable costs and charges most wrongfully sustained. /

/s/ Nicholas Lechmere

E112/915 EBOR 80 – document 8[45]

[44] Parchment, 540mm x 340mm. The document is endorsed at the foot with the *jurat* which is dated 26 May 1707 and signed by Lawrence Kingford and Robert Meade.

[45] Replication of Sir Simon Harcourt, Attorney General 26 April 1707 to 22 October 1708. Parchment, 265mm x 205mm. The document is dated 23 June 1707 [6Anne]. The wording of the replication is identical to that of Sir Edward Northey (document 5 above) and for this reason has not been transcribed.

Attorney General *ex rel.* Reginald Marriott v. Thomas Lord Wharton

Depositions of Witnesses.[46]

E134/6Anne/Mich 44 – document 1[47]

E134/6Anne/Mich 44 – document 2[48]

[membrane i *recto*][49]

Interrogatories exhibitted to be administred to witnesses produced sworne and examined in a cause / depending in her majestyes Court of Exchequer chamber at Westminster betweene Mr Attorney General aswell on the / behalfe of her majesty as on the behalfe and att the relacion of Reginald Marriott esquire complainant and the right / honourable Thomas Lord Wharton now Earle of Wharton Phillip Wharton esquire & others defendants on the parte of / the complainant /

[1:] Imprimis[50]
Doe you know the relator Mr Marriott and the said defendants the Lord Wharton and Phillip Wharton his sonn and did you know Phillip late Lord Wharton deceased father to / the now defendant the Lord Wharton in his life time or any & which of them and how long is it since the late Lord Wharton dyed. Declare etc /

46 In TNA, the following depositions are catalogued as E134/6Anne/Mich 44.

47 Parchment, 255mm x 125mm. Commission of *dedimus potestatem* written in Latin which gave the court's authority to four commissioners to take depositions in the country. The document is badly worn and only partly legible. The commissioners are named as Thomas Thomson gent., Thomas Gyll gent., Robert Hilton gent. and Thomas Waite Esq. The commission is dated 2 July 1707 and signed by the Queen's Remembrancer [Henry] Ayloffe. The document follows closely the formula given by Bryson, 1975, p. 134. It has not been transcribed.

48 This bundle of documents consists of eighteen large parchment membranes held together with a piece of string passed through a hole in the top left-hand corner of each sheet. The first membrane consists of the interrogatories prepared by Marriott's attorneys and is followed by fifteen membranes containing the depositions that witnesses made in response to these questions. This first set of depositions is in turn followed by a list of counter-interrogatories prepared by Lord Wharton's lawyers and the depositions that were made in response to them. Once all the depositions had been engrossed on parchment, they were bound together into a single bundle together with the interrogatories and counter-interrogatories. The sheets were then numbered i to xviij at the top of each sheet and it is this numbering which is followed here. The *verso* of the final sheet (membrane xviij) is endorsed with the signatures of the four commissioners together with the certificate of delivery and the oath, the *liberatur*, which is signed by the chief baron, Sir Edward Ward. The *liberatur* is written in Latin and follows the formula reproduced by Bryson, 1975, p. 119. The date of the *liberatur*, 25 October 1707, is the date on which the depositions were delivered to the court by the messenger, George Crabtree.

49 Parchment, 580mm x 570mm. Marginal tears in the parchment have been repaired with paper which has become semi-opaque with age making the underlying text difficult to read, especially the line endings in the upper right-hand corner of the *verso*.

50 In the original, this and the headings 'Item 2–17' appear in the margin.

[2:] Item
Doe you know the mannor or reputed mannor of Grinton in the county of Yorke, and Doe you know West Grinton Harkaside & Whitaside or any & which of them [x x x][51] / and how long have you knowne them [x x x][52] to what mannor or mannors <doe> they belong; who ha's been the reputed lords of Grinton dureing your / knowledge, and ha's wrought the mines, & taken upp the waifes & estrayes therein; Doe you know the mannor of Helaugh, what ha's been the ancient and knowne boundary betwixt / the said mannors of Grinton & Helaugh Did the present Lord Wharton or any of his ancestors ever worke any mines or claime any royalties in West Grinton or Harkaside or within any / other place within the said mannor of Grinton, Declare your knowledge herein and how you came to know the same; /

3: Item
Doe you know or cann sett forth the severall meetes or boundaries betwixt the said mannors or supposed mannors of Grinton and Helaugh; how doe the said mannors <lye> adjoine and / bounder upon each other, how is the said mannor of Grinton bounder'd northwards and what is the ancient and knowne boundary betwixt the said mannors of Grinton & Helaugh; Doe / you know or can sett forth how the said mannor of Grinton is bounderd on the east, west, & south sides thereof; are there any other & what mannor or mannors that lye north on the river / Swale, & adjoine upon the said mannor of Grinton, what is the reputed bounder or bounders of the said mannor or mannors; Sett forth the same according to the best of your knowledge / judgment & beleife and the reasons thereof; /

4: Item
Doe you knowe or cann sett forth the names of the waists or commons of the townes hamletts vills or places within the said mannor or supposed mannor of Grinton or within the / feilds & territories thereunto belonging; name, describe & sett forth the same perticularly; Doe you knowe the places of West Grinton & Harkaside are the same townes vills or hamletts or / some few houses; & how many standing at any and what distance from each other; Doe you know or cann sett forth whether the same or any & which of them are within or parcell of or / reputed to be parcell of any & what mannor or mannors; By what rents tenures and customes are the lands holden that lye in West Grinton & Harkaside & of whome; perticularly what / freehold or other estate hath the said defendant the Lord Wharton in each of the said places; ha's the freeholders there ever done suite and service at any courts leet or baron, where & when / that you have knowne or heard off Sett forth the same & the reasons of such your knowledge or hearsay; /

[5: Item]
Doe you knowe or cann sett forth whether all the lands that lye at or near Harkaside & West Grinton have a right of common upon the moors & commons of / Grinton Harkaside and West Grinton, or are <there> any other lands there that have a right of common upon any other moors or commons / belonging to any other and what Mannour [x x x][53] has the owners or possessors of the same <lands> ever had or enjoyed or claimed / right of common for the same; where & upon what moores

[51] Approximately one third of a line of text has been erased.
[52] Approximately one third of a line of text has been erased.
[53] Approximately one third of a line of text has been erased.

wastes or commons; Doe the said lands lye intermixed with the fields of West Grinton or Harkaside; or how, & where does the same / lye; doe they lye & adjoine upon any & what river, are they reputed or said to be left there by any & what river altering or changing its current, or how otherwise; are there any banks sandbeds / or other appearances that shew the probability of the same, or what other reasons are given in the neighbourhood for their having a right of common in another mannor; Declare your knowledge hearsay / and beleife herein with the reasons for the same; /

6: Item
Doe you know or can sett forth whether the said mannor or reputed mannor of Grinton did formerly belong to any & what religious house perticularly what lands & tenements in / Harkaside & West Grinton belongd or ha's been reputed to belong to any & what religious house; what hath been or is the reputation in the country & neighbourhood about the same, / Declare your knowledge & hearsay & beleife herein; /

7: Item
Doe you know the dale or vale called Swaledale & how long have you knowne the same; of what extent doe you apprehend the same to be; was the said dale at any time dureing your / knowledge ever said to be a mannor or lordshipp, or does it containe within the limitts thereof any & what mannor or mannors lordship or lordships, what are or is the name or names thereof & / who are the owners & proprietors or reputed owners & proprietors or lords of the same & all the said dale called Swaledale & the mannors or lordshipps therein contained within the limitts / and bounds of one or more parishes, whats the name of the cheife towne of Swaledale wherein the parish church stands Declare your knowledge; /

8: Item
Doe you know have heard or can sett forth whether the lead mines within the mannor of Grinton was ever in lease from the Crowne before the relator had the same; & when; what was the name / of such lessee or lessees; & whether such lessee or lessees ever worked & enjoyed the same, & for what terme, whether were there any & what mines or veines of lead or lead ore, and in what / place <in> perticular, formerly claimed or worked by Sir Solomon Swale or who else by name; where does the same mines lye; name the place & places perticularly; Doe you know or can / sett forth whether the said place or places or any & which of them be within the territories & feilds of Grinton Whitaside & Harkaside or any & which of them & within the meetes / & boundaries of any & what mannor or mannors & whether such places be or are reputed to be any parte of any & what mannor or mannors, How long did the said Sir Solomon Swale / worke the said mines & for how long had they been worked before, & by whome; was the said mines ever very rich & yeilded great quantities of ore or otherwise; & how long / is it since the said Sir Solomon Swale hath been outed of the same, & upon what occasion, & by whome, & who hath since had or enjoyed the same, & whether such person hath since enjoyed / & wrought the same, or any other, or new mines, at or in the same place or places, or in what other places, & for how long, Declare your knowledge etc /

9: Item
What is the boundary or reputed boundary or boundaries that distinguish separate

& divide the mannors or lordshipps of Grinton & Harkaside aforesaid [x x x][54] / [x x x][55] Is there any river betwixt them & what river by name; Is it a capitall river or a small brooke or rivulett, what have you heard fishers and / other antient & what men that used the said river say of its being a boundary betwixt any & what mannors, Doe you know the said river to be the reputed boundary of any / other & what mannors, Declare your knowledge hearsay and beleife herein; /

10: Item
Doe you know have heard or beleife dureing your remembrance any & what waifes & estrays have been found & taken upon any & what parte of the moores & waistes & commons / of Grinton West Grinton Harkaside & Whitaside; when and by whome by what bayliffe or servant or otherwise to any & what lord & as lord of any & what mannor was the herbage / or soile of the said moores waists & commons of the said towne of Grinton & the severall places of West Grinton Harkaside & Whitaside ever dureing your remembrance separately / eaten & enjoyed & by which of the said places; did you ever know or have heard say that any of the said respective places claimed any such right of sole commoning their cattle or / enjoying the herbage or soile of the said moores waists & commons exclusive of the rest; did any & which of the said places ever claime such right & when, Doe you know any division / markes or marke stones betwixt the respective feilds moores waists and commons of Grinton West Grinton Harkaside & Whitaside name & describe the same & to what uses distinctions / intents & purposes are the said markes, Declare your knowledge & beleife herein; /

[membrane i *verso*]

11: Item
Doe you knowe the mines in question which now are or lately were claimed by the defendant the Lord Wharton or some other person on his behalfe & who by name; & which the relato[r . . .] / belong to the Crown & himselfe as lessee or farmer under the Crowne name & describe the same perticularly, & the place & places where the same doe lye & whether the said place or [. . .] / of any & what mannors or lordshipps to what mannor or lordshipp the said places doe belong, or are reputed to belong, & by whome the same have been enjoyed & for how long, & doe you kno[w . . .] / whether the said Lord Wharton or any of his ancestors ever were in possession of or claimed any right or title to the same, or any parte thereof & when & upon what account or occasion / since Declare etc /

12: Item
Doe you know the townes vills or places called Harka alias Harkaside & West Grinton & the moores & wastes thereunto adjoining called Harka Hill Harka Moore Grinton Moore Grinton [Greets] / Grinton How or How Groves, are all the said places reputed or said to belong to one or different mannors & to what mannor or mannors; Doe you knowe or have heard whether there were formerly [. . .] / mines sunk in or upon the said severall places or any & which of them, or whether such mines were ever worked, or wrought, & how long doe you know, have heard or believe they were / sunk & worked, & by whome, were they then worked, & whether the said Lord Wharton or any of his ancestors ever worked the same, or were in possession of or claimed to be

54 One word has been erased.
55 Approximately a quarter of a line of text has been erased.

lords or [owners of . . .] / Harkaside & West Grinton or either of them, or the moores or wastes belonging to them or either of them, & when first they claimed the same; or who hath had the possession or [. . .] / hath been lords or owners or reputed lords or owners of the same; and of the soile thereof; Declare etc /

13: Item
Doe you know or have heard whether any person or persons & who by the order of the Lord Wharton (or as they soe pretended) did ride or pretend to ride the bounder or bounders of / any & what mannor of Harkaside & West Grinton & when; Doe you know whether the bounder they pretended or did soe ride was a true bounder of such mannor or of Harkaside / & West Grinton; was the said bounder ever rid before, when how often & by whome & whether the same did take in any parte or parts of any other & what mannor or mannors / place or places, Sett forth some perticular instances (if you can) wherein the said bounder was mistaken; Declare etc /

14: Item
Doe you know whether there was any certificate <or> writeing made or signed of the pretended bounders in the precedent interrogatorie mencioned, & when, & by whome were you or / other persons, & who by name, desired or asked, & by whome, to signe the same, was the same offered or tendered to you or to any other person & who by name soe to doe; Did you / signe the same or why, or for what reason did you or others refuse the signing thereof, Declare etc /

15: Item
Were you at any time & when a joint lessee, or otherwise, & of whome; with any & what partners, & who by name of any & what lead mines within the said mannor of / Grinton had you or any of your partner lessees any dependance upon the late Lord Wharton or how otherwise related concerned or akquainted with the said lord, was the / said Lord Wharton privy to your takeing the said lease, or then pretended any right or title to the said mines, or were you at any time & when a miner workman or lead / carryer at any & what mines within the said mannor of Grinton who were your masters for whome you carryed lead & wrought the same; By what right did your / then masters pretend to worke the same, Declare etc /

16: Item
Are you a freeholder or inhabitant in Helaugh or within the towneshipp or mannor of Helaugh or within any parte thereof or are you owner farmer or occupyer of any messuage house or lands lying / within the said mannor of Helaugh or within the complainants said mannor or mannors in Swaledale or where Doe you know or did you lately dwell have you right of Common upon any the moores commons or / wasts within the said mannor of Helaugh or within any parte & parcell thereof or upon any the moores waists or commons of West Grinton or Harkaside Doe you put your sheep beasts or horse upon / the said common or any of them to graze & depasture or use your right of common there or if you had any cattle have you liberty or right of common within the said mannor to put them there / or can you dig for or gett turves ling or peates within the said mannor & what right [or title] have you or doe you claime within the said mannor Declare /

17: Item
Doe you know or can set forth any other matter or thing for the benefitt & advantage of the Crowne or relator or for the further & better discovery of the truth in this cause

set forth & specifye the / same with the reason & cause of your knowledge & beleive therein /

/s/ Sam: Dodd[56]

[membrane ii *recto*][57]

Ex parte quer[ens]

Deposicions of wittnesses <presented> sworne and examined aswell on the part and behalfe of her majestyes Attorney Generall of the relacion of Reginald Marriott esquire plaintiffs as on the part of the right Honourable Thomas / Earle Wharton and the Honourable Phillip Wharton esquire defendants taken att the house of Christopher Bowes of Wensly in the county of Yorke innholder the six and twentieth day of August in the sixth yeare of the / reigne of our soveraigne lady Anne Queene of Great Brittaine France and Ireland, *anno domini* one thousand seaven hundred and seaven by vertue of her majestyes commission issued out of her majestyes Court of Exchequer att / Westminster returnable from the day of St Michael[58] next in one month and directed to Thomas Waite esquire, Robert Hilton gentleman Thomas Thomson gentleman and Thomas Gyll gentleman or any two or more of them.[59] /

Benjamin Purchase[60] of Midlam in the county of Yorke gentleman aged seaventy three yeares and upwards sworne and examined saith as followeth. /

1 To the first interrogatorie this deponent saith that he knowes the relator Mr Marriott and the defendant the Lord Wharton and did know the late Lord Wharton the defendant Lord Whartons late father who dyed about eleaven years since. /

2. To the second interrogatorie this deponent saith that he well knowes the mannor or reputed mannor of Grinton and alsoe the places called West Grinton Harkaside and Whitaside and hath soe knowne them for fifty yeares last past and upwards and saith that the said / places have for all the time of this deponents remembrance beene part and parcell of the said mannor of Grinton and this deponent further saith that for and dureing the time aforesaid (untill the relator Mr Marriott entered and became possessed or seized of / the said mannor and premisses) Sir Solomon Swale grandfather to this present Sir Solomon Swale Sir Henry Swale father of this present Sir Solomon Swale and Simpson and Hillary or some of them were alwayes reputed lords of the said mannor of Grinton and / saith that the said lords by their lessees or assignees workemen and servants have from time to time and for most of the time aforesaid digged and wrought the mynes in the severall places aforesaid without interrupcion and saith that / he very well knowes the mannor of Helaugh in this interrogatorie mencioned and saith that the ancient and knowne bounder[s] betweene the mannors of Grinton and Helaugh are and have alwayes beene accounted and reckoned / from

56 Sir Samuel Dodd – see p. 7 footnote 22.

57 Parchment, 710mm x 570mm. On the *verso*, a tear in the lower right-hand corner has been repaired with paper that has become semi-opaque with age, partially obscuring the deposition of John Simpson.

58 The feast of St Michael and All Angels was held on 29 September.

59 Thomas Gyll of Hartford, near Gilling, was one of two commissioners chosen by Lord Wharton, the other being Thomas Thomson. Robert Hilton and Thomas Waite were chosen by Marriott.

60 In the mid-seventeenth century, Benjamin Purchase was agent for Dr John Bathurst's mines in Arkengarthdale (Tyson 1995). The 1673 Hearth Tax return records that his house in Clints, near Marske, possessed nine hearths, putting him in the top rank of householders in Swaledale and Arkengarthdale (Fieldhouse and Jennings 1978, p. 244). In the 1670s, Benjamin Purchase worked mines on Harka Hill (see the deposition of William Binks, p. 49). In May 1708 he gave evidence for Marriott at the trial in London (LI Misc 536, Ward MSS, notebooks vol. 6, f. 111r).

Arkingarth dale beck up the midd streame of the river Swale to Haverdale beck and this deponent saith that he never knew or ever heard or beleived that the now Lord Wharton or any of his ancestors in all the time of his / remembrance or any for or on their behalfes ever wrought any the mynes in the places aforesaid or ever claimed any royaltyes therein or within any other part of the said mannor of Grinton. /

3. To the third interrogatorie this deponent saith that the mannor of Helaugh lyes on the northside of the river Swale and the mannor of Grinton lyes upon the southside thereof and that the river Swale is and anciently hath beene knowne as the north boundarye / to the said mannor of Grinton to the midd streame there and that the ancient boundarye between the said two mannors of Grinton and Helaugh is the midd streame of the said river Swale and soe always reputed dureing the time of this deponents / remembrance and saith that the east boundarye of the mannor of Grinton is Stallerstone Style Haverdale beck on the west the mannors of Askerigg Nappa and his grace the Duke of Boltons mannors on the southside of the said mannor of Grinton / and saith that the mannors of Fremington and Marricke lye on the northside of the river Swale adjoyneing to the mannor of Grinton and that the midd streame of the said river Swale is the reputed bounder betweene the said mannor of Fremington / and the mannor of Grinton and between the said mannor of Marrick and the said mannor of Grinton aforesaid. /

4 To the fourth interrogatorie this deponent saith that he knowes the moores and wastes within the mannor of Grinton called by the names of Harkahill Grinton how Greenas the Birkes and Grinton moore and alsoe knows the places called West Grinton and / Harkasyde which contains severall st<r>agling houses in distance six score yards as he beleives between one and another to witt West Grinton about three houses and Harkaside about twelve houses and saith that all the time of his this deponents remembrance the / said places of Harkaside and West Grinton and the moores and wastes aforesaid were alwayes reputed parcell of the said mannor of Grinton and saith that Swale Hall parcell of West Grinton was anciently belonging as he has heard and beleives / to the late priory of Bridlington that in Harkaside there are severall tenures as crowne lands copyhold lands and lands formerly held of the said late priory of Bridlington the said copyhold lands consisting of two farmes and severall little closes belonging / to the defendant the Lord Wharton the tenants whereof have used to appeare att his lordshipps courts att Helaugh and the tenants of the said crowne lands have used to appeare att the court leet and court baron held within the mannor of / Arkingarth dale and the other tenants held of the late priory of Bridlington have used to appeare att the sheriffes turne and this deponent saith that he is the better enabled to depose herein for that he this deponent was twenty yeares together / steward of the said court leete and court baron in Arkingarth dale and then held and enjoyed an estate att Harkaside which he this deponent severall yeares agoe did grant and transferr to his <this> deponents eldest son. /

5 To the fifth interrogatorie this deponent saith that all the lands lyeing in or neare Harkaside and West Grinton on the southside of the river Swale being within the mannor of Grinton have and for all the time of this deponents remembrance had / right of common upon the moores commons and wastes of Grinton West Grinton and Harkaside and saith that he this deponent about four and twenty years agoe haveing lands aswell in Reeth in the mannor of Helaugh and alsoe in Helaugh / towneshipp as in Harkasyde in the mannor of Grinton had then the right of common in the moores commons and waste grounds in the mannor of Helaugh in respect of his lands there as well as in the commons and wastes of Grinton Harkaside and West Grinton / in the mannor of Grinton in respect of his lands att Harkaside aforesaid. But that the owners and possessors of land in Harkaside and West Grinton

never claimed any right of common in and upon the commons and wast grounds of Helaugh / nor the owners and possessors of land in Helaugh or Reeth never claimed any right of common in and upon the commons and wast grounds of Harkaside or any the places within the mannor of Grinton aforesaid that he this deponent ever heard of till now of late / since the suites begunn between the defendant the Lord Wharton and the said relator the said Mr Marriott. /

[7] To the seaventh interrogatorie this deponent saith that he knowes the dale called Swaledale and has knowne the same for fifty years and upwards and saith the same extends to six or eight myles in length but beleives the same to be noe mannor of itselfe / but containes severall mannors therein to witt the mannors of Grinton Helaugh Fremington and beleives the mannor of Marrick may be within the same dale of Swaledale but knowes it not of his owne knowledge and beleives that the defendant the Lord / Wharton may have another mannor within Swaledale besides the said mannor of Helaugh and that Sir Solomon Swale Simpson and Hillary some or one of them were formerly the reputed lord or lords of the mannor of Grinton and beleives the heire or heires of / Mr Humphrey Wharton is or are heire or heires of the said mannor of Fremington and that the Lord William Pawlett is lord of the said mannor of Marrick and saith that the said dale called Swaledale and the mannors aforesaid (except Marrick) / are all within the parish of Grinton and that Grinton is the mother church haveing therein one or more chappelles of ease. /

[8] To the eighth interrogatorie this deponent saith that he has heard and beleives that the lead mynes in question within the mannor of Grinton were about sixty years agoe or before that time in lease from the Crowne to Lord Scroope late Earle of Sunderland / for the terme of one and thirty years under the yearly rent of twenty shillings and after the determinacion of that terme he this deponent has heard and beleives that the said mynes were in lease againe from the Crowne to Humphrey Wharton Esq for the like / terme of one and thirty years under the like reserved rent of twenty shillings and beleives the said mynes were enjoyed by the severall lessees and by their respective servants and agents and workmen wrought dureing their respective termes and this deponent / further saith that betwixt thirty and forty yeares agoe this deponent together with Mr Robert Jackson late of Stockton merchant deceased Mr Mathew Smales late of Gilling alsoe deceased and Mr Edward Wyvell late of Richmond alsoe deceased and others were / lessees by lease from old Sir Solomon Swale grand father to this present Sir Solomon Swale of Grintonhow Harkahill Greenas and Birks all within the mannor of Grinton for a terme of about seaven or eleaven years paying a fifth lott of all the lead made of / the oare gott and wrought in the severall mynes and places aforesaid and saith that this deponent and partners did worke and cause to be wrought all the said mynes and places for the terme aforesaid and longer as this deponent remembers by an old / lease or leases obtained from <old> Sir Solomon Swale and this deponent further saith that old Sir Solomon Swale did worke or cause to be wrought severall of the said mynes before such lease or leases by him made to this deponent and his partners and after the / expiracion of the lease or leases aforesaid the said mynes or some of them were wrought of caused to be wrought by Sir Henry Swale and after him by his son the present Sir Solomon Swale and saith that this deponent and his partners had reasonable good gettings / in their time but the [mines] wrought by this present Sir Solomon Swale afterwards were very rich and considerable and saith that the said Sir Solomon Swale was outed or evicted of his possession of and in all the said mynes about tenn yeares agoe by the relator / Mr Marriott by vertue of a lease from the Crowne to him or to some in trust for him or to his use or otherwise for a terme of one and thirty yeares upon which lease and the title thereof the said relator Mr Marriott obtained two verdicts

and <saith > that the said / relator since this recoverye of the said mynes from the said Sir Solomon Swale hath possessed and enjoyed the same peaceably and quietly and hath gott severall great quantityes of lead oare therein <&> converted the same into lead till about two years agoe / that the said defendant the Lord Wharton did enter forceably as this deponent beleives into the lead myne called Harkahill and did about that time sinke and worke two pitts a little west before the place and places the said relator Mr Marriott is now working where his / great gettings now are. /

10. 11. To the tenth eleaventh and fifteenth interrogatories this deponent saith that {never}
15. dureing this deponents remembrance or beleife he <never> knew that the herbage and soyle of the said moores wasts and commons of the towne of Grinton and the severall places / of West Grinton Harkaside and Whitaside or any of them were ever att any time separately eaten and enjoyed by any of the inhabitants of the severall places aforesaid but alwayes joyntly & in common and saith that <there> are severall markes markstones / and gylls betweene the said respective fields moores wastes and commons of Grinton West Grinton Harkaside and Whitaside but cannot particularly sett forth the particular markes markstones and gills by their proper names but saith and beleives the same were / sett as marks to distinguish the rights of the inhabitants of the severall places in getting digging pulling and wynning of peates lyng and turves for their separate uses as used and accustomed (to this deponents knowledge) within severall other mannors / namely the mannors of Arkingarth dale Thoralby and Newbigin and further saith that in all the time of this deponents remembrance and for fifty yeares past he this deponent never heard or knew that the defendant the now Lord Wharton or any of his / ancestors ever were in possession of or made <any> claime of in or to the said mannor of Grinton or any the wastes moores or commons aforesaid (except pulling lyng getting peets and turves for his the said Lord Whartons tenants in Harkaside for the time being[61] and this deponent / saith that he this deponent after he was with other his partners concerned as lessees in the mynes aforesaid called Harkahill Grinton How Greenas and Birks and in or about that time was well accquainted with the late Lord Phillip Wharton and in company severall times with him att / Richmond and att his house in St Gyles in the fields London and having severall discourses with his lordshipp in and about the customes of his mannor of Helaugh and other businesse this deponent never heard the said Lord Wharton make or pretend to have any right title / or interest to any of the mynes in question or any the moores wastes and commons aforesaid except as aforesaid and this deponent further saith that he was very well accquainted with one Mr Phillip Swale late of Hartforth deceased for thirty yeares together in his / life time who <was> as this deponent beleives lessee with this deponent or others of the mynes aforesaid or some of them and the said Mr Phillip Swale was then one of the late Lord Whartons comissioners and the onely person advysing old Sir Solomon Swale in and about the letting and / disposeing of the mynes aforesaid and doth beleive that if the said Phillip Swale had knowne or beleived that the late Lord Wharton had then had any title or pretence of title or claime to any the mynes aforesaid or any the moores wastes and commons aforesaid / the said Phillip Swale would not have intermedled or have beene concerned in any of the said mynes as lessee or with the said Sir Solomon Swale or otherwise. /

[61] This bracket is not closed.

Anthony Alderson[62] of Cogden in the parish of Grinton in the county of Yorke gente aged seaventy yeares and above sworne and examined saith as followeth. /

1 To the first interrogatorie this deponent saith that he very well knowes the relator Mr Marriott and the defendants the Lord Wharton and his son but did not know the late Lord Wharton. /

2 To the second interrogatorie this deponent saith that he doth know the mannor of Grinton as alsoe West Grinton Harkaside and Whitaside and hath knowne the same for fifty six yeares past and has lived mostly all his life time within the said mannor and saith that West Grinton / Harkaside and Whitaside are and have beene for all the time of his remembrance belonging to the reputed or supposed mannor of Grinton and this deponent further saith that for all the time aforesaid Sir Solomon Swale Sir Henry Swale his son the present Sir Solomon Swale /

[membrane ii *verso*]

Mr Roger Hillary and now the relator Mr Marriott have been reputed lords successively of the said mannor of Grinton and saith that about fifty yeares agoe old Sir Solomon Swale did worke the mynes both of lead and coale in Grinton how Harkahill and / Birks all within the mannor of Grinton and the mynes then were of such a considerable value that he has heard his workmen declare that the said mynes of lead were then worth one thousand pounds by the yeare that Sir Henry Swale the son after his fathers death / wrought the said mynes <& after Sir Henrys death the said Sir Solomon wrought the said mines> and Mr Roger Hillary wrought the same aswell in the life time of the said Sir Henry as in the time of the present Sir Solomon though he this deponent hath heard that there have beene some troubles disputes and suites betweene the said Swales and the said / Hillary about the said mynes and further saith that he has heard one Christopher Naylor now dead who was bayliffe and tenant to old Sir Solomon Swale say that he took up a waife sheep upon the common of Grinton and Harkaside and Somer Lodge that about fifty / yeares since one John Raper now dead bayliffe and tenant to Mrs Hillary and her son then a minor told this deponent that he took up waifes for the said Mrs Hillary and her son and particularly this deponent remembers that about the same time when he was a / schoolboy that the said Raper had taken up two waife sheep on Harkaside moore and this deponent helpt him to bring them into the inclosed grounds called Coate Inggs and alsoe knew Mr Roger Hillary when he came of age to have taken up waifes and estrayes / in Grinton Moore and this deponent saith that he knows the mannor of Helaugh and that it lyes on the northside of the river Swale and the mannor of Grinton on the southside of the said river Swale and the middstreame of the said river is the ancient and / knowne bounder between the said mannors and saith that the defendant the now Lord Wharton or any of his ancestors never did (that this deponent knew or heard of) worke any mynes or ever claime any mynes or royaltyes within West Grinton or Harkaside or any other place or / places within the said mannor of Grinton till about two yeares last past. /

3 To the third interrogatorie this deponent saith that the said mannor of Grinton is boundered on the east with Stallerstone gate or Stallerstone style which adjoyns upon the mannor of Ellerton and is boundered on the west with Haverdale beck neare Crack pott which discends / into the river Swale and is boundered on the southside

62 In May 1708 Anthony Alderson, aged 71, gave evidence for Marriott at the trial in London (LI Misc 536, Ward MSS, notebooks vol. 6, f. 111r).

with his grace the Duke of Boltons mannors and saith that that[63] the mannor of Fremington and part of the mannor of Marricke lye upon the northside of the river Swale and adjoyne upon the mannor of Grinton and the / middstreame of the river Swale is the ancient and knowne bounder betweene the said mannors of Fremington and Grinton and part of the mannor of Marrick and the said mannor of Grinton and saith that the other parte of the mannor of Marrick which lyes on the / northside of the river Swale and the mannor of Ellerton which lye on the southside of the said river are boundered alsoe by the middstreame of the river Swale /

4: &: To the fourth and twelve interrogatories this deponent saith that the places called
12 Grinton West Grinton Harkaside Whitaside Somer Lodge Grinton How Harkahill Birks green hill Grinton moore Harkaside moore and Whitaside moore are all within and part and parcell of the / mannor of Grinton and this deponent saith that West Grinton and Harkaside consists of a few scattered houses not twenty houses in number and this deponent saith that part of the lands att Harkaside of about forty pounds per annum are crowne lands and fifty yeares agoe the occupyers / of those lands in Harkaside being tenants to the Crowne did appeare at the court in Arkingarthdale which this deponent very well remembers being then present and that there are other lands in Harkaside aforesaid of the yearly value of thirty pounds which belonged to the priory / of Bridlington and part of the mannor of Grinton and saith the defendant Lord Wharton has in Harkaside freehold lands of about the yearely rent of twenty nobles which he letts to tenants att a rack rent and that the Lord Wharton has alsoe some customary lands in Harkaside aforesaid / consisting of about eight tenants the full yearely value of which said lands are about forty pounds out of which the tenants may pay about five pounds per annum to the Lord Wharton some of the said tenants holding of the said lord att fifteene penny fine and others att twenty penny / fine and that upon the death of the tenant the said lands descend or come to all the sons alike and saith that the said customary tenants in Harkaside doe appeare att the Lord Whartons court att Helaugh and notwithstanding that the said customary tenants in Harkaside aforesaid doe appeare / att the Lord Whartons court yet all the wastes commons and moores in Harkaside doe belong to and are part and parcell of the said mannor of Grinton. /

6 To the sixth interrogatorie this deponent saith that the mannor of Grinton did formerly belong to the priory or dissolved monastery of Bridlington as did severall of the lands in Harkaside as before predeposed. /

7 To the seaventh interrogatorie this deponent saith that he knowes the dale or vale called Swaledale and hath soe knowne it for fifty six yeares and that the same is about twelve myles in length and saith that he never knew or beleives that Swaledale is or ever was reputed / a mannor but that there are four severall mannors within the said dale called Helaugh and Mewkar both which doe belong to Lord Wharton Fremington mannor which belongs to Madam Wharton and the reputed mannor of Grinton now belonging to the relator Mr Marriot and / saith that Swaledale and the mannors or lordshipps therein are all within the parish of Grinton where the parish church stands and one onely chappell of ease within the said parish called Mewkar. /

8 To the eighth interrogatorie this deponent saith that he hath heard and beene a wittnesse att a tryall where a lease was proved to be granted from the Crowne to Mr Humphrey Wharton Esq for a terme of one and twenty yeares and alsoe proved that Mr Wharton wrought the said mynes / att Harkahill or near it by his lessees assignees servants agents and workmen for severall yeares together and saith that old

[63] The word 'that' is repeated in the original.

Sir Solomon Swale about fifty yeares agoe wrought severall mynes and gott severall quantityes of lead oare in the severall places called Grinton How / Harka hill Birks and Greenhills within the said mannor of Grinton and that the said old Sir Solomon Swale by times for twenty yeares and Sir Henry Swale his son after his death wrought the same for some small time and after him the present Sir Solomon and Mr Roger Hillary for / severall yeares and saith that the mynes soe wrought by this present Sir Solomon Swale yeilded very great quantityes of oare and beleives that the said mynes were for some time worth one thousand pounds per annum and saith that the relator Mr Marriott evicted the said Sir / Solomon Swale by a tryall att law forth & out of the said mynes about tenn yeares agoe by lease and title from the Crowne and since the said relator Mr Marriott peaceably enjoyed the said mynes without interrupcion till about two yeares agoe that the defendant Lord / Wharton gott into possession of part of the said mynes. /

10 To the tenth interrogatorie this deponent saith that the commons moores and waste grounds of the towne of Grinton and the severall places of West Grinton Harkaside and Whitaside were never enjoyed separately but the cattle of the severall tennants of the said places used to eate and / enjoy the same <one> with another being not enclosed one from another. /

11 To the eleaventh interrogatorie this deponent saith that he knows the mynes in question now in the possession of Lord Wharton lyeing upon Harkahill and the pitt or pitts lately sunk downe by the Lord Wharton or his agents west upon the mynes in Grinton how now wrought by / the relator Mr Marriott in which he has had and now has great gettings all which are within the mannor of Grinton and never enjoyed by the Lord Wharton or any of his ancestors or that they ever claimed any right or title or any royaltyes there dureing the time of this deponents / remembrance. /

13 To the thirteenth interrogatorie this deponent saith that he was present and in company with the Lord Wharton when he and others about two years agoe ridd his pretended bounder in Harkaside and West Grinton and then he this deponent boldly told his lordshipp it was noe true bounder / and saith that he beleives the said pretended bounder never was ridd before however never within the time of this deponents remembrance and saith that all the lands moores and wastes in which the Lord Wharton ridd his pretended bounder doe lye / within the said mannor of Grinton and for the bounder was mistaken. /

Henry Scott[64] of Preston under Scarr in the county of Yorke yeoman aged seaventy yeares or thereabouts sworne and examined saith as followeth. /

1 To the first interrogatorie this deponent saith that he knowes the relator Mr Marriott but knowes not the defendants nor either of them nor did know the late Lord Wharton. /

2 To the second interrogatorie this deponent saith that he knows the mannor of Grinton in the county of Yorke as alsoe the severall places called West Grinton Harkaside and Whitaside and have knowne the same for upwards of forty yeares and saith that the same dureing the time / of this deponents remembrance were reputed to belong to <and> part and parcell of the said mannor of Grinton and that the Swales and Hillarys were the reputed lords thereof and wrought the mynes both of lead and coale therein and this deponent saith that he this deponent hath att / severall times released waifes and estrayes from the said Swales and Hillaryes and part of them were sheepe of this deponent and saith that he very well knows the mannor of Helaugh and hath

64 In May 1708 Henry Scott, aged 70, gave evidence for Marriott at the trial in London (LI Misc 536, Ward MSS, notebooks vol. 6, f. 113r).

soe knowne the same for all the time aforesaid and dureing his knowledge thereof / the midd streame of the river Swale was alwayes reputed and taken to be the bounder betwixt the two mannors of Grinton and Helaugh and saith that about thirty four yeares agoe he this deponent was fishing in the river Swale on the southside thereof over against the towne of Helaugh / and saith he was then opposed by one Alderson who then lived att Helaugh but John Blades then steward to Sir Solomon Swale bid him fish on he would be his warantt and after that time he this Deponant was never disturbed tho fished often in the said river and further saith that dureing the / time aforesaid he this deponent never <knew or> heard that the present Lord Wharton or any of his ancestors did ever worke or cause to be wrought any mynes or claimed any royaltyes in West Grinton or Harkaside or in any other place within the said mannor of Grinton till of late that some difference / happened betweene the said Lord Wharton and the said relator Mr Marriott. /

Martin Wilson of Carperby in the county of Yorke yeoman aged sixty yeares or upwards sworne and examined saith as followeth. /

[1] To the first interrogatorie this deponent saith that he hath seene the relator Mr Marriott but doth not know the defendants nor did he know the late Lord Wharton. /

2 To the second interrogatorie this deponent saith that he knowes the towne of Grinton and the places called Harkaside and Whitaside and hath knowne the same for forty yeares that he beleives Harkside and Whitaside to be within the mannor of Grinton being soe reputed and / hath heard his ancestors old people say that they are within or belong to the said mannor of Grinton and saith that for all the time aforesaid the Swales and the Hillaryes as he has heard have beene the reputed lords of the said mannor of Grinton and has heard and beleives / they have wrought mynes and coles within the moores or commons belonging to Harkaside and this deponent further saith that he this deponent has sought for his goods upon those moores and commons when they have strayed or that he could not find them and has beene / directed to goe to Swale Hall where one Naylor lived who was steward to Sir Solomon Swale to enquire for them and that he would informe him whether he had taken them up as estrayes but this deponent did not goe thither this deponent saith that he knowes the / towne of Helaugh and that hee has heard ancient men say that the river Swale was the reputed boundarie betwixt the mannors of Grinton and Helaugh and that the Lord Wharton was cheife lord on the northside of the river Swale and Sir Solomon Swale / was the chiefe lord on the southside of the river Swale and this deponent saith that the Lord Wharton nor any of his ancestors did ever in the said time of his knowledge worke any mynes or claime any royaltyes in Harkaside or Whitaside or the moores or commons thereto / belonging that ever this deponent heard of and this deponent saith that all his time he has lived att Carperby aforesaid about four or five myles from Grinton. /

John Simpson of Carperby in the county of Yorke yeoman aged eighty three yeares or thereabouts sworne and examined saith as followeth. /

1 To the first interrogatorie this deponent saith that he knows the relator Mr Marriott but doth not know the defendant the Lord Wharton or his son or either of them nor did know the late Lord Wharton. /

2 To the second interrogatorie this deponent saith that he knowes the mannor or reputed mannor of Grinton in the county of Yorke as alsoe the places called Harkaside and Whitaside and soe knowne the same for sixty yeares and for all the time of this <deponents> remembrance the said / places have beene taken and reputed to lye within and parcell of the said mannor of Grinton and this deponent saith that for all the time of his remembrance Sir Solomon Swale and Mr Roger Hillary have beene the reputed lords of the said mannor of Grinton and Mr Hillary / wrought the

mynes of coles and Sir Solomon Swale the leadmynes within the wastes commons and grounds within the said mannor of Grinton and this deponent saith that he this deponent wanted a hogg sheepe about four and twenty yeares agoe and enquireing of one Myles / Freer of Whitaside after the said sheepe the said Myles accquainted this deponent that he the said Myles had taken up the same sheepe being upon Grinton how for his master Mr Hillary and brought it to Whitaside and this deponent asked the said Myles why he carryed / the hogg soe farr the said Myles Freer replyed and said why soe farr for that he had not carryed it out of the mannor of Grinton and this deponent saith that he hath heard and beleives that the river Swale hath beene the ancient and reputed boundarye / betweene the two mannors of Grinton and Helaugh the said Helaugh mannor lyeing on the northside of the said river Swale and Grinton mannor on the southside thereof and saith that when neighbours had lost any goods it was generally reported that if they were / [. . .] on the southside of the said river than they might enquire for the same of bayliffe for the lord of the mannor of Grinton if left on the northside than of the bayliffe for the lord of Helaugh. And saith that dureing the time of this deponents remembrance the Lord / Wharton nor any of his ancestors did ever worke any mynes or claime any royaltyes in Harkaside or Whitaside or the moores or commons thereto belonging in the said mannor of Grinton. /

[membrane iii *recto*][65]

Metcalfe Hutchinson of Carperby in the county of Yorke yeoman aged seaventy seaven yeares or thereabouts sworne and examined saith as followeth. /

[1] To the first interrogatorie this deponent saith that he knowes the relator Mr Marriott but doth not know the defendants nor did he know the late Lord Wharton. /

[2] To the second interrogatorie this deponent saith that he knowes the mannor of Grinton and West Grinton Harkaside and Whitaside and has soe knowne the same for sixty yeares last past saith that the said places called West Grinton Harkaside and Whitaside are reputed to be / within the said mannor of Grinton saith that dureing the time of this deponents remembrance Sir Solomon Swale and Mr Roger Hillary were the reputed lords of the said mannor of Grinton and saith that he knowes of his owne knowledge that the said Mr Hillary [did] / worke for and gott coles in the wastes of the said mannor of Grinton and that he beleives and hath heard that the Swales have wrought the lead mynes there saith that he hath heard that the said Mr Hillary hath taken upp waifes and estrayes on the commons in the said / mannor of Grinton and that some of his owne neighbours have released their goods from him saith that he knowes the mannor of Helaugh and saith that the river Swale has alwayes dureing this deponents knowledge been reputed to be the boundary betweene / the two mannors of Grinton and Helaugh saith that he this deponent never heard that the present Lord Wharton nor any of his ancestors ever wrought any mynes or claimed any royaltyes in West Grinton or Harkaside or within any other place within the said mannor / of Grinton till within these two yeares that the present Lord Wharton and the relator Mr Marriott fell att variance. /

Henry Chapman of Carperby in the county of Yorke yeoman aged seaventy Seaven yeares or thereabouts sworne and examined saith as followeth. /

[1] To the first interrogatorie this deponent saith that he knowes the relator Mr Marriott

65 Parchment, 740mm x 620mm. The left-hand edge of the *recto* has been repaired with paper. The upper right-hand corner and the bottom part of the *recto* are badly rubbed and dirt-stained.

but knowes not the defendants Lord Wharton nor his son nor did know the late Lord Wharton. /

2 To the second interrogatorie this deponent saith that he knowes the reputed mannor of Grinton the towne of Grinton Harkaside and Whitaside and have knowne them for sixty yeares and saith that Harkaside and Whitaside are part of and belonging to the said mannor / of Grinton and saith that dureing his this deponents knowledge Sir Solomon Swale and Mr Roger Hillary were lords of the said mannor of Grinton and saith that to this deponents knowledge Mr Hillary about thirty yeares or upwards did worke or cause to be wrought in the moores / wastes and commons within the mannor of Grinton cole mynes and hath heard that Sir Solomon Swale did worke the lead mynes within the moores wastes and commons aforesaid and saith that he knowes the river Swale and hath heard and beleives that the said river / is the knowne and ancient boundarye between the said mannors of Grinton and Helaugh and this deponent saith that dureing this deponents remembrance the Lord Wharton nor any of his ancestors did ever worke any mynes or claime any royaltyes in Harkaside / Whitaside or the moores or commons thereto belonging in the said mannor of Grinton and this deponent saith that when any neighbours lost any goods it was generally reported that if lost on the southside of the said river Swale they might enquire for the same of the bayliffes or officers / of the lord of Grinton and if lost on the northside of the said river they were to be enquired for of the bayliffes or officers of the lord of Helaugh. /

John Pratt of Carperby in the county of Yorke yeoman aged fifty three yeares or thereabouts sworne and examined saith as followeth. /

1 To the first interrogatorie this deponent saith that he has seen the said relator Mr Marriott and may have seen Lord Wharton but knowes him not nor his son nor did know the late Lord Wharton. /

2 To the second interrogatorie this deponent saith that he this deponent knowes the mannor of Grinton as alsoe the places called West Grinton Harkaside and Whitaside and has soe knowne the same for forty yeares and saith that the said places of West Grinton Harkaside & / Whitaside are part of the said mannor of Grinton and that Sir Solomon Swale and Mr Hillary were reputed lord or lords of the said mannor of Grinton and for some yeares in time of this deponants remembrance the said Sir Solomon Swale and Mr Hillary wrought the mynes within the / moores and wastes within the mannor of Grinton and saith that he this deponent hath seene severall waifes and strayes and for severall yeares att Mr Hillaryes and Sir Solomon Swale their respective houses taken up by them or their officers but where this deponent knows not and saith / that he knowes the mannor of Helaugh and dureing all the time of his remembrance the river Swale hath beene reputed the ancient and knowne boundarye betweene the said mannors of Grinton and Helaugh and this deponent saith that dureing his remembrance the Lord Wharton / nor any of his ancestors did ever worke any mynes upon Harka or How or any other the places within the said mannor of Grinton or claime any royaltyes within the same. /

Brian Metcalfe of Carperby in the county of Yorke yeoman aged eighty three yeares sworne and examined saith as followeth. /

[1] To the first interrogatorie this deponent saith that he knowes the relator Mr Marriott but knowes not the now Lord Wharton nor his son but hath seen the old Lord Wharton. /

[2] To the second interrogatorie this deponent saith that he knowes the towne of Grinton as alsoe the places called Harkaside and Whitaside and hath knowne them for sixty yeares and saith that the said places of Harkaside and Whitaside are part of the said mannor of Grinton / as alwayes reputed and beleives and this deponent saith that he

never knew any lords of the said mannor of Grinton save Sir Solomon Swale and Mr Roger Hillary who were said and reputed to be lord or lords of the said mannor knowes that Mr Hillary wrought the coale mynes within the wastes of the / said mannor {knowes that Mr Hillary wrought the coale mynes within the wastes of the said mannor} and knowes it the better for that he this deponent hath fetched coles from the pitts there and hath heard and beleives that Sir Solomon Swale did worke the lead mynes within the said mannor and also saith / that he hath heard by severall neighbours and beleives it to be true that both Sir Solomon Swale & Mr Hillary tooke waves[66] and estrayes within the said mannor for as they told this deponent such neighbours had redeemed the same of such lords or their officers and knowes the towne of Helaugh / and that the river Swale runns betweene Grinton and Helaugh and this deponent saith that for the time of his remembrance the Lord Wharton nor any of his ancestors did ever worke or cause to be wrought any the mynes upon the wastes and commons of the mannor of Grinton nor ever made / any claime to the royaltyes thereof that this deponent ever heard of. /

Ralph Robinson of Carperby in the county of York miner aged fifty years <or> thereabouts sworne and examined saith as followeth. /

[1] To the first interrogatorie this deponent saith that he knowes the relator Mr Marriott but does not know the defendants nor did {not} <he> know the late Lord Wharton in his life time. /

[2] To the second interrogatorie this deponent saith that he knowes the reputed mannor of Grinton in the county of Yorke saith that he alsoe knowes the places within the said mannor called Harkaside and Whitaside and has soe knowne the same for forty yeares and says they lye & / belong to the said mannor of Grinton saith that dureing this deponents knowledge of the said mannor of Grinton the Swales and Mr Roger Hillary were the reputed lords thereof saith that old Sir Solomon Swale in his life time wrought the leadmynes in Grinton How within the said mannor of Grinton / and this deponent remembers when he was a boy and in the said Sir Solomons time he wrought one day there himselfe and further saith that the said Mr Hillary alsoe wrought for coales in the waistes of the said mannor of Grinton and that this deponent saw coles sold and the said Mr Hillary[s] / overman receive money for them and saith that he <has> often heard that the said Mr Hillary has taken up waifes and estrayes upon the wastes in the said mannor saith that he knowes the mannor of Helaugh and dureing the time of this deponents knowledge thereof which has been about forty / yeares the knowne and reputed boundarye betweene the mannors of Grinton and Helaugh has been the river Swale saith that for the time aforesaid this deponent never heard till of late that the present Lord Wharton nor any of his ancestors did ever worke any mynes or claime any royaltyes / in West Grinton or Harkaside or within any other place within the said mannor of Grinton although he has beene a miner and workeman in the How parcell of the wastes of the said mannor of Grinton and in severall adjacent mynes to the said Grinton mannor. /

John Metcalfe the younger of Carperby in the county of Yorke yeoman aged forty seaven years sworne and examined saith as followeth. /

[1] To the first interrogatorie this deponent saith that he knowes the relator Mr Marriott but doth not know the defendants the present Lord Wharton nor his son nor did know the late Lord Wharton. /

[2] To the second interrogatorie this deponent saith that he doth know the reputed mannor of Grinton as alsoe the places called Harkaside and Whitaside for upwards

66 i.e. 'waifes'.

of thirty yeares and saith that the said places of Harkaside and Whitaside are in and belonging / to the said mannor of Grinton and this deponent further saith that the reputed lords thereof were Sir Solomon Swale and Mr Roger Hillary and saith that Mr Hillary wrought the cole mynes within the wastes of the said mannor for that he this deponent once fetched coales / from the pitts by Mr Hillary soe wrought and that this deponent about fourteene yeares since had a sheepe that strayed and was taken of Harkaside common by one Naylor and carryed to Swalehall and there this deponent gott the same againe and further saith that he / knowes Helaugh towne and that the river Swale hath comonly been said and reputed to be the bounder betweene the mannors of Grinton and Helaugh and the strayes taken on the north side of the said river Swale did belong to the lord of Helaugh / and such as were taken on the south side of the said river did belong to the lord of the mannor of Grinton and this deponant saith that in all the time of this deponants knowledge the present Lord Wharton nor any of his ancestors did ever worke any the / mynes {in the wastes} in the wastes or commons of the mannor of Grinton or caused the same to be wrought for if he had soe done this deponent would have beene privy thereto being frequently upon the said moores or ever claimed any royaltyes within the same till now of / late. /

John Metcalfe[67] senior of Carperby in the county of Yorke yeoman aged sixty three yeares sworne and examined saith as followeth. /

[1] To the first interrogatorie this deponent saith that he knowes the relator Mr Marriott and hath seen the present Lord Wharton but never see his son nor did know the late Lord Wharton. /

[2] To the second interrogatorie this deponent saith that he knowes the reputed mannor of Grinton as alsoe the places called Harkaside Whitaside West Grinton and Harkahill and hath soe knowne the same for forty yeares & upwards and that the said places have always dureing this deponents / knowledge been reputed in and belonging to the said mannor of Grinton and this deponent saith that dureing the said time the Swales and Hillaryes have been the reputed lords of the said mannor and further saith that the Swales have from time to time wrought the mynes of lead within the / wastes <and> commons of the said mannor and that the Hillaryes have wrought the coale-mynes therein and this deponent hath severall times fetched coles from thence being a penny in a horse load cheaper than other coales though wrought nearer where this deponent then lived and saith that the / said lords of Grinton have frequently taken up waifes and estrayes within the mannor of Grinton and this deponent had cattle taken up as estrayes on both side of the river Swale and those taken up on the north-side of the said river he this deponent did loose the same from the Freers bayliffes to the / Lord Wharton and those taken up on the southside of the said river he this deponent did loose the same from one Naylor bayliffe to the lord of the mannor of Grinton and this deponent saith that he knowes the mannor of Helaugh and knowes that the river Swale is the knowne ancient / boundarye between the mannors of Grinton and Helaugh and further saith that for all the time of this deponents remembrance the now Lord Wharton nor any of his ancestors ever wrought any of the mynes or claimed any the royaltyes within West Grinton Harkahill or Harkaside / or any other the places within the said mannor of Grinton. /

[3] To the third interrogatorie this deponent saith that the said mannor of Grinton is boundered by Stallerstone Style adjoyning upon the mannor of Ellerton on the east

67 In May 1708 John Metcalfe, aged 66, gave evidence for Marriott at the trial in London (LI Misc 536, Ward MSS, notebooks vol. 6, f. 112v).

Haverdale beck on the west and the Duke of Boltons mannors on the south and saith that the mannors of Helaugh / and Fremington both lye on the northside of the river Swale and that the mannor of Grinton lyes on the southside of the said river and that the said river Swale is the <knowne> boundarye aswell betweene the said mannor of Grinton and the mannor of Fremington as betweene the said mannor of / Grinton and the mannor of Helaugh. /

William Humphray of Wensley in the county of Yorke yeoman aged seaventy eight yeares or thereabouts sworne and examined saith as followeth. /

[2] To the second interrogatorie this deponent saith that he knowes the church towne of Grinton and alsoe knowes the places called Harkaside and Whitaside and has knowne them for sixty yeares or thereabouts and saith that about thirty yeares agoe he used to goe upon or travel the moore / of Harkaside very frequently for about tenn yeares together and dureing that time the Hillaryes wrought mynes and gott coales in the said moores and that the Lord Wharton nor any of his ancestors dureing that time ever wrought any mynes within the said moores. /

John Robinson[68] of Preston under Scarr in the county of Yorke gentleman and miner aged sixty two yeares or thereabouts sworne and examined saith as followeth. /

[1] To the first interrogatorie this deponent saith that he knowes the relator Mr Marriott and has seen the present Lord Wharton and the late Lord Wharton his father but never saw the other defendant Wharton /

[2] To the second interrogatorie this deponent saith that he very well knowes the mannor of Grinton in the county of Yorke and he alsoe knowes West Grinton Harkaside Whitaside Harkahill Doves hole[69] Grinton How Harka birks and Green hills and has so knowne the same for forty five / yeares and upwards saith that all the said places belong and are within the mannor of Grinton saith that the reputed lords of the said mannor were for the most part of this deponents remembrance the Swales and Mr Roger Hillary saith that old Sir Solomon Swale and one Ralph Lee who / either wrought or tooke the mynes of the said Sir Solomon about forty yeares since gott considerable quantityes of lead in Grinton how parcell of Grinton mannor and saith that about thirty seaven yeares agoe this deponent saith Sir Solomon Swales workmen and agents worked the mynes / att Harka and soe continued for four or five years together and about thirteene or fourteene yeares agoe he saw the present Sir Solomon Swales agents and servants worke the said mynes att Harka and gott several considerable quantityes of lead to a very great value and soe continued / workeing the said mynes untill the said relator Mr Marriott recovered the said mynes from the said Sir Solomon saith that the said Mr Roger Hillarye wrought mynes and cole pitts in Grinton how about thirty yeares since <&> gott both lead and coale saith that he this deponent / att severall times heard that the said Swales and Hillarye took up waifes and estrayes upon Harkaside and Whitaside as lords of the said mannor of Grinton saith he has knowne the mannor of Helaugh for all the time aforesaid and saith that the middstreame of the river Swale / was alwayes reputed to be the boundarye betwixt the two mannors of Grinton and Helaugh saith he has often heard it soe reputed by many old men that has been discourseing about boundaryes and particularly William Copperthwaite of Grinton Christopher Bell of Preston Bryan / and Richard Braithwaites of Preston who are all since dead saith that he this deponent has been very frequently and often upon the wastes and commons

68 In May 1708 John Robinson, aged 62, gave evidence for Marriott at the trial in London (LI Misc 536, Ward MSS, notebooks vol. 6, f. 113v).

69 More commonly known as 'Devis hole'.

in question for forty five yeares last past and in all that time <the> Lord Wharton nor any of his ancestors ever wrought any mynes in the / said moores and waistes or any part thereof for either lead or coale till about two yeares past and this deponent further saith that he had a small estate in Arkingdale and Swaledale adjacent to the places in question and a great part of his businesse dureing his life time has been / advyseing and workeing in mynes and that about thirty yeares agoe hee this deponent was supervisor for the said Mr Hillary in his workes att Grinton How and was alsoe a supervisor for the Duke of Bolton in severall <of his> workes att Marrick which often occacioned this deponent to see and goe over the moores / in question and dureing the time he was soe supervisor to the said Mr Hillary he began for him a levell in Grinton How and tryed in severall places in the said worke for lead oare to the westward of the said levell but the said Mr Hillarye falling into decay and not long able to pay his workmen / this deponent and the rest were forced to desist notwithstanding the worke was then very hopefull /

[3] To the third interrogatorie this deponent saith that the said mannor of Grinton is boundered on the east by Ellerton mannor at Stallerstone Style and from thence to Sharrow and soe to Snawden Man and from thence to Browne hill and from thence to Ridmer Standard and from thence to Grinton [Greets and from] / [thence to] Gibbon hill and from thence to High Carle which said greets Gibbon hill and High Carle are the boundaryes between the Duke of Boltons mannors and the mannor of Grinton the mannor of Grinton lyeing on the northside of the said boundaryes through Harkahill down to the river / Swale [. . .] and the Duke of Boltons mannors lyeing on the southside of the said boundaryes and this deponent saith the reputed boundarye of the mannor of Grinton on the west is Haverdale beck and this deponent further saith that the reputed boundarye between the mannor of Grinton / [. . .] the middstream of the river Swale and that it is alsoe the boundarye betweene the said mannor of Grinton and Marrick Marricke and Fremington lyeing on the northside of the said river Swale and the said mannor of Grinton lyeing on the southside of the said river and / [this deponent] saith that he hath severall times ridd the boundaryes with the late Duke of Bolton between his graces mannors and the mannor of Grinton /

[membrane iii *verso*]

4 To the fourth interrogatorie this deponent saith that he hath heard that the Lord Whartons estate in Harkaside is about seaven or eight pounds per annum. /

8 & 11 To the eight and eleaventh interrogatories this deponent saith that he hath severall times heard that the mynes moores and wastes in question were formerly granted by lease from the Crowne to Henry Lord Scropp of Bolton afterwards to Humphrey Wharton / esquire who as this deponent <hath likewise> heard {hath} wrought the same and further saith that after the relator Mr Marriott about tenn yeares since recovered the mynes from the present Sir Solomon Swale he the said Mr Marriott has peaceably enjoyed the same until about two yeares / agoe that the Lord Wharton by some means or other gott into <the> possession of part thereof lyeing att Harkaside and sinking downe two or three shafts on the forehead of Grinton How to the west of Mr Marriotts works on the How saith that <that> part of the worke now in the / possession of the Lord Wharton is in Harkaside and How which places are within and doe belong to the said mannor of Grinton. /

Henry Forster of East Bolton in the county of Yorke yeoman aged seaventy six yeares sworne and examined saith as followeth. /

1 To the first interrogatorie this deponent saith he knows the relator Mr Marriott and

hath seen the present Lord Wharton and the late Lord Wharton but never see the son of the present Lord Wharton. /

2 & 12 To the second and twelfth interrogatories this deponent saith that he knowes that which is reputed the mannor of Grinton as alsoe West Grinton Harkaside Whitaside Harkahill Harka moore Grinton Moore Grinton Greets Grinton how or How Groves / and hath knowne the same for sixty yeares and that the said West Grinton Harkaside Whitaside Harkahill Harka moore Grinton moore Grinton Greets Grinton How and How groves are within and belonging to the said mannor of Grinton and saith that the / Swales and Hillaryes were the reputed lords of the said mannor and wrought severall mynes of lead and coale in and about Harkahill <&> other places within the said mannor of Grinton and in this deponents remembrance frequently tooke up estrayes within / the said mannor he this deponent having loosed severall from them and further saith that he knowes the mannor of Helaugh and that the river Swale has been all this deponents time accounted the ancient and knowne boundarye betweene / the mannors of Grinton and Helaugh and this deponent saith that he was well accquainted with the said places and frequently upon these moores and that the present Lord Wharton nor any of his ancestors did ever worke any mynes or claime / any royalties in West Grinton Harkaside or in the moores wastes and commons thereunto belonging till now of late. /

Joseph Robinson of Woden in the county of Yorke yeoman aged thirty six yeares sworne and examined saith as followeth. /

1 To the first interrogatorie this deponent saith that he knowes the relator Mr Marriott as alsoe the now Lord Wharton and he did know the late Lord Wharton but doth not know this Lord Whartons son. /

2 & 12. To the second and twelfth interrogatories this deponent saith that he knowes the reputed mannor of Grinton as alsoe West Grinton Harkaside Whitaside Harkahill Harka moore Grinton Moore and Grinton how or Howgroves and hath knowne the same / for twenty yeares and that the said West Grinton Harkaside Harkahill Harka moore Grinton moore and Grinton how are within and belong to the said mannor of Grinton and saith that the Swales and Hillaryes for all the time of his / remembrance have been the reputed lords thereof till now of late and alsoe saith that they by their agents and servants have wrought mynes of lead {and coale} upon Harkahill and for severall yeares together and had very good / gettings and alsoe wrought mynes of coale in Grinton moore without disturbance and this deponent saith that he has heard and believes that waifes and estrayes have been frequently taken in Sir Solomons time in Harka & carryed to / Swale Hall and knows the mannor of Helaugh and that the river Swale as he hath heard ancient people within the neighbourhood declare is the ancient and knowne bounder between soe much of the mannor of Grinton as from Stollerstone / Style to Scabbawath on the southside of the said river and the mannor of Helaugh on the northside thereof and further saith that he this deponent was a nigh neighbour to the said moores and wastes all the time till now of late and frequently / upon the said moores and saith that the present Lord Wharton nor any of his ancestors did ever worke any mynes or claime any royaltyes in Harkaside or Harkamoore or any other places within the said mannor of Grinton till now of late. /

George Tirry of Ridmer in the county of Yorke yeoman aged seaventy six yeares or thereabouts sworne and examined saith as followeth. /

1 To the first interrogatorie this deponent saith that he knowes the relator Mr Marriott but knowes not the defendants or either of them nor did know the Lord Wharton nor ever see any of them. /

2. 12. To the second and twelfth interrogatories this deponent saith that he knowes the reputed mannor of Grinton as alsoe the places called West Grinton Harkaside Whita-

side Harkahill Harka moore Grinton Moore and Grinton How or How groves and / hath knowne the same for sixty yeares and that the said places of West Grinton Harkaside Whitaside Harkahill Harka moore Grinton moore and Grinton How are within and belong to the said mannor of Grinton and saith that Sir Solomon Swale and / Mr Hillary for all the time of this deponents remembrance have been the reputed lords of the said mannor of Grinton till now of late and have respectively wrought the mynes of lead & coale within the wastes and commons of the said mannor the said / lords haveing respectively wrought mynes of coale within the said wastes nigh unto the mannor or mannors of the late Duke of Bolton and Sir Solomon Swale hath wrought the mynes of lead at Harka hill and Mr Hillary wrought other the mynes of lead upon the / place called Grinton how <saith> Mr Hillary gott not much advantage by his workeing but Sir Solomon Swale had in his time very great gettings and att the time when the relator Mr Marriott evicted him by one or more verdicts att law as this deponent hath heard / and beleives and saith that he hath knowne waifes and estrayes frequently in the time of his remembrance taken up by Sir Solomon Swale and Mr Hillary respectively or their respective bayliffes or servants upon the moores of Whitaside Harkaside / and Grinton moore as this deponent hath heard and believes and this deponent saith that he knowes the mannor of Helaugh and that the river Swale is and for all the time of this deponents knowledge hath beene accounted and reputed the ancient / and knowne bounder between the mannor of Grinton and Helaugh and further saith that dureing all the time of this deponents knowledge the now Lord Wharton nor any of his ancestors did ever worke any mynes or claime any royaltyes / in Harkaside or Harka moore or any other the places there or within the mannor of Grinton. /

Bartholomew Theakstone of Preston under Scarr in the county of Yorke miner aged sixty six yeares sworne and examined saith as followeth. /

1 To the first interrogatorie this deponent saith that he knowes the relator Mr Marriott But knowes not the defendants nor did know the late Lord Wharton yet hath seen the present Lord Wharton at Bolton /

2. 12 To the second and twelfth interrogatories saith that he knowes the reputed mannor of Grinton as alsoe the places called West Grinton Harkaside Harkahill Harka moore Grinton Moore Grinton how or Howgroves and hath knowne the same for / above fifty yeares and that the said places of West Grinton Harkaside Harka moore <Harka hill> Grinton moore and Grinton how are within and belong to the said mannor of Grinton and saith that Sir Solomon Swale and Mr Hillarye have been reputed lords of / the said mannor till of late and the said Sir Solomon wrought mynes of lead upon Harkahill where some times there were considerable gettings and the said Hillarye wrought mynes of lead upon Grinton how and mynes of coale upon Grinton Moore where this / deponent was a workman about fifty yeares agoe and this deponent saith that he knowes Helaugh and that the midd streame of the river Swale all the time of this deponents knowledge hath been accounted and reputed the ancient and knowne / bounder between the mannors of Grinton and Helaugh and further saith that he this deponent from his age of tenn yeares till about four yeares agoe hath wrought in the moores and wastes within the mannor of Grinton and in other adjacent / mannors and frequently hath been upon the moores of the mannor of Grinton and in all this deponents time the now Lord Wharton nor any of his ancestors did ever worke any mynes or claime any royaltyes within the moores of Harkaside / West Grinton or in any place there or within any other place within the said mannor of Grinton till now of late. /

Henry Wheatley of Preston under Scarr in the county of Yorke miner aged seaventy eight yeares sworne and examined saith as followeth /

1 To the first interrogatorie this deponent saith that he knowes the relator Mr Marriott but knowes not the defendants nor did know the late Lord Wharton /

2. 12 To the second and twelfth interrogatories this deponent saith that he knowes the reputed mannor of Grinton as alsoe the places called West Grinton Harkaside Whitaside Harkahill Harka Moore Grinton Moore Grinton Greetes Grinton How or How Groves & has soe / knowne the same for about 50 yeares and that the said West Grinton Harkaside Whitaside harkahill Harka Moore Grinton Moore Grinton Greetes & Grinton How are within <&> belong <to> & parcell of the said mannor of Grinton and alsoe saith that Sir Solomon Swale & Mr Roger Hillary for / all the tyme of <his> this deponents remembrance have beene the reputed lords of the said mannor till now of late and saith that the said Sir Solomon Swale wrought the lead mynes at Grinton How by his lessees agents & workemen and this deponent himself wrought at the / said lead mynes {with} under James Allen the lessee of Sir Solomon and that the said Sir Solomon wrought other <lead> mynes at Harkahill & Green hills within the said mannor and saith that one John Blaides overseer of Sir Solomon Swale's mynes & workes & his bayliffe took up / waifes & estrayes within the mannor of Grinton & this deponent had a mare of his owne that estrayed & taken up by the said John Blaides within the mannor of Grinton & carryed her to his owne house at Swale Hall where this deponent paid the said John Blaides 3^s for her / release & for cryeing the said mare in the markett & other places and this deponent saith that he knowes the mannor of Helaugh & dureing all the tyme of his this deponents knowledge the river Swale from the midd streame thereof hath been alwayes accounted & taken to be / the ancient & knowne bounder between the said mannors of Grinton & Helaugh and also saith that he this deponent hath frequently been upon the moores for 50 yeares past & in all the tyme of this deponents knowledge the now Lord Wharton nor {of} any of his ancestors did ever work / any mynes or claime any royaltyes in Harkaside Harkahill or any places there or any other places within the said mannor of Grinton till now of late. /

William Atkinson of Preston under Scarr in the county of Yorke myner aged 55 yeares sworne & examined saith as followeth /

1 To the first interrogatorie this deponent saith that he knowes the relator Mr Marriott but knowes not the now Lord Wharton (but hath seen him) nor his son nor did know the late Lord Wharton /

2 & 12 To the second & twelveth interrogatories this deponent saith that he knowes the reputed mannor of Grinton in the county of Yorke as alsoe the places called West Grinton Harkaside Whitaside Harka hill Harka Moore Grinton Moore Grinton Greetes Grinton how or How {or} Groves and / hath knowne the same for 45 yeares and that the said West Grinton Harkaside Whitaside Harka hill Harka Moore Grinton Moore Grinton Greetes & Grinton How are within & belonging to the said mannor of Grinton and saith that Sir Solomon Swale & Mr Roger Hillary were / the reputed lords of the said mannor till of late and that Sir Solomon wrought the lead mynes on Harkahill for severall yeares & had very good gettings there & that Mr Hillary wrought the coale mynes on Grinton Moore & the lead mynes on Grinton How for severall yeares / and this deponent was a workeman under the said Mr Hillarye at the coale mines aforesaid for some short tyme but the same did not prove to advantage soe as the said Mr Hillary did desist from the <further> workeing thereof and this deponent saith that the lords of Grinton did take up waifes / & estrayes in the said mannor and this deponent saith that he had two sheep strayed at severall tymes & taken up on some parte of the moores within the mannor of Grinton but upon what parte thereof knows not but one of the said sheep was taken by the officer of the said / Sir Solomon Swale & the other by the officer of the said Mr Hillary & by

this deponent severally released & paid money for the discharge thereof respectively but how much for either of them now remembers not and this deponent saith that he knowes the reputed mannor of Helaugh & / reputed mannor of Grinton & that the midd streame of the river Swale for all the tyme of his knowledge ha's been accounted & reputed the ancient & knowne bounder between the said mannors of Grinton & Helaugh and alsoe saith that for all the tyme of his knowledge the present / Lord Wharton nor any of his ancestors did ever worke any mynes or claime any royaltyes in West Grinton Harkaside or in any the moores or waists thereof or in any other the places within the said mannor of Grinton till now of late. /

Henry Carter of Preston in the county of Yorke {myner} mason aged 62 yeares sworne & examined saith as followeth /

1 To the first interrogatorie this deponent saith that he knowes the relator Mr Marriott but does not know the defendants nor did he know the late Lord Wharton /

2 & 12 To the second & twelveth interrogatories this deponent saith that he knowes the mannor of Grinton in the county of Yorke & the severall places called West Grinton Harkaside Whitaside Harkahill Harka Moore Grinton Moore Grinton Greetes Grinton How or Howgroves & has / soe knowne the same for above forty yeares & saith that all the aforesaid places lye within & belong to the said mannor of Grinton and saith that old Sir Solomon Swale & Mr Roger Hillary were the reputed lords of the mannor of Grinton till of late & that both the / said Sir Solomon Swale & the said Mr Hillarye wrought for lead & coales in Grinton How but beleives their gettings there were not very great and further saith that he hath heard & beleives that the agents of the said Sir Solomon Swale & Mr Hillarye used from tyme to tyme / to take up waifes & estrayes upon Grinton commons & moores and this deponent further saith that he knowes the mannor of Helaugh & ha's soe knowne the same for above these forty yeares & that dureing his knowledge thereof the midd streame of the / river Swale was alwayes the knowne & reputed boundarye betweene the two mannors of Grinton & Helaugh and alsoe saith that for all the tyme of this deponents knowledge the present Lord Wharton nor any of his ancestors did ever worke any mynes or claime / any royaltyes in West Grinton Harkaside or in any the moores or waists thereof or in any other place or places within the said mannor of Grinton till now of late /

[membrane iiij *recto*][70]

Henry Mason[71] the elder of East Bolton in the county of Yorke myner aged fifty nine yeares sworne & examined saith as followeth /

1 To the first interrogatorie this deponent saith that he knowes the relator Mr Marriott & knowes the present Lord Wharton but doth not know his son nor did know the late Lord Wharton /

2 & 12 To the second & twelfth interrogatories this deponent saith that he knowes the mannor or reputed mannor of Grinton as alsoe the places called <West Grinton> Harkaside Whitaside Harka hill Harka Moore Grinton Moore Grinton Greetes Grinton How or how Groves & hath soe knowne the same for nigh fifty yeares and saith that / all the places aforesaid lye within & belong to & are parcell of the mannor of Grinton and saith that old Sir Solomon Swale & Mr Roger Hillary were

70 Parchment, 700mm x 570mm. Tears and other damage to the parchment have been repaired with paper but without loss of legibility.

71 In May 1708 Harry Mason, aged 60, gave evidence for Marriott at the trial in London (LI Misc 536, Ward MSS, notebooks vol. 6, f. 113r).

the reputed lords of the said mannor of Grinton till of late and saith that the said Sir Solomon Swale wrought the lead mynes in Harka and the coale mynes on the / south east of Grinton How & this deponent did fetch coales from thence about 46 yeares agoe as did severall other persons and beleives that Mr Hillary did worke some other coale mynes in Grinton Moore and this deponent further saith that the said Sir Solomon Swale & Mr Hillarye did from tyme to tyme take up waifes & / estrayes within the said mannor of Grinton and particularly saith that about 22 yeares since he this deponent had lost a sheepe & repaired to Mr Hillaryes house at Grinton to enquire after what estrayes were taken up by him who then was not at home but his wife told this deponent that they had several estrayes which / were ringed about the neck & in a little garth & she went with this deponent to shew him the same & said if any of them were his this deponents she directed him to take it but there were five estrayed sheep but none of them were belonging to this deponent and saith that he knowes the mannor of Helaugh and that / the midd streame of the river of Swale for all the tyme of this deponents knowledge hath been {the} reputed to be the ancient & knowne bounder betweene the mannors of Grinton & Helaugh and saith that dureing the tyme of this deponents knowledge he has from tyme to tyme frequently been upon the moores / & waists of the mannor of Grinton & never knew or heard that the present Lord Wharton or any of his ancestors did ever worke any mynes or claime any royaltyes in West Grinton or Harkaside Harka or any the waists there or in any other place within the said mannor of Grinton till about two yeares since /

[3] To the third interrogatorie this deponent saith that the mannor of Grinton is bounde<r>d on the east with the mannor of Ellerton by Staller Stone Style bounderd on the west by Haverdale beck & on the south by the mannors of the now Duke of Bolton and saith that he this deponent knowes the mannors of / Fremington & Marrick which lye on the north side of the river Swale & that the mannor of Fremington doth adjoyne upon the mannor of Grinton & that the midd streame of the said river Swale is the ancient reputed bounder betweene the said mannors of Grinton & Fremington as it is betweene the / mannors of Grinton & Helaugh /

[7] To the seaventh interrogatorie this deponent saith that he knowes the dale or vale of Swaledale for about 50 yeares & that the same is of about 12 myles extent and saith that this deponent never in all the tyme aforesaid ever knew or heard that Swaledale was {ever} any mannor or ever soe reputed but saith / that there are four mannors contained therein namely Meukar Lord Wharton the lord thereof the mannor of Helaugh Lord Wharton lord of the same the mannor of Fremington belonging to Madam Wharton & the mannor of Grinton the relator Mr Marriott now lord / thereof and saith that the whole dale or vale of Swaledale & all the said mannors are contained within one parish called Grinton which is the cheife towne thereof /

James Thompson of Ridmer in the county of Yorke yeoman aged 63 yeares sworne & examined saith as followeth /

1 To the first interrogatorie this deponent saith that he knowes the relator Mr Marriott but knows not the Lord Wharton or his son nor did know the late Lord Wharton /

2 & 12 To the 2d & 12 interrogatories this deponent saith that he knowes the reputed mannor of Grinton as alsoe the places called West Grinton Harkaside Whitaside Harkahill Grinton Moore {Grinton} Grinton how or how Groves & hath knowne the same for about for<ty> yeares & saith that all the places aforesaid / doe belong to & are parcell of the mannor of Grinton as [he][72] believes & saith that old Sir Solomon Swale & Mr Roger Hillary were the reputed lords thereof and saith that the said lords

72 Word missing in the original.

have from tyme to tyme taken up waifes & estrayes within the mannor of Grinton as this deponent hath heard amongst the / neighbours and saith that the river Swale hath been generally reputed to be the bounder between the mannors of Grinton & Helaugh and further saith that he this deponent hath from tyme to tyme been frequently upon the waists & moores within the mannor of Grinton & in all the tyme of this / deponents knowledge the present Lord Wharton nor his ancestors did ever worke any mynes or claime any royalties in West Grinton Harkaside or any the wasts or commons there or in any other place within the said mannor of Grinton /

Henry Mason the younger of East Bolton in the county of Yorke miner aged 56 yeares or thereabouts sworne & examined saith as followeth /

1 To the first interrogatorie this deponent saith that he knowes the relator Mr Marriott but does not know the defendants nor the late Lord Wharton /

2 & 12 To the 2^{d} & 12 interrogatories this deponent saith that he very well know's the mannor or reputed mannor of Grinton & know's the severall places called West Grinton Harkasyde Whitasyde Harkahill Harka Moore Grinton Moore Grinton Greetes Grinton How or Howgroves & hath soe knowne / the same for 40 yeares & upwards saith that all the aforesaid places lye within & belong to the said mannor of Grinton saith that dureing this deponents remembrance till of late the reputed lords of the said mannor of Grinton were the Swales & the Hillaryes saith that severall of this deponents / neighbours who are since dead were workemen for old Sir Solomon Swale & wrought upon Harka & Harka Moore for lead & got lead ore there for the said Sir Solomon & saith that the said Mr Hillarye wrought for coales & gott coales in other partes of the said waists eastward belonging to / the mannor of Grinton saith that he this deponent has been at Swalehall severall tymes looking for estrayes of his owne but mett with none but ha's seen severall estrayes there which as the neighbours (who were old Sir Solomons Swales bayliffs) told this deponent were waifes & / estrayes taken up upon the <waists &> moores belonging to Grinton mannor and this deponent alsoe saith that he often heard & beleives that the antient boundarye between the Lord Whartons mannor of Helaugh & Sir Solomons Swales mannor of Grinton was & is the midd streame of the / river Swale and further saith that for these 40 yeares last past he never heard that the present Lord Wharton nor any of his ancestors ever wrought any mynes or claimed any <of the> royalties in West Grinton or Harkaside or in any of the waists or other places within the said mannor / of Grinton /

3 To the 3^{d} interrogatorie this deponent saith that the mannor of Grinton is bounded on the east with Ellerton mannor by Stallerston Style & on the west by Haverdale beck & on the south by the mannors of the Duke of Bolton & the reason of this deponents knowing the said boundaryes is that he / ha's severall tymes sought for goods upon the moores & waists belonging to Grinton mannor & ha's observed the aforesaid boundaryes & sayes the high Carle Gibbon hill Bolton & Grinton Greetes are the boundaryes on the south between the Duke of Boltons mannors & the said mannor of Grinton /

Thomas Watson of East Bolton in the county of Yorke myner aged 54 yeares or thereabouts sworne & examined saith as followeth /

1 To the first interrogatorie this deponent saith that he knowes the relator Mr Marriott but does not <know> the defendants nor did he know the late Lord Wharton /

2 To the 2^{d} <& 12> interrogatories this deponent saith that he knowes the reputed mannor of Grinton & the places called West Grinton Harkaside Whitaside Harkahill Harka Moore Grinton Moore Grinton Greetes Grinton How or Howgroves & ha's knowne them for 44 yeares or thereabouts & that the / said places are & have been commonly reputed dureing the tyme of his remembrance to belong to the mannor

of Grinton saith that Swales & Hillaryes have been the reputed lords of the mannor of Grinton & Swales wrought the mynes of lead upon Harka hill & Hillaryes / wrought mynes of lead upon Grinton how & also mynes of cole upon the east side thereof which said mynes were for sometyme of a considerable vallue he this deponent haveing wrought at the said mynes for sometyme saith that both the Swales & Hillaryes took up waifes & estrayes / in the said mannor as this deponent ha's heard this deponent further saith that he know's the mannor of Helaugh & that the midle parte of the river Swale ha's been alwaise reputed to be the antient boundarye betwixt the mannor of Grinton & the mannor of Helaugh saith that he / very well knowes & ha's been frequently upon the said moores & commons & that the present Lord Wharton nor any of his ancestors did ever worke any mynes or claime any royaltyes in West Grinton or Harkaside or any of the said moores or commons dureing his tyme except of late /

3 To the third interrogatorie this deponent saith that the said mannor of Grinton is bounderd on the east by the mannor of Ellerton {by} at Stallerston Style & is bounded on the west by Haverdale beck & is alsoe bounderd on the south side by the Duke of Boltons mannors saith he was borne at Grinton & lived there / for twenty five yeares together and further saith that he know's the mannor of Fremington which lyes upon the northside of the said river Swale & adjoynes on that side of the river upon the mannor of Helaugh & alsoe adjoynes upon the said mannor of Grinton the said river Swale divideing / the same & is the reputed boundarye betwixt the said mannors of Fremington & Grinton /

Robert Maitchel of Ridmer in the county of Yorke yeoman aged sixty three yeares sworne & examined saith as followeth /

1 To the first interrogatorie this deponent saith that he know's the relator Mr Marriott but know's not the now Lord Wharton or his son nor did know the late Lord Wharton /

2 & 12 To the second and twelfth interrogatories this deponent saith that he knowes the mannor or reputed mannor of Grinton as alsoe the places called West Grinton Harkaside Whitaside Harkahill Harka Moore Grinton Moore Grinton Greetes Grinton how or Howgroves & hath knowne the / same above forty yeares & saith that the places aforesaid have been commonly reputed to lye within & belong to the said mannor of Grinton saith that Sir Solomon Swale & Mr Hillarye were the reputed lords of the said mannor & that old Sir Solomon Swale by his servants & workemen wrought / the lead mynes at Grinton How about forty yeares since where the gettings were then very ordinarye and that Hillarye wrought the mynes of coale on the east side of Grinton How about halfe a mile from which pitts this deponent hath fetched coales severall tymes and this deponent saith / that he hath heard and believes that the lords of the mannor of Grinton have used to take up waifes & estrayes within the same although this deponent never had any occasion to enquire for any either for himselfe or any his neighbours and this deponent saith that dureing all the tyme of this deponents knowledge / the midd streame of the river Swale hath been accounted & reputed the antient & knowne bounder between the said mannor of Grinton & the mannor of Helaugh and this deponent further saith that in all the tyme of this deponents knowledge & since he was a boy he this deponent hath frequently been / upon the moores & waists within the mannor of Grinton & the now Lord Wharton nor any of his ancestors did ever worke the mynes or claime {the} any royaltyes within West Grinton Harkaside Harkahill or the moores & waists there or in any other place within the said mannor of Grinton / till now of late /

Thomas Young of Preston under Scarr in the county of Yorke myner aged one & forty yeares sworne & examined saith as followeth /

1 To the first interrogatorie this deponent saith that he know's the relator Mr Marriott but know's not the defendants /

2 & 12 To the second & twelfth interrogatories this deponent saith that he know's the reputed mannor of Grinton & the severall places called Harkaside Whitaside Harka hill Harka Moore Grinton Moore Grinton How or How Groves & hath knowne the same for twenty yeares last past & that all the said severall places / have been commonly reputed {&} to belong to & lye within the mannor of Grinton & that the Swales & Hillaryes have been the lords of the said mannor dureing the tyme of his knowledge till of late & know's that Hillarye work't mynes of coale upon Grinton Moore & heard amongst his neighbours / that Swales & Hillaryes tooke up waifes & estrayes of their goods within the said mannor saith he know's the towne of Helaugh & that the river Swale ha's been commonly taken to be the boundarye betwixt the said mannors of Grinton & Helaugh & ha's heard his grandfather about twenty / yeares agoe (who was then about four score yeares of age & dyed about sixteene years agoe) say that the river Swale was the boundarye betwixt the said mannors this deponent further saith that he ha's dureing the tyme aforesaid been frequently upon the said moores & ha's wrought in mynes / adjoyneing thereto & that he never heard or believes {that he never heard or believes} that the Lord Wharton or any of his ancestors ever wrought any mynes upon the said moores /

3 To the third interrogatorie this deponent saith that the said mannor of Grinton is bounderd upon the mannor of Ellerton on the east & upon the Duke of Boltons mannors on the south /

Mathew Braithwaite of Preston in the county of Yorke yeoman aged fifty eight yeares sworne & examined /

1 To the first interrogatorie this deponent saith that he knowes the relator Mr Marriott but does not know the defendants nor did not know the late Lord Wharton /

2 & 12 To the second & twelfth interrogatories this deponent saith that he know's the reputed mannor of Grinton & the severall places called Harkaside Whitaside Harkahill Harka Moore Grinton Moore Grinton How or Howgroves & ha's knowne the same for forty yeares last past & that all the said / severall places have been commonly reputed to belong to & lye within the mannor of Grinton & that the Swales & the Hillaryes have been dureing this deponents knowledge till of late the reputed lords of the said mannor of Grinton & saith that old Sir Solomon Swale wrought the / lead mynes in Harka & gott lead ore & that Mr Hillarye wrought for coales & gott coales in Snawdon Man within the said mannor of Grinton saith that the knowne reputed boundarye between the two mannors of Grinton & Helaugh is the river Swale saith that / he this deponent never heard for these forty yeares by past that the present Lord Wharton or any of his ancestors ever wrought any mynes or claimed any of the royaltyes in Grinton or Harkaside or within any of the moores or wasts within
3 the said mannor of Grinton till / about two yeares last {past} To the third inter-
rogatorie this deponent saith that the said mannor of Grinton is bounderd on the east on Ellerton mannor by Stallerston Style & on the south on the mannors of the Duke of Bolton saith that the river Swale is the reputed boundarye betweene / the mannors of Grinton & Fremington Grinton lyeing on the south side the river Swale & Fremington on the north /

Richard Hamond of Preston under Scarr gentleman aged sixty six yeares & upwards sworne & examined saith as followeth /

1 To the first interrogatorie this deponent saith that he know's the relator Mr Marriott but know's noe other person in this interrogatorie named /

2 & 12 To the second & twelfth interrogatories {interrogatories} this deponent saith that he know's the mannor of Grinton & the severall places called West Grinton Harkaside

Whitaside Harkahill Harka Moore Grinton Moore Grinton Greetes Grinton how or How Groves & ha's soe knowne the same for fifty yeares & / upwards & that all the said severall places are reputed to belong & lye within the said mannor of Grinton & saith that old Sir Solomon Swale & Mr Roger Hillarye were till of late the reputed lords of the said mannor of Grinton saith that both old Sir Solomon Swale & the said Mr Hillarye wrought for & got / coales in Snawden Man on the parte of the waists adjoyneing on the Duke of Bolton & belonging to the mannor of Grinton & that one of the pitts is to this day called Swales Pitt & the other Hillaryes Pitt and further this deponent saith that the river Swale was alwayse reputed to be the boundarye / between the two mannors of Grinton & Helaugh and further this deponent saith that for fifty yeares last past he never heard that the present Lord Wharton nor any of his ancestors ever claimed any of the mynes or royaltyes in Grinton or Harkaside or within any of the moores or waists within the mannor / of Grinton till about two yeares last past /

3 To the third interrogatorie this deponent saith that the said mannor of Grinton is bounderd on the east on {the} Ellerton mannor by Stallerton Style & on the south on the Duke of Boltons mannors & that the Greets Gibbon Hill & high Car<le> are the boundarye marke betweene the Duke of Boltons mannors & / Grinton mannor saith that he know's the reputed mannor of Fremington lyeing on the north side of the river Swale & saith that the river Swale is the reputed boundarye between the said Fremington mannor & the <said> mannor of Grinton lyeing on the southside of the said river Swale /

[membrane iiij *verso*]

Alexander Hibert of East Bolton in the county of Yorke yeoman aged seaventy nine yeares sworne & examined saith as followeth /

2 & 12 To the second & twelfth interrogatories this deponent saith that he know's the mannor of Grinton & know's the severall places of West Grinton Harkaside Whitaside Harka hill Harka Moore Grinton Moore Grinton Greetes Grinton How or How groves & ha's soe knowne the same for above / fifty yeares saith that the said severall places doe belong & are parcell of the said mannor of Grinton saith that the Swales & the Hillaryes were the reputed lords of the said mannor of Grinton & that old Sir Solomon Swale wrought & gott lead ore on Harka hill & saith Mr Roger / Hillary both wrought & gott lead ore & coale lead on Harkahill & the coale in Grinton Moore nigh the Duke of Boltons mannor saith that he this deponent had a neighbour that had a sheep strayed & was taken upon Grinton common by one Christopher Naylor a servant or / steward of the Swales & saith that this deponents son went for the sheep & had it delivered by the said Naylor as this deponent was informed by his son saith that he knowes the river Swale & the mannors of Grinton & Helaugh & ha's soe knowne the same for about fifty yeares / saith that dureing this deponents knowledge thereof the midd streame of the said river ha's been the reputed boundarye between the said mannors of Grinton & Helaugh never knew or heard dureing the tyme aforesaid that ever the present Lord Wharton or any of his / ancestors ever wrought any mynes or claimed any royaltyes in West Grinton or Harkaside or in any other place within the said mannor of Grinton till of late /

Henry Waggett[73] of Marrick in the county of Yorke miner aged sixty five yeares sworne & examined saith as followeth /

73 In May 1708 Harry Waggen (sic), aged 60, gave evidence for Marriott at the trial in London (LI Misc 536, Ward MSS, notebooks vol. 6, f. 111v).

1 To the first interrogatorie this deponent saith that he know's the relator Mr Marriott but know's not the defendants nor did he know the late Lord Wharton /

2 & 12 To the second & twelfth interrogatories this deponent saith that he know's the mannor of Grinton & alsoe know's Harkaside & Whitaside & alsoe that place called West Grinton & likewise know's Harkahill Harka Moore Grinton Moore Grinton How or Howgroves & ha's knowne them / all for fifty yeares & for all the said tyme the said places called Harkaside West Grinton Whitaside Harkahill Harka Moore Grinton Moore Grinton How or Howgroves have been commonly reputed & taken to be within the mannor of Grinton saith that old Sir Solomon Swale & Roger / Hillary were the reputed lords of the said mannor & that about fifty yeares agoe this deponent wrought at the lead mynes upon Grinton How & Harkahill the places in question under the said Sir Solomon & that the said Sir Solomon by his agents {&} or lessees did for many yeares together worke / mynes of lead & coale upon Grinton How & Harkahill & had considerable gettings saith that the said Roger Hillarye wrought alsoe mines of coale at east end of Grinton How & for some tyme wrought mynes of lead upon Harka hill saith that he knew that the / bayliff's or agents of both the said Sir Solomon & Roger Hillarye frequently tooke up waifes & estrayes upon the said moores & particularly this deponent loosed or released one of his owne sheep from them that had estrayed upon the said moores & paid one shilling for it saith that / he alsoe know's the mannor of Helaugh & that the river Swale ha's for all the tyme of this deponents remembrance Vizt fifty yeares been reputed the antient boundarye betwixt the mannor of Helaugh & the said mannor of Grinton saith that the Lord Wharton nor any of his ancestors / did ever dureing all the said tyme worke any mynes of lead or coale or claime any royaltyes within Harkaside or West Grinton or the said moores or commons thereunto belonging or within any other place within the said mannor of Grinton /

3 & 9 To the third & ninth interrogatories this deponent saith that the said mannor of Grinton is bounderd on the east by the mannor of Ellerton at Stallerston Style & on the south by the mannors of the Duke of Bolton saith that he ha's for all the said tyme knowne the mannors of Fremington / & Marrick both which lye on the north Side of the river Swale Fremington adjoyneing to the said mannor of Grinton which lyes on the south side of the said river & parte of the mannor of Marrick adjoyneing likewise to the said mannor of Grinton saith that the river Swale ha's / for all his tyme been reputed the antient boundarye betwixt the said mannors of Fremington & Grinton & that parte of the mannor of Marrick & the mannor of Grinton saith that the river Swale is a large river and further this deponent saith that he hath severall tymes / heard antient fishermen who lived in Reeth & since dead say that they claimed liberty of fishing to the midd streame of the said river saith that the said river Swale ha's been all his tyme the reputed boundarye betwixt parte of the mannor of Marrick & the mannor of Ellerton / & alsoe betweene the other parte of the said mannor of Marrick & the mannor of Stainton & alsoe between the mannor of Ma<r>ske & the mannor of Downham and this deponent remembers particularly that about twenty yeares agoe there was a dispute betwixt the Duke of / Bolton lord of the mannor of Downham & Mr Hutton lord of the mannor of Marsk about the right to two little islands lyeing about the middle of the said river between the said two mannors & the wood growing thereon & at last they agreed one of them to take one of the said / islands & the other to take the other island /

11 To the eleaventh interrogatorie this deponent saith that he know's the mynes in question upon Harka hill or Harka Moore lately claimed by the defendant the Lord Wharton & alsoe claimed by the relator Mr Marriott to belong to the Crown & himselfe as lessee or farmer under the / Crown saith that the same have all his {life}

tyme been reputed to belong to the mannor of Grinton & have for fifty yeares last past been constantly enjoyed by the Swales & the Hillaryes & the relator Mr Marriott & in all that tyme this deponent never heard that / the present Lord Wharton or any of his ancestors were in the possession of or claimed any right or tytle to the same or parte thereof till about two yeares agoe that the defendant the Lord Wharton entred upon the same /

15 & To the fifteenth & seaventeenth interrogatories this deponent saith he this deponent
17 was a workeman or myner in the said mynes for about one yeare and Benjamin Purchase Mr Philip Swale of Hartford Mr Mathew Smales of Gilling & Mr Robert Jackson who were lessees or farmers / of the said mynes under Sir Solomon Swale and the said lessees wrought the said mynes & gott ore & lead under the right of the said Sir Solomon saith that he the said Phillip Swale all the tyme he was a partner & wrought the said mynes as aforesaid was a steward to or under the late Lord Wharton & / beleives that if the said Lord Wharton had any right to the mynes or moores or commons in question the said Phillip Swale would have promoted or asserted his lordshipps right to them & not have been concerned as a partner with others as aforesaid for that he the said Phillip was alsoe at that tyme / a partner or had some share with his lordshipp in his mynes in Swaledale as well as his steward further saith that the said Phillip Swale did remove this deponent from the mynes in question & employed him to worke in mynes at the old Gang on the north side of the said river in the mannor of Helaugh /

Robert Richardson of Craven Kell nigh Greenahill in the county of Yorke myner aged sixty yeares of thereabouts sworne & examined saith as followeth /

2 & 12 To the second & twelfth interrogatories this deponent saith that he knows the mannor of Grinton in the county of Yorke & alsoe know's Harkaside Whitaside Harka hill Harka Moore Grinton Moore Grinton Greets Grinton How or Howgroves & ha's soe knowne the same for nigh / fifty yeares and saith that the said severall places doe belong & are parte & parcell of the reputed lordshipp of Grinton saith that old Sir Solomon Swale Mr Roger Hillarye & the present Sir Solomon Swale were the reputed lords of the said reputed lordshipp of Grinton & wrought / for both lead & coale within the said lordshipp of Grinton saith that the said Mr Hillarye wrought for coale on the Coale Moore adjoyneing on the Duke of Boltons mannor & that both old Sir Solomon Swale & the present Sir Solomon wrought & got lead upon / Harkaside being the place in question & within the lordshipp of Grinton & gott considerable profitt & the present Sir Solomon continued workeing there till the relator Mr Marriott outed him saith that waifes & estrayes were frequently taken upon Harkaside Moore / by one John Blaides a servant to old Sir Solomon Swale and saith that he this deponent himselfe ha's taken up waifes upon Harkaside Moore & delivered them to Blaides as servant to the said Sir Solomon who detained the same untill the owners thereof came & paid for the releaseing / of them saith that dureing this deponents knowledge of the lordshipp of Grinton & Harkaside Moores which lyes as he beleives within the said lordshipp he never knew that the late Lord Wharton the present Lord Wharton nor any of their ancestors ever wrought any mynes or claimed / any royalties in West Grinton or Harkaside or in any other place or places within the said lordshipp of Grinton & say's he beleives if any such workeing had been he would have either knowne or heard of it /

William Binks of the parish of Marsk in the county of Yorke yeoman aged Eighty two or thereabouts sworne & examined saith as followeth /

1 To the first interrogatorie this deponent saith that he know's the relator Mr Marriott & the present Lord Wharton & did know Phillip late Lord Wharton /

2 & 12 To the second & twelfth interrogatories saith he know's the mannor or reputed mannor of Grinton & alsoe all those places called Harkaside West Grinton Whitaside Harkahill Harka Moore Grinton Moore Grinton Greetes Grinton How or Howgroves & ha's knowne the same for about sixty yeares this deponent being born at a place called Moore / houses in the parish of Marsk about four miles from the said places & lived all his tyme there saith that all the said places dureing his knowledge thereof have been accounted to be within & belong to the said reputed mannor of Grinton & that the Swales & the Hillaryes were the reputed lords of the said mannor & for many / yeares wrought mynes there the Swales wrought mynes of lead upon the said moore of Harkahill & Hillarye wrought mynes of coale upon Grinton Moore & the said Swales their agents lessees or farmers some tymes gott considerably there saith he know's the mannor of Helaugh & that the midd streame of the river / Swale ha's been alwayes accounted dureing his tyme to be the ancient boundarye betwixt the said mannor of Helaugh & the mannor of Grinton further saith that he never knew or heard that the present Lord Wharton or any of his ancestors wrought any mynes or claimed any royaltyes in <West Grinton> Harkaside or any the moores or /
15 commons thereunto belonging till about two yeares agoe To the fifteenth interrogatorie this deponent saith that he this deponent about thirty yeares since wrought as a myner for severall yeares together upon Harkahill or Harkaside Moore and Mr Purchase Robert Jackson Phillip Swale & Mathew Smales who were lessees under / Sir Solomon Swale & that they gott considerable quantityes of lead ore saith that the said Phillip Swale was dureing all that tyme a steward to or under the late Lord Wharton for his mynes in Swaledale & after they the said Purchase Jackson Swale & Smales gave over working at Harka hill the said Phillip Swale carried this deponent / & employed him to worke at the Lord Whartons mynes called old Gang upon the north side of the river Swale /

Francis Ascough[74] of Marrick in the county of Yorke yeoman aged seaventy yeares and upwards sworne & examined saith as followeth /

1 To the first interrogatorie this deponent saith that he know's the relator Mr Marriott & has seen the present Lord Wharton but do's not know his son nor did know the late Lord Wharton /

2 & 12 To the second & twelfth interrogatories this deponent saith that he know's the reputed mannor of Grinton in the county of Yorke & knows the severall places within the said mannor called West Grinton Harkaside Whitaside Harka hill Harka Moore Grinton Moore Grinton Greetes Grinton How or Howgroves / & ha's soe knowne the same for nigh sixty yeares saith that dureing the tyme of this deponents knowledge thereof the said severall places were alwayes reputed to belong & to be parcell of the said mannor of Grinton saith that the Swales & Hillaryes were alwayes the reputed lords of the said mannor of Grinton saith that to / this deponents knowledge both old Sir Solomon Swale & the present Sir Solomon Swale {both} wrought & gott lead on Harkahill or Harkaside the place in question & that the present Sir Solomon Swale continued workeing for lead there untill the relator Mr Marriott recovered the said mynes from him saith that Mr Roger Hillarye / wrought for both lead & coale on Grinton How say's that he ha's heard & beleives that the servants or agents of the Swale's tooke up waifes & estrayes upon Harkaside Moore in the right of the said Swale's being lords of the mannor of Grinton saith that the ancient & reputed boundaryes between the two mannors of / Helaugh & Grinton is the midd streame

74 In May 1708 Francis Askew, aged 70, gave evidence for Marriott at the trial in London (LI Misc 536, Ward MSS, notebooks vol. 6, f. 111v).

of the river Swale saith that he never heard nor doe's he beleive that dureing these sixty yeares last past being the tyme of this deponents remembrance that ever the present Lord Wharton or any of his ancestors ever wrought for either lead or coale upon Harkaside Moore the place / in question or in any other place within the said mannor of Grinton till of late saith that he this deponent was borne at Reeth within the mannor of Helaugh {& ha's all} & ha's lived all his life tyme nigh the said mannors of Helaugh & Grinton therefore beleives noe such workeing could have
3 been but he should have / heard thereof To the third interrogatorie this deponent
saith that the said mannor of Grinton is bounderd on the east side thereof on Ellerton mannor by Stallerston Style on the south on the Duke of Boltons mannors & on the west on Haverdale beck and further saith that on the northside of the river Swale ly's / the mannors of Helaugh & Fremington & adjoyn's upon the said mannor of Grinton and saith that the midd streame of the river Swale is the reputed boundarye between the said mannors of Fremington & Grinton as alsoe between parte of the mannors of Marrick & Grinton & between other parte of the / said mannor of Marrick & the mannor of Ellerton & is alsoe the reputed boundarye betweene other parte of the said mannors of Marrick & Stainton & alsoe between the two mannors of Downhame & Marsk /

7 & 9 To the seaventh & ninth interrogatories this deponent saith that he very well know's the dale called Swaledale & ha's soe knowne the same for nigh sixty yeares saith the same doth extend in length about twelve miles saith he never heard the said dale reputed to be either a mannor or lordshipp of it selfe / but hath therein four severall mannors to witt the mannors of Helaugh & Mukar belonging to the present Lord Wharton the mannor of Fremington belonging to Mrs Wharton or her son the mannor of {Fremington} Grinton formerly belonging to the Swales or Hillaryes & now to the relator / Mr Marriott saith that all the said dale called Swaledale & all the mannors therein are within the parish of Grinton & that the parish church stands {within} in the towne of Grinton and this deponent saith that when he was a boy & lived at Reeth he went & fished in the river Swale & gott / a fish but was threatned by a man that was fishing on the parte of the river Swale on which Harkaside adjoyn's to take the said fish from him but this deponent thereupon left off fishing & carryed the said fish home & telling of his being soe threatned George Chalder William / Race Anthony Race & Michael Garth who were great fishers severally told this deponent that he might fish to the midd streame of the said river Swale as well as the people of Grinton & Harkaside for that they had right to fish noe further /

11 & 17 To the eleaventh & seaventeenth interrogatories this deponent saith that he know's the mynes in question on Grinton How & Harka hill claimed by the present Lord Wharton as alsoe claimed by the relator Mr Marriott as lessee to the Crowne & ha's soe knowne the same for nigh sixty / yeares saith the same doe lye & are reputed to be within the mannor of Grinton saith the said mynes dureing this deponents knowledge of them have belonged & been wrought {by} & enjoyed by the Swale's & Hillary's & the present relator Mr Marriott untill about two yeares agoe that the defendant the present / Lord Wharton caused a myne or mynes to be opened in the said places in question & till then this deponent never heard or knew dureing his knowledge of the said mynes that ever the said Lord Wharton or his ancestors ever claimed any right to the said mynes or ever wrought for either lead or coale in the place or moores / in question /

John Fawcett[75] of Marrick in the county of Yorke gentleman aged seaventy two yeares & upwards sworne & examined saith as followeth /

1 To the first interrogatorie this deponent saith that he know's the relator Mr Marriott & did know the late Lord Wharton but do's not know the defendants /

2 & 12 To the second & twelfth interrogatories this deponent saith that he know's the reputed mannor of Grinton & the severall places within the said mannor called West Grinton Harkaside Whitaside Harka hill Harka Moore Grinton Moore Grinton Greetes Grinton How or How Groves ha's soe known the / same for sixty yeares or thereabouts saith that the said severall places belong to & are within the said mannor of Grinton & that the Swales & the Hillary's were the reputed lords thereof dureing this deponents knowledge thereof saith that old Sir Solomon Swale grandfather of the present Sir Solomon wrought / & gott lead both in Grinton How & Harka hill and saith that he being an agent at Marrick Mills lead ore was brought there from Grinton How & Harka hill & the which was gott by old Sir Solomon to be smelted at Marrick Mills & after the same was smelted at Marrick this deponent carryed the same / to Stockton & sold the said lead to one Mr William Peers[76] a merchant in Stockton & that for two or three yeares together this deponent agreed for & bought what lead ore was by old Sir Solomon wrought & gott at Grinton How & Harkahill & smelted the same at Marrick & then sold the same at Stockton saith / that he this deponent hath often heard that both Sir Solomon & Mr Roger Hillary took up waifes & estray's upon Harkaside Moore & saith that one Henry Wyott a neighbour of this deponents told this deponent that he had a sheep estrayed into Grinton mannor & that Mr Roger Hillary's servant tooke it as a waife & / made him pay one shilling for the releaseing of it and this deponent further saith that for sixty yeares last past the common & knowne boundarye between the two mannors of Helaugh & Grinton was alwaies reputed to be the midd streame of the river Swale saith that till within this two yeares / last past this deponent for sixty yeares together never heard that the present Lord Wharton nor any of his ancestors ever wrought any myne or mynes or claimed any of the royalties upon West Grinton or Harkaside or in or upon any of the places moore waists or commons within the said mannor of Grinton saith / he this deponent was borne within a mile & one halfe of Grinton & has lived all his tyme within one mile & one halfe of Grinton aforesaid /

[membrane v *recto*][77]

3 & 9 To the third & ninth interrogatories this deponent saith that the mannor of Helaugh lyes on the north side of the river Swale & that the mannor of Grinton lyes on the southside of the said river & that the boundarye betwixt them is the midd streame of the river Swale & that the said mannor of Grinton is bounderd on the east side / thereof on Ellerton mannor by Stallerstone Style on the south side of the {said mannor} Duke of Boltons mannors & on the west side on Haverdale beck and further saith that on the north side of the river Swale lyes the mannors of Helaugh & Fremington & adjoin's upon the said mannor of Grinton & saith that the midd streame / of the river Swale is the reputed boundarye between the said mannors of Fremington & Grinton as alsoe between parte of the <mannors of> Marrick &

75 In May 1708 John Fawcet, aged 73, gave evidence for Marriott at the trial in London (LI Misc 536, Ward MSS, notebooks vol. 6, f. 112r).

76 In the 1660s, a Richard Piers or Pierce was an agent purchasing lead at Stockton – see Tyson 1989, pp. 23–4.

77 Parchment, 730mm x 470mm.

Grinton & between other parte of the said mannor of Marrick & the mannor of Ellerton & is alsoe the reputed boundarye betwixt the other parte of the said mannor of Marrick & Stainton & / alsoe betweene the two mannors of Downham & Marsk and saith that he this deponent ha's knowne the said severall mannors & places for nigh sixty yeares last past /

11 To the eleaventh interrogatorie this deponent saith that the mynes in question lately claimed by the defendant the Lord Wharton as alsoe by the relator Mr Marriott under the Crowne lye on Grinton How & Harkahill & are reputed & were enjoyed as parte & parcell of the reputed mannor of Grinton & have been for sixty yeares last / past enjoyed by the Swales & Hillary's & by the relator Mr Marriott untill about two yeares agoe that the present Lord Wharton entred upon the same but by what meanes the said Lord Wharton gott into possession this deponent know's not but saith that till then this deponent never heard that the present Lord Wharton nor any / of his ancestors ever were in possession of or claimed any right or tytle to the same or any parte thereof /

John Loftus[78] of Marrick in the county of Yorke yeoman aged fifty seaven yeares sworne & examined /

1 To the first interrogatorie this deponent saith that he knows the relator Mr Marriott the present Lord Wharton & did know the late Lord Wharton /

2 9 & 12 To the second ninth & twelfth interrogatories this deponent saith that he know's the reputed mannor of Grinton as alsoe West Grinton Harkaside & Whitaside & the moores & commons called Harka hill Harka Moore Grinton Moore Grinton Greetes Grinton How or Howgroves & ha's soe knowne the same for about forty yeares / last past & saith the same doe belong to the said reputed mannor of Grinton & that Sir Solomon Swale & the Hillary's were the reputed lords thereof and that the said Sir Solomon wrought for lead for many yeares upon Grinton How & Harka Moore & continued working there till the relator Mr Marriott recovered the / mynes from him saith the said Sir Solomon had very good gettings whilst he continued workeing & saith the reason of such his knowledge is that he being a carryage man often did & was employed in the carrying of lead ore from the said mynes to Marrick & Clints mills to be smelted there saith / that he know's the reputed mannor of Helaugh & that for the tyme aforesaid the reputed boundarye between the two mannors of Helaugh & Grinton was the midd streame of the river Swale saith that he this deponent did never hear nor do's he beleive that the present Lord Wharton nor any of his ancestors ever wrought any / mynes or claimed any royalties in West Grinton Harkaside Harka hill Harka Moore Grinton Greetes Grinton How or Howgroves within the mannor of Grinton or within any other place within the said mannor till now of late saith that he this deponent hath heard his father say & severall other old men to witt / John Parke James Bowson two antient fishermen both dead & severall others that the midd streame of the river Swale was the boundarye between the mannors of Marrick & Ellerton between the mannors of Marrick & Stainton & between the mannors of Marrick & Downham /

3 To the third interrogatorie this deponent saith that the mannor of Helaugh lyes on the north side of the river Swale & that the mannor of Grinton lyes on the south side thereof & that the boundaryes of those two mannors are reputed to be as followeth the midd streame of the river Swale is the / north boundarye between Grinton mannor & the mannor of Helaugh the Duke of Boltons mannors the south

[78] In May 1708 John Loftus, aged 57, gave evidence for Marriott at the trial in London (LI Misc 536, Ward MSS, notebooks vol. 6, f. 113r).

boundarye Ellerton mannor by Stallerston Style the east boundarye & hath frequently heard that Haverdale beck is the west boundarye /

11 To the eleaventh interrogatorie this deponent saith that he know's the mynes in question now claimed by the defendant the present Lord Wharton as alsoe by the {defendant} relator Mr Marriott as lessee or farmer under the Crowne saith the same lye on the moores or commons called Harkahill on Harkaside Moore parcell / of & lyeing within the said reputed mannor of Grinton & have been dureing the tyme that this deponent knew the same which is above forty yeares enjoyed & wrought by old Sir Solomon Swale his son the present Sir Solomon's father Roger Hillarye the present Sir Solomon & the relator Mr Marriott / untill about two yeares agoe that the present Lord Wharton got into the possession thereof /

Ralph Hillary of Marrick in the county of Yorke myner aged fifty six yeares or thereabouts sworne & examined deposeth as followeth /

1 To the first interrogatorie this deponent saith that he knows Mr Marriott & the present Lord Wharton & alsoe knew the late Lord Wharton but know's not the present Lord Wharton's son /

2 & To the second & twelfth interrogatories this deponent saith that he know's the mannor
12 or the reputed mannor of Grinton as alsoe Harkaside West Grinton <&> Whitaside & the moores or commons called Harkahill Harka Moore Grinton Moore Grinton Greetes <Grinton How> or How Groves & ha's knowne the same for forty yeares saith that all the said / moores & commons are accounted to be within & belong to the reputed mannor of Grinton & that the Swales & Hillary'{s} have been <the> reputed lords thereof dureing his knowledge saith that the Swales wrought mynes of lead upon Harka hill or Harka Moore the place in question & some tyme gott considerably / there & the present Sir Solomon Swale continued to worke there till the relator Mr Marriott recovered the same against him and saith that Hillary did for some tyme worke mynes of coale upon Grinton Moore further saith that he knows <the> mannor of Helaugh & that the river Swale ha's been all his tyme / reputed to be the ancient boundarye betwixt the said mannors of Helaugh & Grinton & ha's heard old people account it soe saith that he never heard that the present Lord Wharton or any of his ancestors dureing his tyme wrought any mynes or claimed any royaltyes within Harkaside or West Grinton / or the moores or commons thereunto belonging till about <two> yeares last past /

3 & 9 To the third & ninth interrogatories this deponent saith that the said mannor of Grinton is bounderd on the east by the mannor of Ellerton at Stallerstone Style & on the south by the Duke of Boltons mannors in Wensleydale saith that the mannor of Fremington the mannor of Marrick & the mannor / of Marsk lye on the northside & adjoyne upon the river Swale & that the said river is reputed to be the ancient boundarye betwixt the mannor of Fremington & mannor of Grinton & alsoe betwixt parte of the mannor of Marrick & the said mannor of Grinton and the said river is likewise reputed / to be the ancient boundarye betwixt another parte of the mannor of Marrick & the mannor of Ellerton which mannor of Ellerton lyes east on the said mannor of Grinton & adjoynes on the south upon the said river and saith that the said river is alsoe reputed to be the ancient boundarye / betwixt the other parte of the mannor of Marrick & the mannor of Stainton the said mannor of Stainton lyeing & adjoyneing upon the south side of the said river further saith that the said river is alsoe reputed to be the ancient boundarye betwixt the mannor of Downham & the mannor of Marsk / one of them lyeing on the north & the other lyeing of the south side of the said {ri} capitall river /

7 To the seaventh interrogatorie this deponent saith that he know's Swaledale & beleives it extends about twelve miles in length & saith that there are four mannors

or reputed mannors in Swaledale Vizt the mannor of Grinton the mannor of Helaugh the mannor of Mukar & the mannor of Fremington & that Mr Marriott / is the reputed lord of the mannor of Grinton & Madam Wharton or her son lord of the mannor of Fremington & the present Lord Wharton lord of the mannors of Helaugh & Mukar & saith that all the said Swaledale & the said several mannors lye within the parish of Grinton & that Grinton is the [. . .] of the towne in Swaledale & the parish church stands in <Grinton> /

11 To the eleaventh interrogatorie this deponent saith that he know's the mynes in question upon Harka hill or Harka Moore which the present Lord Wharton claimes & which the relator Mr Marriott alsoe claimes to belong to the Crowne & himselfe as lessee under the Crowne saith that the same hath been for all his tyme reputed to be / parte & parcell of & to lye within the said mannor of Grinton & that the relator Mr Marriott was in possession thereof as aforesaid till the present Lord Wharton entred upon him about two yeares agoe /

Thomas Hudson of Marrick in the county of Yorke yeoman aged seaventy six yeares or thereabouts sworne & examined /

1 To the first interrogatorie this deponent saith that he know's the relator Mr Marriott & the defendant the Lord Wharton & did know the late Lord Wharton /

2 & 9 To the second ninth & twelfth interrogatories this deponent saith that he know's the
& 12 mannor or reputed mannor of Grinton in the county of Yorke as alsoe know's West Grinton Harkaside Whitaside Harka hill Harka Moore Grinton Moore Grinton How or Howgroves & ha's soe knowne the same for sixty yeares & upwards saith that all the / said severall places are reputed to belong to & be parte & parcell of the <said> mannor of Grinton & saith that the Swale's the Hillary's & the relator Mr Marriott have been dureing all the tyme of this deponents knowledge the reputed lords of the said mannor of Grinton & that the Swals' both wrought & gott considerable quantitys of / lead on Harka Moore in Harkaside & that the present Sir Solomon continued workeing untill the defendant Mr Marriott recovered the possession of the said mynes from him say's that he ha's often heard that waifes & estray's have been frequently taken up upon Harkaside Moore by one Naylor a servant or tenant to the / Swale's & beleives they paid for haveing their goods released say's he know's the mannor of Helaugh & that the reputed boundarye dureing this deponents knowledge between the two reputed mannors of Helaugh & Grinton was the midd streame of the river Swale saith that he this deponent never heard or beleives that / ever the present Lord Wharton or any of his ancestors wrought at any mynes for lead or coale or claimed any royalties within or upon any of the moores waists or commons in West Grinton or Harkaside or in any other place within the said mannor of Grinton saith that he hath often heard severall antient / fishermen say that the midd streame of the said river Swale was not onely the antient boundarye between the two mannors of Helaugh & Grinton but the antient boundarye betweene the mannors of Fremington & Grinton <&> of the mannors of Marrick & Grinton & between another parte of the mannors / of Marrick & Ellerton which lyes east of the mannor of Grinton & alsoe another parte of the mannor of Marrick & the mannor of Stainton & between the mannor of Downham & the mannor of Marsk /

11 To the eleaventh interrogatorie this deponent saith that he know's the mynes in question claimed by the present Lord Wharton as alsoe by the relator Mr Marriott as lessee or farmer under the Crowne saith that the same lyes on Harka Moor in Harkaside within the reputed mannor of Grinton & ha's dureing this / deponents knowledge thereof been enjoyed by the Swales & the relator Mr Marriott & never was dureing the tyme aforesaid ever wrought or enjoyed by the present Lord Wharton or

any of his ancestors till about two yeares since that the present Lord Wharton entred upon the mynes in question /

Peter Hemsley[79] of Marrick in the county of Yorke yeoman aged sixty four or thereabouts sworne & examined /

1 To the first interrogatory this deponent saith that he knows the relator Mr Marriott & alsoe the present Lord Wharton, but knows not his son nor did he know the late Lord Wharton deceased; /

2: 12: To the second & twelfth interrogatories this deponent knows the mannor or reputed mannor of Grinton in the county of Yorke & knows alsoe the places called West Grinton Harkaside & Whitaside, & the moors & commons called Harka Hill Harka Moor Grinton Moore / Grinton Greets Grinton How or How Groves all which said places have been commonly reputed all his time to belong to & be parte & parcell of the said mannor of Grinton & ha's knowne the same for forty yeares, saith that the Swales & Hillaries have dureing his remembrance been the / reputed lords of Grinton as alsoe the relator Mr Marriott of late, and that old Sir Solomon Swale & the present Sir Solomon Swale wrought mines of lead and cole upon Grinton How & Harka Moore the place in question and gott considerably there & this deponent well remembers that about a / dozen or fourteen yeares agoe hee this deponent was present upon Harka Moor the place in question when Mr William Cowling & Mr George Cowling in the name & on the behalfe of John Hillary discharged or warned the workmen that were then working the mines upon Harka Moore / under the present Sir Solomon from proceeding to worke there, and the workmen then declared to the said Mr Cowling's that if the said John Hillary would make them Satisfacttion for the oare they had then gott which was about thirty or forty pounds value they would deliver the said / oare to him & likewise deliver him possession of the said mines & endeavour to keep possession thereof for him, further saith that hee hath heard that both the Swales & the Hillaryes have taken up waifes & estrayes within the liberty or mannor of Grinton & that severall of his / neighbours have told him that their sheep ha's been taken upp there by one Naylor steward or agent to Sir Solomon Swale, saith he alsoe knows the manor of Helaugh, and that the river Swale ha's been alwaies dureing his time reputed to be the boundary betwixt / the said mannors of Grinton & Helaugh and that the present Lord Wharton nor any of his ancestors did ever (that this deponent heard of) work any mines or claime any royalties in West Grinton or Harkaside or in any the said moors or commons or in any other place within / the said manor of Grinton till about two yeares agoe, saith that the relator Mr Marriott wrought the said mines upon Harka Moore till of late that the Lord Wharton enterd upon him; /

3 & 9: To the third & ninth interrogatories this deponent saith that the said manor of Grinton is bounder'd on the east by the manor of Ellerton at Stallerstone Stile & on the south by the Duke of Boltons manors & has heard & beleives that Haverdale Beck is the boundary thereof on the / west saith that the manors of <Fremington &> Marrick lye on the north side of the river Swale & adjoine to the said manor of Grinton & that the said river or the mid stream thereof has been alwaies accounted the boundary betwixt the said manors saith that he has been severall / times with the Duke of Bolton & Lord Powlett or their agents & great many of their neighbours riding his boundaries of the said manor of Marrick about twenty yeares & about three yeares agoe and the said mid streame of the river Swale was then reputed & /

[79] In May 1708 Peter Hemsley, aged 55, gave evidence for Marriott at the trial in London (LI Misc 536, Ward MSS, notebooks vol. 6, f. 112v).

taken to be the boundary betwixt the said manor of Grinton & parte of the manor of Marrick & betwixt the manor of Ellerton & another parte of the manor of Marricke & betwixt the manor of Stainton & another part of the manor of Marrick & betwixt parte of the manor / of Downham & the other parte of the manor of Marrick & that the said river Swale is alsoe the reputed boundary betwixt the other parte of the said manor of Downham & the manor of Marske & that about fifteen yeares agoe there was some dispute betwixt the /

[membrane v *verso*]

the Duke of Bolton & Mr Hutton about the right to two little islands lying about the midle of the said river & they agreed to divide the same islands betwixt them; /

7: To the seaventh interrogatory this deponent saith that he knows the dale {or} called Swaledale & that the same is reputed a lordshipp, that the extent thereof is about twelve miles & that the manors of Mukar & Helaugh belonging to the Lord Wharton & the manor of Fremington belonging / to Madam Wharton & the manor of Grinton belonging to Mr Marriott are all within the extent of Swaledale that the said dale lies within the parish of Grinton & that the parish church stands at the town of Grinton which is the cheife towne in the said dale /

11: To the eleaventh interrogatory this deponent saith that hee very well knows the mines in question upon Harka Moore claimed by the Lord Wharton & which the relator Mr Marriott alsoe claimes to belong to the Crowne & himself as lessee under the Crown / which this deponent beleives to lye within the manor of Grinton that the same were enjoyed by the Swales & the Hillaryes till about tenn yeares agoe that Mr Marriott recovered the same and the said Mr Marriott enjoyed the same & gott great / quantities of lead there till about two yeares agoe that the present Lord Wharton enter'd upon him, and that the present Lord Wharton nor any of his ancestors did ever claime any right of title thereto till about two yeares agoe that ever this deponent heard of; /

Charles Willis[80] of Marrick in the county of Yorke aged seaventy three or thereabouts sworne and examined on the parte of the relator saith /

[1] To the first interrogatory this deponent saith that hee knows the relator Mr Marriott but knows not the defendants /

2: & 12: To the second and twelfth interrogatories this deponent saith that he knows the manor or reputed manor of Grinton & alsoe West Grinton Harkaside & Whitaside & the moors & commons call'd Harkahill Harka Moor Grinton Moore Grinton Greets Grinton How or / how Groves & has knowne the same about fifty yeares and for all that time West Grinton Harkaside & Whitaside & all the said moors & commons were commonly taken and reputed to belong to & be parte & parcell of the said manor of Grinton & that the Swales & / Hillaries have for all the said time been the reputed lords of the said manor till Mr Marriott has had the same saith that for fifty yeares last past & downwards old Sir Solomon Swale & the present Sir Solomon Swale wrought mines of lead & coale upon / Harka Hill or Harka Moore the place in question & sometimes gott considerably there, and the present Sir Solomon Swale continued to worke there till Mr Marriott recovered the same saith that the Hillaries wrought mines of lead & cole as he has heard / and beleives both upon the said Harkaside Moore and Grinton Moore, and that one John Blades formerly an agent

80 In May 1708 Charles Willis, aged 74, gave evidence for Marriott at the trial in London (LI Misc 536, Ward MSS, notebooks vol. 6, f. 112r).

or steward to old Sir Solomon Swale used to take upp waifes & estrays there, saith he knows the manor of Helaugh and that for / all the time of his remembrance the mid stream of the river Swale ha's been alwaies taken to be the ancient & knowne boundary betwixt the said manor of Helaugh & manor of Grinton, and that the present Lord Wharton nor any of his ancestors / did ever worke any mines or claime any royalties (as this deponent beleives) in West Grinton or Harkaside or any other place within the said manor of Grinton till about two yeares agoe; /

3: & 9: To the third & ninth interrogatories this deponent saith that for all his time the reputed boundaries of the said manor of Grinton <was> the manor of Ellerton at Stallerstone Stile on the east & Haverdale Beck on the west & the Duke of Boltons manors of Wensley / dale on the south & the said river Swale on the north saith that there is the manor of Fremington lying on the north side of the river Swale & adjoines upon the said manor of Grinton and that the said river has been all along the reputed boundary / betwixt the said manors of Fremington Grinton and that the said river has been alsoe the reputed boundary betwixt parte of the manor of Marrick and the said manor of Grinton and alsoe betwixt another parte of the said manor of Marrick and the / manor of Ellerton and alsoe betwixt another parte of the said manor of Marrick & the manor of Stainton and alsoe betwixt another parte of the manor of of Marrick and the manor of Downham & alsoe betwixt the manor of Marske & Downham saith / that this deponent ha's often fished in the said river and that the mid streame amongst such as used to fish has been always looked upon to be a boundary as aforesaid; /

7: To the seaventh interrogatory this deponent saith that he knows the dale called Swaledale, and beleives it extends twelve miles in length and saith there are four manors or reputed manors in Swaledale the mannors of Helaugh & Mukar belonging / to the Lord Wharton the manor of Fremington belonging to Madam Wharton and the said manor of Grinton belonging to Mr Marriott that all the said dale lies within one parish called Grinton & that Grinton is the cheife towne thereof; /

Thomas Lamb of Marrick aged forty one yeares or thereabouts sworne & examined deposeth as followeth; /

1: To the first interrogatory this deponent saith that he know's the relator Mr Marriott but knows not the said Lord Wharton or his sonn; /

2: & To the second & twelfth interrogatories this deponent saith that he knows the manor
12: or reputed manor of Grinton & alsoe the places called West Grinton Harkaside & Whitaside & alsoe knows the moors & commons called Harkahill Harka Moor Grinton Moore Grinton How or How Groves & has / knowne the same for near thirty yeares & the said places moors & commons have for all the time of his knowledge been commonly reputed & taken to belong to & be parte & parcell of the said manor of Grinton, and that for all that time Sir Solomon Swale Mr Hillary & Mr Marriott / have been called & esteemed the lords thereof and that the said Mr Hillary Sir Solomon & Mr Marriott have respectively wrought mines of lead upon Harkahill or Harka Moor the place in question till about two yeares agoe & Sir Solomon & Mr Marriott had sometimes / good gettings there saith that the said Sir Solomon & Mr Hillary tooke upp waifes & estrays there which he has seen in the custody of their agents and perticularly this deponent has loosed or released one sheep of his owne that was taken upp by one Naylor agent to the / said Sir Solomon; further saith that hee knows the manor of Helaugh and that for all his time the midd streame of the river Swale has been reputed & commonly taken in the neighbourhood to be the ancient & knowne boundary betwixt the said manors of Grinton & Helaugh and saith / that he verily beleives the present Lord Wharton nor any of his ancestors did ever worke any mines or claime any royalties in West Grinton or Harkaside or the

moores or commons thereunto belonging or in any other place within the manor of Grinton / till of late; /

3: & 9: To the third & ninth interrogatories this deponent saith the common & reputed boundaries of the said manor of Grinton is by the manor of Ellerton on Stallerstone Stile on the east & Haverdale Beck on the west & by the Duke of Boltons manors of Wensley Dale on the south and saith / the manor of Fremington lies on the north-side of the said river Swale & adjoines upon the said mannor of Grinton & the mid streame of the said river Swale is & has been the reputed boundery betwixt the said manors of Fremington & Grinton & the said river is alsoe the reputed / boundery betwixt parte of the manor of Marrick & the said manor of Grinton and is alsoe the reputed boundery betwixt another parte of the manor of Marrick & the manor of Ellerton and alsoe betwixt another parte of the manor Marrick & the manor of Stainton / and alsoe betwixt another parte of the manor of Marrick & the manor of Downham & alsoe betwixt the manor of Marske & the manor of Downham and ha's alsoe heard severall ancient men say that the said river is & alwaies was a boundery as aforesaid; /

11: To the eleaventh interrogatory this deponent saith that hee knowes the mines in question upon Harka Hill or Harkaside Moore & that the same have for all the time of [his] remembrance been reputed to lye within the said manor of Grinton saith the same have for all the time / aforesaid been enjoyed by Mr Hillary & Sir Solomon Swale till about tenn yeares agoe that Mr Marriott recovered the same who has since enjoyed them till about two yeares agoe that the present Lord Wharton enterd upon him and that hee never heard / nor does hee beleive that the present Lord Wharton or any of his ancestors claimed any right or title thereto till about two yeares agoe; /

Christopher Kay of Marrick in the county of Yorke yeoman aged fifty eight years or thereabouts sworne & examined saith; /

1 To the first interrogatory this deponent saith that hee know's the relator Mr Marriott /

2: & 12: To the second & twelfth interrogatories this deponent saith that he knows the manor of Grinton & alsoe the places called West Grinton Harkaside & Whitaside & alsoe know's the moors & commons called Harka Hill Harka Moore Grinton Moore Grinton How or How Groves / and has knowne the same for forty yeares last past & saith the said places moores & commons have for all the time of his knowledge been reputed & taken to be parte & parcell of the said manor of Grinton and that for all that time the present Sir Solomon Swale old Sir Solomon / Roger Hillary & the relator Mr Marriott have been the reputed lords of the said manor & saith that old Sir Solomon Swailes lett & bargained forty yeares since with Christopher Chambers Ralph Lee & John Blaides all now dead to digg & worke mines for lead within the / liberties of Sir Solomon in the waists & commons within the said manor of Grinton the said Sir Solomon was to have every tenth pigg that the said Chambers Lee & Blaides should winn & gett, & that in pursuance of the said agreement they did worke & gett considerable / quantities of lead upon the said moores & waists saith that hee this deponent has often heard that the said John Blaides frequently tooke upp waifes & estray's upon the said Harkaside Moore; further saith that he well knows the manor of Helaugh & dureing his knowledge / thereof the mid streame of the river Swale has been reputed to be the ancient & knowne boundary betwixt the said manors of Grinton & Helaugh and the reason of such this deponents knowledge is that from his childhood hee this deponent has been brought upp & lived at / Marrick within a mile & a halfe of Grinton & never heard it otherwise reported and saith that the present Lord Wharton the late Lord Wharton nor any of their ancestors did ever that this deponent heard of worke or caused to be wrought any mine or mines for lead or / cole within or upon any of the waists & commons of Harkaside & West Grinton or in any other place within the said manor of Grinton till about two yeares since; /

3: & 9: To the third & ninth interrogatories this deponent saith the the manor of Helaugh lyes on the north side of the river Swale and the manor of Grinton on the south side thereof & the ancient & knowne boundary between the said two manors are as follows Stallerstone Stile / on the east parte thereof the Duke of Boltons manors of Wensley Dale on the south Haverdale Beck on the west & the mid streame of the said river Swale on the north and further saith that the manor of Fremington lies on the north side of the said river / Swale & adjoines upon the said manor of Grinton and that the mid streame of the said river Swale is & has been the reputed boundary betwixt the said manors of Fremington & Grinton & the said river is alsoe the reputed boundary between parte of the manor / of Marricke and the said manor of Grinton & is alsoe the reputed boundary between other parte of the said manor of Marrick & Ellerton and alsoe between other parte of the said mannor of Marrick & Stainton & alsoe between other parte of the manor of Marrick / & Downham and alsoe between the said manor of Marske & the manor of Downham, and ha's alsoe heard severall ancient men say the said river is and always was a boundary as aforesaid and this deponent about tenn yeares since did goe with Harry Water / Notts[81] to fish and persons belonging to Ellerton hinder'd this deponent & those that were with him for comeing any further to fish than the mid streame of river Swaile and told them the mid streame was the boundary & they ought to fish noe further; /

[membrane vi *recto*]

7: To the seaventh interrogatory this deponent saith that he know's the dale or vale called Swaledale & has soe knowne the same for above forty yeares, saith it is in length about thirteen or fourteen miles never heard it reputed a lordshipp but saith within the said dale there are four manors / or lorshipps to witt the manor or lordshipp of Helaugh and Mukar belonging to the present Lord Wharton the mannor or lordshipp of Fremington belonging to Madam Wharton of Gilling Wood and the manor or lordshipp of Grinton belonging to the relator Mr Marriott saith the said dale is within / the parish of Grinton and the parish church is at Grinton; /

11: To the eleaventh interrogatory this deponent saith that he knows the mines in question upon Harka Hill or Harkaside Moore and that the same have been for all the time of his remembrance reputed to lye within the said mannor of Grinton saith the same has been for / all the time aforesaid enjoyed by the Swales or Hillaries till about tenn yeares agoe that the relator Mr Marriott recovered the same & ha's since the recovery of them been in the peaceable possession thereof till about two years agoe that the present Lord Wharton enterd / upon him & saith that hee this deponent never heard nor does he beleive that the present Lord Wharton or any of his ancestors ever claimed any right or title to the mines in question till about two yeares since; /

Thomas Waggett of Marricke in the county of Yorke yeoman aged fifty yeares or thereabouts sworne & examined saith /

1. To the first interrogatory this deponent saith he knows the relator Mr Marriott /

[81] The text as written does not appear to make sense and it may be that the phrase as originally intended was 'Harry Water's netts'. It may be relevant here that the words 'Harry Water' have been re-written (but by the same hand), possibly to correct a mistake, but that in doing so further errors were inadvertently introduced. In the context, 'the nets of Harry Water' would be an appropriate interpretation.

2: & To the second & twelfth interrogatories this deponent saith that he knows the manor
12. of Grinton & alsoe the places called West Grinton Harkaside & Whitaside & alsoe knowes the moors & commons called Harka Hill Harka Moore Grinton Moor Grinton How or How Groves / & ha's soe knowne the same for thirty seaven yeares or thereabouts & saith the said places moors & commons have for all the time of this deponents knowledge been reputed & taken to be parte & parcell of the said manor of Grinton & for all that time the present Sir Solomon / Swale old Sir Solomon Roger Hillary & the relator Mr Marriott have been the reputed lords of the said mannor and that the present Sir Solomon Swaile wrought for & gott considerable quantity's of lead upon Harka Hill or Harkaside & continued working there / untill the relator Mr Marriott recovered the said mines from him saith that the said Roger Hillary wrought for & gott lead on Grinton How but cannot be positive as to his working <on> Harka hill, saith that hee this deponent has frequently heard that / John Blaides a servant or tennant to Sir Solomon Swaile tooke upp waifes & estrayes upon Harkaside Moore; further saith that hee well know's the manor of Helaugh {thir} <and> for thirty seaven yeares last past the mid streame of the river Swaile has been / reputed to be the ancient & knowne boundary betwixt the said manor of Grinton & the manor of Helaugh saith that the late Lord Wharton the present Lord Wharton nor any of their ancestors did ever that this deponent heard off worke or cause to be wrought any mine / or mines for lead or coale within or upon any of the waists moors & commons of Harkaside & West Grinton or in any other place within the said manor of Grinton; /

3: & To the third & ninth interrogatories this deponent saith that the manor of Helaugh
9: lyes on the north side of the river Swaile & the manor of Grinton on the south side thereof & that the mannor of Grinton is bounderd as followeth [x x x][82] Stallerstone Stile / on the east Haverdale Beck on the west the Duke of Boltons two manors of Wensley on the south & the mid streame of the river Swaile on the north, and further saith that the mannor of Fremington lyes on the north side of the river Swaile & adjoines / upon the said manor of Grinton & that the mid streame of the river Swaile is & ha's been the reputed boundary between the said manor of Fremington & Grinton & alsoe between parte of the manor of Marrick & Grinton & between other parte of the said / manor of Marrick & Ellerton manor & other parte of Marrick & Stainton manor & between other parte of Marrick & Downham manor & also between the manor of Marske & Downham & alsoe heard severall ancient men say the said mid streame of the / said river Swale is & always was the reputed boundary as aforesaid and saith that about twelve yeares since Mr Timothy Blackburne of Grinton deceased would not suffer this deponent who lived at Marrick to come any further to fish than the mid / streame of the river Swale and told this deponent that the said mid streame was the boundary between Marrick & Grinton; /

7: To the seaventh interrogatory this deponent saith that for twenty seaven yeares last past he has knowne the dale called Swaledale & that the same is in length about twelve miles never heard it reputed a lordshipp but saith in the said dale there is four / mannors or lordshipps the manor or lordshipp of Helaugh & Mukar belonging to the present Lord Wharton the manor or lordshipp of Fremington belonging to Madam Wharton & the manor of Grinton belonging to the relator Mr Marriott saith that / Swaledale is all in the parish of Grinton & that the parish church stands at Grinton /

11: To the eleaventh interrogatory this deponent saith that hee well knows the mines in

82 Some few words have been erased.

question upon Harka Hill or Harkaside Moore & that dureing his knowledge thereof the same have been reputed to lye within the said manor of Grinton & have been enjoyed / by the Swales & the relator Mr Marriott who recovered the same from the present Sir Solomon Swaile about tenn yeares since & saith that the relator Mr Marriott till about two yeares since peaceably enjoyed the same beleives the present Lord Wharton nor any / of his ancestors ever claimed any right or title to the mines in question till the time aforesaid; /

Phillip Greathead[83] of Marrick aged forty seaven or thereabouts sworne & examined deposeth as followeth; /

1: To the first interrogatory this deponent saith that he knows the relator Mr Marriott and alsoe the present Lord Wharton but doe not know his sonn nor did he know Phillip late Lord Wharton; /

2. & 12: To the second & twelfth interrogatories this deponent saith that he knows the manor or reputed manor of Grinton & the hamletts called West Grinton Harkaside & Whitaside and alsoe the moors or commons called Harkahill Harka Moore Grinton Moor Grinton How or How Groves / and has knowne the same for thirty six yeares & upwards and for all his time all the said hamletts places & commons have been comonly reputed to belong to and be parte & parcell of the said manor of Grinton saith that Sir Solomon Swale and the Hillaries and of late / Mr Marriott have been all along the reputed lords of the said manor of Grinton and that Sir Solomon has wrought mines of lead dureing his time upon Harka Hill or Harkaside Moore the place in question & gott considerably there till Mr Marriott recovered the / same against him about ten yeares agoe and that the said Mr Marriott ha's since wrought the same & had very good gettings there till about two yeares agoe that the present Lord Wharton enterd upon him saith that Mr Hillary wrought mines of lead upon / Grinton How, saith that the Swales & Hillary's have often taken upp waifes & estrays within the said manor of Grinton to this deponents knowledge & that Christopher Naylor & James Naylor agents for Sir Solomon Swale used to take upp waifes & estrays / & carryed them to Swaile Hall but in what particular place they tooke them within the said manor this deponent cannot say saith that he knows the manor of Helaugh & that for all his time the river Swale has been reputed the ancient boundery betwixt the / said manor of Helaugh & the said manor of Grinton and saith that he never knew or heard that the present Lord Wharton or any of his ancestors did ever worke any mines of lead or claime any royalties in the said moors or commons belonging to West Grinton / or Harkaside till of late /

3: & 9: To the third & ninth interrogatories this deponent saith that the said manor of Grinton is & ha's been reputed to be bounderd on the east by the manor of Ellerton at Stallerstone Stile & on the west by Haverdale beck and on the south by the Duke of Boltons manors & on / the north by the river Swale from the manor of Helaugh as aforesaid saith that the manor of Fremington is on the north side of the river Swale & adjoines upon the said manor of Grinton and that the mid streame of the said river is the ancient / & knowne boundery betwixt the said manors and that the said river is alsoe the knowne boundary betwixt parte of the manor of Marrick & the manor of Grinton & alsoe betwixt another parte of the manor of Marrick & the manor of Ellerton and / further saith that <there was> about twenty five yeares agoe a dispute betwixt the Duke of Bolton (then Marquess of Winchester) lord of the manor of Downham & Mr Hutton lord of the manor of Marske about the right of two little

[83] In May 1708 Philip Greathead (no age given), gave evidence for Marriott at the trial in London (LI Misc 536, Ward MSS, notebooks vol. 6, f. 113r).

islands lying about the / midle of the said river & they at last divided the same saith that the mid stream of the said river has been alsoe accounted amongst fisher men to be the boundery as aforesaid; /

7: To the seaventh interrogatory this deponent saith that hee knows Swaledale that hee beleives the extent thereof to be about twelve miles saith that there are four manors within Swaledale the manor of Helaugh & Mukar belonging to the Lord / Wharton the manor of Fremington belonging to Madam Wharton & the said manor of Grinton belonging to the relator Mr Marriott that all Swaledale ly's within the parish of Grinton & that the towne of Grinton is the cheife town / therein and the parish church stands in Grinton /

11: To the eleaventh interrogatory this deponent saith that he knows the mines in question lying upon Harka or Harkaside Moore & that the same has dureing his knowledge been severally reputed & taken to lye within the said manor of Grinton and that / the same have been enjoyed by the Swale's & Hillary's till about tenn yeares agoe & since by Mr Marriott till about two yeares agoe that the present Lord Wharton enterd upon him saith that he never heard or beleives that the present Lord / Wharton claimed any right or title to the mines or moores or commons in question till of late; /

George Arnold of Ellerton Abbey gentleman aged forty five or thereabouts sworne & examined saith /

1 To the first interrogatory this deponent saith that he know's the relator Mr Marriott & the present Lord Wharton but know's not his son nor did he know the late Lord Wharton deceased /

8 To the eight interrogatory this deponent saith that he this deponent haveing a smelting mill at Ellerton about eleaven yeares agoe had severall quantity's of lead from Sir Solomon Swale which were gott by him at that tyme out of the mynes of Harkahill saith that within one yeare & a month's tyme he this deponent smelted for / the said Sir Solomon one hundred & fifty fothers of lead & beleives Mr Blackburne & Mr Carrington who had the said smelting mill before this deponent farmed the same did also smelt considerable quantity's at the said place for the said Sir Solomon for that he this deponent haveing a right to some share of the profitts of / the said smelting mill had an account of the profitts of twenty fothers thereof & verily beleives that they the said Mr Blackburne & Mr Carrington did smelt about one hundred & fifty fothers of lead gott by the said Sir Solomon out of the said mynes upon Harkahill saith that the said Sir Solomon Swale / enjoyed the said mynes till about nine or tenn yeares agoe that Mr Marriott recovered the same against him & the said Mr Marriott since enjoyed the same till about two yeares agoe that the said Mr Marriott entred upon him /

9 To the ninth interrogatory this deponent saith that the midd streame of the river Swale is the ancient & known boundarye betwixt the mannor of Ellerton & the mannor of Marrick & Marrick Abby & the lords of both the said mannors or their agents as well as all the neighbourhood doe allow the / midd streame of the said river to be the said boundarye & ha's heard & verily beleives the said river is alsoe a boundarye to the adjacent mannors of Stainton & Marrick saith that he this deponent did at the request of Sir Solomon Swale about eleaven yeares agoe & alsoe about two yeares agoe ride with him / & others parte of the boundarye of the mannor of Grinton saith that they begun at Stallerstone Style which is the ancient & knowne boundarye betwixt the said mannor of Grinton & the mannor of Ellerton & this deponent ridd with the said Sir Solomon as far as the mannor of Ellerton [which] / extended & joyned upon the said mannor of Grinton but how far the said Sir

Solomon proceeded to ride further this deponent know's not he the said Sir Solomon then laying claime to the mannor of Grinton /

Henry Fawcett of Marrick in the county of Yorke yeoman aged sixty three yeares or thereabouts sworne & examined saith /

2 & 12 To the second & twelfth interrogatories this deponent saith that he know's the reputed mannor of Grinton & alsoe the places called West Grinton Harkaside & Whitaside & the severall moores & commons thereto belonging called Harka hill Harka Moore Grinton Moore Grinton Greetes Grinton How or Howgroves & ha's / soe knowne the same for fifty yeares saith the said places are reputed to {belong or} be within & to belong to the reputed mannor of Grinton & that dureing his knowledge the Swales Hillary's & the relator Mr Marriott have been the reputed lords of the said mannor of Grinton saith that the present Sir / Solomon wrought for severall yeares the mines and gott considerable quantity's of lead within & upon Harka hill on Harkaside Moore the place in question & continued workeing there till Mr Marriott gott possession & that Roger Hillarye wrought & gott coale on Grinton Moore saith that / waifes & estray's have been severall tymes taken up upon Harkaside Moore by Sir Solomon Swale his servants & agents & ha's heard that the Hillary's alsoe took up waifes & estray's there saith that he know's the mannor of Helaugh & ha's soe knowne the same for the tyme aforesaid & dureing this deponents / knowledge thereof the midd streame of the river Swale has alwayes been the reputed boundarye between the two mannors of Helaugh & Grinton as alsoe between the mannors of Marrick & Downham mannor & other parte of the mannor of Marrick & the mannor of Stainton & other parte of the mannor of Marrick / & the mannor of Ellerton and this deponent further saith that dureing the tyme aforesaid the present Lord Wharton the late Lord Wharton nor any of their ancestors did ever that this deponent knew or heard of worke any myne or mynes or claimed any royalties within or upon West Grinton or Harkaside commons / or in any other place within the said mannor of Grinton till of late and this deponent further saith his servants for severall yeares in the present Sir Solomon's tyme carryed lead ore from the mynes on Harkahill to Marrick mill to be smelted & that once this deponents selfe his servants being not well carryed lead / ore from the said mynes to the said Marrick mill to be smelted as aforesaid & saith that dureing his being employed as aforesaid the present Lord Wharton nor any of his ancestors ever claimed any right to the said mynes in question that ever this deponent knew or heard off /

11 To the eleaventh interrogatorie this deponent saith that he know's the mynes in question now claimed by the defendant the Lord Wharton as alsoe claimed by the relator Mr Marriott as lessee under the Crowne say's the same doe lye & are on Harkahill & on Harkaside Moore & are reputed to be parte & parcell of the said reputed mannor / of {enjoyed} Grinton & have been enjoyed dureing this deponents remembrance by the Swale's the Hillary's & the present relator Mr Marriott till within these two yeares last past & saith that the present Lord Wharton the late Lord Wharton nor any of their ancestors dureing this deponents knowledge of the said mynes ever were in the possession / thereof or claimed any right thereto that this deponent ever knew or heard of till of late /

George Spenceley of Downham yeoman aged seaventy two or thereabouts sworne & examined saith /

1 To the first interrogatorie this deponent saith that he know's the relator Mr Marriott & the defendant Lord Wharton but know's not his son nor did he know Phillip late Lord Wharton deceased /

2 & 12 To the second & twelfth interrogatories this deponent saith that he know's the mannor or reputed mannor of Grinton & the places called Harkaside West Grinton &

Whitaside & the moores & commons thereunto belonging & hath knowne the same for near fifty yeares & that for all the tyme of his knowledge the said / places have been reputed to lye within the said mannor of Grinton saith that Sir Solomon Swale & Roger Hillarye have been the reputed lords of Grinton & that for all the tyme of his remembrance the river Swale ha's been reputed to be the ancient boundarye betwixt the said mannors of Grinton & Helaugh further / saith that he never knew or beleives that the present Lord Wharton or any of his ancestors did worke any mynes upon any of the said moores or commons of West Grinton or Harkaside till of late /

Thomas Buckton of Oxnipp gentleman aged fifty nine yeares or thereabouts sworne & examined deposeth as followeth; /

1: To the first interrogatory this deponent saith that he knows the relator Mr Marriott & beleives hee knows the present Lord Wharton but does not know his sonn /

2: & 12: To the second & twelfth interrogatories this deponent saith that he knows the reputed manor of Grinton & the places called West Grinton Harkaside & Whitaside & the moors & commons thereunto belonging & ha's knowne the same for about forty six yeares saith that the Swales & the Hillarys have been / the reputed lords of Grinton further saith that he knows the manor of Helaugh & that the river Swale has been generally accounted the boundery betwixt the said manors of Helaugh & Grinton & is now perticularly induced to beleive the same for that this deponents father purchased / some lands in Whitaside which he settled upon this deponents brother and the said river Swale haveing by the current or running thereof left some parte of the said lands upon the north side of the said river his said brother was & is obliged to appear at the Lord Whartons court at Helaugh / for and in respect of the said lands left on the north side of the said river but is not obliged to appear there for or in respect of his said estaite on the south side of the said river; /

[membrane vi *verso*]

9 To the ninth interrogatory this deponent saith that hee knows the manor of Ellerton which adjoines upon the south side of the river Swale & that the said river is the ancient & knowne boundery betwixt the said manor of Ellerton & parte of the manor of / Marrick & alsoe a boundery betwixt the mannor of Stainton & another parte of the manor of Downham & another parte of the manor of Marrick & alsoe betwixt the other parte of the mannor of Downham & the manor / of Marsk; /

John Willis of Marrick in the county of Yorke yeoman aged forty two yeares sworne & examined saith /

2 To the second interrogatory this deponent saith that he know's the reputed manor of Grinton & the places called West Grinton Harkaside & Whitaside & ha's soe knowne the same for twenty eight yeares last past saith the said places are reputed to belong / & be within the said reputed manor of Grinton & that Sir Solomon Swaile & Roger Hillary were the reputed lords thereof saith that the said Sir Solomon wrought lead mines on Harka Hill the place in question & continued working there till / the relator Mr Marriott recovered the possession from him saith that he knows the manor of Helaugh & that the mid streame of the river Swale is the reputed boundery between the two manors of Helaugh & Grinton & this deponent / saith that the mid streame of the river Swaile is the boundery between the manors of Marrick & Ellerton & between other parte of the manor of Marrick & Stainton & other parte of the manor of Marrick & Downham & betweene / the manor of Marsk & Downham saith that this deponent was a fishing in the river Swaile between the manors of Marrick & Mr Blackburne steward of the manor of Ellerton discharged this deponent & the rest that were fishing with him to come noe / further than the mid streame

for that was the boundary and if they did not keep within their boundery he would not only take their netts from them but would alsoe indict them /

Mathew Watson[84] of Marrick aged fifty three or thereabouts sworne & examined saith /

2 & 12 To the second & twelfth interrogatories this deponent saith that hee knows the manor or reputed manor of Grinton & the places called Harkaside West Grinton & Whitaside & Harka Hill Harkaside Moore & Grinton How & Grinton Moore & ha's knowne them about forty yeares / saith that for all his time the said places have been reputed to belong to & lye within the said manor of Grinton and that Sir Solomon Swale wrought mines of lead upon Harka or Harkaside Moore and gott considerably till Mr Marriott recovered the same against / him and that Mr Hillary wrought alsoe mines upon Grinton How saith that one Naylor used to take upp waifes & estray's there and the said Sir Solomon and Mr Hillary were dureing his time reputed lords of the said manor of Grinton; further saith that hee / knows the manor of Helaugh and that the river Swale ha's been all along the reputed boundary betwixt the said manors of Grinton & Helaugh & beleives the same to be the boundery there as it is generally accounted & taken to be the boundery / betwixt the manor of Ellerton (which adjoines upon the said manor of Grinton) & parte of the manor of Marrick and alsoe betwixt the manor of Stainton & parte of the said manor of Marrick & alsoe betwixt the manor of Downham & another parte of / the manor of Marrick & alsoe betwixt Downham & the manor of Marske saith hee never heard or beleives that the present Lord Wharton or any of his ancestors did ever worke any mines of lead or cole upon any of the said moors or commons of West Grinton / or Harkaside till about two yeares agoe that the said Lord Wharton enter'd upon the complainant Marriott /

7 To the seaventh interrogatory this deponent saith that there are four mannors lying in Swaledale the mannors of Helaugh & Mukar belonging to the said Lord Wharton the mannor of Fremington belonging to Madam Wharton & the said mannor of Grinton belonging to / Mr Marriott /

11 To the eleaventh interrogatory this deponent saith that hee knows the mines in question lying upon Harkaside Moor that Sir Solomon Swaile enjoyed the same as aforesaid this deponent haveing often carryed lead from thence & that Mr Marriott / enjoyed the same (after the said Sir Solomon) till the Lord Wharton enterd thereon about two yeares agoe; /

James Place of Brunton upon Swale in the county of Yorke yeoman aged forty four years or thereabouts sworne & examined; /

2 To the second interrogatory this deponent saith that he know's the reputed manor of Grinton & knows West Grinton Harkaside & Whitaside & the moors & waists thereto belonging & ha's soe knowne the same for above thirty yeares, saith dureing his knowledge / thereof the said places moors & waists were reputed to be parcell & to belong to the said reputed manor of Grinton saith that Sir Solomon Swaile & Roger Hillary were the reputed lords of Grinton and that the said Sir Solomon wrought lead mines / on Harkaside Moor that he knows the manor of Helaugh saith that dureing this deponents time the river Swale was & is the reputed boundary between the two manors of Grinton & Helaugh saith that hee this deponent was borne at Grinton and / hath heard severall people say that the river Swaile was the said boundary and that Stallerstone Stile was the east boundarye of Grinton; /

84 In May 1708 Mathew Watson, aged 54, gave evidence for Marriott at the trial in London (LI Misc 536, Ward MSS, notebooks vol. 6, f. 113r).

Roger Harland of Downham in the county of Yorke yeoman aged sixty five years or thereabouts sworne & examined /

2 To the second interrogatory this deponent saith that he knows the reputed manor of Grinton & alsoe the places called West Grinton Harkaside & Whitaside and the moors & waists thereto belonging & ha's knowne the same for about fifty yeares saith the Swales / & the Hillaries have been comonly called & reputed lords of Grinton further saith that hee knows the manor of Helaugh & that the river Swale has been all along accounted to be the boundary betwixt the said manors of Grinton & Helaugh / as it is the boundary betwixt the manors of Downham <Marrick & Downham> & Marske and alsoe betwixt the manor of Ellerton (which adjoines upon the said mannor of Grinton & parte of the said manor of Marrick & alsoe betwixt the manor of Stainton & another parte of / the said manor of Marrick saith that he never heard that the Lord Wharton or any of his ancestors laid any claime to the moors & comons of Harkaside or West Grinton or the mines there till of late; /

Edward Ellerton of Downham in the county of Yorke yeoman aged eighty three yeares or thereabouts sworne & examined saith /

2 To the second interrogatory this deponent saith that he knows the reputed manor of Grinton & the places called West Grinton Harkaside & Whitaside & the comons & waists thereto belonging & ha's soe knowne the same for sixty yeares last past saith the said places / were reputed to belong to the reputed manor of Grinton & that the Swales & the Hillaries were the reputed lords thereof that the said Sir Solomon wrought mines & got lead on Harka & that Hillary wrought & gott coles on Grinton Moore has heard / that waifes & estrays have been taken upon the said Harkaside Moore by the said Sir Solomon knows the manor of Helaugh & saith that the reputed boundery between the said manors of Grinton and Helaugh is the mid streame of the river Swale / and this deponent beleives it to be the boundary & the reason of such his beleife is that the said mid streame of the said river is the boundery between the manor of Downham & Marske & Ellerton & parte of the manor of Marrick & Stainton and / parte of the manor of Marrick & parte of the manor of Downham & parte of the manor of Marrick saith that dureing this deponents remembrance the late Lord Wharton the present Lord Wharton nor any of their ancestors did ever worke / any mine or mines or claime any royalties in West Grinton or Harkaside or within any other place within the said manor of Grinton that ever this deponent knew or heard off; /

John Clerkson of Leyburn in the county of Yorke yeoman aged seaventy nine or thereabouts sworne & examined saith /

[1] To the first interrogatory this deponent saith that he knows the relator Mr Marriott & alsoe knew the late Lord Wharton but does not know the defendants /

[2] To the second interrogatory this deponent saith that he knows the manor or reputed manor of Grinton & alsoe knows the places called West Grinton Harkaside & Whitaside & the moors & commons thereunto belonging & has known the same / for above sixty yeares & that for all the time of his remembrance the said places have been reputed to belong to & lye within the said manor of Grinton & that the Swales <& Hillarys> have been the reputed lords of the said manor saith that both the said / Swales & Hillarys used to take upp waifs & estrays as lords there further saith that he knows the manor of Helaugh & that the mid streame of the river Swale ha's been all along the reputed boundery betwixt the said mannors of Helaugh / & Grinton & this deponent ha's the greater reason to beleive the said mid streame of the said river to be the said boundery for that about sixty seaven yeares agoe this deponent very well remembers that hee being a boy was then serving or helping / his father at some worke about the mill at Swale Hall which mill that time wanted water & Mr

Swale then owner thereof told this deponent & his father (who is now dead) that if he had as much water above as he had / below he should have water enough for that he had right to the water to the mid streame of the river Swale the said river running a little below the said mill; /

Thomas Wilson of Gilling weaver aged ninety one years or thereabouts sworne & examined saith /

1: To the first interrogatory this deponent saith that he knows the relator Mr Marriott & the present Lord Wharton & did know the late Lord Wharton /

8: To the eighth interrogatory saith that about seaventy three or seaventy four years agoe this deponent being an apprentice to a weaver & his master dying one Henry Blenkiron a smelter at Gilling mill came to this deponents mother & told her till shee could gett / this deponent another master if he would come & worke at the said mill hee would give him six pence a day & hee should be loose whenever shee could gett him another master whereupon this deponent did goe thither to worke and saith that lead / ore was then wrought for & gott on Grinton Moor as the carryer of the said lead told this deponent by old Mr Humphrey Wharton & afterwards by his sonn and it was then said by severall people by a lease they had from the / Crowne & saith Leonard Imeson & Robert Pickering were the jaggers that brought the said ore to be smelted & that James Crathorne & Clement Chambers then were miners at the said Grinton mines as this deponent verily beleives & / the reason of such his beleife is that hee this deponent did see Francis Fawcett steward for the said Mr Wharton pay the said Crathorne & Chambers their wages for working at the said Grinton mines; /

Christopher Dawson of Marske in the county of Yorke aged seaventy two yeares or thereabouts sworne & examined saith; /

2 To the second interrogatory this deponent saith that he knows the manor or reputed manor of Grinton & alsoe knows the places called Harkaside West Grinton & Whitaside & the moors & commons thereunto belonging & ha's knowne the same / for forty yeares saith that the said places have been reputed to belong to the said manor of Grinton & that the Hillarys have been called & taken to be lords of the said manor ha's alsoe heard that one Simpson pretended to be lord thereof / saith that hee knows the manor of Helaugh & that he beleives the river Swale is the boundery betwixt the said manors of Helaugh & Grinton for that the said river is & alwaies was the ancient boundery betwixt the manors / near thereunto adjoining Vizt betwixt the manor of Ellerton & parte of the manor of Marrick & alsoe betwixt the manor of Stainton & parte of the manor of Marrick & alsoe betwixt parte of the manor of Downham & another / parte of the manor of Marrick & alsoe betwixt Downham & the manor of Marsk and further saith that about thirty yeares agoe there was a dispute betwixt the lords of Downham & Marsk about two little islands lying about / the midle of the said river betwixt their own two manors & they agreed to divide them; /

Peter Hutchinson of Kirby Hill black smith aged sixty four years or thereabouts sworne & examined saith /

1 To the first interrogatory this deponent saith that hee has seen the relator Mr Marriott & knows the defendant the present Lord Wharton & did know the old Lord Wharton deceased; /

2 To the second interrogatory saith he knows the reputed manor of Grinton in the county of Yorke & knows the places called West Grinton Harkaside & Whitaside & the moores & commons thereunto adjoyneing & ha's soe knowne the same for forty years last past & / that the said places are reputed to belong & are within the said reputed mannor of Grinton & that Sir Solomon Swale & Roger Hillarye were the reputed lords of the said mannor & hath heard that Sir Solomon gott lead upon Harkaside moore & alsoe this deponents father being an old man told / this deponent

that Mr Humphrey Wharton by vertue of a lease from the Crowne gott lead upon the moores belonging to the mannor of Grinton saith he know's the mannor of Helaugh & that the river Swale dureing this deponents remembrance was reputed to be the / boundarye between the said mannors of Grinton & Helaugh and further saith that neither the late Lord Wharton the present Lord Wharton nor any of their ancestors did ever worke any myne or mynes or claime any royalties upon Harkaside or West Grinton or in any / other place within the said mannor of Grinton that ever this deponent knew or heard of till within these two yeares last past /

Anthony Coates of Marsk in the county of Yorke aged seaventy two yeares or thereabouts sworne & examined deposeth /

2 To the second interrogatory this deponent saith that he know's the mannor or reputed mannor of Grinton & the places called Harkaside West Grinton & Whitaside & the moores & commons thereunto belonging & ha's knowne the same about fifty years that the / Swale's & Hillary's have been in his tyme called the lords there saith he know's the mannor or that which is called the mannor of Helaugh & that he beleives the river Swale ha's been the reputed boundarye betwixt the said mannors of Helaugh & / Grinton because the said river is comonly taken to be the boundarye betwixt the adjacent mannors of Ellerton & Marrick Stainton & Marrick Downham & Marsk & soe downwards & towards Richmond /

[membrane vij *recto*][85]

Samuell Musgrave of Skelton in the county of Yorke aged fifty four yeares or thereabouts sworne & examined saith /

2 To the second interrogatory this deponent saith that he know's the mannor or reputed mannor of Grinton & know's West Grinton Harkaside & Whitaside & the moores & waists thereto belonging called Harka hill Harka Moore Grinton Moore Grinton How or Howgroves & ha's soe knowne the said places moores & / waists for forty yeares last past & saith the same are reputed to belong to the said reputed mannor of Grinton & saith the reputed lords of the said mannor were Sir Solomon Swale & Roger Hillarye & now the relator Mr Marriott saith that one Thomkinson[86] & his partner wrought & gott lead ore on Harkaside / under Sir Solomon Swale & brought the same to Marrick mill to be smelted & that Hillary wrought & gott coale on Grinton moore saith that he ha's heard about twenty or thirty yeares since that the Swale's & Hillary's took up waifes & estray's upon Harkaside moore saith that he know's the mannor of Helaugh / & that for the tyme aforesaid the reputed boundarye betweene the said mannors of Grinton & Helaugh ha's been the river Swale saith that forty yeares last past the present Lord Wharton the late Lord Wharton nor any of their ancestors did ever worke any myne or mynes or claime any royalties upon Harkaside or / West Grinton or in any other place within the said mannor of Grinton that ever this deponent knew or heard of till within these two yeares that the present Lord Wharton entred upon the mynes in question /

11 To the eleaventh interrogatory this deponent saith that he know's the mynes in question claimed by the present Lord Wharton as alsoe by the relator Mr Marriott under the Crowne as this deponent ha's heard & ha's knowne the same for forty <yeares> last past saith the same lye on Harka hill or Harkaside & are reputed parcell of the

85 Parchment, 750mm x 630mm. Tears in the right-hand edge have been reinforced with paper which has darkened with age leading to significant loss of legibility on both the *recto* and *verso*.

86 The reading of this name is uncertain. It appears to have been erased and then rewritten.

[mannor or] / reputed mannor of Grinton saith that dureing the tyme aforesaid the said mynes & places in question have been enjoyed by the Swales & the present relator Mr Marriott & never heard that the present Lord Wharton nor any of his ancestors ever enjoyed or were in the possession thereof or any parte thereof till within these 2 yeares [past] /

George Cowling[87] of Richmond in the county of Yorke alderman aged fifty six or thereabouts sworne & examined deposeth /

1 1 To the first interrogatory this deponent saith that he know's the relator Mr Marriott & alsoe the defendant the Lord Wharton & alsoe <knew> the late Lord Wharton deceased /

2 To the second interrogatory this deponent saith that he know's the mannor or reputed mannor of Grinton & know's the place called Harkaside & the moore & common called Harkaside moore & ha's knowne the same for about thirty yeares saith that the said Harkaside & the moores or commons thereof have been reputed to lye within the / said mannor of Grinton & that Sir Solomon Swale & the Hillary's have been the reputed lords of the said mannor for the said tyme and this deponent particularly remembers that about sixteene yeares agoe he this deponent went with his brother & Peter Hemsley & others on the behalfe & at the request of Mr Hillarye to discharge [&] / warne the workemen then working on the said Harka hill or Harkaside moore from proceeding to worke upon that place for that the said Mr Hillarye claimed the same to belong to him as lord of the said mannor & all the said workemen were then willing to desist (except one Galloway) if they might be con[tent &] / satisfyed for the worke they had done further saith that this deponent was <about> twelve or fourteene years agoe goeing to hunt about Grinton & Helaugh & this deponent being on the north side of the river Swale one James Fryar (now dead) then a bayliffe or steward under the Lord Wharton charged this deponent not to hunt on that side / but told him afterwards that night or the next morneing that if he went over Swale Vizt the south side of the said river to hunt it was out of his lords libertye's & he could not prevent him or had nothing to say to him or to that effect for the same belonged to the Hillarys further saith that the river Swale is the knowne bounderye betwixt Richmond [. . .] /

Ralph Close of Richmond in the county of Yorke butcher aged fifty seaven years or thereabouts sworne & examined saith /

1 To the first interrogatory this deponent saith that he know's the relator Mr Marriott & the present Lord Wharton & did know the late Lord Wharton /

2 & 12 To the second <& 12º> interrogatories this deponent saith that he very well know's the reputed mannor of Grinton & the places called Grinton Harkaside or West Grinton & Whitaside & the moores & waists thereunto belonging & ha's soe knowne the same for forty yeares saith the said places moores waists & comons are reputed to lye & be within the / said mannor of Grinton & saith that Sir Solomon Swale & Mr Hillarye were the reputed lords of the said Grinton mannor & saith that the said Mr Hillarye wrought & gott lead of Grinton How & gott considerable quantity's of lead there & that about thirty yeares agoe there was a suite betwixt the said Sir Solomon Swale & the aforesaid Mr Hillarye [about] / the royalty's of Grinton mannor & that Mr Hillarye as this deponent was informed gott a verdict & was in the possession of the mynes & royalty's belonging to Grinton & gott lead there & tooke up waifes & estray's upon the commons thereto belonging & beleives the late Lord Wharton

[87] In May 1708 George Cooling, aged 60, gave evidence for Marriott at the trial in London (LI Misc 536, Ward MSS, notebooks vol. 6, f. 112r).

was in the country when the said tryall was & saith that [he] / know's the mannor of Helaugh & dureing his remembrance & knowledge thereof the midd streame of the river Swale ha's been reputed & this deponent beleives it to be the boundarye between the mannors of Grinton & Helaugh never dureing this deponents tyme the late Lord Wharton the present Lord Wharton nor any of their ancestors did / ever worke any myne or mynes or claime any royalties upon Harkaside or West Grinton or in any other place within the said mannor of Grinton nor do's this deponent beleive that ever the present Lord Wharton pretended any right thereto till within the said two yeares /

Robert Maltus[88] of East House <near Grinton> aged seaventy five or thereabouts sworne & examined saith /

1 To the first interrogatory this deponent saith that he know's the relator Mr Marriott & alsoe the present Lord Wharton /

2 & 12 To the second & twelfth interrogatories this deponent saith that he know's the mannor or reputed mannor of Grinton & alsoe know's the places called West Grinton Harkaside & Whitaside & the moores & commons called Harka hill Harka Moore Grinton Moore Grinton Greetes Grinton How or Howgroves & ha's knowne the same for sixty yeares saith that / for all the tyme of his knowledge the said places moores & commons have been generally reputed & accounted to lye within & belong to the said mannor of Grinton and that the Swales & the Hillaries have been called & reputed lords of the said manor saith that old Sir Solomon Swale & the present Sir Solomon Swale have continued to / worke mines of lead upon Harkaside moore the place in question & the present Sir Solomon had great gettings there & continued to worke till the relator Mr Marriott recovered the same against him about tenn yeares agoe and from that time the said Mr Marriott enjoyed the same till the Lord Wharton [entred] / upon him about two yeares <agoe> saith that the Hillaries wrought mines of cole upon Grinton moore saith that the Swales as lords used to take upp waifes & estray's & perticularly remembers that about sixty yeares agoe old Mr Swale owner of Swale Hall & the lands thereunto belonging (before he [sold] / the same to old Sir Solomon) tooke upp a ewe & two lambs of this deponents fathers as estray's & this deponent went to fetch them away who presently lett him take them his said father then liveing at East House within the manor of Grinton saith he knows the manor of Helaugh and that for all his / time the river Swale has been reputed the ancient boundery betwixt the said manor of Helaugh & the said manor of Grinton from Haverdale Beck down to Stallerstone stile & beleives the said river is alsoe a boundery betwixt other manors downwards saith that the present Lord Wharton nor any [of his] / ancestors did ever (that this deponent heard or beleives) worke any mines or claime any royalties in West Grinton or Harkaside or the said moors or commons till about two yeares agoe; /

3: & 9: To the third & ninth interrogatories saith that the said manor of Grinton is bounderd on the east by the manor of Ellerton at Stallerstone Stile & on the west by Haverdale Back & on the south by the Duke of Boltons manors & on the north by the river Swale as aforesaid; /

4: To the fourth interrogatory this deponent saith that the said places called West Grinton & Harkaside are little hamletts containing some few houses about the number of twelve or fourteen lying at some distance from each other /

[88] In May 1708 Robert Maltus, aged 76, gave evidence for Marriott at the trial in London (LI Misc 536, Ward MSS, notebooks vol. 6, f. 112v).

6: To the sixth interrogatory this deponent saith that the said manor of Grinton is & ha's been all his time reputed & taken to belong to the Priory of Bridlington; /

7: To the seaventh interrogatory this deponent saith that he ha's knowne the dale called Swaledale for about sixty yeares & that the same is in length about twelve miles that there are four manors lying within the said dale Vizt the manors of Helaugh & Mukar belonging to the Lord Wharton / the manor of Fremington belonging to Madam Wharton & the said manor of Grinton belonging to Mr Marriott all lying within the parish of Grinton & that the town of Grinton (where the church stands) is the cheife towne therein /

8: To the eighth interrogatory this deponent saith that he this deponent ha's formerly heard from ancient men that one Wharton of Gillingwood had formerly a lease from the Crowne of the mines within the said manor of Grinton in the place in question & that he wrought & enjoyed the [same] /

10: 10:To the tenth interrogatory this deponent saith that inhabitants of Grinton & Harkaside have a rake for their cattle to goe upon each others moors or commons without disturbance but beleives they have & use a separate right in taking turfe and ling in their severall commons [. . .] / & that they know their severall divisions & boundaries for that purpose But this deponent cannot perticualarly sett forth the same; /

11: To the eleaventh interrogatory this deponent saith that he knows the mines in question lying upon Harkaside moore which the Lord Wharton claimes & which the relator Mr Marriott alsoe claimes under the Crowne & that the same have for all his time been comonly reputed [. . .] / amongst the neighbourhood (hee this deponent being alsoe a near neighbour thereto) to lye within & to belong to the said manor of Grinton; /

Roland Snow of Nappa Scarr in the county of Yorke gentleman aged six & fifty yeares or thereabouts sworne & examined saith /

1: To the first interrogatory this deponent saith that he knows the relator Mr Marriott & the present Lord Wharton & has seen the late Lord Wharton /

2: & 12: To the second & twelfth interrogatories this deponent saith that he well know's the manor of Grinton & alsoe knows the places called West Grinton Harkaside & Whitaside & the moores & waists called Harka Moore Grinton Moore Grinton Greets Grinton How or How Groves & has [soe] / knowne the same for thirty five yeares or thereabouts & saith the said places moors & waists doe belong & lye within the said reputed manor of Grinton & that the Swailes & the Hillaries have been the reputed lords of the said manor till the same was recovered by the said relator Mr / Marriott saith that the present Sir Solomon Swaile wrought the mines on Harkahill the place in question & got considerable quantities of lead ore there and that the relator Mr Marriott since his entry upon the said mines ha's gott very great quantities of lead <ore> & continued / working at the said mines untill the present Lord Wharton entered upon him about two yeares since saith that Roger Hillary wrought & gott cole on Grinton Moore saith that he know's the manor of Helaugh & dureing this deponents knowledge thereof which is for the time [. . .] / the river Swale has been the reputed boundery between the two manors of Grinton & Helaugh saith for the time aforesaid the late Lord Wharton the present Lord Wharton nor any of their ancestors did ever worke any mine or mines or claime any royalties within or upon the moores [&] / waists of Harkaside & West Grinton or in any other place within the said manor of Grinton that ever this deponent heard of till within these two yeares that the present Lord Wharton enterd upon Harka Hill; /

3: & 9: To the third & ninth interrogatories this deponent saith that the river Swale is reputed to be the ancient boundery northwards from Haverdale Beck to Stallerstone Stile between the two manors of Grinton & Helaugh & that the manor of Grinton is bounderd on the east on the said Staller / =stone Stile on the west on the said Haverdale Beck & on the south on the Duke of Boltons manors saith that the said river Swale is alsoe a boundery between severall other manors that is to say between the manors of Fremington & Grinton Marrick & Ellerton & beleives it is soe / downwards saith the said river Swaile is a large river & that hee this deponent hath heard severall ancient men say the same was a boundery as aforesaid; /

7: To the seaventh interrogatory this deponent saith that he never heard Swailedale reputed a lordshipp of it selfe but saith there are four manors in the said dale to witt Helaugh & Mukar belonging to the present Lord Wharton Fremington to Madam Wharton of Gillingwood & Grinton manor [. . .] / Mr Marriott saith that the said dale is about twelve miles in length & all ly's within the parish of Grinton & that the parish church stands at Grinton; /

11: To the eleaventh interrogatory this deponent saith that he well knows the mines in question claimed by the present Lord Wharton as alsoe by the relator Mr Marriott as lessee under the Crowne saith the said mines lye on Harka Hill on Harkaside & are reputed to be & lye within the / reputed manor of Grinton & to be parte & parcell thereof & have been enjoyed dureing the time aforesaid by the Swales & the relator Mr Marriott untill of late that the present Lord Wharton entered thereon saith the present Lord Wharton the late Lord Wharton nor any of their / ancestors did ever that this deponent knew or heard of claime any right or had any possession of the said mines in question till within these two years as aforesaid;

Thomas Wood[89] of Grinton yeoman aged sixty six or thereabouts sworne & examined saith; /

2: & 12: To the second & twelfth interrogatories this deponent saith he well knows the manor or reputed manor of Grinton & the places called West Grinton Harkaside & Whitaside & alsoe the moors & commons called Harka Hill Harka Moore Grinton Moore Grinton Greets Grinton How or How Groves / & ha's knowne the same for fifty yeares; hee this deponent having wrought at the mines upon Harka Hill or Harkaside Moore under old Sir Solomon & the present Sir Solomon Swale saith that for all the time of his knowledge the said places moors & commons have been generally reputed / to lye within & belong to the said manor of Grinton & that the Hillarys & Swales have been called & reputed lords of the said manor saith that old Sir Solomon Swale & the present Sir Solomon wrought mines of lead upon Harka Hill or Harkaside Moore the place in question & the present Sir Solomon / had great gettings and continued to worke there till the relator Mr Marriott recovered the same against him about tenn yeares agoe & since that time the said Mr Marriott enjoyed the same till the Lord Wharton enterd upon him about two yeares since saith that the Hillaries wrought mines of [. . .] / upon Grinton Moor and that both Hillarys & Swales as lords used to take upp waifes & estrays within the said moors & commons saith he knows the manor of Helaugh & that for all his time the river Swale has been reputed the ancient boundery betwixt the said manor of Grinton and the manor / of Helaugh from Haverdale Beck down to Stallerstone Stile & beleives the said river is alsoe a boundery betwixt the other manors downwards & that fishermen used to observe the mid streame of the said river to be a boundery as aforesaid saith that the present Lord Wharton or any of / his ancestors did ever (that this deponent heard

89 In May 1708 Thomas Wood, aged 66, gave evidence for Marriott at the trial in London (LI Misc 536, Ward MSS, notebooks vol. 6, f. 112v).

or beleives) work any mines or claime any royalties in West Grinton or Harkaside or the said moors or commons till about two yeares agoe; /

3: & 9: To the third & ninth interrogatories this deponent saith that the said manor of Grinton is bounderd on the east by the manor of Ellerton at Stallerstone Stile & on the west by Haverdale Beck & on the south by the Duke of Boltons manors & on the north by the river Swale as aforesaid /

4: To the fourth interrogatory this deponent saith that the said places called West Grinton & Harkaside is a little hamlett containing some few houses about the number of fourteen lying at some distance from each other & are & have been reputed as aforesaid to lye within the manor of Grinton /

5: & 7: To the fifth & seaventh interrogatories this deponent saith that he ha's known the dale called Swaledale for about fifty yeares & that the same is in length about twelve miles that there are four manors lying within the said dale Vizt the manors of Helaugh & Mukar belonging to the said Lord Wharton / the manor of Fremington belonging to Madam Wharton & the said manor of Grinton belonging to Mr Marriott all lying within the parish of Grinton & that the town of Grinton (where the church stands) is the cheife towne in the said dale; /

10: To the tenth interrogatory this deponent saith that the inhabitants of Grinton & Harkaside have a rake for their cattle to goe upon each others moors or commons and that the boundery lately rid by the present Lord Wharton is comonly taken & reputed to be only the bounderies for div[iding . . .] / their moors & commons; /

11: To the eleaventh interrogatory this deponent saith that he knows the mines in question lying upon Harkaside moor which the said Lord Wharton claimes & which the relator Mr Marriott alsoe claimes under the Crown & that the same have for all his time been comonly reputed to lye within & b[elong to the said] / manor of Grinton; /

13: To the thirteenth interrogatory this deponent saith that hee never knew any boundery rid between Grinton Harkaside West Grinton & Whitaside till of late by the present Lord Wharton & that he does not beleive the same to be the boundery of the manor of Helaugh but only a boundery of [. . .] / turbary as aforesaid; /

Francis Robinson[90] of Fremington in the county of Yorke yeoman aged forty three yeares or thereabouts sworne & examined saith: /

1: To the first interrogatory this deponent saith that he knows the relator Mr Marriott & the present Lord Wharton & did know the late Lord Wharton /

2: & 12: To the second & twelfth interrogatories this deponent saith that he knows the reputed manor of Grinton & knows the places caled West Grinton or Harkaside & the moors & commons thereto belonging called Harkahill Harka Moore Grinton Moore Grinton Greets Grinton How or Howgroves / & has soe knowne the said manor places moores & commons for about thirty yeares last past & saith West Grinton Harkaside & the moors & commons are reputed to lye & be parte & parcell of the said reputed manor of Grinton saith the Swales & the Hillaries & the relator Mr Marriott [have been] / reputed lords of the said manor dureing this deponents knowledge thereof saith that Sir Solomon Swale wrought the mines of Harka Hill the place in question & got considerable quantitys of lead there & continued working till the relator Mr Marriott about <eleaven> years since recovered [the same] / saith that Roger Hillary wrought & gott coale on Grinton Moore saith that this deponent ha's seen severall times Christopher Naylor & his sonns take upp waifes & estrays on

90 In May 1708 Francis Robinson (no age given) gave evidence for Marriott at the trial in London (LI Misc 536, Ward MSS, notebooks vol. 6, f. 112v).

Harkaside Moore as servants & agents for Sir Solomon Swaile & ha's heard that one Thomas Pratt [. . .] / Hillary has likewise taken upp waifes & estrays upon the said Harkaside Moore saith that hee knows the manor of Helaugh & dureing the time aforesaid the reputed boundary between the said manors of Grinton & Helaugh has been the mid streame of the river Swale / saith the late Lord Wharton the present Lord Wharton nor any of their ancestors did ever that this deponent knew or heard of worke any mine or mines or claime any royalties upon Harkaside or West Grinton or in any other place within the said manor of Grinton till [. . .] /

3: & 9: To the third & ninth interrogatories this deponent saith that the manor of Grinton is bounderd on the east on Ellerton manor by Stallerstone Stile on the west on Haverdale Beck & on the south on the Duke of Boltons manors & on the north on the river Swale as aforesaid saith that the [river Swale] / is alsoe reputed a boundery between the manors of Fremington & Grinton & between the manors of Marrick & Ellerton & parte of the said manor of Marrick & Stainton & soe downwards as this deponent beleives as farr as the river Swaile goes saith that he this deponent [has] / heard both his grandfather & uncle ancient men & fishermen severall times say that the said river Swale was a boundery as aforesaid /

[membrane vii *verso*]

4: To the fourth interrogatory this deponent saith that he knows the places of West Grinton & Harkaside saith it is a hamlett consisting of fourteen or fifteen houses & the which stands at some considerable distance from each other & beleives the same to be within the [boundaries of] / Grinton saith that there is both copyhold & freehold lands lying & belonging to Harkaside & West Grinton Beleives the defendant the present Lord Wharton ha's of both tenures there beleives there are other freeholders there but what perticular rents are [paid by] / them to the Lord Wharton this deponent know's not or at what court they all appeare this deponent does not know, saith that the present Sir Solomon Swaile has lands in Reeth for which he used to appeare at Helaugh court as this deponent beleives saith that about fourteene yeares [ago] / the old Sir Solomon Swaile sold the said lands at Reeth to one Whip but whether the said Sir Solomon Swale has appeared or been called at Helaugh court since this deponent knows not; /

6th To the sixth interrogatory this deponent saith that hee has heard it reputed that the manor of Grinton formerly belonged to the Priory of Bridlington but what perticular parcells in Harkaside or West Grinton belongs to the said Priory this deponent knows not;

7: To the seaventh interrogatory this deponent saith that hee well knows the dale called Swaledale saith it is in length about twelve miles never heard it reputed a lordshipp but saith there are four manors in the said dale the manors of Helaugh & Mukar belonging to the Lord [Wharton the] / manor of Fremington to Madam Wharton & the reputed manor of Grinton to the relator Mr Marriott saith all the said dale is in the parish of Grinton & that the parish church stands at Grinton; /

10: To the tenth interrogatory this deponent saith that he this deponent knows the boundary rid by the present Lord Wharton & never heard the same reputed to be a boundery to divide Harkaside & West Grinton from Grinton but only to divide the common rights between the said places [for the] / getting of ling & turfes & saith that hee this deponent has been disturbed from graveing of turfs on Harkaside Moore he living at Fremington & not asking leave; /

11: To the eleaventh interrogatory this deponent saith that he knows the mines in question on Harkaside Moore which the Lord Wharton claimes & the which the relator Mr Marriott alsoe claimes under the Crowne & that the same have for all this deponents time been / reputed to lye within & belong to the manor of Grinton /

13: To the thirteenth interrogatory this deponent saith that hee never knew any boundery rid between Grinton Harkaside West Grinton & Whitaside till of late that the present Lord Wharton rid the same & that he does not beleive the same to be the boundery of the [Mannnor of] / of Helaugh but only the boundary of the commons for turbary as aforesaid /

17: To the seaventeenth interrogatory this deponent saith that the present Lord Wharton has copyhold lands lying on the east side of Arkell beck & there is another parcell belonging to Mr Covill lying on the east side of Fremington both which are in the manor of Fr[emington . . .] / & appeare at Helaugh court; /

Francis Hall of Fremington in the county of Yorke miner aged forty five yeares or thereabouts sworne & examined saith /

1: To the first interrogatory this deponent saith that he knows the relator Mr Marriott & alsoe knows the present Lord Wharton but does not know the sonn nor did he know the late Lord Wharton deceased /

2: & 12: To the second & twelfth interrogatories this deponent saith that he knows the manor or reputed manor of Grinton & alsoe knows the places called West Grinton Harkaside & Whitaside & the moors & commons called Harka Hill Harka Moor Grinton Moor Grinton Greets [Grinton How or] / How Groves & hath knowne the same for thirty yeares and for all that time the said places moors & commons were generally reputed & taken to lye within & belong to the said manor of Grinton & that the Hillarys & Swales have been the reputed lords of the [said manor] / for all the time of his remembrance & this deponent for severall yeares wrought under Sir Solomon Swale in the mines at Harka Hill where hee had great gettings of lead ore & continued to worke there till about tenn yeares agoe that Mr Marriott [recovered the] / same against him & that Mr Hillary wrought mynes of cole upon Grinton Moore saith that the Hillarys & the Swales used as lords to take upp waifes & estray's upon the said moors & commons saith he alsoe knows the manor of Helaugh & that [dureing] / the time of his remembrance the river Swale has been always accounted & taken to be the ancient boundery betwixt the said manor of Grinton & mannor of Helaugh saith that the present Lord Wharton nor any of his ancestors did ever worke [any] / mines or claime any royalties in the said places moors or commons till about two yeares agoe that this deponent ever heard of hee this deponent being frequently working upon the said moor's; /

Christopher Hetherington of Fremington in the county of Yorke miner aged fifty seaven yeares or thereabouts sworne & examined saith /

1: To the first interrogatory this deponent saith that hee knows the relator Mr Marriott the present Lord Wharton & did know the late Lord Wharton /

2: & 12: To the second & twelfth interrogatories this deponent saith that he very well knows the reputed manor of Grinton & the places called West Grinton Harkaside & Whitaside as alsoe the moors waists & commons called Harka Hill Harka Moor Grinton Moor [Grinton] / Greets Grinton How or How Groves & ha's soe knowne the same for forty six yeares last past saith the same dureing this deponents knowledge thereof have been reputed to belong & to be parte & parcell of the said manor of Grinton and that the Swales [&] / Hillaries & the relator Mr Marriott have been the reputed lords of the said manor saith that about forty five or forty six yeares agoe when this deponent was a boy he wrought at the mines at Harka hill the place in question under Sir Solomon / Swale and from that time continued working there about two years together saith that then the said mines were very rich and afterwards this deponent for severall year's together was a workman under Mr Humphrey Wharton at other / mines at Copperthwaite above Marrick saith that about eleaven years since this deponent was a workman under the relator Mr Marriott in the mines in question & that the said Mr Marriott had very good gettings there saith that the / Hillaries wrought & gott coales on Grinton Moore saith that John Blaides as steward to Sir Solomon Swaile tooke upp waifs & estrays upon Harkaside moor saith that hee this deponent ha's seen the said Blaides take upp waifs & estrayes / & that he this deponent ha's carryed in waifs & estray's from the said moors <to> the said Blaids who paid him four pence for every sheep & twelve pence for a horse or beast saith that one Francis Place of Grinton used to take up waifes & estrays for the Hillarys upon Grinton [. . .] / saith that he this deponent know's the mannor of Helaugh & that the river Swale for all his remembrance was & is the reputed boundary between the two mannors of Grinton & Helaugh saith that for all the tyme of this deponents knowledge of West Grinton & Harkaside the said Lord Wharton [nor any] /of his ancestors that ever this deponent knew or heard of did worke any mine or mynes within or upon the aforesaid places on the mynes[91] & commons thereto belonging or in any other place within the said mannor of Grinton till of late /

3 & 12 To the third & twelfth interrogatories this deponent saith that the said mannor of Grinton is bounderd on the north by the river Swale from Haverdale beck to Stallerstone Style <on the east on Ellerton mannor by the said Stallerstone Stile> on the west by the said Haverdale beck & on the south by the Duke of Boltons mannors saith the said river Swale [is] / the boundarye of the mannors of Fremington & Grinton as alsoe Marrick & Grinton /

4 To the fourth interrogatory this deponent saith that he know's the places called West Grinton & Harkaside saith that the same are hamletts consisting of some few {acres} houses lyeing at a considerable distance from each other saith that West Grinton & Harkaside / consist of about eighteen or twenty houses saith that the said places were for all the tyme of this deponents remembrance reputed to belong & to be within the said mannor of Grinton /

7 To the seaventh interrogatory this deponent saith that he know's the dale called Swaledale & ha's soe known the same dureing his remembrance saith that the said dale is about twelve miles in length & has four mannors in it the mannors of Helaugh & Mukar belonging to / the Lord Wharton the mannor of Fremington to Madam Wharton & the mannor of Grinton to the relator Mr Marriott saith the said dale is all in the parish of Grinton & that the parish church stands in Grinton /

10 & To the tenth & 13th interrogatories this deponent saith that about two yeares agoe
13 the Lord Wharton caused a boundary to be ridd (as this deponent beleives seeing them ride parte thereof) between the moores & commons of Grinton West Grinton & Harkaside saith he verily beleives [the] / boundary is onely for the divideing the common rights of each other for the getting of peates turves & ling & claimed to be such a division for common right by the Harkaside men the said moores not being fenced & the said Grinton men haveing <a> rake in the said Common [. . .] /West Grinton & Harkaside men. /

To the eleaventh interrogatory this deponent saith that the mynes in question lye on
11 Harka Moor & are reputed to be within the said mannor of Grinton & have been enjoyed by the Swale's & Hillary's & the said Mr Marriott dureing this deponents remembrance & that the Lord Wharton [nor any] / of his ancestors ever were in the possession thereof or claimed any tytle thereto that ever this deponent knew or heard

[91] The word 'mynes' is clearly an error for 'moores'.

of till about two yeares since that the present Lord Wharton entred thereon but by what title or by what meanes this deponent knows not /

17 To the seaventeenth interrogatory this deponent saith that the present Lord Wharton ha's coppyhold lands in Whitaside Harkaside & Fremington & that his tennants there appeare at his lordshipps court at Helaugh for the said lands & that Mr Covill one of his Lordships tennants he [verily beleives does] / appeare at Helaugh court & that the Fremington people used to see the Lord Whartons tennants at the said Helaugh court /

Richard Etherington[92] of Fremington in the county of Yorke aged sixty yeares or thereabouts sworne & examined saith /

1 To the first interrogatory this deponent saith that he know's the relator Mr Marriott & the present Lord Wharton /

2 & 12 To the second & twelfth interrogatories this deponent saith that he know's the mannor or reputed mannor of Grinton & alsoe know's the places called West Grinton Harkaside & Whitaside & the moores & commons called Harkahill Harka Moore Grinton Moore Grinton Greetes Grinton How or [How Groves] / & has known the same for about fifty yeares & that for all his tyme the said places & commons have been reputed to belong to & lye within the mannor of Grinton & that the Hillary's & Swales & of late Mr Marriott have been the reputed lords of the said mannor and that [old Sir Solomon] / Swale & the present Sir Solomon wrought mynes of lead upon Harkahill the place in question for many yeares together & gott considerably there & has heard that Hillarye some tyme wrought mynes of lead alsoe there the said Hillary & Sir Solomon disputeing the right [. . .] / that Sir Solomon continued workeing there till Mr Marriott recovered the same against him about ten yeares agoe & that Mr Marriott continued to worke there & had considerable gettings till about two yeares agoe that the Lord Wharton entred upon him saith that both Sir Solomon & Hill[ary] / used to take up waifes & estray's as lords upon the said moores & commons of West Grinton & Harkaside saith he know's the mannor of Helaugh & that the river Swale ha's been all his tyme reputed to be the antient boundarye betwixt the mannors of Grinton [&] / Helaugh from Haverdale beck down to Stallerstone Style saith that the present Lord Wharton nor any of his ancestors did ever worke any mynes or claime any royalties in the moores & commons of West Grinton & Harkaside till about two yeares agoe /

3 & 9 To the third & ninth interrogatories this deponent saith that the said mannor of Grinton is bounderd on the east by the mannor of Ellerton at Stallerston Style & on the west by Haverdale beck & on the south by the Duke of Boltons mannors about Gibbon Hill & on the north [by the] / river Swale as aforesaid saith that the river Swale ha's been alsoe generally taken to be the boundarye betwixt the mannors of Fremington & Grinton Marrick & Ellerton Marrick & Stainton Marrick & Downham & soe downwards saith he hath heard old fishermen / say that the Harkaside & West Grinton men often hindred them from fishing further then the midd streame of the said river & would not suffer them to come on their side unless they gave them some of their fish & the said fishermen used to account the midd streame [of] / the said river to be a boundarye as aforesaid /

4 To the fourth interrogatory this deponent saith that the Lord Wharton hath coppyhold lands in West Grinton & Harkaside the tennants whereof appeare at his lordshipps courts at Helaugh & that he hath alsoe some coppyhold lands in Whitaside

92 In May 1708 Richard Etherington, aged 60, gave evidence for Marriott at the trial in London (LI Misc 536, Ward MSS, notebooks vol. 6, f. 112r).

& Fremington his tennants whereof app[eare] / at Helaugh saith that Sir Solomon Swale owner of Swale hall & the lands thereunto belonging had alsoe some lands in Reeth which he sold about fourteen or fifteene yeares agoe & saith that the said Sir Solomon or his tennants used to appeare at Helaugh court for the lands at [. . .] / but beleives he ha's not appeared there since he sold the said lands at Reeth & this deponent is the more able to depose therein for that this deponents unckle was tennant to the said Sir Solomon at Reeth & this deponent lived with him /

7 To the seaventh interrogatory this deponent saith that there are four mannors in Swaledale Vizt the mannors of Helaugh & Mukar belonging to the Lord Wharton the mannor of Fremington belonging to Madam Wharton & the mannor of Grinton belonging to Mr Marriott all lyeing within [the mannor of] / Grinton & that Grinton (where the church stands) is the chiefe towne therein /

10 To the tenth interrogatory this deponent saith that the inhabitants of Grinton West Grinton & Harkaside used to have a rake for their cattle to goe {within} upon each other moores but have a known boundarye or division marks to distinguish or divide the right of their severall commons from [each] / other as to getting of ling peates or turves & that the White Stone on the How & some other marks thereabouts are & have been the reputed markes for the division of their commons as aforesaid /

11 To the eleaventh interrogatory this deponent saith that he know's the mynes in question lyeing upon Harka or Harkaside Moore & that the same have been enjoyed by the Swale's Hillary's & Mr Marriott as by him predeposed till the Lord Wharton entred upon the same about [. . .] /

13 To the thirteenth interrogatory this deponent saith that he never knew or heard of any boundarye ridd betwixt Grinton West Grinton & Harkaside or Whitaside till about two yeares agoe he heard the Lord Wharton ridd some boundarye thereabouts /

Samuell Robinson of Fremington in the county ofYorke yeoman aged fifty four yeares or thereabouts sworne & examined saith /

1 To the first interrogatory this deponent saith that he know's the relator Mr Marriott & the present Lord Wharton & did know the late Lord Wharton /

2 & 12 To the second & twelfth interrogatories this deponent saith that he knows the reputed mannor of Grinton & the places called West Grinton Harkaside & Whitaside & the moores & commons called Harka hill Harka Moore Grinton Moore Grinton Greetes Grinton How or How groves & ha's known the said / places moores & waists for fifty six years[93] & upwards saith the same dureing this deponents knowledge ha's been reputed to belong & be within the said mannor of Grinton & that the Swale's the Hillary's & Mr Marriott have been the reputed lords thereof saith that Sir Solomon Swaile [wrought] / mynes on Harka Moore the place in question for severall yeares together & had good gettings saith that he continued workeing the said mynes untill the relator Mr Marriott recovered them from him saith this deponent wrought at the said mynes under Sir Solomon at the tyme [Lord Wharton] / took possession of them & afterwards wrought in the said mynes for the said Mr Marriott who had very good gettings there saith the said Hillarye wrought for & gott coale on Grinton How saith that one John Blaides tooke up waifes & estrayes for Sir Solomon Swale [upon Harkaside] / Moore saith that Roger Hillarye took waifes & estray's upon Grinton Moore saith that he know's the mannor of Helaugh & that dureing the tyme of his remembrance which is for above fifty six yeares the river Swale ha's been the reputed boundarye from [Haverdale beck] / to Stallerstone Style between the two mannors of

93 Possibly an error for 'forty six years' in view of the stated age of the deponent.

Grinton & Helaugh saith that dureing the tyme aforesaid the present Lord Wharton nor any of his ancestors did ever worke any myne or mynes or claime any royalties within the said places of West Grinton & Harkaside [or] / any of the moores & waists thereto belonging or in any other place within the said mannor of Grinton that ever this deponent knew or heard of till about two yeares last past /

3 & 9 To the third & ninth interrogatories this deponent saith that the said mannor of Grinton is bounderd on the east on Ellerton mannor at Stallerstone Style on the west by Haverdale beck & on the south by the Duke of Boltons mannors & on the north on the said river Swale as aforesaid [&] / that the said river Swale is the antient reputed boundarye between Fremington & Grinton & soe downwards as this deponent beleives saith that he hath heard severall antient men who are since dead say that the said river Swale was a boundarye as aforesaid /

[membrane viij *recto*][94]

4 To the fourth interrogatory this deponent saith that he know's the places of West Grinton & Harkaside saith the same are one hamlett consisting of about eighteene or twenty houses lyeing at a considerable distance from each other & are reputed to lye within the said mannor of Grinton saith that the Lord [Wharton] / hath coppyhold lands in West Grinton & Harkaside the tennants whereof appeare at his lordshipps court at Helaugh & that he hath alsoe some coppyhold lands in Whitaside & Fremington the tenants whereof alsoe appeare at Helaugh court saith that the present Sir Solomon Swale had some coppyhold lands / at Reeth for which his tennants appeared at Helaugh court & that this deponent about forty yeares since being tennant to the said lands appeared at the said court saith that about fourteene yeares agoe the said Sir Solomon sold the said coppyhold lands to one Whip saith that since the selling of that land this deponent do's [not] / know of the said Sir Solomon's being called or of his tennants appeareing for him at the said Helaugh court /

7 To the seaventh interrogatory this deponent saith that he know's Swaledale & ha's soe knowne the same for fifty six yeares last past saith it's in length about twelve miles & ha's in it four mannors to witt the mannors of Helaugh & Mukar belonging to the Lord Wharton the mannor of Fremington to Madam / Wharton & the mannor of Grinton to the relator Mr Marriott saith the said dale is all in the parish of Grinton & that the parish church stands there /

11 To the eleaventh interrogatory this deponent saith that he know's the mynes in question lyeing upon Harkaside Moore & that the same ha's been enjoyed dureing this deponents remembrance by Sir Solomon Swale's & Mr Marriott till the Lord Wharton entred upon the same about two yeares agoe /

Michael Waggett[95] of Downholme clerke aged fifty <five> yeares or thereabouts sworne & examined saith /

2 & 12 To the second & twelfth interrogatories this deponent saith that he know's the mannor or reputed mannor of Grinton, and alsoe < knows the places called West Grinton Harkaside & Whitaside> & has knowne the same for above thirty yeares & that the said places have dureing the tyme of his knowledge thereof been reputed to belong to the mannors of Grinton & Harkaside / & Swale's have been reputed lords of the

94 Parchment, 760mm x 630mm. The sheet is discoloured and the edges are torn and dirt stained. Paper used to repair damaged areas has become partly opaque leading to loss of legibility, particularly in the top left-hand corner of the verso and down the right-hand sides of both *recto* and *verso*.

95 In May 1708 Michael Waget, clerke, aged 55, gave evidence for Marriott at the trial in London (LI Misc 536, Ward MSS, notebooks vol. 6, f. 113v).

royaltyes there & that he ha's never heard or beleives that the Lord Wharton or any of his ancestors claimed any royalties there till about two yeares agoe saith he know's the mannor of Helaugh & that the river Swale has been dureing his tyme reputed to be the antient boundery betwixt the said mannors of Grinton & [Helaugh] /

3 & 9 To the third & ninth interrogatories this deponent saith that the boundery of the said mannor of Grinton is commonly taken to be by the mannor of Ellerton at Stallerstone Style on the east & by the Duke of Boltons mannors on the south & on the north by the river Swale as aforesaid saith that the said river is & hath been dureing / his tyme alwaies taken & accounted to be the boundery betwixt the mannor of Fremington & parte of the mannor of Grinton & parte of the mannor of Marrick & Grinton down to Stallerstone Style & alsoe betwixt the mannors of Marrick & Ellerton & Marrick & Stainton & Marrick & Downham & Marsk & Downham /

7 To the seaventh interrogatory this deponent saith that he know's Swaledale & that he ha's never heard it to be a mannor of itselfe but that there are four mannors within the said dale Vizt the mannors of Helaugh & Mukar belonging to the Lord Wharton the mannor of Fremington belonging to Madam Wharton & the mannor of Grinton belonging to Mr Marriott & that / they all lye within the parish of Grinton /

Bryan Appleby of Grinton in the county of Yorke yeoman aged twenty Eight yeares or thereabouts sworne & examined saith /

2 & 12 To the second & twelfth interrogatories this deponent saith that he know's the mannor or reputed mannor of Grinton & knows the places called West Grinton Harkaside & Whitaside & the moores & commons thereto belonging called Harka hill Harka Moore Grinton Moore Grinton Greetes Grinton How or Howgroves & ha's soe knowne / the same for eighteen yeares last past say's they are reputed to belong & be within the said mannor of Grinton & that Sir Solomon Swale Roger Hillarye & Mr Marriott have been the reputed lords thereof saith the present Sir Solomon Swale wrought & gott lead on Harka hill or Harkaside Moore the place in question & continued / workeing there till the relator Mr Marriott recovered the mynes from him saith Mr Marriott ha's since wrought the <said> mynes & had good gettings there untill the present Lord Wharton entred upon the same about two yeares since saith Sir Solomon Swale & Roger Hillarye took up waifes & estray's upon Grinton Moore saith / he know's the mannor of Helaugh & that for these eighteene yeares last past the river Swale ha's been the reputed boundery betweene the two mannors of Grinton & Helaugh saith the Lord Wharton nor any of his ancestors dureing this deponents knowledge did ever worke or claime any of the myne or my[nes or any] / of the royaltyes in West Grinton or Harkaside or in any other place within the said mannor of Grinton till of late /

3 & 9 To the third & ninth interrogatories this deponent saith that the river Swale is the north boundarye betweene the two mannors of Grinton & Helaugh from Haverdale beck to Stallerstone Style saith the said mannor of Grinton is bounderd on the east by Ellerton mannor at Stallerstone Style on the west by Haverdale beck & on the so[uth] / by the Duke of Boltons mannors & saith that the said river Swale ha's dureing this deponents knowledge been taken & accounted to be the boundery between the mannor of Fremington & parte of the mannor of Grinton & parte of the mannor of Marrick & Grinton & soe downwards /

7 To the seaventh interrogatory this deponent saith Swaledale is reputed to be in length about 12 myles & ha's in't four mannors to witt Helaugh Mukar Fremington & Grinton the Lord Wharton is Lord of Helaugh & Mukar Madam Wharton of Fremington & Mr Marriott of Grinton saith the whole dale is in Grinton parish & that the parish church stands att Grinton /

11 To the eleaventh interrogatory this deponent saith he know's the mynes in question

on Harka hill or Harkaside saith they are reputed to belong to the mannor of Grinton & have dureing his tyme been enjoyed by Sir Solomon Swale & Mr Marriott never knew for the tyme aforesaid the present Lord Wharton nor any of his ancestors in the possession thereof / claime the same till about two yeares since /

William Pratt of Grinton myner aged forty or thereabouts sworne & examined saith /

2 & 12 To the second & twelfth interrogatories this deponent saith that he know's the mannor or reputed mannor of Grinton & alsoe know's the places called West Grinton Harkaside & Whitaside & the moores & commons called Harkahill Harka Moore Grinton Moore Grinton Greetes Grinton How or Howgroves & ha's knowne / them all for neare thirty yeares & that they have for all his tyme been accounted & taken to lye within the said mannor of Grinton & that the Hillary's & Swales & of late Mr Marriott have been called & reputed the lords of the said mannor saith that Sir Solomon Swale wrought mynes of lead upon Harka hill / or Harkaside Moore & this deponent wrought under him & this deponent also wrought waists of lead ore upon Harkahill under the said Roger Hillarye he claimeing the said feilds as his owne but do's not know that the said Hillarye wrought any mynes there under ground but the said mynes proved very rich in the [time] / of the said Sir Solomon & he continued to worke there till Mr Marriott recovered the same & turned him out of possession about ten or eleaven yeares agoe & the said Mr Marriott after that continued to worke there & had very good gettings till the Lord Wharton entred upon him about two yeares agoe saith that both Sir Solomon & Roger / Hillarye <used to> take up waifes & estray's upon the said moores or commons betweene Stallerstone Style & Haverdale beck saith that he know's the mannor of Helaugh & that the river Swale ha's been dureing his tyme the reputed & knowne boundery betwixt the said mannor of Grinton & Helaugh from Haverdale beck / to Stallerstone Style saith that the Lord Wharton nor any of his ancestors did ever that this deponent heard or beleives dureing his tyme worke any mynes or claime any royalties in the moores or commons of West Grinton & Harkaside till about two yeares agoe /

3 & 9 To the third & ninth interrogatories this deponent saith that the boundary's of the mannor of Grinton dureing his tyme has been commonly reputed to be as followeth (he this deponent haveing for most parte of his tyme since he was a boy wrought upon the said moores & commons) Vizt by the mannor of Ellerton / at Stallerstone Style on the east & by Haverdale beck on the west & by the Duke of Boltons mannors on the south the said boundery thereof begining at Stallerston Style & soe going up to White Stone in Sharrow & soe to Snowden Man & thence to Ridmer Standard thence to Grinton Greetes [thence to] / Gibbon Hill & soe to high Carle currack & then to Pickersgill currack & then to the hurrack of Windyates & soe to the currack of Somerlodge Tarne & then to bulls head & soe to bloody wall & soe down to Haverdale beck further saith that the said mannor of Grinton is bounderd on the north [by the] / river Swale from the mannor of Helaugh aforesaid saith that the said river Swale ha's been alsoe the reputed boundery betwixt the mannor of Fremington (which adjoynes eastward on the said mannor of Helaugh) & the said mannor of Grinton & alsoe betwixt the mannors of Marrick & Ellerton & beleives the same to be alsoe a boundery betwixt [. . .] /

4 To the fourth interrogatory this deponent saith that the said West Grinton & Harkaside is a little hamlett containeing about sixteen houses lyeing at some distance from one another saith the Lord Wharton hath some coppyhold lands in Harkaside & West Grinton & alsoe in Fremington for which his tenannts appeare at Helaugh court but know's not the particular [rents . . .] /

6 To the sixth interrogatory this deponent saith that he ha's heard & beleives that the mannor of Grinton did formerly belong to the priory of Bridlington /

7 To the seaventh interrogatory this deponent saith he know's Swaledale & that the same ha's not been reputed a mannor saith there are four mannors or reputed mannors in Swaledale Vizt the said mannor of Grinton belonging to Mr Marriott the mannor of Fremington belonging to Ma[dam] / Wharton & the mannors of Helaugh & Mukar belonging to the Lord Wharton & that the towne of Grinton is the cheife towne therein /

10 To the tenth interrogatory this deponent saith that the inhabitants of <Grinton> Harkaside & West Grinton used to have their cattle goeing upon one anothers commons without interrupcion but the Harkaside & West Grinton men claime a distinct right exclusive of the Grinton men to gett stones peates turfe & ling upon the moores of Harkaside/

Henry Place[96] of Grinton in the county of Yorke myner aged forty three years or thereabouts sworne & examined saith /

2 & 12 To the second & twelfth interrogatories this deponent saith that he know's the reputed mannor of Grinton & the places called West Grinton Harkaside or Whitaside & the moores & waists thereto belonging called Harkahill Harka Moore Grinton Moore Grinton Greetes Grinton How or Howgroves & has known / the same for thirty five yeares last past saith that the said places moores & waists are reputed to belong & be within the said mannor of Grinton & that the Swale's the Hillary's & the relator Mr Marriott were & are the reputed lords of the said mannor of Grinton saith that Sir Solomon Swale wrought / the mynes on Harka hill on Harkside Moore for severall yeares & gott considerable quantity's of lead ore there & continued workeing there untill the relator Mr Marriott recovered the said mynes from him saith the said Mr Marriott from his being in possession of the said mynes (untill / two yeares since that the Lord Wharton entred into possession thereof) wrought the said mynes in question & had very great gettings there & the reason of such this deponents knowledge is that he was a myner & workeman at the said mynes for the said Mr Marriott saith that he was informed / that about twenty years agoe the said Mr Hillarye wrought the waist ore that Sir Solomon left on Harkaside moore & gott lead ore out of it but sunk noe shafts that this deponent knew of saith that he this deponent himselfe ha's taken up waifes & estray's upon Harkaside Moore both for / Sir Solomon Swaile & the said Mr Hillarye & ha's carryed the same to their repective servants & had for his pains for every sheep four pence & for every horse & beast one shilling saith that he know's the mannor of Helaugh & that the river Swale ha's been the ancient & reputed boundarye / between the two mannors of Grinton & Helaugh saith the present Lord Wharton nor any of his ancestors did ever that this deponent knew or heard of worke any myne or mynes or gett any lead ore in West Grinton or Harkaside or in the moores or waists thereto belonging save that about two / yeares since the present Lord Wharton ent<r>ed upon the mynes in question /

3 & 9 To the third & ninth interrogatories this deponent saith that the manor of Grinton is bounderd on the north by the river Swale as farr as from Haverdale Beck to Stallerstone Stile saith the said mannor of Grinton is bounderd on the east by Ellerton manor at Stallerstone Style on the west by the / Haverdale Beck & on the south by the Duke of Boltons manors the boundery whereof begining at Snowden man from

[96] Henry Place swore affidavits on 18 May 1706 (E207/112/1) and 5 June 1707 (E207/114/2). He was present on 16 and 17 May 1706, and 5 May 1707, when Mathew Sisson and Edward Smithson served court injunctions on Lord Wharton's miners. In May 1708 Henry Place, aged 44, gave evidence for Marriott at the trial in London (LI Misc 536, Ward MSS, notebooks vol. 6, f. 111v).

thence to High Browsitt from Browsitt to Redmire Standard or Grinton Cross & from thence to the Height of Grinton Greets & from thence to Gibbon Hill & from thence to / the High Carle Currock & from thence to Pickershill Rigg Hurrock & from thence to Tarnsant & from thence to Wind Yate Hurrack & from thence to Summer Lodge Tarn Hurrack & from thence to Bull head & from thence to Bloody Wall & from thence down Blagill to Haverdale Beck / the reason of this deponents knowledge of these boundarys is that he has lived nigh the moores ever since he was borne & ha's wrought at lead mines at his age of nine yeares saith that the said river Swale is a boundary between the manors of Fremington & Grinton; the manor of Marrick / & Grinton & the manor of Marrick & Ellerton & between the manors of other parte of Marrick & Stainton & soe downwards as this deponent beleives saith that this deponent ha's heard his father an old man now dead say that the river Swale was the boundary between Helaugh & Grinton from / Haverdale Beck to Stallerstone Stile /

4: To the fourth interrogatory this deponent saith that he knows the places called West Grinton & Harkaside, saith the same are a hamlett consisting of about twenty houses which stands at a distance from each other some nearer & some farther of saith that the present Lord Wharton hath copyhold lands / at West Grinton & Harkaside & in Whitaside Reeth & Fremington saith his lordshipps copyhold tennants appeare at Helaugh court but what rents duty's or services the said Lord Whartons tennants pay or doe this deponent knows not, saith that Sir Solomon Swale hath copyhold lands at Reeth / which as this deponent has heard he was called & his tennants appeared at Helaugh court /

6: To the sixth interrogatory this deponent saith that dureing his time the manor of Grinton has been reputed to belong to the priory of Bridlington /

7: To the seaventh interrogatory this deponent saith that he well knows the daile called Swaledale saith that he does not know it to be a mannor of itselfe but that there are four manors in it to witt the manors of Helaugh & Mukar belonging to the Lord Wharton the manor of / Fremington belonging to Madam Wharton & the manor of Grinton belonging to the relator Mr Marriott saith that all Swaledale is in the parish of Grinton & that the parish church is at Grinton; /

8: To the eighth interrogatory this deponent saith that since the relator Mr Marriott recovered the mines on Harkahill from Sir Solomon Swaile the said Mr Marriott sunk nineteen new shafts upon the said Harka Hill & had great gettings there untill about two yeares since that the Lord / Wharton entred upon them since which time my Lord Wharton sunk three new shafts near Grinton Greets upon Harkaside Moore the which this deponent beleives to be within the manor of Grinton; /

10: To the tenth interrogatory this deponent saith that the people of West Grinton Harkaside & Grinton used to have their cattle goeing together on Harkaside Moor saith the inhabitants of West Grinton & Harkaside claime right upon the said moors for getting of peats turfe & ling in their / own right; /

11: To the eleaventh interrogatory this deponent saith that he knows the mines in question claimed by the Lord Wharton as alsoe by the relator Mr Marriott as lessee to the Crowne saith that the same lye on Harka Hill on Harkaside Moore & have dureing this deponents remembrance been [reputed to belong to] / Solomon Swaile & Mr Marriott & never by the said Lord Wharton nor his ancestors that ever this deponent heard of till within these two years last past that the Lord Wharton entred thereon; /

13: To the thirteenth interrogatory this deponent saith that he never knew or heard of the riding of any boundary between the places & moors of West Grinton & Harkaside & Grinton till about two yeares since that the Lord Wharton caused a boundary to be rid there; /

John Haykin of Grinton in the county of Yorke aged forty years or thereabouts sworne & examined saith /

2. 8 & 12: To the second eighth & twelfth interrogatories this deponent saith that he knows the manor or reputed manor of Grinton & alsoe the places called Harkaside West Grinton & Whitaside & the moors & commons called Harka Hill Harka Moore Grinton Moore Grinton Greets Grinton How or / How Groves & ha's known them all for about thirty yeares & that they have been dureing the time of his remembrance reputed to lye within & belong to the manor of Grinton & that the Swales & the Hillarys & of late Mr Marriott have been the reputed lords thereof saith that Sir / Solomon Swale wrought mines of lead upon Harka Hill or Harkaside Moore for many yeares as hee has heard & had very great gettings & continued to worke there till Mr Marriott recovered the same against him & the said Mr Marriott continued to worke as hee has / heard till about two yeares agoe that the Lord Wharton entred upon him & that Hillary's wrought the waist hillocks upon the How for lead ore saith that Sir Solomon used to take upp waifs & estray's upon the moors & commons belonging to Harkaside & West Grinton & ha's heard / the Hillaries used <alsoe> to take upp waifes & estrays there saith that hee knows the manor of Helaugh & that the river Swale has been reputed to be a <boundery> betwixt the said mannor of Helaugh & the manor of Grinton saith hee never heard that the Lord Wharton or any of his / ancestors wrought any mines upon any of the said moors of Harkaside & West Grinton till about two yeares agoe /

3 & 9: To the third & ninth interrogatories this deponent saith that the boundery of the said mannor of Grinton has been reputed to begin at the Swale & soe on the east at Stallerstone Stile & goes upp southward to the Grey Stone in the Sharrow & soe to Snowden Man thence to the height of Brown [sitt . . .] / to Redmere Standard thence to Grinton Greets thence westwards to Gibbon Hill Currock thence to High Carle thence to Pickerstin Rigg Currock thence to Wind Yate Currock thence to Summerlodge tarn Currock thence to Bull head thence down to the Bloody Wall & soe to Haverdale Beck /

7: To the seaventh interrogatory saith hee knows the dale or vale called Swaledale & that there are four reputed manors in the said dale vizt. the manors of Helaugh & Mukar belonging to Lord Wharton the manor of Fremington belonging to Madam Wharton & the said manor of Grinton / belonging to Mr Marriott; that they all lye in the parish of Grinton, that Grinton is the cheife towne in the said dale; /

10 & 13: To the tenth & thirteenth interrogatories this deponent saith that the inhabitants of Grinton West Grinton & Harkaside used to have their cattle goe upon one anothers commons & this deponent has heard that there are some marks or mark stones or divisions betwixt the moores & commons / of Grinton & West Grinton & Harkaside & that the Grey Stone in the How is reputed one of the said marks to divide the said moors & commons for the said inhabitants to gett ling turfe peats & stone within their own moors; /

Christopher Chambers of Grinton in the county of Yorke miner aged fifty yeares or thereabouts sworne & examined saith; /

2. 8. & 12: To the second eighth & twelfth interrogatories this deponent saith that he very well knows the reputed manor of Grinton & the places called West Grinton Harkaside & Whitaside & the moors waists & commons called Harka Hill Harka Moore Grinton Moore Grinton Greets Grinton [How or] / How groves saith he hath knowne the said places moors & waists for nigh forty yeares last past saith that during this deponents knowledge thereof the same were reputed to belong & be within the said mannor of Grinton saith that the Swales the Hillarys & the relator Mr Marriott have [been] / the reputed lords of the said manor saith that he has known old Sir Solomon

Swaile his sonn Sir Henry & the present Sir Solomon all worke mines for lead on Grinton How & Harka on Harkaside moore & in the Birkes all in the manor of Grinton saith that old Sir Solomon had [. . .] / good gettings on Grinton How & Harka that Sir Henry did not worke much, but that the present Sir Solomon wrought many yeares both on Grinton How & Harka on Harkaside Moore & ha's had very good gettings & soe continued working untill the relator Mr Marriott [recovered] / the said mines from him which was about eleaven yeares agoe since which time the said relator Mr Marriott has enjoyed the said mines both on Grinton How & Harka on Harkaside Moore & has had very great gettings in both places & has peaceably enjoyed the said mines untill /

[membrane viij *verso*]

about two yeares agoe that the present Lord Wharton entred upon the mines on Harka in Harkaside and the reason of such this deponents knowledge is that about thirty foure yeares agoe he this deponent wrought as a miner under old Sir Solomon Swale both on Grinton How & Harka[. . .] / side moores that he wrought for many years as a miner in the said places for persons who tooke the said mines of the present Sir Solomon & that he ha's also wrought & been a miner in the said places for some years under the relator Mr Marriott; saith that both [. . .] / & the Hillarys had estrays & waifes brought to them that were taken upp on Harkaside Moore; /

3 & 9 To the third & ninth interrogatories this deponent saith that the manor of Helaugh lyes on the north side of the river Swale & that the manor of Grinton lys on the southside thereof saith that the mid streame of the river Swale dureing this deponents remembrance has always [been] / boundary between the said two mannors & that the said manor of Grinton is boundered on the north by the said river Swaile as farr as from Haverdale Beck to Stallerstone Style on the east by Ellerton mannor at Stallerstone style on the west [by] / Haverdale Beck on the south by the Duke of Boltons manors adjoining on Grinton Moors saith that the manors of Fremington & Grinton are alsoe boundered by the said river Swaile & parte of the manor of Marrick & Grinton are soe boundered & saith that hee this [deponent] / has heard severall ancient men often say that the said river Swale was a boundery as aforesaid; /

4: To the fourth interrogatory this deponent saith that he knows the places called West Grinton & Harkaside saith the same are a hamlett consisting of about twenty houses most of them standing at a distance from each other saith the same are reputed to belong & be in the [said manor] / of Grinton saith the Lord Wharton hath copyhold lands at West Grinton Harkaside Whitaside & Fremington saith that the lords tennants usually appeared at his lordshipps courts at Helaugh but where the freeholders there appeares or what rents suits or services they pay [this] / deponent knows not saith that one Christopher Naylor Sir Solomon Swales tennant at Swale Hall who usually tooke upp waifes & estrays for him & who was this deponents master told this deponent that he was goeing to Helaugh court to appeare for his master Sir Solomon Swale but [. . .] / it was for lands at Grinton or Reeth this deponent knows not or whether Sir Solomon was then in the possession of lands in Reeth this deponent knows not; /

7: To the seaventh interrogatory this deponent saith that he knows the dale called Swailedale which is in length about twelve miles & has within it four manors to witt Helaugh & Mukar belonging to the Lord Wharton the manor of Fremington to Madam [Wharton] / & Grinton manor belonging to the relator Mr Marriott saith all the dale is in the parish of Grinton & that the parish church stands at Grinton /

13: To the thirteenth interrogatory this deponent saith that he has heard that the Lord Wharton about two years agoe caused a boundary to be rid between Harkaside West

Grinton & Grinton but before that time never knew or heard of any boundary rid bet[ween] / the said places nor doth he beleive there is any such boundary beleives there may be severall bounder stones & places that may be for the dividing the moors & waists between them as to the getting of peates ling & turfe & not otherwise as the elme tree Jinny Wrenn [cow] / house the Ox Intack head & the White or Grey Stone in the How & soe to Grinton Greets & severall others which this deponent cannot sett forth; /

Christopher Fothergill of Reeth in the county of Yorke miner aged fifty six yeares or thereabouts sworne & examined saith /

2: 8: & To the second eighth & twelfth interrogatories this deponent saith that he well knows
12: the reputed manor of Grinton & the places called West Grinton Harkaside & Whitaside & the moors waists & commons called Harka Hill Harka Moore Grinton Moore Grinton Greets / Grinton How or How Groves & has soe knowne the said places & commons for about forty five years last past saith the same are reputed to belong <to> & be within the said manor of Grinton & that the Swales the Hillaries & the relator Mr Marriott have been [the] / reputed lords of the said mannor dureing the time aforesaid saith that this deponent has heard & verily beleives that old Sir Solomon Swaile wrought & gott lead on Harka on Harkaside Moore the place in question saith that his brother was a miner there in [. . .] / time & that this deponents selfe has seen miners worke there for old Sir Solomon Swaile and as his brother & the other workmen told this deponent saith that the present Sir Solomon Swaile wrought the mines in question upon Harka for severall years & had very [good] / gettings & saith that the present Sir Solomon continued working there & that this deponent himselfe was a miner under him untill the relator Mr Marriott recovered the said mines from him about eleaven yeares agoe & this deponent further saith that he continued [a] / miner there with the said Mr Marriott from that time for the most parte till the present Lord Wharton about two years since entred upon the mines on Harka on Harkaside the place in question saith that the relator Mr Marriott sunk as he beleives about fifteen [shafts] / on Harka the place in question & had very great gettings there & beleives the said mines would have continued good if the said Mr Marriott could have peaceably wrought the same; saith that this deponent has <known> waifes & estrays <taken> up on Grinton Moore for the Swailes & Hillaries / as lords thereof but never heard that any waifes & estrayes were taken upp by any other lords saith that he knows the manor of Helaugh & that dureing the time aforesaid the reputed boundary between the manors of Grinton & Helaugh has always been the mid [streame] / of the river Swale; saith that the present Lord Wharton nor any of his ancestors did ever worke any mine or mines or claime any royalties within Harkaside & West Grinton or in any of the moors & commons thereto belonging that ever this deponent knew or [heard of] / till within these two yeares that the Lord Wharton entred upon the mines in question /

3: & 9: To the third & ninth interrogatories this deponent saith that Helaugh lys on the north side of the river Swaile & that the manor of Grinton lys on the southside thereof and saith that the mid streame of the river Swale ha's always dureing this deponents knowledge been the reputed / boundary between the said two mannors of Grinton & Helaugh & that the said manor of Grinton is bounderd on the north by the said river from Haverdale Beck to Stallerstone Stile & on the east by Ellerton manor at Stallerstone Stile on the west by the said Haverdale beck / on the south by the Duke of Boltons mannors adjoining on Grinton Moore saith that the said river Swaile is alsoe a boundary between the manors of Fremington & Grinton & between the manors of Marrick & Ellerton & soe downwards saith that he / has heard ancient men say that the mid streame of the river Swaile was a boundary & that the people of each side ought to fish noe further than the mid streame /

4: To the fourth interrogatory this deponent saith that he knows Harkaside & West Grinton saith the same is a hamlett consisting of some few houses but how many this deponent knows not saith the same lye at a distance from each other & beleives the said places to be within the said manor / of Grinton; /

7: To the seaventh interrogatory saith that he knows Swaledale and does not know it to be a manor saith there are four manors in it and that the whole dale is in Grinton parish; /

10: To the tenth interrogatory this deponent saith that dureing his remembrance the moores & commons of West Grinton Harkaside & Grinton as to the herbage & soile thereof have been enjoyed jointly by the inhabitants their sheep & beasts / went thereon without any disturbance of each other that this deponent knows or heard of saith that hee ha's heard that a division & marke stones as to the divideing & knowing of each other[s][97] rights as to the getting of peats turfe & ling but not as to the herbage / thereof; /

11: To the eleaventh interrogatory this deponent saith that he knows the mines in question claimed by the Lord Wharton as alsoe by the relator Mr Marriott under the Crown saith the same lye on Harka on Harkaside Moore & are reputed to be within the manor of Grinton saith [the] / same have been dureing this deponents knowledge in the possession & enjoyed by the Swailes & the relator Mr Marriott & never till of late by the Lord Wharton or any of his ancestors; /

William Race of Fremington miner aged sixty yeares or thereabouts sworne & examined saith /

2: 8: & 12. To the second eighth & twelfth interrogatories this deponent saith that he knows the manor or reputed manor of Grinton & alsoe knows the places called West Grinton Harkaside & Whitaside & the moors & commons called Harka Hill Harka Moor Grinton Moore Grinton [Greets] / Grinton How or How Groves & has knowne the same for fifty yeares & that they have been all his time reputed to belong to <and> be parte & parcell of the said manor of Grinton & that the Hillaries & Swales & of late Mr Marriott have been commonly called or reputed / lords of the manor of Grinton & that old Sir Solomon Swale & the present Sir Solomon Swale for many yeares wrought mines of lead upon Harkaside Moore the place in question & had great gettings & the present Sir Solomon continued to worke there till Mr Marriott [recovered] / the same against him & from that time Mr Marriott enjoyed the said mines & had as this deponent has heard & beleives very great gettings till the Lord Wharton enterd upon him about two yeares agoe saith that the Swales & Hillarys have used to take upp waifes & / estrays upon the said moors & commons saith he knows manor of Helaugh & that the mid streame of the river Swale has for all his time been <the> reputed ancient boundary betwixt the said manors of Grinton & Helaugh from Haverdale Beck to Stallerstone / Stile saith he has every yeare for about fifty years been frequently upon the said moors & commons of Harkaside & West Grinton & that the present Lord Wharton or any of his ancestors (that this deponent heard or beleives) never did worke any mines of lead [. . .] / upon any of the said moors or commons belonging to Harkaside or West Grinton or any other of the moors or commons aforesaid belonging to the said manor of Grinton till about two yeares agoe; /

3: & 9: To the third & ninth interrogatories this deponent saith that the commons & reputed

[97] The word 'other' is written in the singular.

boundaries of the said manor of Grinton for all his time has been on the east by the manor [of][98] Ellerton at Stallerstone Stile & on the west by Haverdale Beck & on the south by the Duke of Boltons manors / & on the north by the river Swale from the manor of Helaugh from Haverdale Beck to Stallerstone Stile saith the said mid streame of the river Swale is alsoe the common reputed boundary in the neighbourhood betwixt the manor of Fremington (which adjoines [. . .] / the said manor of Helaugh) & the said mannor of Grinton & alsoe betwixt the manors of Marrick & Ellerton (Ellerton adjoining upon & lying east upon the said mannor of Grinton) and the said mid streame of the said river is alsoe the reputed ancient boundary [between] / the manors of Marrick & Stainton & Marrick & Downham & Marsk & Downham & beleives it is alsoe downwards saith that this deponents grandfather (who was an old fisher man & dy'd about forty yeares agoe & was above one hundred yeares [old when] / he dyed) frequently told this deponent that the said river Swale was alway's accounted a boundary as aforesaid & this deponents father & himselfe who were alsoe fisher men & used to fish in the said river used to observe the said river to be a boundary betwixt the / mannors from Haverdale Beck downwards as aforesaid and this deponent has been often disturbed in fishing in the said river & told that he must observe the mid streame of the said river to be the boundary & that he must goe noe further; /

4: To the fourth interrogatory this deponent saith that West Grinton & Harkaside is a hamlett consisting of a few houses lying at some distance from each other & that the same has been reputed for all his time to lye within the said mannor of Grinton /

7: To the seaventh interrogatorie this deponent knows the dale or vale called Swaledale saith he never knew it called a mannor of it selfe but that there are four mannors therein the mannors of Helaugh & Mukar belonging to the Lord Wharton / of Fremington belonging to Madam Wharton & the said mannor of Grinton belonging to the said Mr Marriott & that Grinton (where the church stands) is the cheife towne there & all of them lye in the parish of Grinton; /

10: & 13. To the tenth & thirteenth interrogatories this deponent saith that the inhabitants of Grinton & Harkaside & West Grinton used to have their cattle goe by rake in one anothers commons saith hee never heard of or knew any boundery rid between / Harkaside & West Grinton or the moors or commons thereunto belonging till about two years agoe he has heard some boundery was rid there by the present Lord Wharton; /

11: To the eleaventh interrogatory this deponent saith that he knows the mines in question both those that lye upon Harka Hill & that mine alsoe lately sunck by the present Lord Wharton which lys upon Harka & lys neare to a mine lately [. . .] / by Mr Marriott upon Grinton How & beleives all the said mines upon Harka Hill & the said mine soe sunck by the said Lord Wharton lyes within the said mannor of Grinton; /

John Stoddart of Fremington in the county of Yorke miner aged fifty six yeares or thereabouts sworne & examined saith /

2: To the second interrogatory this deponent saith that he knows the reputed manor of Grinton & the places called West Grinton Harkaside & Whitaside & the moors waists & commons thereto belonging & has soe knowne them for about forty six yeares / and saith that the said places moors & waists are reputed to belong & be within the said mannor of Grinton & that the Swailes the Hillary's & the relator Mr Marriott have been the reputed lords of the said mannor saith that Sir Solomon [Swaile] / wrought the mines on the How & Harka the places in question for many

98 The word 'of' is omittted in the original.

yeares & had very considerable gettings & that he continued workeing there untill the relator Mr Marriott about eleaven yeares agoe recovered the said mynes from the said Sir Solomon since which time the relator Mr Marriott [. . .] / the said mines & had great gettings both on the How & Harka the place in question without the least disturbance that ever this deponent heard of till about two yeares since that the present Lord Wharton entred upon the mynes on Harka the place in question saith that the said Roger Hillarye wrought / for coales on the How & wrought the waists of lead on Harka on Harkaside Moore saith this deponent ha's knowne all the said mynes for forty yeares last past & lived in the neighbourhood most of his tyme & ha's been a myner & wrought in the said mynes for the relator Mr / Marriott saith that he ha's heard that both the Swale's & the Hillary's took up waifes & estray's on Grinton Moor saith that he know's the mannor of Helaugh & that the midd streame of the river Swale dureing this deponents tyme ha's been the reputed boundarye betweene the / mannors of Grinton & Helaugh saith the present Lord Wharton the late Lord Wharton nor any of their ancestors did ever that this deponent heard off worke any myne or mynes or claime any royalties within or upon West Grinton or Harkaside or in any the moores & waists thereto [belonging] / or in any other place within the said mannor of Grinton /

3 & 9 To the third & ninth interrogatories this deponent saith that Grinton mannor is bounderd on the north by the river Swale from Haverdale Beck to Stallerstone Style on the south on the Duke of Boltons mannors on the west on the said Haverdale beck & on the east on Ellerton mannor at Stallerstone Style saith / that the said river Swale is the boundery between Fremington mannor & Grinton mannor & between parte of Marrick & Grinton & betweene Marrick & Ellerton /

7 To the seaventh interrogatory this deponent saith that he know's Swaledale & that it's in length about twelve myles know's it not to be a lordshipp but saith there are four mannors in it to witt Helaugh & Mukar the Lord Wharton's Fremington Madam Whartons of Gillingwood & Grinton / the relator Mr Marriott's saith that all the dale is in Grinton parish & that the church stands at Grinton /

11 To the eleaventh interrogatory this deponent saith that hee know's the mynes in question claimed by the Lord Wharton as alsoe by the relator Mr Marriott saith they lye on Harka on Harkaside Moore within the said mannor of Grinton & have been dureing this deponents remembrance in the possession / of & enjoyed by the Swale's Hillary's & the relator Mr Marriott & that dureing the tyme aforesaid the present Lord Wharton nor any of his ancestors ever were in possession of or claimed any right thereto till two yeares since that ever this deponent knew or heard of /

Francis Breckenfield of Fremington myner aged forty yeares or thereabouts sworne & examined saith /

2: 11 & 12 To the second eleaventh & twelfth interrogatories this deponent saith that he know's the mannor or reputed mannor of Grinton & alsoe know's the places called Harkaside & West Grinton & the moores & commons thereunto belonging & hath knowne the same for twenty yeares & dureing the [time] / of his knowledge the said places moores & commons have been reputed & taken to lye within & belong to the said mannor of Grinton & that the Hillary's & Swale's & of late Mr Marriott have been usually called & reputed lords of the said mannor saith that Mr Hillary's wrought mynes of coale / upon Grinton How & Sir Solomon Swale <& his ancestors> wrought mynes of lead upon Harka or Harkaside Moore the place in question & had very good gettings there & continued to worke the same till the said Mr Marriott recovered the same against him about eleaven yeares agoe & after that the said Mr Marriott enjoyed / the same & this deponent was at the sinking of one new shaft upon the said place (& beleives he sunk severall others) & had very great gettings till the Lord Wharton entred upon him about two yeares agoe & saith the Lord Wharton hath since sunck one shaft upon the upper end of the said Harka Moore saith / the Swale's & Hillary's used to take up waifes & estray's further saith that he know's the mannor of Helaugh & that the midd streame of the river Swale ha's for all his tyme been reputed the antient boundary betwixt the said mannors of Grinton & Helaugh from Haverdale Beck downewards to / Stallerstone Style Sayth he this deponent ha's wrought in the mynes belonging to the Lord Wharton & his father upon the northside of the said river Swale for about twenty yeares by tymes & for all that tyme the Lord Wharton nor any of his ancestors did ever worke any mynes of lead or coale upon the / said Harka or Harkaside Moore till about two yeares agoe that this deponent ever knew or heard off /

3 & 9 To the third & ninth interrogatories this deponent saith that the said mannor of Grinton is bounderd on the east by the mannor of Ellerton at Stallerstone Style & on the west by Haverdale Beck & on the south by the Duke of Boltons mannors & on the north by the midd streame of the river Swale from the mannor / of Helaugh as aforesaid saith that the said river Swale is alsoe & ha's been the reputed & knowne boundery betwixt the mannors of Fremington & Grinton & Marrick & Grinton & Marrick & Ellerton & beleives it is soe between the other mannors downwards that this deponent ha's been disturbed in / fishing in the said river between the mannors of Marrick & Ellerton & ha's been told he must observe the midd streame of the said river to be a boundery as aforesaid /

7 To the seaventh interrogatory this deponent saith that he know's the vale or dale called Swaledale that it ha's never been accounted a mannor of it selfe but that there are four mannors lyeing in the said dale Vizt the mannors of Helaugh & Mukar belonging to the Lord Wharton the mannor of Grinton belonging to / Mr Marriott & the mannor of Fremington belonging to Madam Wharton all lyeing in the parish of Grinton /

[membrane ix *recto*][99]

John Lickley of Grinton in the county of Yorke myner aged seaventy three yeares or thereabouts sworne & examined saith /

2 11 & 12. To the second eleaventh & twelfth interrogatories this deponent saith that he know's the mannor or reputed mannor of Grinton saith he knows the places called West Grinton Harkaside & Whitaside & the waists & moores thereto belonging called Harka hill Harka Moore Grinton Moore Grinton Greetes Grinton How or How groves / & has soe knowne the same for forty yeares last past saith the same are reputed to belong & be parcell of the said mannor of Grinton & that the Swale's & Hillary's & the relator Mr Marriott have been the reputed lords of the said mannor of Grinton saith that for severall yeares together Sir Solomon Swale wrought / mynes for lead on Grinton How & Harka on Harkaside Moore the place in question saith that the said Sir Solomon had very good gettings of lead on Harka & that he continued workeing there untill the relator Mr Marriott recovered them from him which was about eleaven yeares agoe & since that the said / relator Mr Marriott continued workeing the said mynes both on the How & Harka & ha's had very great gettings in both places till the present Lord Wharton entred upon the mynes at Harka the place in question about two yeares agoe saith that Mr Roger Hillarye wrought &

99 Parchment, 730mm x 650mm. Parts of the sheet are heavily discoloured with age and the top left-hand corner of the *verso* is badly rubbed.

gott coales on Grinton / Moore & wrought the waists of lead on Harkaside Moor for two yeares together which waists were left by Sir Solomon Swale above ground saith that the mynes on Harka now claimed by the Lord Wharton as alsoe by the relator Mr Marriott lyeing on Harka or Harkaside Moore are / reputed to belong to the mannor of Grinton & have been for forty yeares last past in the possession of & enjoyed by the Swale's the Hillarys & the relator Mr Marriott & in all that tyme the present Lord Wharton the late Lord Wharton nor any of their ancestors ever were in the possession of or enjoyed or / wrought any myne or mynes or claimed any royalties in West Grinton or Harkaside or in the moores or commons thereto belon<ging> till about two yeares last past as aforesaid that ever this deponent knew or heard of saith that the reason of such this deponents knowledge as aforesaid is that he was a workeman at / the lead mynes both on the How & Harka the place in question & on Harka at the workeing the waists & on Grinton Moore at the collerye as well as under Sir Solomon Swale as under Mr Hillarye & the relator Mr Marriott /

3 & 9 To the third & ninth interrogatories this deponent saith that the mannor of Helaugh lyes on the north side of the river Swale & that the mannor of Grinton lyes on the south side thereof & saith the river Swale is reputed boundarye between the said two mannors & that the said mannor of Grinton is bounderd by the / said river Swale on the north from Haverdale Beck to Stallerston Style & on the south by the Duke of Boltons mannors & on the west by the said Haverdale Beck & on the east by Ellerton mannor att Stallerston Style saith that the said river is alsoe a boundarye betweene the mannors of Fremington & Grinton / & parte of the mannors of Marrick & Grinton & Marrick abbey & Ellerton & Marrick abbey & Stainton saith that one Robert Dunn an ancient fisherman now dead severall yeares told this deponent that the said river Swale was a boundery as aforesaid /

7 To the seaventh interrogatorie this deponent saith that he know's Swaledale & that the extent of it is in length about twelve myles he ha's heard it called a lordshipp but not a mannor saith there are four mannors in the said dale to witt the mannors of Helaugh & Mukar the Lord Whartons the mannor / of Fremington Madam Whartons of Gillingwood & the mannor of Grinton belonging to the relator Mr Marriott saith all the whole dale is in the parish of Grinton & that the church stands at Grinton /

Charles Jaques of Arkingarth Dale myner aged fifty nine yeares & upwards {saith} sworne & examined saith

2 11 & 12 To the second eleaventh & twelfth interrogatories this deponent saith he know's the mannor or reputed mannor of Grinton & alsoe know's the places called West Grinton Harkaside & Whitaside & the moores or commons called Harka or Harkaside Moore & ha's knowne the same about forty yeares & that dureing all the / said tyme the same have been accounted to lye within the said mannor of Grinton & that the Hillary's & Swales & of late Mr Marriott have been the reputed lords of the said mannor & this deponent particularly remembers that when he was a boy about forty yeares agoe there was some difference or dispute betwixt Sir Solomon / Swale & the late Duke of Bolton then called Lord St John's about some rights belonging to their mannors there adjoineing to each other & was then told by his neighbours that they ridd some boundarye thereabouts but this deponent cannott particularly sett forth the same saith that old Sir Solomon Swale used to worke mynes / upon Harka or Harkaside Moore & this deponent about thirty four or thirty five yeares agoe made a bargaine with the said Sir Solomon Swale for workeing some mynes of lead upon Harka hill & one George Hall then a steward or agent to the Lord Wharton for the mynes upon the north side of the said river was then present at the / makeing of the said bargaine between this deponent & the said Sir Solomon saith that the present Sir

Solomon Swale wrought the said mynes for severall yeares upon Harkahill the place in question & had very good gettings till about eleaven yeares agoe that Mr Marriott recovered the same against him & after that Mr Marriott enjoyed the same till the / Lord Wharton entred upon him saith that he ha's been a near neighbour to the said places most of his tyme & never knew or heard that the Lord Wharton or any of his ancestors ever workt any mines of lead or coale upon the said Harkaside Moore or claimed any royalties there or upon any the moores or commons belonging to the mannor of Grinton till about two yeares agoe /

Thomas Haukin[100] of Grinton in the county of Yorke myner aged sixty two yeares or thereabouts sworne & examined saith /

2 11 & To the second eleaventh & twelfth interrogatories this deponent saith that he well
12 know's the mannor or reputed mannor of Grinton & the places called West Grinton Harkaside & Whitaside & the moores waists & commons called Harka hill Harka Moore Grinton Moore Grinton Greetes Grinton How or Howgroves / saith he ha's knowne the same for above forty yeares last past & dureing the tyme aforesaid the said places moores & waists have been reputed to belong & be within the said mannor of Grinton saith that about forty yeares agoe old Sir Solomon Swale wrought & gott lead on the How & on Harka the place in / question & had some tymes good gettings & other sometimes not soe good knew Sir Henry Swale son of old Sir Solomon & father of the present Sir Solomon doe's not remember his workeing the mynes but saith the present Sir Solomon wrought the mynes both on the How & Harka the place in question & gott very / considerable quantity's of lead in both places saith the mynes in question lyes on Harka or Harkaside Moore & are reputed to belong to the said mannor of Grinton & that the present Sir Solomon wrought the mynes there untill about eleaven yeares agoe that the relator Mr Marriott recovered them from him / since which tyme the said Mr Marriott wrought the said mynes & sunck eighteene shafts & had very good gettings there & continued such his workeing till about two yeares since that the present Lord Wharton entred upon the said mynes on Harka on Harkaside Moore & till then the present Lord Wharton nor any of his / ancestors ever wrought or claimed any of the myne or mynes on West Grinton or Harkaside Moore or in any of the moores waists or commons thereunto belonging or in any other place within the said mannor of Grinton that ever this deponent knew or heard of saith that he hath heard that the Swale's & the / Hillary's tooke up waifes & estray's on Harkaside Moore /

3 & 9 To the third & ninth interrogatories this deponent saith that he very well know's the mannor of Helaugh & that it ly's on the northside of the river Swale & that the mannor of Grinton ly's on the south side thereof saith that he ha's heard that the mannor of Grinton is bounderd on the north by the midd streame of the river / Swale from Haverdale beck to Stallerstone Style on the south by the Duke of Bolton's mannors on the west on the said Haverdale beck & on the east by Ellerton mannor at Stallerstone Style saith that the division betweene Grinton mannor & the Duke of Bolton's mannor are as followeth from Snowden man or / Ridmer Standard to Grinton Greetes & from thence to Gibbon Hill & from thence to High Carle Harrock soe to Pickerstin Harrock soe to Wingates Harrock soe to Somerlodge Car Harrock soe to Bulhead soe to bloody Wall soe downe the Gill to Haverdale beck /

10 To the tenth interrogatory this deponent saith that the inhabitants of Grinton & Harkaside have their cattle goeing on each others commons by way of a rake & that the inhabitants of West Grinton & Harkaside lay claime to some boundary's & stones

100 Thomas Hawkins of Grinton, a miner, swore an affidavit on 5 June 1707 (E207/114/2).

to divide the said common's & begins at the elme tree & goe's to the Oxe / Intack nook & to the White Stone in the How but saith that he never tooke the said places or stones to be boundary's but marks to distinguish their common rights onely /

13 To the thirteenth interrogatory this deponent saith that he ha's heard that the present Lord Wharton about two yeares since ridd a boundarye between West Grinton Harkaside & Grinton saith till then for above {fifty} forty yeares last past he never knew of any boundarye ridd betweene the said places nor do's he beleive there is any such boundarye /

John Metcalfe[101] of Grinton waller aged sixty seaven yeares or thereabouts sworne & examined saith /

2, 8, 11 & 12 To the second eight eleaventh & twelfth interrogatories this deponent saith that he well know's the mannor or reputed mannor of Grinton & alsoe know's the places called West Grinton Harkaside & Whitaside & the moores & commons called Harka hill Harka Moore Grinton Moore Grinton Greetes Grinton How or Howgroves / & ha's knowne the same for above fifty yeares & for all the said tyme the said places moores & commons have been generally reputed & accounted to lye within & belong to the said mannor of Grinton & that the Hillary's & Swale's & of late the Mr Marriott have been called & taken to be lords of the said mannor saith that Hillarye / wrought mynes of coale upon Grinton How & old Sir Solomon Swale wrought mynes of lead about forty eight years agoe upon Harka hill or Harkaside Moore the place in question & had great gettings & wrought many years & after the present Sir Solomon wrought mynes of lead upon the said Harka hill & had alsoe great / gettings & continued to worke there untill about eleaven yeares agoe the relator Mr Marriott recovered the same against him & from that tyme the said Marriott wrought the said mynes & enjoyed the same till about two yeares agoe that the Lord Wharton entred upon him saith that both Hillary's & Swales used as lords to take up / waifes & estray's upon the said moores & commons & this deponent ha's taken up estray's upon Harkaside Moore for the said Hillarye saith that he know's the mannor of Helaugh & that the midd streame of the river Swale ha's for all his tyme Vizt above fifty yeares been reputed the antient boundary betwixt the / said mannors of Grinton & Helaugh from Haverdale beck downe to Stallerstone Style saith that he ha's for all his tyme lived at Grinton & that he never knew or heard that the present Lord Wharton nor any of his ancestors ever wrought any mynes of lead or coale or claimed any royalties in Harkaside or West Grinton / or the moors or commons thereunto belonging till about two yeares agoe /

3 & 9 To the third & ninth interrogatories this deponent saith that the said mannor of Grinton is bounderd on the east by the mannor of Ellerton at Stallerstone Style & on the south by the Duke of Bolton's mannors & on the west by Haverdale beck on the north by the midd streame of the river Swale from the said mannor of / Helaugh as aforesaid saith the same have not onely been commonly reputed the {Man} boundarye of the said mannor but this deponent very well remembers that about forty eight yeares agoe old Sir Solomon Swale ridd all the said boundary's of the said mannor & this deponent went a great parte thereof with him he the said / Sir Solomon then began at the midd streame of the said river & went from thence to Stallerstone Style & from thence to the Gray Stone o'th Sharrow (soe far the boundarye betwixt the mannors of Grinton & Ellerton) from thence to Snowden man (& there begins the boundery's betwixt the Duke of Boltons / mannors & the said mannor of Grinton)

[101] In May 1708 John Metcalfe, aged 66, gave evidence for Marriott at the trial in London (LI Misc 536, Ward MSS, notebooks vol. 6, f. 112v).

from thence to Ridmer Standard thence to Grinton Greetes thence to Gibbon Hill & soe as heaven water deales to Somerlodge Gill & soe down to Haverdale saith the said river Swale is alsoe accounted the antient boundary betwixt the mannors of Grinton and / Fremington Grinton & Marrick & Ellerton & Marrick & Stainton & Marrick & beleives it to be a boundery downwards saith that he has seen the said Hillarye formerly fish with netts in the said river upon the southside within the said mannor of Grinton & ha's heard the said river by ancient men formerly / be reputed a boundary as aforesaid /

4 To the fourth interrogatory this deponent saith that West Grinton & Harkaside is a hamlett consisting of a few houses lyeing at some distance from one another /

7: To the seaventh interrogatory this deponent saith that he knows the dale or vale called Swaledale about twelve miles long that it is not accounted a manor of it selfe but there are four manors within it Vizt the manor of Grinton belonging to Mr Marriott the manors of Helaugh & Mukar belonging / to the Lord Wharton the manor of Fremington belonging to Madam Wharton all lyeing in the parish of Grinton /

10: To the tenth interrogatory this deponent saith that the inhabitants of Grinton Harkaside & West Grinton used to have their cattle goe upon each others moors & commons by way of a rake & that the Harkaside & West Grinton men claime a right of getting peats turf & ling / within their owne moors & commons soe farr as the White Stone in the How exclusive of the inhabitants of Grinton /

13: To the thirteenth interrogatory this deponent saith that he saw some parte of a boundary rid by the Lord Wharton or his agents between Grinton Harkaside & West Grinton about two yeares agoe but cannot perticularly describe the same nor did he ever know that any / boundary was rid betwixt the said Grinton Harkside & West Grinton before that time /

William Clarkson[102] of Feetham in the county of Yorke yeoman aged fifty four yeares and upwards sworne & examined saith /

2: 8:11: & 12: To the second eighth eleaventh & twelfth interrogatories this deponent saith that he knows the manor or reputed manor of Grinton & the places called West Grinton Harkaside & Whitaside & the moors & waists called Harka Hill Harka Moor Grinton Moore Grinton Greets Grinton How or How / Groves & has soe knowne the same for about forty yeares last past saith the same are reputed to belong & be parte & parcell of the said manor of Grinton & that the reputed lords thereof were the Swailes & the Hillaries & the relator Mr Marriott saith the present Sir Solomon Swale wrought mines / both on the How & Harka the place in question & had very good gettings there & continued working on Harkaside Moore the place in question till the relator Mr Marriott recovered the said mines from him saith that the said Mr Marriott from that time wrought the said mines & / had very good gettings & that he sunck five new shafts on Harka & that he continued working there untill about two yeares agoe that the present Lord Wharton enterd upon the said mines saith that for these forty yeares last past the present Lord Wharton the late Lord Wharton nor any of their ancestors / were ever in the possession or did ever worke or claime any right to the mines in question on Harka or Harkaside Moore reputed to belong & <to> be within the manor of Grinton or to any royalties either in West Grinton & Harkaside or in any other place within the said manor of Grinton till of late that / ever this deponent knew or heard off saith that Roger Hillary wrought for coles on

102 In May 1708 William Clarkson, aged 55, gave evidence for Marriott at the trial in London (LI Misc 536, Ward MSS, notebooks vol. 6, f. 113v).

Grinton Moore saith that one Christopher Naylor had frequently waifes & estrays brought to him to Swaile Hall in Sir Solomon Swailes right from Harkaside Moore as this deponent has heard & been told / & saith that he knows above thirty yeares agoe that one Anthony Close tooke upp four estrayed beasts on Harkaside Moore & carryed them as waifes to the said Christopher Naylor to Swaile Hall; /

3: & 9: To the third & ninth interrogatories this deponent saith that he knows the manors of Helaugh & Grinton & has soe knowne the same for above forty yeares last past saith that Helaugh manor ly's on the north side of the river Swaile & that Grinton manor lys on the south side thereof & that the / mid streame of the river Swaile is the reputed ancient boundary between the said two manors & that the said river is the north boundary of the said manor of Grinton from Haverdale Beck to Stallerstone Stile & that the said Grinton manor is bounderd on the east by Ellerton / manor at Stallerstone Stile on the west by Haverdale Beck & on the south by the Duke of Boltons manors at Redmer Standard Gibbon Hill High Carle Pickerston Stone & soe downe to Haverdaile Beck saith that he this deponent has heard old men say that the Lord Wharton claimed all the river Swaile / but the lords of Grinton would let them come noe further than the mid streame which the old men said was the boundary as aforesaid; /

4: To the fourth interrogatory this deponent saith that he knows the places called West Grinton & Harkaside & saith that the same are a small hamlett consisting of some few houses at a distance from each other saith the said places dureing this deponents knowledge were reputed to belong & <to> be parcell / of the said manor of Grinton; /

7: To the seaventh interrogatory this deponent saith he knows the vale called Swaledale & that its in length about twelve miles & has four mannors in it to witt Helaugh Mukar Fremington & Grinton & that the whole dale is in Grinton parish & that the parish church is at Grinton /

10: To the tenth interrogatory this deponent saith that the inhabitants of Grinton West Grinton & Harkaside had a rake of comoning in one anothers moors without disturbance; /

13: To the thirteenth interrogatory this deponent saith that hee heard that the late Lord Wharton agents about two years since caused a boundary to be rid between the moors and commons of West Grinton Harkaside & Grinton saith that forty years together till then this deponent never heard of / the riding of any boundary between the said moores waists & commons;

John Clerkson of Whitaside husbandman aged sixty five or thereabouts sworne & examined saith /

2: 8:11: & 12: To the second eighth eleaventh & twelfth interrogatories this deponent saith that he knows the manor or reputed manor of Grinton & alsoe the places called West Grinton Harkaside & Whitaside & the moors & commons called Harkahill Harka Moore Grinton Moore Grinton Greets Grinton / How or How Groves & has knowne the same for fifty yeares & that for all that time the said places moors & commons have been reputed to belong to & lye within the said manor of Grinton & that the Hillaries & Swales & of late Mr Marriott have been the reputed lords of the said / manor saith the Hillarys used to worke mines upon Grinton Moore & the present Sir Solomon Swale wrought mines of lead upon Harkahill or Harkaside Moor for severall yeares & had considerable gettings till about eleaven yeares agoe Mr Marriott recovered the same / against him & after that Mr Marriott continued to worke there & sunck severall new shafts & had good gettings till about two yeares agoe the present Lord Wharton enterd upon him saith Hillarys used to take upp waifes & estrays upon Grinton Moore & Swales used to take upp waifes & / estrays both upon Harkaside & Whitaside Moors saith he knows the manor of Helaugh & for all his time the mid streame of the river Swale has been reputed the ancient boundary betwixt the said manors of Grinton & Helaugh from Haverdale Beck down to near Grinton / & so to Stallerstone Stile saith he never knew that the Lord Wharton or any of his ancestors ever wrought any mines of lead or cole or claimed any royalties in West Grinton & Harkaside or the moors & commons thereof till about two yeares agoe; /

3: & 9: To the third & ninth interrogatories this deponent saith that the said manor of Grinton is bounderd on the east by the manor of Ellerton beginning at the mid streame of the river Swale & soe to Stallerstone Stile thence to White Stone on Sharrow thence to Snowdon Man (where begins / the boundary betwixt the Duke of Boltons manors & the said manor of Grinton) & soe proceeds to Redmere Standard thence to Grinton Greets thence to Gibbon Hill thence to High Carle thence to Pickerstin Rigg or Currock & from thence the said boundary of Grinton manor goes to / Wind Yate thence to Tarnsant thence to Summerlodg Tarn Currack thence to Bullhead thence to Bloody vale or Wall thence down Blagill to Haverdale Beck which is the boundary on the west & the said manor of Grinton is bounderd on the north by the river Swaile from the manor / of Helaugh as aforesaid saith the said boundaries as aforesaid are comonly reputed to be the said boundarys in the neighbourhood all his time; /

4: To the fourth interrogatory this deponent saith that West Grinton & Harkaside is a hamlett consisting of a few houses standing at some distance from each other saith the same have been all the time of his remembrance taken to be within the said mannor of Grinton /

Nicholas Hall of Booze in the county of Yorke gentleman aged sixty six years or thereabouts sworne & examined saith /

[2] To the second interrogatory this deponent saith that he knows the manor or reputed manor of Grinton & knows the places called West Grinton Harkaside & Whitaside & the moors waists & commons thereto belonging & has knowne the same for fifty yeares & upwards & saith the same / have been dureing this deponents knowledge reputed to belong & be parte & parcell of the said manor of Grinton saith that the Swales Hillarys & the relator Mr Marriott have been the reputed lords thereof saith that the Swailes have wrought lead mines on Harka & that the / Hillarys have wrought for coale on Grinton Moore as this deponent has heard saith that he knows the mannors of Grinton & Helaugh & that the river Swale has been the reputed boundary between the said two mannors saith that the present Lord Wharton nor any of / ancestors did ever dureing this deponents remembrance worke any mine or mines or claime any royalties in West Grinton or Harkaside or in any other place within the said manor of Grinton that he knows or ever heard off till about two yeares since; /

[3] To the third interrogatory this deponent saith that the manor of Helaugh lyes on the north side of the river Swale & the manor of Grinton on the southside thereof & that the river Swaile is the reputed boundary between the said two mannors as aforesaid saith that it is / north boundary of the mannor of Grinton is the river Swale from Haverdale Beck to Stallerstone Stile & that the said mannor is bounderd on the e east by Ellerton mannor at Staller Stone Stile on the west by the said Haverdale Beck & on the south by the Duke of Boltons mannors /

[membrane ix *verso*]

saith above forty years agoe this deponent was at the riding of a boundary between old Sir Solomon Swale & the late Duke of Bolton saith the footmen begun the said boundary at the river Swale & came to Stallerstone Stile & from thence the horsemen and / footmen rid & went as farr as the cole pitts on the topp of the moore between Wensleydaile & Swaledaile where the Duke & Sir Solomon met & had some discourse & soe parted & the boundary was at that time rid noe further; /

John Coates[103] of Grinton miner aged forty nine years or thereabouts sworne & examined saith /

2: 8:11: To the second eighth eleaventh & twelfth interrogatories this deponent saith that he
& 12: knows the manor or reputed manor of Grinton & alsoe knows the places called West Grinton Harkaside & Whitaside & the moors & commons called Harka Hill Harka Moore Grinton / Moore Grinton Greets Grinton How or How Groves & has soe known the same for near forty years & that for all his time the said places moors & commons have been reputed to lye within & belong to the said manor of Grinton & that Sir Solomon Swale & Mr / Roger Hillary [&][104] of late Mr Marriott have been the reputed lords of the said manor saith the Sir Solomon Swaile wrought mines of lead upon Harka hill or Harkerside Moore the place in question & had good gettings there till about eleaven yeares agoe / that Mr Marriott recovered the same against him & Mr Marriott afterwards sunck severall new shafts upon the said place & had good gettings & enjoyed the same till the Lord Wharton enterd upon him about two years agoe and that the said Sir / Solomon & <Mr> Roger Hillary used to take upp waifes & estrays upon Grinton Moore & Whitaside Moore to this deponents knowledge saith that hee knows the manor of Helaugh & that for all his time the river Swale (as hee has heard said by fishermen / the mid streame of the said river) has been reputed & accounted the ancient boundery betwixt the said manors of Grinton & Helaugh from Haverdale Beck downe to neare Grinton & soe to Stallerstone Stile saith that hee never knew or / heard that the Lord Wharton or any of his ancestors ever wrought any mines of lead or coale or claimed any royalties within the moors or commons of West Grinton or Harkaside hill till about two yeares agoe; /

3 & 9: To the third & ninth interrogatories this deponent saith that the said manor of Grinton is reputed to be bounderd on the east by the manor of Ellerton beginning at the river Swale & soe to Stallerstone Stile thence to White Stone on Sharrow thence to / Snowdon man (where begins the boundary betwixt the Duke of Boltons manors & the said manor of Grinton) & soe proceeds to Redmere Standard thence to Grinton Greets thence to Gibbon Hill thence to High Carle thence to Pickerstin Rigg / & from thence the said boundary of Grinton manor goes to Tarnsant thence to Wind Yate Currock thence to Sumerlodge Tarn Currock thence to Bullhead thence to Bloody Wall soe down Blagill to Haverdale Beck which is the boundary on the / west & the said manor of Grinton is bounderd on the north from the manor of Helaugh by the river Swale as aforesaid saith hee has heard one Robert Dunn a fisherman who dyed about twelve years agoe & was about eighty yeares of age / say that the mid streame of the river Swale was a boundary as aforesaid saith that the said river Swale is alsoe a common reputed boundary betwixt the manors of Fremington & Grinton Marrick & Grinton Marrick & Ellerton / & beleives it is soe downwards /

4: To the fourth interrogatory this deponent saith that West Grinton & Harkaside is a hamlett consisting of some few houses lying at some distance from one another & are as deposed by this deponent within the said manor of Grinton /

6: To the sixth interrogatory this deponent saith that hee has often heard as a report

103 In May 1708 John Coates, 'near 50', gave evidence for Marriott at the trial in London (LI Misc 536, Ward MSS, notebooks vol. 6, f. 112v).

104 Word omitted in the original.

in the neighbourhood that the said manor of Grinton did formerly belong to the Priory of Bridlington /

7: To the seaventh interrogatory this deponent saith that Swaledale was never called (as he has heard) a manor of it selfe but saith there <are> four manors within the same Vizt the manor of Grinton belonging to Mr Marriott the manors of / Helaugh & Mukar belonging to Lord Wharton & the manor of Fremington belonging to Madam Wharton /

10: To the tenth interrogatory this deponent saith that the inhabitants of Grinton Harkaside & West Grinton have used to have their cattle going upon one anothers commons by way of a rake without disturbance to each other saith that / the Harkaside & West Grinton men claime a right of getting ling turfe & peats upon the commons belonging to Harkaside & West Grinton so farr to a place called Sevy More which lys above the White Stone of the How from Grinton / moors or commons; /

George Hall of Fremington in the county of Yorke miner aged forty nine years or thereabouts sworne & examined saith as followeth /

2: To the second interrogatory this deponent saith that he knows the reputed manor of Grinton & the places called West Grinton Harkaside & Whitaside & the moors & waists thereto belonging & has soe knowne the same for about thirty five years saith they are reputed to belong & be / parcell of the said manor of Grinton & that the Swales the Hillarys & the relator Mr Marriott were the reputed lords of the said manor saith that Sir Solomon Swaile wrought mines for lead on Harka the place in question & continued working there till the relator / Mr Marriott recovered the said mines from him saith that Roger Hillary wrought for cole on Grinton Moore; has heard that the Swailes & the Hillaries both tooke upp waifes & estrays on Harkaside for the time aforesaid never heard that the Lord Wharton / or any of his ancestors ever wrought any mine or mines or claimed any royalties in West Grinton or Harkaside or in the moors or waists thereto belonging till about two years since /

3: To the third interrogatory this deponent saith that he knows the manor of Helaugh & that the mid streame of the river Swale is reputed to be the division between the said manor of Grinton & Helaugh saith that the manor of Grinton is bounderd on the east by / Ellerton manor at Stallerstone Stile & on the south by the Duke of Boltons manors; /

4: To the fourth interrogatorie this deponent saith that he knows the places called West Grinton & Harkaside; that they consist of a few stragling houses lying at a distance from each other & are reputed to be within the manor of Grinton /

7: To the seaventh interrogatorie this deponent saith that he knows Swaledale; that it is about twelve miles in length & has four mannors in it to witt Grinton Helaugh Mukar & Fremington that the whole dale is in the parish of Grinton & that the parish church / stands in Grinton; /

Richard Raw of Fremington yeoman aged sixty years or thereabouts sworne & examined saith /

2: To the second interrogatory this deponent saith that hee knows the manor or reputed manor of Grinton & alsoe knows the places called West Grinton Harkaside & Whitaside & the moore or common belonging to Harkaside & has knowne the same for about forty yeares / saith that for all the time of his knowledge the said places have been reputed to belong to & lye within the said manor of Grinton & that the Hillarys & Swales & of late Mr Marriott have been the reputed lords of the said manor & that Mr Hillary wrought mines / of cole upon Grinton Moore & Sir Solomon Swale wrought mines of lead upon Harkaside Moore the place in question for severall years & had good gettings there till about tenn or eleaven yeares agoe that Mr Marriott recovered the same against him & after / that Mr Marriott enjoyed the same & had good gettings till about two yeares agoe that the Lord Wharton enterd upon him saith that hee ha's heard that the Swales & Hillarys used to take upp waifs & estrays as lords within the said manor saith he / knows the manor of Helaugh & that for all the said time of his knowledge the mid streame of the river Swale has been reputed to be the boundary betwixt the said manors of Grinton & Helaugh & beleives the same to be soe as it is comonly taken / to be the boundary betwixt the said manor of Fremington & Grinton saith hee has lived at Fremington ever since he was twenty one years of age & a neare neighbour to the said moors & commons of West Grinton & Harkaside & never heard that the Lord / Wharton or any of his ancestors ever wrought any mines of lead or cole or claimed any royalties within West Grinton & Harkaside or the said moors or commons till about two years agoe; /

William Spencley[105] of Whitaside in the county of Yorke miner aged sixty years & upwards sworne & examined saith; /

2: 11: & 12: To the second eleaventh & twelfth interrogatories this deponent saith that hee well knows the manor or reputed manor of Grinton & the places called West Grinton Harkaside & Whitaside & the moors & wasts & commons called Harka hill Harka Moor Grinton Moor / Grinton Greets Grinton How or How Groves & has so knowne the said places moors & waists for above forty yeares last past saith the same are & were dureing the time aforesaid reputed to belong to the said manor of Grinton saith that the present Sir Solomon Swale / wrought the mines for lead on the How & Harka the place in question for many years & had good gettings in both places saith that he continued working there untill about eleaven years agoe that the relator Mr Marriott recovered them from him / saith that the said Mr Marriott has since that time wrought the said mines & had very good gettings & sunck two or three new shafts on Harka the place in question & reputed parcell of the said manor of Grinton & the which has been dureing / this deponents remembrance enjoyed by the Swales & the relator Mr Marriott & never that <ever> this deponent heard of either in the possession or been enjoyed by the present Lord Wharton or any of his ancestors till about two years since that the / present Lord Wharton enterd upon Harka the place in question & saith that the present Lord Wharton nor any of his ancestors did ever worke any mine or mines or claime any royalties either in Harkaside or West Grinton or in any / the moors & waists thereto belonging that ever this deponent heard of save as aforesaid saith that Roger Hillary wrought for cole in Grinton moore; saith that John Lickly tooke upp waifes & estrays for Roger Hillary and brought / them to Miles Freer as this deponent beleives either on Grinton or Harkaside Moore; /

3: & 9: To the third and ninth interrogatories this deponent saith that he knows the manor of Helaugh & that dureing the time aforesaid the mid streame of the river Swale has been the reputed division between the two mannors of Helaugh & Grinton / & that the said manor is reputed to be bounderd on the north by the said river as farr as from Haverdale Beck to Stallerstone Stile & on the east by the manor of Ellerton beginning at the said river Swale & soe to Stallerstone Stile thence to / White Stone oth Sharrow thence to Snowdon Man (where begins the boundary betwixt the Duke of Boltons mannors & the said mannor of Grinton southwards) & soe proceeds to Redmere Standard thence to Grinton Greets thence to Gibbon / Hill thence to High Carle thence to Pickerstone Rigg & from thence the said boundary of Grinton manor goes to Tarnsant thence to Windyate Currock thence to Sumerlodge tarn

105 In May 1708 William Spencely, aged 'above 60', gave evidence for Marriott at the trial in London (LI Misc 536, Ward MSS, notebooks vol. 6, f. 111v).

Currock thence to Bulhead thence to Bloody Wall soe down Blagill / to Haverdale Beck which is the west boundary of the said mannor; /

4: To the fourth interrogatory this deponent saith that he knows the places of West Grinton & Harkaside saith the same are a hamlett consisting of some few houses some parte whereof stands together & others of them at a considerable distance from / each other & saith the said places are reputed to lye & belong to the said manor of Grinton; /

6: To the sixth interrogatory this deponent saith that the manor of Grinton was comonly reputed to belong to the Priory of Bridlington but cannot sett forth what perticular lands in West Grinton & Harkaside doe or was reputed to belong to the / said Priory; /

7: To the seaventh interrogatory this deponent saith that he knows Swaledale & that the same is in length about twelve miles & that there is in the said dale the manor of Grinton belonging to Mr Marriott the manors of Helaugh & Mukar to the Lord / Wharton the manor of Fremingon to Madam Wharton that the whole dale is in Grinton parish & that the church stands at Grinton; /

10: To the tenth interrogatory this deponent saith that the inhabitants of Grinton Harkaside & West Grinton used to have their goodes going on each others commons by way of a rake & that the Harkaside & West Grinton men claime right upon Harka / =side Moore for the getting peats turfe & ling separately from the Grinton men as farr as the White Stone on the How to Grinton Greets; /

Francis Greathead of Marrick miner aged fifty one years or thereabouts sworne & examined saith; /

2: 8:11: & 12: To the second eigth eleaventh & twelfth interrogatories this deponent saith that he knows the manor or reputed manor of Grinton & the places called West Grinton Harkaside & Whitaside & the moors & commons called Harka Hill Harka Moore Grinton / Moore Grinton Grinton How or How Groves & has knowne the same for about thirty five yeares & that for all the time of his knowledge the said places moors & commons have been reputed to lye within & belong to the said manor of Grinton & that the / Hillarys & Swales & of late Mr Marriott have been called & reputed lords of the said manor saith that Mr Hillary wrought mines of cole upon Grinton Moore that old Sir Solomon & the present Sir Solomon wrought mines of lead for many / years upon Harkahill or Harkaside Moore the place in question & had good gettings & the present Sir Solomon continued to worke there till Mr Marriott recovered the same against him & after Mr Marriott wrought the said mines upon the said / Harka & had very good gettings & continued to worke there till the Lord Wharton enterd upon him about two years agoe saith he knows the manor of Helaugh & that the mid streame of the river Swale has for all his time been comonly reputed the / boundary betwixt the said manor of Grinton & Helaugh from Haverdale Beck down to neare Grinton over against Arkle Beck & soe to Stallerstone Stile saith he never knew or heard that the Lord Wharton or any of his ancestors wrought any mines of / lead or cole or claimed any royalties in West Grinton & Harkaside or any the moors & commons aforesaid till about two years agoe; /

3: & 9: To the third & ninth interrogatories this deponent saith that the said manor of Grinton is & has been all this time comonly accounted & taken to be bounderd on the east by Ellerton manor at Stallerstone Stile & on the west by Haverdale Beck & on the south by the Duke of / Boltons manors & on the north by the river Swale as aforesaid saith that the said river Swale has been alsoe the reputed boundary betwixt the manors of Fremington & Grinton Marrick & Ellerton Marrick & Stainton Marrick & Downham Marske & Downham & / beleives it is alsoe soe downwards;

& that the midd streame of the said river has been observed by the fishermen in drawing their netts to be a boundary as aforesaid; /

4: To the fourth interrogatory this deponent saith that West Grinton & Harkaside is a hamlett consisting of betwixt tenn & twenty houses lying stragling at some distance from one another; & comonly reputed to lye within the said manor of Grinton as aforesaid; /

7: To the seaventh interrogatory this deponent saith that he knows Swaledale that it is not accounted a manor of it selfe but there are four manors within the said dale Vizt the mannors of Helaugh & Mukar belonging to the Lord Wharton the said manor of Grinton belonging / to Mr Marriott & the manor of Fremington belonging to Madam Wharton & that they all lye in the parish of Grinton /

[membrane x *recto*][106]

William Fryer[107] of Downham in the county of Yorke yeoman aged forty years or thereabouts sworne & examined saith /

2: 8:11: & 12: To the second eigth eleaventh & twelfth interrogatories this deponent saith that he well knows the reputed manor of Grinton & the places called West Grinton Harkaside & Whitaside & the moors commons & waists thereto belonging & ha's soe knowne the same for twenty years / last past saith the Swailes & the relator Mr Marriott have been the reputed lords of the said manor saith the the said places moors & waists are reputed to belong & be in the said manor of Grinton saith that the present Sir Solomon Swaile wrought mines both of Grinton / How & Harka Hill the place in question & had good gettings of lead in both places & continued working in Harka till the relator Mr Marriott recoverd the mines from him who has since his entry wrought the lead mines on Harka & has had very good /gettings there & soe continued to work untill about two years since that the Lord Wharton entred upon the said mines in question saith waifes & estrays have as this deponent has heard been taken upp on Harkaside Moore & carried to the servant of the said Sir / Solomon who was then the reputed lord of Grinton saith that dureing the time of this deponents remembrance being for the time aforesaid the present Lord Wharton nor any of his ancestors did ever worke any mine or mines or claime any the royalties / within West Grinton or Harkaside or in any of the waists thereto belonging that ever this deponent knew or heard of till of late; /

3: & 9: To the third & ninth interrogatories this deponent saith that he knows the manor of Helaugh & that the mid streame of the river Swaile is the reputed boundary between the said manor of Grinton & the said manor of Helaugh from Haverdale Beck to Ellerton manor / saith the said manor of Grinton is bounderd on the north on the said river Swale on the east by Ellerton mannor at Stallerston Stile on the south by the Duke of Boltons manor & on the west by the said Haverdale Beck saith that the manor of Fremington & Grinton / is bounderd by the river Swale & parte of the manor of Marrick & Grinton is alsoe bounderd by the said river saith that hee this deponent has heard his master Gyles Metcalfe who was an ancient man & is since dead say the said river was a / boundary as aforesaid; /

4: To the fourth interrogatory this deponent saith that hee knows the places called West

[106] Parchment, 630mm x 520mm.

[107] In May 1708 William Fryer, aged 20 [?], gave evidence for Marriott at the trial in London (LI Misc 536, Ward MSS, notebooks vol. 6, f. 113v).

Grinton & Harkaside that they are a hamlett consisting of a few stragling houses lying at a distance from each other & are reputed to lye within the said manor of Grinton /

7: To the seaventh interrogatory this deponent saith that he knows Swaledale & that the same is in length about twelve miles & has in it four manors to witt Grinton manor Fremington manor & the manors [of][108] Helaugh & Mukar & saith all the said daile is in the / parish of Grinton & that the parish church stands at Grinton; /

10: To the tenth interrogatory this deponent saith that the inhabitants of Grinton West Grinton & Harkaside usually had a rake of comoning in each others comons without interrupcion that ever this deponent heard of; /

Mathew Horseman of Milfeild near Marsk yeoman aged fifty years or thereabouts sworn & examined saith /

2:11: & 12:. To the second eleaventh & twelfth interrogatories this deponent saith that hee knows the manor or reputed manor of Grinton & alsoe knows the places called West Grinton Harkaside & Whitaside & the moore or common of Harkaside or Harkahill & / has knowne the same for above thirty years & that the said places & commons have been for all the said time reputed to lye within & belong to the said manor of Grinton & that the Swales & the Hillarys & of late Mr Marriott have for all / his time been the reputed lords of the said mannor saith that Sir Solomon Swaile wrought mines of lead upon Harka Hill or Harkaside Moore the place in question & had good gettings there till Mr Marriott recoverd the same against him and / that afterwards Mr Marriott enjoyed the same till the Lord Wharton enterd upon him saith hee knows the manor of Helaugh & beleives the river Swale to be a boundary betwixt the manors of Helaugh & Grinton for that the said / river has been comonly reputed a boundary betwixt the manors of Marsk & Downham Marrick & Downham Marrick & Stainton Marrick & Ellerton & Fremington & Grinton saith hee never knew or heard that the present / Lord Wharton or any of his ancestors ever wrought any mines in the moors & commons of West Grinton & Harkaside till of late; /

3 & 9: To the third & ninth interrogatories this deponent saith that the said manor of Grinton is commonly reputed to be bounderd on the east by the manor of Ellerton at Stallerstone Stile & on the south by the Duke of Boltons manors & has heard fishermen often / talke of observing the mid streame of the river Swale to be a boundary; /

John Hutchinson of Cherry Tree house in Skelton felmonger aged sixty three years or thereabouts sworne & examined saith /

2 & 11: To the second & eleaventh interrogatories this deponent saith that hee knows the manor or reputed manor of Grinton & the places called West Grinton Harkaside & Whitaside & the moors & commons called Harka Hill Harkaside Moore & Grinton / Moore & has soe knowne the same for about thirty seaven years saith that for all the time aforesaid the said places & commons have been reputed to lye within & belong to the said manor of Grinton & that the Swales & the Hillarys & / of late Mr Marriott have been the reputed lords of the said manor saith hee knows the manor of Helaugh and that the river Swale has been for all his time reputed a boundary betwixt the said manors of Grinton & Helaugh / & this deponent ha's often heard fisher men say the same to be a boundary further saith that about thirty seaven years agoe this deponent having occasion to goe over the said moors hired a guide one Ralph Collier then an / old man who lived at Grinton & since dead & as this deponent & the said Ralph were travelling upon Grinton How some discourse happening betwixt them about the said moors hee the said Ralph bid this deponent turne his

[108] Word omitted in the original.

horse & there / shewed him some mines lying upon the west some distance from the said place & told him that they were rich mines & alsoe then shewed him a place northward which hee called Helaugh & which hee said belonged to Lord Wharton / & another place which hee called Fremington & which he said belonged to Wharton of Gillingwood & said the river Swale was the boundary betwixt the said moors & commons & the said manors of Helaugh & Fremington /

3: & 9: To the third & ninth interrogatories this deponent saith that the said manor of Grinton is commonly reputed to be bounderd on the east by the manor of Ellerton at Stallerstone Stile & on the west by Haverdale Beck & on the south by the Duke of Boltons / manors saith the said river Swale is comonly reputed & taken to be the boundary's betwixt the mannors of Fremington & Grinton Marrick & Ellerton Marrick & Stainton Marrick & Downham Marsk & Downham & alsoe betwixt the places of Aplegarth & / Thorpe & this deponent hath heard Marmaduke Binks of Thorpe who was about eighty years old & lately dead & alsoe one William Dawson of Thorpe alsoe dead say that the said river Swale was a boundery betwixt the said places of Applegarth & Thorpe and the said Binks / alsoe told this deponent that he had designed to build a mill near unto the river Swale but was purswaded by his neighbours against it because he could have noe advantage of the river noe further then the {river Swale} midd streame /

John Milner of Skelton in the county of Yorke yeoman aged seaventy two years sworne & examined saith /

2 To the second interrogatory this deponent saith that he knows the reputed mannor of Grinton & the places called West Grinton Harkaside & Whitaside & the moors waists & commons thereto belonging & ha's soe knowne the same for forty years last past saith the said / places moores & waists are reputed to belong & be parcell of the said reputed mannor of Grinton & that the Swale's the Hillary's & the relator Mr Marriott have been the reputed lords of the said mannor saith that the Swale's for many years together wrought / lead mynes on Harka the place in question & that the present Sir Solomon had good gettings there & continued workeing untill the relator Mr Marriott recovered the said mynes from him since which tyme the relator Mr Marriott wrought the said mynes in / question & ha's heard he had greate gettings there & continued soe workeing untill about two years agoe that the present Lord Wharton entred upon the said mynes saith that Roger Hillarye wrought for both lead & coale on Grinton Moore saith that he this deponent ha's / heard that both the Swale's & Hillary's have frequently taken up waifes & estray's on the moores & commons aforesaid saith that dureing this deponents remembrance the river Swale ha's been the reputed boundary betweene the manors of Grinton & Helaugh / & is alsoe the boundarye betweene Fremington & Grinton & Marrick & {Grinton} Ellerton Marrick & Downham Ma<r>sk & Downham betweene the places of Applegarth & Thorpe & soe downwards saith that forty years agoe he was a jagger for three years / together for the Lord Wharton & in that tyme nor ever since did the present Lord Wharton nor any of his ancestors ever worke any myne or mynes or claime any royalties in West Grinton or Harkaside or in the moores or waists thereto belonging that ever this deponent ever heard of till / about two years last past /

William Jeofferson of Marrick in the county of Yorke myner aged sixty five years or thereabouts sworne & examined saith /

2 & 9 To the second & ninth interrogatories this deponent saith that he know's the reputed mannor of Grinton & the places called West Grinton Harkaside & Whitaside saith that he ha's knowne the same for fifty years & that the said places were dureing the tyme aforesaid reputed / to {be} belong & to be parcell of the said mannor of

Grinton & the Swale's & the Hillary's have been the reputed lords of the said mannor saith that they both wrought mynes for lead on Grinton Moore & that he this deponent smelted lead on the How for Roger Hillarye / saith they both tooke up waifes & estray's on Grinton Moore as lords of Grinton saith he know's the mannor of Helaugh & for fifty years last past the mid streame of the river Swale ha's been the reputed boundarye betweene the two mannors of Grinton & / Helaugh saith that he ha's heard antient fishermen say that the midd streame of the said river devides the two mannors & that the people of either side used to fish noe further than the said midd streame saith that for the tyme aforesaid the present / Lord Wharton nor any of his ancestors did ever worke any myne or mynes for lead or coale or claime any royalties in West Grinton or Harkside Moores that ever this deponent heard of till about two years since that the Lord Wharton entred upon the mynes in question. /

Ralph Scott of Whitaside husbandman aged sixty six sworne & examined saith /

2 To the second interrogatory this deponent saith that he know's Grinton & alsoe the places called Harkaside & the moores & commons thereunto belonging & ha's knowne the same for about fifty years & that Harkaside & the said moores & commons have been reputed dureing his tyme to / belong to & be within Grinton & that the Swales have been reputed lords & alsoe ha's heard that the Hillary's have claimed to be lords thereof saith that he ha's heard in the neighbourhood that Sir Solomon Swale wrought mynes upon Harka & ha's heard that / one Miles Fryar tooke up waifes & estray's under Hillarye saith that he ha's heard that the south side of the river Swale belonged to the Lord Wharton tho' the said Lord Wharton has some lands in Whitaside saith he ha's been frequently upon Harka Moore / & ha's for all his tyme lived in Whitaside except two years at Crackpott & never knew or heard {of} that the Lord Wharton or any of his ancestors ever wrought any mynes there till of late /

James Spenceley[109] of Whitaside in the county of Yorke rough mason aged sixty seaven yeares sworne & examined saith /

2,11 & 12 To the second eleaventh & twelfth interrogatories this deponent saith that he know's the mannor of Grinton & the places called Grinton Harkaside & Whitaside & the moors & waists called Harka hill Harka Moore Grinton Moore Grinton Greetes Grinton How or Howgroves / & ha's knowne the same for fifty seaven years last past saith the said places moores & waists are reputed to belong & be parte & parcell of the said mannor of Grinton saith that Sir Solomon Swale wrought the mynes on Harka the place in question for lead & continued / workeing there by tymes untill the relator Mr Marriott recovered the said mynes from him & the said Mr Marriott hath since wrought the said mynes in question & ha's had very good gettings never knew or heard till of late that ever the present Lord Wharton or any / of his ancestors ever were in possession of or wrought any myne or mynes for lead or coale upon Harka or Harkaside Moore the place in question or claimed any royalties within or upon the said Harkaside Moore saith that the Hillary's wrought & gott both / lead & coale on Grinton Moore saith that the Swale's took up waifes & estray's on Harkside Moore saith that about thirty yeares agoe one John Sympson told this deponent that one Mr Swale an ancestor of the present Sir Solomon caused to be taken up as waife / upon Whitaside a faune or young deer calfe & ordered it to be brought to him to Swale Hall & kept the same as being lord of Grinton mannor saith that dureing this

109 In May 1708 James Spencely, aged 67, gave evidence for Marriott at the trial in London (LI Misc 536, Ward MSS, notebooks vol. 6, f. 112r).

deponents remembrance the midd streame of the river Swale ha's been reputed to be the division / between the said mannor of Grinton & the mannor of Helaugh /

3 & 9 To the third & ninth interrogatories this deponent saith that the mannor of Grinton is bounderd on the east by the mannor of Ellerton from the midd streame of the river Swale to Stallerston Style from thence up Ellerton pasture side & from thence to the White Stone o'th / Sharrow thence to Snowden Man thence to Brownsett (where begins the boundery of the Duke of Boltons mannors from the said mannor of Grinton) thence to Ridmer Standard thence to Grinton Greetes thence to Gibbon Hill thence to High Carle thence to / Pickerstone Riggs & from thence the bounder of Grinton mannor proceeds to Tarnsant thence to Windyate Currack thence to Summerlodge Tarne Currock thence to Bullhead thence to Bloody vale or Wall thence downe {to} Blagill to Haverdale beck the / west boundery of the said mannor & from thence downwards to Stallerston Style the river Swale lyeing on the north saith the said boundery dureing this deponents remembrance ha's been a reputed boundery as aforesaid /

4 To the fourth interrogatory this deponent saith that he know's Harkaside a small hamlett consisting of about twenty houses some few standing together & others at a considerable distance from each other saith that the Lord Wharton ha's copyhold & freehold lands at Harkaside & alsoe coppyhold lands at / Whitaside the tennants of all which appeare at his lordshipps court at Helaugh /

6 To the sixth interrogatory this deponent saith that the mannor of Grinton ha's been & is reputed to belong to the Priory of Bridlington /

7 To the seaventh interrogatory this deponent saith there are four mannors in Swaledale to witt Helaugh & Mukar the Lord Whartons Fremington Madam Whartons Grinton the relator Mr Marriotts all the said dale is in the parish of Grinton & Grinton where the church stands is the cheife towne therein /

10 & 13 To the tenth & thirteenth interrogatories saith that the inhabitants of Grinton Harkaside & Whitaside used to have their cattle goe on each others moores by way of a rake saith that Harkaside men claime a right of getting lyng peate & turfe on their owne moores / & that the place called Browny Gill is a division marke betweene the said moores & commons of Harkaside & Whitaside for that purpose & not a division marke for divideing any mannor saith he never knew or heard of any boundery ridd by the present Lord Wharton nor / any of his ancestors betweene Grinton & Harkaside till about two years last past/

[membrane x *verso*]

17 To the seaventeenth interrogatory this deponent saith that he this deponent about forty seaven years agoe was a workeman & helped to build a cowhouse for one Butterfeild who lived on the north side of the river Swale & beleives the stones were gott out of the river Swale / for that as this deponent ha's heard there was a suite comenced against the said Butterfeild for the said stones by Anthony Close who lived in Whitaside who told this deponent that Butterfeild paid him twenty four shillings for an agreement & the said Butterfeild accused this deponent for / informing the said Close /

John Garth of Low Spenceley in the county of Yorke yeoman aged ninety one years or thereabouts sworne & examined saith /

2 & 12 To the second & twelfth interrogatories this deponent saith that he know's the mannor or reputed mannor of Grinton & the places called <West Grinton> Harkaside & Whitaside & the moores & commons thereunto belonging & ha's knowne the same for about three score yeares saith that for / all his tyme they have been reputed

to lye within & belong to the said mannor of Grinton & that the Swal's & Hillary's & of late Mr Marriott have been reputed the lords of the said mannor saith that the Swales have taken up waifes & estray's & Hillary's / have alsoe as this deponent ha's heard laid claime to waifes & estray's there saith he know's the mannor of Helaugh & that the river Swale ha's been accounted a division betwixt the said mannors of Grinton & Helaugh /

3 & 9 To the third & ninth interrogatories this deponent saith that he <has> known the bounder markes following for above sixty years & that they have for all that tyme been comonly reputed the bounder marks betwixt the said mannor of Grinton & the Duke of Boltons mannors & from / thence downe Haverdale beck (that is to say) Redmer Standard & soe to Gibbon Hill & thence to High Carle & soe to Pickerstin rigg & soe to Tarnsant & soe to Windyate thence to Somerlodge Tarn Currack soe to Bull head so to bloodywall thence down Blaygill to Haverdale <beck> /

4 To the fourth Interrogatory this deponent saith that the hamlett of Harkaside & West Grinton consists of a few houses standing straggling & at some distance from one another saith that the Lord Wharton hath coppyhold lands in Harkaside & hath alsoe coppyhold lands in Whitaside & that the tennants / thereof appear at his lordshipps court at Helaugh /

6 To the sixth interrogatory this deponent saith that the mannor of Grinton ha's been comonly reputed to belong to the Priory of Bridlington formerly /

7 To the seaventh interrogatory this deponent saith that there are four mannors within the dale of Swaledale Vizt Helaugh & Mukar belonging to the Lord Wharton the mannor of Grinton belonging to Mr Marriott & the mannor of Fremington belonging to the Madam Wharton /

10 & 13 To the tenth & thirteenth interrogatories this deponent saith that inhabitants of Grinton & Harkaside & Whitaside have used to have their cattle goe from one common to another to each others commons by way of a rake & that there is a division or boundery marke called Browna Gill / which is & alway's ha's been onely accounted a division marke of their commons for getting peates ling & turfe /

Ralph Close of Whitaside in the county of Yorke miner aged eighty years & upwards sworne & examined saith /

2: To the second interrogatory this deponent saith that he knows the manor of Grinton & the places called West Grinton Harkaside & Whitaside & has soe knowne the same for sixty years last past & saith the same are reputed to belong to the said manor of / Grinton & that the Swales & the Hillarys have been the reputed lords thereof & that the Swales wrought mines for lead on Harka the place in question that for some short time this deponent was a miner at the said mines under the present Sir Solomon / Swale & when this deponent wrought at the said mines the gettings were pretty good saith that he has known the waifes & estrays taken upp upon Whitaside Moore & carryed to Swale Hall did not dureing the time aforesaid ever know that the present / Lord Wharton nor any of his ancestors did ever worke any mine or mines for lead or cole upon West Grinton or Harkaside Moores /

4: To the fourth interrogatory this deponent saith that he knows the places called West Grinton & Harkaside that they consist of a few stragling houses some standing together & others standing at a distance from each other that they lye in Grinton /

7: To the seaventh interrogatory this deponent saith that Swaledale is in length about twelve miles & has four manors in it to witt Helaugh Mukar Fremington & Grinton & that the whole dale is in the parish of Grinton & that the church stands at Grinton /

17: To the seaventeenth interrogatory this deponent saith that the present Lord Wharton has copyhold lands at Whitaside & that the tennants thereof appears at the Lord

Whartons courts at Helaugh does not know of any other person but his own tenannts that / appears there; /

John Simpson of Grinton in the county of Yorke yeoman aged sixty eight years or thereabouts sworne & examined saith /

2: 8: & 12: To the second eigth & twelfth interrogatories this deponent saith that hee knows the manor or reputed manor of Grinton & alsoe knows the places called West Grinton Harkaside & Whitaside & the moors & commons called Harkahill Harka Moore Grinton Moore / Grinton Greets Grinton How or How Groves & has knowne the same for above fifty years saith that for all that time the said places moors & commons have been reputed to lye within & belong to the said manor of Grinton & that the Swales have been / the reputed lords of the said manor & the Hillarys alsoe claimed to be lords thereof saith he knew old Sir Solomon Swale wrought mines of lead upon Harkaside Moore & this deponent about forty eight yeares agoe wrought there for some tyme under / him & that the Hillarys wrought mines of cole near Grinton moore saith that he has heard that the present Sir Solomon wrought mines of lead upon Harkaside moore (but did not see the same) & that hee had great gettings & enjoyed the / same till Mr Marriott recovered the same against him saith that the Swales used to take upp waifes & estrays upon the said moors & commons & the Hillarys sometimes alsoe tooke upp waifes & estrays there saith hee knows the manor of Helaugh & / that hee has lived fifty six years at Grinton and for all the time of his remembrance the mid streame of the river Swale has comonly been reputed the boundary betwixt the said mannors of Grinton & Helaugh & this deponent has frequently / fished in the said river Swale & it was alwaies accounted amongst the fishermen that the mid streame of the river Swale was a boundary all along from Haverdale Beck downe to Stallerstone Stile soe downe betwixt the severall mannors / lying on each side of the said river as farr as Richmond bridge saith that he never knew that the Lord Wharton or any of his ancestors wrought any mines or claimed any royalties in Harkaside or West Grinton or the moors or / commons thereof till of late; further this deponent saith that hee this deponent & one George Hall of Helaugh about thirty eight years agoe tooke a lease of old Sir Solomon Swale of the mines upon Harka or Harkaside Moore & they were to give the said Sir / Solomon two parts of ten out of the gettings & this deponent & his said partner continued to worke there for about one yeare & halfe but the said mines not proving good they gave over as they had liberty soe to doe by their lease further / saith that about thirty years agoe Mr Purchas Mr Smales & Phillip Swales (who was then a steward under the Lord Wharton) tooke a lease from old Sir Solomon Swales of the mines & wrought at Grinton How but whether they / wrought upon Harkaside moore or that the same was in their lease this deponent knows not /

3: & 9: To the third & ninth interrogatories this deponent saith that the said manor of Grinton is bounderd on the east by the manor of Ellerton at Stallerstone Stile that about fifty <four> years agoe there was some dispute betwixt old Sir Solomon & one Frankland / lord of Ellerton about the boundaries & they agreed upon a joint rideing & mett then together & begun at the mid streame of the river Swale & went upp to Stallerstone Stile & thence to Ellerton pasture nooke thence to the Broad / stone on the Sharrow from thence to Snowdon Man & this deponent being then a boy went along with them further saith that the said manor of Grinton is bounderd on the south by the Duke of Boltons manors & on the west / by Haverdale Beck; /

7: To the seaventh interrogatorie this deponent saith that Swaledale was never reputed a mannor of itselfe but that there are four manors lying within the same Vizt the manor of Grinton belonging to Mr Marriott the manor of / Fremington belonging to Madam Wharton & the manors of Helaugh & Mukar belonging to Lord Wharton all lying in the parish of Grinton & that Grinton (where the church stands) is the cheife towne therein; /

13: To the thirteenth interrogatory this deponent saith that he never knew or heard that the present Lord Wharton nor any of his ancestors ever rid or pretended to ride any boundary betwixt Grinton West Grinton or Harkaside till about two years / agoe; /

Thomas Atkinson of Fremington in the county of Yorke yeoman aged ninety two years or thereabouts sworne & examined saith /

2: To the second interrogatory this deponent saith that he knows the places & townes called Grinton & Harkaside & Whitaside & has soe knowne the same for eighty years & upwards to what mannor they belong this deponent knows not nor who has been the / reputed lords thereof saith that severall persons wrought for lead on Harka the place in question for the present Sir Solomon Swale saith that waifes & estrays were taken upp upon Grinton Moore & carryed to Swale Hall for the / use of the said Sir Solomon & sometimes were taken upp for Roger Hillary saith that hee knows the Lord Whartons manor of Helaugh & beleives the mid streame of the river Swale is the boundary between Grinton & Helaugh / but this deponent has not till about two years since heard it called a boundary saith that dureing this deponents remembrance which is for eighty years last past the late Lord Wharton the present Lord Wharton nor any of their ancestors / did ever worke any mine or mines for lead or cole or claime any royalties in Grinton or Harkaside or in any of the moors & waists thereto belonging that ever this deponent knew or heard of till about two years since; /

Christopher Loftas of Fremington in the county of Yorke miner aged forty five or thereabouts sworne & examined saith /

2: 11: & 12: To the second eleaventh & twelfth interrogatories this deponent saith that hee knows the manor or reputed manor of Grinton & alsoe knows the places called West Grinton Harkaside & Whitaside & the moors & commons called / Harka Hill Harka Moore Grinton Moore Grinton How or How Groves & ha's knowne the same for about thirty six yeares & for all the said time the said places moors & commons have been reputed to lye within and / belong to the said manor of Grinton & that the Swales & the Hillarys have been the reputed lords thereof saith that the present Sir Solomon Swale wrought mines of lead upon the said Harkaside Moore & this deponent for / some time wrought under him & the said Sir Solomon had very good gettings & continued to enjoy the same till Mr Marriott recovered the same against him & after the said Mr Marriott under whome this / deponent was alsoe a workman had good gettings & enjoyed the same till about two years agoe that the Lord Wharton enter'd upon him & that Hillarys wrought mines of cole upon the How saith that hee has knowne / both the Hillarys & Swales take upp waifes & estrays upon the said moors further saith that hee knows the manor of Helaugh & that the river Swale has for all his time been accounted the division betwixt the said / manors of Grinton & Helaugh and that hee never knew or heard that the present Lord Wharton or any of his ancestors wrought any mines or claimed any royalties in West Grinton or Harkaside or the said / moors or commons till about two years agoe; /

Anthony Atkinson of the Hagg in the county of Yorke yeoman aged sixty years or thereabouts sworne & examined saith /

2: To the second interrogatory this deponent saith that hee knows the reputed manor of Grinton & the places called West Grinton Harkaside & Whitaside & has soe knowne the same for fifty years last past saith the same dureing / this deponents knowledge ha's been reputed to belong to the said mannor of Grinton saith that the present Sir Solomon Swale wrought lead mines on Harkaside Moore & continued workeing there untill the relator Mr Marriott recovered them from / him saith that the said Mr Marriott ha's wrought the said mynes since & as he ha's heard had very good gettings there saith he ha's heard that waifes & estray's has been taken up upon Harkaside Moore & carryed to Swale Hall to Sir Solomons / steward saith that he ha's heard that Hillarye wrought for coale on Grinton Moore know's the mannor of Helaugh saith that dureing this deponents tyme the river Swale ha's been reputed to be the division betweene the said mannors of / Grinton & Helaugh & is soe in many other places downwards where the lordshipp lyes <not> on both sydes never knew or heard that ever the present Lord Wharton or any of his ancestors ever wrought for either lead or coale on West Grinton or / Harkaside Moore till about two years since /

[membrane xi *recto*][110]

Absolom Petty[111] of Fremington in the county of Yorke yeoman aged forty five yeares or thereabouts sworne & examined saith /

2: 8:11: To the second eight {eight} eleaventh & twelfth interrogatories this deponent saith
& 12: that he know's the mannor or reputed mannor of Grinton & alsoe the places called West Grinton Harkaside & Whitaside & the moores & commons called Harka hill Harka Moore Grinton Moore Grinton / Greetes Grinton How or How Groves & ha's knowne the same about thirty two yeares & that dureing all the said tyme the said places moores & commons have been reputed to belong & to lye within the said mannor of Grinton & that the Swale's & Hillary's & of late Mr Marriott / have been the reputed lords of the said mannor saith that Hillary's wrought mynes of lead upon the How & alsoe mines of cole upon Grinton Moore & that Sir Solomon Swale wrought mynes of lead for severall years upon Harka hill or Harkaside Moore the place in / question & had good gettings there & continued to worke & enjoyed the same till about ten or eleaven yeares agoe that Mr Marriott recovered the same from him & after that Mr Marriott sunck severall new shafts & had good gettings there till about two years agoe that the Lord Wharton / entred upon him saith that he ha's heard that the Swale's & Hillary's used as lords to take up waifes & estray's upon the said moores & commons further saith that he know's the mannor of Helaugh & that the midd streame of the river Swale ha's been for all his tyme reputed & / accounted to be the boundary betwixt the said mannors of Grinton & Helaugh & that he this deponent ha's for all his tyme lived a near neighbour to Harkaside & never knew or heard that the present Lord Wharton or any of his ancestors wrought any mynes of lead or coale or / claimed any royalty's within the said moores or commons of Harkaside or West Grinton till about two yeares agoe /

3: & 9: To the third & ninth interrogatories this deponent saith that the said mannor of Grinton is bounderd on the east by the mannor of Ellerton at Stallerstone Style & on the south by the Duke of Boltons mannors at the Marks called Grinton Greetes & soe by Gibbon Hill & is bounderd on the west / by Haverdale beck & on the north by the said midd streame of the river Swale from Haverdale beck down to Stallerstone Style as aforesaid saith that the said midd streame of the river Swale ha's been commonly reputed in the neighbourhood & by antient men to be the boundery / betwixt the mannors of Fremington (which adjoyn's eastward of the said mannor of Helaugh) & the said mannor of Grinton the mannor of Fremington lyeing on the north & the mannor of Grinton on the south side of the said river & alsoe the

110 Parchment, 640mm x 520mm.

111 In May 1708 Absolom Petty, aged 45, gave evidence for Marriott at the trial in London (LI Misc 536, Ward MSS, notebooks vol. 6, f. 113v).

boundery betwixt the / mannors downwards Vizt the mannors of Marrick & Ellerton Marrick & Stainton & beleives it is soe downwards /

4: & To the fourth & seaventeenth interrogatories this deponent saith that the hamlett of
17: West Grinton & Harkaside consists of a few houses standing at some distance from one another saith the Lord Wharton ha's some copyhold lands there as he ha's alsoe in the mannor of / Fremington all the tennants whereof appeare & doe suite & service at his lordshipps court at Helaugh /

7: To the seaventh interrogatory this deponent saith that the said dale or vale called Swaledale is about twelve miles long that he never heard it called a mannor or lordshipp of it selfe but saith there are four mannors within the same Vizt the mannors of Helaugh & Mukar / belonging to the Lord Wharton the mannor of Grinton belonging to Mr Marriott & the mannor of Fremington belonging to Madam Wharton all lyeing in one parish & that Grinton (where the church stands) is the cheife towne there /

10: & To the tenth & thirteenth interrogatories this deponent saith that the inhabitants
13: of Grinton & Harkaside have used to have their cattle goeing from each others commons without interruption saith that there are certain bounder marks called the corner of the Ox / intang[112] & the White Stone oth the How which were rid about two years agoe on the behalfe of the Lord Wharton betwixt the said moors & commons of Grinton & Harkaside & this deponent has great reason to beleive that the same was never thought to be a / boundary of any manor as was pretended at the said riding but that the same is only the bounder marks betwixt their said severall commons for getting ling peats & turfs separately from each other for that the West Grinton & Harkaside men / have for all this deponents time claimed the same to be the boundary marks for that purpose only & beleives they doe alsoe claime other marks for that purpose to divide their said moors from Whitaside; /

Leonard Lamb of Marrick in the county of Yorke cooper aged sixty eight years sworne & examined saith; /

2: To the second interrogatory this deponent saith that he knows the manor or reputed manor of Grinton & he knows the places called Grinton or West Grinton Harkaside & Whitaside & ha's soe knowne the same for forty years last past saith that dureing this deponents knowledge the said places / have been reputed to belong & be parcell of the said manor of Grinton & the Swales & the Hillarys have been the reputed lords of the said manor dureing this deponents time saith hee has heard & beleives that Sir Solomon Swaile wrought & got lead both on the How & Harka the place / in question saith he knows the manor of Helaugh & for these forty years last past the mid streame of the river Swale has been accounted & reputed to be the boundary between the said manors of Grinton & Helaugh never in all this deponents time heard or knew that / the present Lord Wharton or any of his ancestors ever wrought any mine or mines for lead or coale upon West Grinton or Harkside Moors till about two years since /

9: To the ninth interrogatory this deponent saith that he has heard severall of his ancient neighbours say that the river Swale was a boundary between the said manors of Grinton & Helaugh & saith <it> is alsoe accounted & reputed to be a boundary between the manors of / Fremington & Grinton Marrick & Ellerton Marrick & Stainton Marrick & Downham Marsk & Downham & soe downwards as far as Richmond /

112 i.e. 'intack'.

Edmund Lamb of Marrick in the county of Yorke yeoman aged seaventy two years or thereabouts sworne & examined saith /

2: & 9 To the second & ninth interrogatories this deponent saith that hee knows the places called West Grinton & Harkaside & Harkaside Moore & has known the same for above fifty years saith that the Swales & Hillarys were called & reputed lords of Grinton & that the said places / were for all his time reputed to belong to them as lords saith he has many years wrought as a cowper about the said places & it was always reckoned & accounted that the Lord Whartons had never any right of working any mines or claiming any royalties in West Grinton or / Harkaside & never heard they did soe till of late but that the same was in the Swales & Hillarys saith the Swales wrought mines of lead there & that hee has heard both Swales & Hillarys used to take upp waifes & estrays there saith that he knows the manor of Helaugh / & that the mid streame of the river Swale has been for all the time of his remembrance the reputed boundary betwixt the said manor of Grinton & Helaugh & that the mid streame of the said river is & has alsoe <been> comonly taken & accounted to be the boundary betwixt / the adjacent manors Vizt the manors of Fremington & Grinton Marrick & Ellerton Marrick & Stainton Marrick & Downham Marsk & Downham & the places of Thorp & Aplegarth & beleives it is soe downwards /

John Hudson of Marrick in the county of Yorke wood wright aged sixty six yeares sworne & examined saith /

2: & 9: To the second & ninth interrogatories this deponent saith that he knows Grinton & Harkaside & has soe knowne the same for above forty years & has heard they belong & are parcell of Grinton & saith that the Swales & Hillarys have been reputed & called lords of / Grinton saith that he knows Helaugh & that hee never heard to the contrary but that the river Swale was the boundary between Grinton & Helaugh & this deponent beleives it to be soe for that he knows the said river is the boundary between the manors of / Marrick & Ellerton Marrick & Stainton Marrick & Downham & soe downwards; /

Anthony Hopps of Marrick in the county of Yorke husbandman aged sixty three years or thereabouts sworne & examined saith /

2: To the second interrogatory this deponent saith that he knows the manor or reputed manor of Grinton & alsoe knows the places called West Grinton & Harkaside & the moors & commons thereunto belonging & has soe knowne the same for fifty years saith the said places moors & commons / belong to & lye within the said manor of Grinton & that the Swales & Hilarys & of late Mr Marriott have been the reputed lords of the said manor saith the Swales have wrought mines of lead upon the moors & commons & that both Swales & Hillarys have used to take / upp waifes & estrays there further saith hee knows the manor of Helaugh & that the mid streame of the river Swale has for all the time of this deponents remembrance been reputed the ancient boundary betwixt the said manors of Grinton & Helaugh & has / been alsoe always accounted the main boundery betwixt the adjacent manors of Marrick & Ellerton Marrick & Stainton Marrick & Downham & beleives it is soe downwards saith that hee has lived all his time in the neighbourhood & that the Lord / Wharton nor any of his ancestors did ever worke any mines of claime any royalties in West Grinton or Harkaside or the moors or commons thereunto belonging till about two years agoe that ever he knew or heard of; /

Bryan Harland of Marsk in the county of Yorke yeoman aged eighty eight years or thereabouts sworne & examined saith /

2: To the second interrogatory this deponent saith that he knows the manor or reputed manor of Grinton as alsoe the places called West Grinton Harkaside & Whitaside & has soe knowne the same for seaventy years last past saith that West Grinton Harkaside & Whitaside / are reputed to belong to the said manor of Grinton & that the Swales & the Hillarys were the reputed lords of the said manor saith that he knows the manor of Helaugh & that for the time aforesaid the river Swale has been reputed to be the boundary between the said / manors of Grinton & Helaugh & alsoe saith that hee beleives it to be the boundary as aforesaid for that the said river is the boundary between severall other manors downwards as farr as Marsk saith that for seaventy years last past the present Lord Wharton / nor any of his ancestors did ever work any mine or mines for lead or cole on West Grinton or Harkaside moors that ever this deponent knew or heard of till about two years since; /

Richard Walker of Marrick yeoman aged fifty two years or thereabouts sworne & examined saith /

2: & 9: To the second & ninth interrogatories this deponent saith that hee knows the manor or reputed manor of Grinton & alsoe knows the places called West Grinton Harkaside & Whitaside & the moors & commons thereunto belonging & has known the same for above / thirty years & that for all the time of this deponents knowledge the same have been comonly reputed to lye within & belong to the said manor of Grinton & that the Swales & the Hillarys have been the reputed lords thereof saith that Sir Solomon Swale / wrought mines of lead upon Harkaside Moore & had good gettings till Mr Marriott recovered the same against him & Mr Marriott afterwards enjoyed the same till the Lord Wharton enterd upon him about two years agoe saith Sir Solomon / Swale used to take upp waifes & estrays upon the said moors & commons & this deponent hath been present at the looseing of some from his bailiffe or agent saith hee knows the manor of Helaugh & that the mid streame of the river Swale has for all this / deponents remembrance [been] comonly accounted & reputed to be the boundary between the said manors of Grinton & Helaugh & that the Lord Wharton nor any of his ancestors did ever worke any mines or claime any royalties in West Grinton or / Harkaside or the moors or commons thereunto belonging that ever this deponent knew or heard of till about two years agoe saith that the mid streame of the river Swale is & has been alsoe generally accounted to be the bounderies betwixt the adjacent / manors of Fremington & Grinton Marrick & Grinton Marrick & Ellerton Marrick & Stainton Marrick & Downham Marsk & Downham & beleives it is soe downwards saith hee remembers about twenty years agoe that there was / some dispute betwixt the lords of the manors of Marsk & Downham about two little islands lying about the mid streame of the said river Swale & at last they agreed to divide the said islands betwixt them; further saith that hee this deponent / has been frequently fishing in the said river with old fishermen & the said fishermen used to observe & account the mid streame of the said river Swale to be a boundary betwixt all the said manors aforesaid /

Simon Close of Red Hurst in the county of Yorke yeoman aged forty eight years or thereabouts sworne & examined saith /

2: & 9: To the second & ninth interrogatories this deponent saith that hee knows the reputed manor of Grinton & that hee knows the places called West Grinton Harkaside Whitaside & has soe knowne the same for thirty two years last past saith the / said places were dureing the time aforesaid reputed to belong & be parcell of the said manor of Grinton & that the Swales & Hillarys were the reputed <lords> thereof & tooke upp waifes & estrays on Grinton Whitaside & Harkaside moors as / this deponent has severall times heard saith that hee knows the manor of Helaugh & has soe knowne the same for thirty two years last past & saith that dureing that time the river Swale has been the reputed boundary between the two / manors of Grinton & Helaugh saith that hee often heard his father Parcivall Close[113] who was an old man near seaventy when he dyed say that the said river Swale was a boundary as aforesaid & this deponent beleives it to be a / boundary having heard severall of his ancient neighbours say the said river was a boundary betwixt the manors of Fremington & Grinton Marrick & Ellerton & soe downwards saith that dureing the time aforesaid the / present Lord Wharton nor any of his ancestors did ever worke any mine or mines for lead or coale upon the moors & waists of West Grinton or Harkaside or in any other place within the said manor of Grinton that ever this deponent knew or heard / of till about two yeares last past; /

Stephen Orton of Thorpe in the county of Yorke yeoman aged forty six years or thereabouts sworne & examined saith; /

2: & 12: To the second & twelfth interrogatories this deponent saith that he knows the manor or reputed manor of Grinton & alsoe knows the places called West Grinton Harkaside & Whitaside & the moors & commons called Harka hill Harka Moore Grinton Moore Grinton / Greets Grinton How or How Groves & has knowne the same for about thirty five years saith that for all the time of his knowledge all the said places moors & commons have been comonly reputed & taken to belong to & be parte of the said manor / of Grinton & that the Swales & Hillarys & of late Mr Marriott have been usually called & reputed lords thereof saith that Mr Hillary wrought mines of lead & cole upon Grinton How & got pretty well there & the present Sir Solomon / wrought mines of lead upon Harka or Harkaside Moore the place in question & had great gettings & continued to enjoy the same till about tenn or eleaven years agoe that Mr Marriott recovered the same from him and after that Mr Marriott / enjoyed the same till about two years agoe that the Lord Wharton enterd upon him saith that the said Sir Solomon used to take upp waifes & estrays upon the said moors & commons & has heard that Mr Hillary alsoe tooke upp waifes & estrays / there saith that hee knows the manor of Helaugh & that for all the said time of his remembrance the mid streame of the river Swale has been comonly reputed & looked upon to be the boundery betwixt the said manors of Grinton & Helaugh & / that hee never knew or heard that the present Lord Wharton or any of his ancestors ever wrought any mines or claimed any royalties in Harkaside or West Grinton or the said moors or commons till about two yeares agoe; /

[membrane xi *verso*]

3 & 9 To the third & ninth interrogatories this deponent saith that the said manor of Grinton is bounderd on the north by the river Swale from Haverdale Beck all along downwards to Stallerstone Stile that Haverdale Beck is the west boundery / of the said manor & Stallerstone Stile the east boundery thereof & the Duke of Boltons manors is the boundary thereof on the south saith that the mid streame of the river Swale is alsoe comonly reputed & accounted to be the / boundary betwixt the adjacent manors of Fremington & Grinton Marrick & Grinton Marrick & Ellerton Marrick & Stainton Marrick & Downham & the places of Thorp & Aplegarth and more perticularly / this deponent has reasons to beleive it is a certain boundery betwixt Thorp & Aple Garth because about two years agoe Mr Dinckney steward to Colonell Byerley lord of Aplegarth for the satisfaction of the inhabitants of both

113 Parcivall Close of Redhurst, carpenter, aged 74, deposed on behalf of the Crown on 1 June 1697, in the case of the Attorney General v. Sir Solomon Swale in the Court of Exchequer (E134/9Wm3/Trin18).

places as to their / common rights & other priviledges rid a boundary there (as hee believes by his lords orders) to the mid streame of the said river & had then an old booke which he declared to be a boundary booke & that it was therein sett / downe the mid streame of the said river to be a boundary betwixt them as aforesaid that there were a great many of the inhabitants of both places then present & were satisfyed therewith & alsoe this deponent has reason to beleive it a / boundary as aforesaid because hee remembers two persons at severall times were found drowned in the said river one about twenty & the other about twelve years agoe betwixt Thorpe & Aplegarth & then all the inhabitants were / satisfyed that the charge thereof should lye upon that place on which side of the mid streame the said dead bodies were found and further this deponent saith that the mid streame of the said river is & has been alsoe the comon reputed boundery / betwixt the lordshipps or mannors of Hudswell & Richmond & beleives it is soe downwards and hee this deponent is the better able to depose herein as aforesaid because he has in his time lived in most of the places there abouts & has been very / well acquainted with them all from Haverdale Beck downe to Richmond as aforesaid & has knowne severall of the lords of the said manors ride the boundarys of the said manors & they made their claime to the mid streame of the / said river and this deponent remembers that there was a dispute betwixt the lords of Marsk & Downham about twenty five years agoe about two little islands lying in the midst of the river betwixt their said manors & they are at last agreed to divide / the same; /

4: & 17: To the fourth & seaventeeth interrogatories this deponent saith that the Lord Wharton has copy hold lands in Harkaside & alsoe in Whitaside & Fremington all the tennants whereof as hee has heard & beleives appeare at his lordshipps / court at Helaugh /

7: To the seaventh interrogatory this deponent saith that the dale or vale of Swaledale is about twelve miles long that he does not beleive it to be a manor of itselfe but that there are four manors therein Vizt the manors of Helay / & Mukar belonging to Lord Wharton the manor of Grinton belonging to Mr Marriott & the manor of Fremington belonging to Madam Wharton that they all lye in Grinton parish & that Grinton (where the church stands) / is the cheife towne in the said dale; /

Ralph Orton of Hellwith in the county of Yorke jagger aged fifty nine years or thereabouts sworne & examined saith /

2: 9: 11: & 12: To the second ninth eleaventh & twelfth interrogatories this deponent saith that hee knows the manor or reputed manor of Grinton & the places called West Grinton Harkaside & Whitaside & the moors commons & waists called Harka hill Harka / Moore Grinton Moore Grinton Greets Grinton How or How Groves & has soe knowne the said manor places moors & waists for about forty seaven years last saith the said places moors & waists have for all the time aforesaid been / reputed to belong & to be parcell of the said manor of Grinton & that the Swales the Hillarys & the relator Mr Marriott have been the reputed lords thereof saith that Roger Hillary who was a reputed lord of Grinton both wrought / for lead & cole on Grinton Moore that hee this deponent being then a jagger brought lead from the said mines to Richmond to one Dorothy Murfod who lived then in Richmond saith that hee beleives that Sir Solomon / Swale wrought for & got considerable quantities of lead upon Harkaside Moore & that the said Sir Solomon continued working there untill the relator Mr Marriott recovered the said mines from him who has since / wrought the same & has had good gettings there & soe continued to worke untill about two years since that the Lord Wharton entred upon the same never heard that the present Lord Wharton nor any other person (but the Swales the Hillarys / & the relator Mr Marriott) that ever wrought any mine or mines for either lead or cole upon any of the said places moors & commons or in any other place within the said manor of Grinton or that the Lord Wharton or any of / his ancestors were ever in the possession of the said mines till about two years since that the Lord Wharton entred upon the mines in question saith that hee this deponent knows the mines in question saith that hee this deponent knows the manor of Helaugh & for the time aforesaid the / mid streame of the river Swale has been the reputed boundary between the said two manors of Grinton & Helaugh and further this deponent saith that the said mid streame of the river Swale has been dureing the time aforesaid / reputed to be the boundary between the manors of Marrick & Grinton Marrick & Ellerton Marsk & Downham Marrick & Downham <& soe downwards> saith the reason of such this deponents knowledge is that hee has lived most of his time / within three miles of Grinton & being a jagger has frequently used the moors & daily when rideable crossed the said river Swaile in most parts thereof in the parts before mencioned and further this deponent saith that hee has / severall times knowne differences arise between fishers for comeing to fish on either side of the mid stream of the river Swale between Ellerton & Marrick & saith that about thirty years agoe a man was drowned / high upp in Swaledale & the strength of the river brought him downe as farr as Thorpe & that he was taken upp on the Thorpe side of the river & Mr Hutton the lord of the Marske side would be at noe charge with the said / drowned corps by reason he was taken upp beyond the mid streame of the river Swale on the other side; /

Robert Hawkins of New Forest in the county of Yorke yeoman aged forty years or thereabouts sworne & examined saith /

2: 9: 11: To the second ninth eleaventh & twelfth interrogatories this deponent saith that he
& 12: knows the manor or reputed manor of Grinton & alsoe knows the places called West Grinton Harkaside & Whitaside & the moors or commons called Harka Hill Harka / Moore Grinton Moore Grinton How or How Groves & has knowne the same neare thirty years & that all the said places have been during his time reputed to belong to & be parcell of the said manor of Grinton & that Mr Roger Hillary & Sir Solomon / Swale were reputed to be lords of the said manor saith hee knew the said Mr Hillary worke mines of lead & cole upon the How & Grinton Moore & this deponent carryed severall quantities from his said mines & that Mr Marriott wrought / mines of lead upon Harkaside Moore after hee recovered them from the said Sir Solomon till about two years agoe that the Lord Wharton enterd upon him and saith hee has heard that waifes & estrays has been taken upp by Sir Solomons / agents his neighbours who had left[114] their cattle having told this deponent that they fetcht them from Swale Hall saith that he knows the mannor of Helaugh & that for all the time of his remembrance the mid streame of the river Swale / has been accounted the boundary betwixt the said manors of Grinton & Helaugh & that it is & ha's been alsoe the boundary betwixt the manors of Fremington & Grinton Marrick & Grinton Marrick & Ellerton Marrick & Stainton / Marrick & Downham Marsk & Downham & beleives it is soe downwards; further saith that hee for most of his time has been frequently upon the said moores & commons & never knew or heard that the present Lord Wharton or any / of his ancestors ever wrought any mines of lead or cole or claimed any royalties in West Grinton & Harkaside or any of the said moors or commons till about two yeares agoe saith hee hath often heard old fishermen talk / of the mid streame of the said river being a boundary as aforesaid & that they used to observe the same & either gett leave of the severall lords on both sides or that they have been threatened to have their netts or rods taken / from them /

[114] Probably a scribal error for 'lost'.

Richard Nelson[115] of Ellerton in the county of Yorke yeoman aged fifty one years or thereabouts sworne & examined saith /

2: 11: & 12: To the second eleaventh & twelfth interrogatories this deponent saith that he well knows the manor or reputed manor of Grinton & alsoe knows the places called West Grinton Harkaside & Whitaside & the moors & wasts called Harka hill / Harka Moore Grinton Moore Grinton Greets Grinton How or How Groves & has soe knowne the said manor places moors commons & waists for forty years & upwards saith that for the time aforesaid the said places moores waists & / commons have been reputed to belong & to be parcell of the said manor of Grinton & that the Swales the Hillarys & the relator Mr Marriott have been the reputed lords thereof saith that above thirty years since old Sir Solomon / Swale let mines to be wrought for lead on Harkaside Moore to Thomas Nelson John Blaides & John Richardson who wrought the said mines under the said Sir Solomon & this deponent saith that at the same time he this / deponent carryed wood from Thorpe to the said mines on horse back where the said workmen had sunk shafts & wrought for lead on Harkaside Moore aforesaid saith that the present Sir Solomon Swale wrought & got lead on / Harkaside Moore & continued working <there> untill the relator Mr Marriott recovered the said mines saith that hee this deponent was a jagger under the said Sir Solomon & carryed ore from the mines on Harka the place / in question to Ellerton mill to be smelted saith the present relator since he recoverd the said mines wrought for & had good gettings on Harka the place in question saith that Hillary wrought for cole on Grinton / Moore saith that the present Lord Wharton nor any of his ancestors ever wrought or were in the possession of any mine or mines or got either lead or cole upon West Grinton or Harka on Harkaside Moore the place / in question or in any other place within the said manor of Grinton that ever this deponent knew or heard of till about two yeares since that the present Lord Wharton entred upon the mines in question; saith that both the Swales / & the Hillaries {& been told by themselves as} as this deponent has heard <& been told by themselves> used to take upp waifes & estrays on Harkaside Moore & that hee this deponent about twenty years since loosed estrayed sheep from Hillary which / as hee this deponent was told was taken upp upon Grinton Moore saith that he knows the manor of Helaugh & that dureing this deponents remembrance the mid streame of the river Swale has been the reputed / boundary between the said manors of Grinton & Helaugh; /

9: To the ninth interrogatory this deponent saith that the manor of Grinton is bounderd by the said river Swale on the north & on the east on Ellerton manor by Stallerstone Stile & from thence goes to the White Stone on the / Sharrow & from thence to Snowdon Man where the Duke of Boltons manors begins & this deponent rid the said boundary soe farr beginning at Stallerston Stile & the foot men came upp from the river / Swale to Stallerstone Stile & then the horse & foot proceed{ed} as above said saith this boundary was rid by Sir Solomon Swale & notice thereof was given to Mr Arnold the lord of Ellerton who mett & rid the said / boundary & agreed thereto; /

10: To the tenth interrogatory this deponent saith that the inhabitants of West Grinton Harkaside & Grinton used to have a rake upon each other commons for their cattle & sheep without any interrupcion of each other / that ever this deponent heard of saith there are severall bounder marks that divide the commons of Grinton West

[115] In May 1708 Richard Nelson, aged 50, gave evidence for Marriott at the trial in London (LI Misc 536, Ward MSS, notebooks vol. 6, f. 112r).

Grinton & Harkaside as to the getting of ling peate & turfe but the perticular of such bounder / marks this deponent cannot sett forth; /

Christopher Binks of Waitwith in the county of Yorke aged seaventy four or thereabouts sworne & examined saith /

2: & 12: To the second & twelfth interrogatories this deponent saith that he knows the manor or reputed manor of Grinton & the places called West Grinton Harkaside & Whitaside & alsoe the moors or commons thereunto belonging & has knowne the same for about / forty five years & that all the said places moors & commons have for all the said time been reputed to belong to & be parte of the said manor of Grinton & that Mr Hillary & Sir Solomon Swale & of late Mr Marriott have been the reputed lords of the / said manor saith that the said Mr Hillary (as this deponent has heard) has wrought some mines upon Harka & this deponent has seen mines of lead wrought upon Harka Hill & Harkaside Moore as he has frequently passed by there as a carrier & beleives / the same were wrought by the said Sir Solomon & has heard that both the said Mr Hillary & Sir Solomon used to take upp waifes & estrayes upon the said moors & commons & that this deponent was familiarly acquainted with Phillipp Swale a steward or agent under the / Lord Wharton & had frequent business with him & never heard him lay any claime to the mines or any other right or royalties upon Harkaside Moore upon the Lord Whartons behalfe & verily beleives the Lord Wharton or any of his ancestors never did worke or was in possession / of any mines upon any the moors & commons aforesaid saith he knows the manor of Helaugh & that for the time of his knowledge the said river Swale has been reputed to be the boundery betwixt the said manors of Grinton & Helaugh & beleives it to be soe as it is reputed to be betwixt the manors / of Marrick & Ellerton & Marsk & Downham & soe downwards /

[membrane xii *recto*][116]

Leonard Spenceley of Hallgaites in the county of Yorke gentleman aged forty nine years or thereabouts sworne & examined saith /

2: To the second interrogatorie saith that he knows the reputed manor of Grinton & the places & moors called West Grinton Harkaside & Whitaside & Grinton Moore & has soe knowne the same for about forty years saith the said places & moors dureing the time aforesaid have been reputed to belong & lye within the said / manor of Grinton & that the Hillarys & Swales have been the reputed lords of the said manor & as this deponent beleives & has heard they wrought for both lead & coal Swailes on Harka the place in question & Hillary upon Grinton Moore saith he has heard they both got waifes & estrays upon some / of the said moors saith hee knows the manor of Helaugh & for the time aforesaid the river Swale has been the reputed boundary between the said mannors of Grinton & Helaugh saith that during this deponents remembrance the present Lord Wharton nor any of his ancestors did ever worke any mine or / mines upon West Grinton or Harkaside Moors or claime any royalties there or in any other place within the said manor of Grinton that ever this deponent knew or heard about till about two years last past; /

John Terry of Yorke gentleman aged forty nine years or thereabouts sworne & examined saith /

2: 9: & 12: To the second ninth & twelfth interrogatories this deponent saith that hee knows the manor or reputed manor of Grinton & the places called West Grinton Harka-

116 Parchment, 720mm x 430mm.

side & Whitaside & the moors & commons thereunto belonging & has knowne the same for about twenty five years or thereabouts & that all the / said places moors & commons have during the said time been reputed to lye within & belong to the manor of Grinton & that the Swales & Hillarys have been the reputed lords of the said manor saith he hath heard that the Swales wrought mines of lead upon Harka or Harkaside Moore & knows / that Hillary wrought mines of cole upon Grinton Moore & has alsoe heard that hee wrought mines of lead there saith that hee knows the manor of Helaugh & that for the said time the mid streame of the said river Swale has been accounted to be the boundery betwixt the said manors of Grinton / & Helaugh and this deponent is the more induced to beleive the same because it is alsoe comonly reputed to be the boundary betwixt the adjacent manors Vizt the manors of Fremington & Grinton Marrick & Ellerton Marrick & Stainton Marske & Downham Marsk & Downham / & soe downwards & saith that there was some contest betwixt the lords of Marsk & Downham about some parcell of ground lying about the mid streame of the said river which this deponent has heard was agreed betwixt them saith hee has lived most of his time neare to Grinton / & that hee never knew or heard that the Lord Wharton or any of his ancestors ever wrought any mines or claimed any royalties in West Grinton or Harkaside or the said moors & commons till about two yeares agoe further saith that hee has often fished in the said river & ha's / been severall times threatned by one Mr Blackburne who lived at Grinton & alsoe was concerned for the lordshipp of Ellerton that hee & others that fished in the said river should not goe further than the mid streame of the said river and this deponent & other fishermen have / alsoe observed the mid streame of the said river to be a boundary as aforesaid; /

John Williamson[117] of Feetham in the county of Yorke miner aged forty four years or thereabouts sworne & examined saith

2: & To the second & last interrogatories this deponent saith that hee knows the manor or
17: reputed manor of Grinton & the places called West Grinton Harkaside & Whitaside & the moors waists & commons thereunto belonging & has soe knowne the same for twenty eight years last past saith the / same during this deponents knowledge were reputed to be parte & parcell of the said manor of Grinton & that the Swales & Hillarys have been the reputed lords of the said manor & that the present Sir Solomon Swale wrought for lead on Harka on Harkaside Moore the place / in question & continued soe working untill the relator Mr Marriott recovered the said mines saith that Mr Marriott since his entry wrought the said mines & has had very good gettings saith that Hillary wrought mines for cole on Grinton Moore saith that hee this deponent / has seen waife sheep brought in to old Christopher Naylor deceased steward or servant to Sir Solomon Swale at Swale Hall & that the persons that brought in the said sheep then said they tooke them upp as waife <up>on Harkaside Moore knows the manor of Helaugh saith that during / this deponents knowledge the mid streame of the river Swale has been the reputed boundary between the said manor of Grinton & Helaugh & that hee has severall times heard his uncle George Burbeck & Margaret his wife who were two ancient people & are now dead say / that the said mid streame of the said river was a boundary as aforesaid as farr as from Haverdale Beck to Stallerstone Stile saith that about tenn years since Ralph Freer an

[117] John Williamson was one of Lord Wharton's miners who was served with an injunction by Mathew Sisson on 17 April 1706. After Sisson claimed in his affidavit (E207/112/1) that Williamson said he 'vallued not the injunction', the Court ordered his arrest (Order, E127/25, f. 156r). In May 1708 John Williamson, aged 45, gave evidence for Marriott at the trial in London (LI Misc 536, Ward MSS, notebooks vol. 6, f. 113v).

agent of the present Lord Wharton & this deponent got stones in the said river for the makeing of a fence on the northside of the / said river by the side thereof of William Freers ground lying between Haverdale Beck & Scabba Wath and this deponent saith that the said Freer said they could goe noe further than the mid streame to gett stones for that was the boundary between the north & south sides saith that hee ha's by times / been a miner under the Lord Wharton for about twenty eight years last past & for that time beleives the present Lord Wharton nor any of his ancestors did ever worke any mine or mines for lead or cole upon West Grinton or Harkaside Moors till about two years since; /

Richard Garth of Crackpott in the county of Yorke yeoman aged fifty five years or thereabouts sworne & examined saith /

2: 11: & To the second eleaventh & twelfth interrogatories this deponent saith he knows the
12: manor or reputed manor of Grinton & alsoe knows the places called West Grinton Harkaside & Whitaside & the moors & commons called Harka Hill Harka Moore Grinton Moore <Grinton Greets> Grinton How or How Groves / & has knowne the same for forty years saith that for all the said time the said places moors & commons have been generally accounted & reputed to belong to & be parte of the said manor of Grinton & that the Swales & Hillarys & of late Mr Marriott have been all his time reputed to / be the lords of the said manor & that Hillary used to worke mines upon the How & Sir Solomon Swale wrought mines of lead upon Harka or Harkaside Moore & had for some time good gettings & enjoyed the same till Mr Marriott about tenn or eleaven years agoe / recovered the same from him & afterwards the said Mr Marriott enjoyed the same & continued in possession thereof till about two years agoe the Lord Wharton enterd upon him saith hee never heard that the Lord Wharton or any of his ancestors claimed any royalties in / the said places moors or commons or wrought any mines there till after hee had soe enterd upon the said Mr Marriott; /

7: To the seaventh interrogatory this deponent saith he knows the dale or vale called Swaledale which is in length about twelve or thirteen miles that the same is not a manor of it selfe but that there <are> four manors in the said dale Vizt the manor of Grinton belonging to Mr Marriott / the manors of Helaugh & Mukar belonging to Lord Wharton & the manor of Fremington belonging to Madam Wharton all lying in the parish of Grinton & that Grinton (where the church stands) is the cheife towne therein; /

10: To the tenth interrogatory this deponent saith that the inhabitants of Grinton Harkaside & Whitaside used to have their cattle goe from one anothers commons by way of a rake without interruption; /

James Maddison of Thorpe in the county of Yorke yeoman aged seaventy seaven years or thereabouts sworn & examined saith /

2: & 9: To the second & ninth interrogatories this deponent saith that hee knows the manor or reputed manor of Grinton & the places called West Grinton Harkaside & Whitaside & the moors & commons thereto belonging called Harka Hill Harka Moore Grinton Moore Grinton Greets Grinton / How or How Groves & has soe knowne the said places Moors waists & commons for above fifty years last past & saith that dureing that time the same have been reputed to belong & be parcell of the said manor of Grinton & that the Swailes & the Hillarys have / been the reputed lords of the said mannor saith that hee this deponent has been a fisherman himselfe for many yeares & saith that dureing his remembrance the mid streame of the river Swale was the division between the lords that had lands lying on each side / thereof & that hee this deponent was borne & brought upp at Thorpe which lyes on the southside of the said river Swale & that he usually fished to the mid streame of the said river that he never was disturbed by any of the lord or lords or their servants that lay on / the north side of the said river; /

John Garth junior of Whitaside in the county of Yorke yeoman aged forty five years or thereabouts sworne & examined saith /

2: 11: To the second eleaventh & twelfth interrogatories this deponent saith that he knows
& 12: the manor or reputed manor of Grinton & alsoe the places called West Grinton Harkaside & Whitaside & the moors & commons called Harka Hill Harka Moore Grinton Moore / Grinton Greets Grinton How or How Groves & has knowne the same for about thirty years saith the said places moors & commons have all the said time of his remembrance been reputed to lye within & belong to the said manor of Grinton & that / the Swales & Hillary & of late Mr Marriott were the reputed lords of the said manor saith Hillary wrought mines of lead upon the How & Sir Solomon wrought mines of lead upon Harka Hill or Harkaside Moore & continued to worke there till the said / Mr Marriott recovered the same from him and afterwards the said Mr Marriott wrought there & sunck severall new shafts & had the possession of & enjoyed the same till about two years agoe that the present Lord Wharton enterd upon him saith that both Hillarys & Swales / as lords used to take upp waifes & estrays upon the said moors & commons & hath seen waifes taken upp both on Harkaside & Whitaside & carryed to both Hillary & Sir Solomon saith hee has lived all his time in Whitaside & never heard that the present Lord / Wharton or any of his ancestors ever wrought any mines or claimed any royalties in West Grinton or Harkaside or any of the said moors or commons till about two yeares agoe; /

4: & To the fourth & seaventeenth interrogatories this deponent saith that the hamletts of
17: West Grinton & Harkaside consists of a few houses lying at some distance from each other saith the Lord Wharton ha's some copyhold lands in Harkaside & has alsoe copyhold lands in Whitaside / the tennants of all which appeare at his lordshipps court of Helaugh; /

7: To the seaventh interrogatory this deponent saith that the dale or vale called Swaledale is not reputed a manor of it selfe but there are four manors within the said dale Vizt the manors of Helaugh & Mukar belonging to the Lord Wharton the manor of Grinton belonging to Mr / Marriott & the manor of Fremington belonging to Madam Wharton all lying in the parish of Grinton & that Grinton (where the church stands) is the cheife towne therein. /

10: To the tenth interrogatory this deponent saith that the inhabitants of Grinton West Grinton & Harkaside used to have their cattle goeing from one common to another by way of a rake without interruption saith that the moors & commons of West Grinton & Harkaside are divided / from the moors & commons of Whitaside by the division or boundery marke of Browna Gill as to getting ling peates & turfes separately upon their own commons; /

George Jefferson of Hudswell in the county of Yorke yeoman aged sixty years or thereabouts sworne & examined saith /

2: 9: & To the second ninth & twelfth interrogatories this deponent saith that hee knows the
12: manor or reputed manor of Grinton & alsoe knows the places called West Grinton Harkaside & Whitaside & the moors & commons thereunto belonging & ha's knowne the same for about / thirty five years & that the said places moors & commons have for all his time been reputed to belong to the manor of Grinton & that the Swales & Hillarys have been the reputed lords of the said manor saith that he knows the manor or reputed manor of / Helaugh & beleives the river Swale to be the boundary betwixt the said manors of Grinton & Helaugh & the reasons of his beleife thereof are that this deponent well remembers that about twelve years agoe there was a girle found drowned in the said river / betwixt Thorpe & Applegarth & this deponent

being a constable of Hudswell (to which constablery Thorpe belongs) was forct to pay the coronor his fees because the dead body was taken upp on that side from the midd streame of the said river and alsoe / this deponent has heard of some riding of the boundary betwixt the lords of Hudswell & Richmond & that the mid streame of the said river was then agreed to be the boundery betwixt them and alsoe for that the mid streame of the said river is commonly / reputed to be the boundery betwixt Lord & Lord of all the manors from Grinton down the said river as far as Catterick bridge and further this deponent saith that hee was a miner & wrought at the Old Gang under the Lord Wharton about thirty five years agoe for about three / months & never heard that the said Lord Wharton or any of his ancestors wrought any mines or claimed any royalties in West Grinton or Harkaside till about two yeares agoe; /

[membrane xii *verso*]

James Blaides[118] of Crackpott in the county of Yorke yeoman aged fifty six years or thereabouts sworne & examined saith /

2: 11: & To the second eleaventh & twelfth interrogatories this deponent saith that hee knows
12 the manor or reputed manor of Grinton & the places called West Grinton Harkaside & Whitaside & the moors commons & waists thereto belonging called Harka Hill Harka / Moore Grinton Moore Grinton Greets Grinton How or How Groves & has soe knowne the same for forty years last past saith that for the time aforesaid the said places moors & waists have been reputed to be parcell & to belong to the said manor of Grinton & / that the Swales the Hillarys & the relator Mr Marriott have been the reputed lords thereof that the Hillarys have wrought & got both lead & cole on Grinton Moore & that Sir Solomon Swale wrought mines & got lead on Harka on Harkaside Moore / the place in question & had pretty good gettings there & continued soe working untill the relator Mr Marriott recovered the said mines from him since which time the relator Mr Marriott wrought the said mines & has had very good gettings on Harka the / place in question untill about two years since that the present Lord Wharton enterd upon the said mines saith that James Naylor a servant to Sir Solomon Swaile told this deponent that hee had a ewe & a lambe brought to him to Swale Hall as waife from Harka / =side Moore & has heard that both the Swales & the Hillarys tooke upp waifes & estrays upon Grinton Moore never heard the present Lord Wharton nor any of his ancestors ever were in the possession of or wrought any mines upon the moors waists & commons / of West Grinton or Harkaside or claimed any royalties there till about two yeares last past that the present Lord Wharton enterd upon the mines in question on Harka on Harkaside Moore; /

7: To the seaventh interrogatory this deponent saith that he knows Swaledale that there are four manors in it to witt Mukar & Helay belonging to the Lord Wharton Fremington to Madam Wharton & Grinton to Mr Marriott saith the said dale is about twelve miles long / & all in the parish of Grinton & that the parish church stands at Grinton; /

10: To the tenth interrogatory this deponent saith that the inhabitants of Grinton West Grinton Harkaside & Whitaside have a rake for each others cattle on their severall moores & commons without interruption & that West Grinton & Harkaside claime sole right on / their commons for the getting of ling peate & turfe & that Browna

118 In May 1708 James Blades, aged 56, gave evidence for Marriott at the trial in London (LI Misc 536, Ward MSS, notebooks vol. 6, f. 112v).

Gill is <one of> the boundarys of the said commons between Harkaside & Whitaside commons for that purpose; /

4: & To the fourth & seaventeenth interrogatories this deponent saith that the present Lord
17: Wharton has copyhold lands at Harkaside & Whitaside the tennants whereof appears at his lordshipps court at Helaugh /

Thomas Garth of Whitaside in the county of Yorke yeoman aged forty eight years or thereabouts sworne & examined saith /

2: 11: To the second eleaventh & twelfth interrogatories this deponent saith that hee knows
& 12: the manor or reputed manor of Grinton & the places called West Grinton Harkaside & Whitaside & the moors waists & commons waists thereto belonging called Harka Hill Harka Moore / Grinton Moore Grinton How or How Groves & has soe knowne the same for thirty four years last past saith the same for the time aforesaid have been reputed to belong & to be parcell of the said manor of Grinton & that the Swales the Hillarys & the relator Mr Marriott / have been the reputed lords of the said manor & that the Hillarys wrought for lead & cole on Grinton Moore & that the present Sir Solomon Swale wrought mines & got considerable quantitys of lead on Harka the place in question & continued working there untill the / relator Mr Marriott recovered the said mines from him saith Mr Marriott since his entry has wrought the said mines & has had good gettings untill the present Lord Wharton about two years since entred upon the said mines in question saith that hee this deponent ha's / seen sheep brought in as waifes to one Christopher Naylor a servant to Sir Solomon Swale to Swale Hall which Naylor told this deponent they were taken upp on Harkaside Moor saith hee never heard that the present Lord Wharton nor any of his ancestors ever / wrought any mines or claimed any royalties in West Grinton or Harkaside Moors till about two yeares since; /

4: & To the fourth & seaventeenth interrogatories this deponent saith that the present
17: Lord Wharton has copyhold & freehold lands in Harkaside as alsoe copyhold lands in Whitaside the tennants of which lands appeare at his lordshipps courts at Helaugh /

7: To the seaventh interrogatory this deponent saith that Swaledale has four manors or lordshipps in it but is not reputed to be a manor of it selfe saith its about twelve miles in length & all in Grinton parish; /

10: To the tenth interrogatory this deponent saith that the inhabitants of Whitaside Harkaside & Grinton have a rake for each others cattle on each others commons & that West Grinton & Harkaside people claime a sole right on their own moores for / the getting of peats turfe & ling & that Browna Gill is a boundary marke between the said commons & Whitaside for that purpose; /

Edward Bell[119] of Whitaside in the county of Yorke miner aged thirty six or thereabouts sworne & examined saith /

2: & To the second & twelfth interrogatories this deponent saith that hee knows the manor
12: or reputed manor of Grinton & alsoe knows the places called West Grinton Harkaside & Whitaside & the moors & commons called Harka Hill Harka Moore Grinton Moore Grinton Greets Grinton How or How Groves / & has knowne the same for fifteen years & upwards & during the time of his knowledge thereof the said places moors & commons have been reputed & accounted to belong to & be parte of the said manor of Grinton & that the Swales & Hillarys & of late Mr Marriott have been / the reputed lords of the said manor saith that the Hillarys wrought mines of

[119] Edward Bell swore affidavits on 5 June 1707 and 25 October 1708 (E207/114/2; E207/118/1). On the second occasion he accused Lord Wharton's workmen of unlawfully removing timber from a 'house' built by Marriott on 'Harkey' [i.e. Harka] Hill.

cole upon Grinton Moore & Sir Solomon Swale wrought mines of lead upon Harka <hill> or Harkaside Moore & had good gettings & continued to worke & enjoyed the same till the said Mr Marriott recovered the / same from him & after that Mr Marriott wrought there & sunck severall new shafts & had good gettings & continued in possession thereof till about two years agoe that the present Lord Wharton enterd upon him saith that hee hath heard both Hillarys & Swales used to take upp / waifes & estrays upon the said moors & commons further saith that hee knows the manor of Helaugh & that he beleives the river Swale to be the boundary betwixt the said manors of Grinton & Helaugh saith hee hath lived fifteen years at Whitaside & never knew or / heard that the Lord Wharton or any of his ancestors wrought any mines or claimed any royalties in West Grinton or Harkaside or the said moors or commons till about two years agoe; /

3: To the third interrogatory this deponent saith that the said manor of Grinton is bounderd on the east by the manor of Ellerton at Stallerstone Stile & on the west by Haverdale Beck & on the south by the Duke of Boltons manors & that the marks called Grinton Greets Gibbon Hill & High / Carle are some of the reputed boundery marks betwixt the Duke of Boltons manors & the said manor of Grinton saith hee has alsoe heard that the mid streame of the said river Swale is alsoe the reputed boundary betwixt the adjacent manors of Fremington & Ellerton /

4: & 17: To the fourth & seaventeenth interrogatories this deponent saith that the Lord Wharton hath copyhold lands in Harkaside & Whitaside the tennants of all which (as this deponent beleives) appeare at his lordshipps court at Helaugh; /

Edward Bland[120] of Whitaside in the county of Yorke miner aged fifty two years or thereabouts sworne & examined saith /

2: 8 11 To the second <eighth> eleaventh & twelfth interrogatories this deponent saith that
& 12. he well know's the mannor or reputed mannor of Grinton & the places called West Grinton Harkaside & Whitaside & the moors & commons called Harka hill Harka Moore Grinton Moore Grinton Greetes Grinton / How or Howgroves & ha's soe knowne the said places moores & commons for forty years last past saith the same are reputed to belong to & to be parcell of the said mannor of Grinton saith that the Swale's the Hillary's & the relator Mr Marriott have been the reputed lords of the said / mannor saith that Sir Solomon Swale wrought mynes for lead on the How & Harkaside the place in question & on Green hill & Birks in Whitaside & that he had sometymes good gettings & sometymes but indifferrent saith that Mr Marriott ha's wrought mynes for lead / both on the How & Harka & ha's had good gettings in both places saith that Mr Hillarye wrought for both lead & coale on Grinton Moore saith that he this deponent ha's himselfe taken up waife sheep on Whitaside & carryed them to John Blaides an agent for Sir / Solomon Swale saith that he beleives waifes & estray's were taken up on Grinton Moore & carryed to the Servants & agents both of Sir Solomon Swale & Mr Hillarye saith that the river Swale is the boundery or division between Grinton & Helaugh & this deponent beleives the said / river to be a boundery as aforesaid by reason the same is a boundery between Ellerton & Marrick saith that he this deponent ha's been for most of the tyme aforesaid frequently upon the said moores but never in all that tyme heard or knew that the present Lord Wharton or / any of his ancestors did ever worke any myne or mynes for lead or coale upon any the moores waists or commons of West Grinton or Harkaside or in any other place within the said mannor of Grinton till about two yeares since /

120 In May 1708 Edward Bland, aged 50, gave evidence for Marriott at the trial in London (LI Misc 536, Ward MSS, notebooks vol. 6, f. 110v).

3 & 9 To the third & ninth interrogatories this deponent saith that the mannor of Grinton is bounderd on the east by Ellerton mannor at Stallerston Style from thence to the Gray Stone in the Sharrow from thence to Snowden Man (& there begins the boundery betweene the Duke of Boltons / mannor & the mannor of Grinton) & from thence to a place called Brownsitt & from thence to Ridmer Standard & from thence to Grinton Greetes & from thence to Gibbon Hill soe to High Carle & from thence to Pickerstin rigg & from thence to Windgates & from thence / to Tarne Currack & from thence to Bullhead & soe to Blagill & from thence to Haverdale beck the west boundery betweene the said mannors & from thence to the river Sale being the north boundery or division between the said mannor of Grinton & Helaugh /

4 & 17 To the fourth & seaventeenth interrogatories this deponent saith that he know's West Grinton & Harkaside saith they are a small hamlett consisting of some few stragling houses some few whereof stand{ing} together others at a considerable distance from each other & are reputed / to be within the said mannor of Grinton saith the present Lord Wharton hath both freehold & coppyhold lands in Harkaside as alsoe coppyhold lands in Whitaside the tennants whereof appear at his lordships court at Helaugh /

6 To the sixth interrogatory this deponent saith that the mannor of Grinton ha's been dureing this deponents tyme reputed to belong to the Priory of Bridlington /

[membrane xiij *recto*][121]

7 To the seaventh interrogatory this deponent saith that he know's Swaledale & that it's in length about twelve miles & ha's in it four mannors to witt Helaugh & Mukar belonging to the Lord Wharton Fremington to Madam Wharton & Grinton to Mr Marriott saith the whole dale is in Grinton parish & that the church stands at Grinton /

10 To the tenth interrogatory this deponent saith that the inhabitants of Harkaside Whitaside & Grinton have a rake for each others cattle through their severall moores & commons without interrupcion of each other saith they all claime severall rights in their owne common's for the getting / of peates turfe & ling & that Brouna Gill is a divison marke or boundery betweene the said commons of West Grinton Harkaside & Whitaside for that purpose /

Anthony Sympson[122] of Whitaside in the county of Yorke aged sixty six yeares or thereabouts sworne & examined saith /

2 11 & 12 To the second eleaventh & twelfth interrogatories this deponent saith that he know's the mannor or reputed mannor of Grinton & alsoe the places called West Grinton Harkaside & Whitaside & the moores & commons called Harka hill Harka Moore Grinton Moore Grinton Greetes Grinton How or / Howgroves & ha's knowne the same for fifty yeares saith that for all the said tyme the same hath been reputed to lye within & be parte & parcell of the said mannor of Grinton & that the Swale's & Hillary's & of late Mr Marriott have been the reputed lords of the said mannor saith that Hillary / wrought mynes of coale upon Grinton Moore & the Swale's wrought mynes of lead upon Harka or Harkaside Moore the place in question & that the present Sir Solomon had good gettings there & continued to enjoye them till about ten or eleaven years agoe the said Mr Marriott recovered them from / him & after that the said Mr Marriott had the possession thereof & continued to worke or enjoye

[121] Parchment, 730mm x 650mm.

[122] In May 1708 Anthony Sympson, aged 57 (error for 67?), gave evidence for Marriott at the trial in London (LI Misc 536, Ward MSS, notebooks vol. 6, f. 113r).

the same till about two years agoe the Lord Wharton entred upon him saith that both the Hillary's & Swale's as lords used to take upp waifes & estray's upon the said moores & commons saith that he know's the mannor of Helaugh / & that for all his tyme the midd streame of the river Swale ha's been the reputed division betwixt the said mannors of Grinton & Helaugh & the rather beleives it to be soe because the inhabitants of both sides Vizt Whitaside & about Helaugh have used to observe the midd streame of the said river as to their / getting stones in the said river upon their owne side for their owne use saith he ha's lived all his tyme at Whitaside & never heard that the Lord Wharton or any of his ancestors wrought any mines or claimed any royalties in West Grinton or Harkaside or any of the said moores or commons till about two yeares agoe /

3 & 9 To the third & ninth interrogatories this deponent saith that the said mannor of Grinton is bounderd on the east by the mannor of Ellerton begining at the midd streame of the <said> river Swale & soe to Stallerston Style & thence to Ellerton nooke thence to Gray Stone in the Sharrow thence to Snawdon Man & there begins / the boundery betwixt the said mannor of Grinton & the Duke of Boltons mannors & proceeds southward to Ridmer Standard thence to Grinton Greetes soe to Gibbon Hill where the said mannor of Grinton is bounderd on the south by the Duke of Boltons mannors & soe go's to High Carle thence westward to Pickerstin / Rigg thence to a Tarne & soe to Windyate Currack thence to Sumerlodge Tarne thence to Middle Tongue thence to Bull head & soe downe to Haverdale beck & soe to the midd streame of the said river Swale which is the west boundary of the said mannor of Grinton /

10 To the tenth interrogatory this deponent saith that the inhabitants of Grinton Harkaside & Whitaside used to have their cattle going from one common to another by way of [a][123] rake without interrupcion but the inhabitants of Harkaside & Whitaside claime a separate right of getting ling peates & turfe upon their / commons & that the marke called Browna Gill is a boundery or division marke betwixt them for that purpose /

Mathew Sympson of Whitaside miner aged thirty eight years or thereabouts sworne & examined saith /

2 11 & 12. To the second eleaventh & twelfth interrogatories this deponent saith that he know's the mannor or reputed mannor of Grinton the places called West Grinton Harkaside & Whitaside & the moores & waists commonly called Harka hill Harka Moore Grinton Moore Grinton Greetes Grinton How or Howgroves / & ha's soe knowne the same for twenty five years last past saith that dureing the tyme aforesaid the same have been reputed to belong & to be parcell of the said mannor of Grinton and that the Swale's the Hillary's & of late Mr Marriott have been the reputed lords of the said mannor saith that / the present Sir Solomon Swale wrought mynes {of} <for> lead on Harka or Harkaside Moore the place in question & continued workeing there untill the relator Mr Marriott recovered the said mynes from him since which tyme the said Mr Marriott wrought the said mynes & ha's had very good gettings & / continued workeing the same untill the present Lord Wharton about two years since entred upon him saith that waifes & estray's have been taken up for the present Sir Solomon Swale on Whitaside Moore know's the mannor of Helaugh & saith that the reputed boundarye or division betweene the / mannors of Grinton & Helaugh is the midd streame of the river Swale saith that for the tyme aforesaid this deponent

123 Indefinite article omitted in the original.

[n]ever[124] knew or heard that ever the present Lord Wharton nor any of his ancestors ever wrought any myne or mynes for lead or coale upon West Grinton or Harkaside Moores or / upon any other place within the said mannor of Grinton till about two years last past /

3 & 9 To the third & ninth interrogatories this deponent saith that the reputed mannor of Grinton is bounderd on the north by the midd streame of the river Swale & soe to Ellerton mannor by Stallerstone Style on the east & from thence to the Gray Stone on the Sharrow & from thence to Snawden Man (which divides the / Duke of Boltons mannor & the said mannor of Grinton) & is the south boundery from Snawden Man to Brownsett & from thence to Ridmer Standard & from thence to Grinton Greetes & from thence to Gibbon Hill & from thence to High Carle & from thence to Pickerstin rigg & from thence to Haverdale Tarne / & from thence to Wind Yates & from thence to Tarne Currack & from thence to bloody wall & from thence to Bullhead & from thence to Blagill & from thence to Haverdale beck which is the west boundery & soe downe to the river Swale /

4 & 17 To the fourth & seaventeenth interrogatories this deponent saith that he know's West Grinton & Harkaside saith the same are a hamlett consisting of a few houses some standing together & others of them at a considerable distance from each other & are reputed to lye & belong to the said mannor of Grinton / saith the defendant the Lord Wharton ha's coppyhold lands in Harkaside Whitaside & Fremington the tennants whereof appeares at his lordshipps court at Helaugh /

7 To the seaventh interrogatory this deponent saith that he knows Swaledale that it's in length about twelve miles know's it not to be a mannor but saith there are four mannors in it to witt Mukar & Helaugh belonging to the Lord Wharton Fremington to Madam Wharton & Grinton to Mr Marriott / saith the whole dale is in Grinton parish & that the parish church stands there /

10 To the tenth interrogatory this deponent saith that the inhabitants of Whitaside West Grinton Harkaside & Grinton have a rake for each others cattle on each others moores & that they goe together & graze on each others commons without interrupcion saith that as to the getting of peates ling / & turfe they gett the same severally on their owne {bounder} moores & do's not claime right soe to doe on each others /

William Blaides of Crackpott in the county of Yorke yeoman aged fifty seaven years or thereabouts sworne & examined saith /

2 & 11 To the second & eleaventh interrogatories this deponent saith that he know's the mannor or reputed mannor of Grinton & alsoe the places called West Grinton Harkaside & Whitaside & the moores & commons thereunto belonging & ha's knowne the same for about forty five yeares & that dureing the said tyme / the said places moores & commons have been reputed to lye within & belong to the said mannor of Grinton saith that the Swale's wrought mynes of lead upon Harka or Harka <side> Moore Sir Solomon Swale had good gettings there till about ten or eleaven years agoe that Mr Marriott recovered the same from him / & after that the said Mr Marriott enjoyed the same till about two years agoe that the Lord Wharton entred upon him saith that for all that tyme of his said knowledge Vizt twenty five years he never knew or heard that the present Lord Wharton or any of his ancestors wrought any mynes of lead or coale / or claimed any royalties in West Grinton or Harkaside or any the said moores or commons till about two yeares agoe /

124 Written 'ever' in the original.

James Fothergill[125] of Grinton miner aged forty nine years sworne & examined saith /

2 11 & 12 To the second eleaventh & twelfth interrogatories this deponent saith that he know's the mannor or reputed mannor of Grinton saith he know's the places called West Grinton Harkaside & Whitaside & the moors waists & commons thereto belonging called Harka hill Harka Moore Grinton Moore Grinton Greetes Grinton How / or Howgroves & ha's knowne the said places moores & waists for forty years last past saith the same are within & reputed to be parte & parcell of the said mannor of Grinton & that the Swale's the Hillary's & Mr Marriott have been the reputed lords of the said mannor saith that old Sir Solomon Swale grandfather / to the present Sir Solomon wrought mynes for lead on Harka or Harkaside Moore the place in question & saith that he this deponent was a workeman under the said Sir Solomon at the said mynes thirty five years since saith that the present Sir Solomon wrought mynes for lead on Harka the place in question / & continued workeing there untill the relator Mr Marriott recovered the said mynes from him about eleaven years agoe since which tyme the said Mr Marriott ha's wrought the said mynes untill about two years since that the defendant the present Lord Wharton entred upon him saith that he this deponent carryed oare / off Harka the place in question which was wrought by Roger Hillarye out of the waist oare when he was in the possession of the said mynes saith he this deponent tooke up a waife sheep on Grinton Moore & carryed the same to old Sir Solomon Swale & afterwards it was delivered to his servant Naylor / saith he alsoe heard that severall waifes & estray's have been taken upon the said moores & commons & carryed to the servants both of the Swale's & Hillary's saith that dureing this deponents remembrance the midd streame of the river Swale ha's been reputed to be the division or parting betweene / the two mannors of Grinton & Helaugh saith that he this deponent ha's lived about forty yeares in Grinton & in all that tyme never knew or heard that the present Lord Wharton nor any of his ancestors ever wrought any myne or mynes for lead or coale upon West Grinton or Harkaside Moore / or in any other place within the said mannor of Grinton till about two yeares since that the present Lord Wharton gott into possession of the mynes in question /

3 & 9 To the third & ninth interrogatories this deponent saith that the mannor of Grinton is bounderd by the midd river of Swale on the north on the east on Ellerton mannor by Stallerstone Style & from thence to the Gray Stone on the Sharrow & from thence to Snawden Man (where the Duke of Boltons / mannors begin to joyne on the said mannor of Grinton) on the south from Snawden Man to Brownsitt & from thence to Ridmer Standard & from thence to Grinton Greetes from thence to Gibbon Hill from thence to High Carle which are the south bounderys between the Duke of Bolton's & Grinton / mannors & the which is bounderd on the west by Haverdale beck & saith that the said river Swale is the boundery betweene the mannors of Fremington & Grinton Marrick & Grinton & Marrick & Ellerton /

4 & 17 To the fourth & seaventeenth interrogatories saith that West Grinton & Harkaside is a hamlett consisting of some few stragling houses standing at a distance from each other saith that the present Lord Wharton ha's coppyhold lands lyeing at Harkaside Whitaside & Fremington the tennants whereof appear at his lordshipps court at Helaugh /

6 To the sixth interrogatory saith that this deponent ha's heard that the mannor of Grinton did formerly belong to the Priory of Bridlington /

125 In May 1708 James Fothergill (no age given) gave evidence for Marriott at the trial in London (LI Misc 536, Ward MSS, notebooks vol. 6, f. 113r).

7 To the seaventh interrogatory saith that he know's Swaledale & that it's in length about twelve myles is not a mannor of it selfe but ha's within it four mannors to witt Mukar & Helaugh belonging to the Lord Wharton Fremington to Madam Wharton & Grinton to Mr Marriott / saith the whole dale is in Grinton parish & that the church stands at Grinton the cheife towne in the dale /

10 To the tenth interrogatory this deponent saith that the inhabitants of West Grinton Harkaside Whitaside & Grinton have a rake for each others cattle on their severall moores & are not disturbed by each other saith the inhabitants of Harkaside claime a right of getting / peates turfe & ling on their owne moores as far as the Ox intack & the White Stone in the How being a division marke betweene them & Grinton Moores & commons /

Nicholas Stewardson of Fremington {of Fremington} in the county of Yorke miner aged fifty eight years or thereabouts sworne & examined saith /

2 8 11 & 12 To the second eight eleaventh & twelfth interrogatories this deponent saith that he know's the mannor or reputed mannor of Grinton & alsoe knows the places called West Grinton Harkaside & Whitaside & the moores & commons called Harka hill Harka Moore Grinton Moore Grinton Greetes Grinton / How or Howgroves & ha's knowne the same for forty five years & dureing all the said tyme the said places moores & commons have been commonly reputed & taken to belong to & be parte & parcell of the said mannor of Grinton & that the Swale's & Hillary's & of late Mr Marriott have been the reputed lords / of the said mannor saith that Hillary's wrought mynes of coale upon Grinton Moore & the Swale's wrought mynes of lead upon Harka or Harkaside Moore the place in question & Sir Solomon Swale had good gettings & continued to worke there till about eleaven years agoe that Mr Marriott / recovered the same from him & after that Mr Marriott wrought there & sunck about fourteene or fifteene new shafts & enjoyed the same till about two years agoe the present Lord Wharton entred upon him saith that both Hillary's & Swale's used as lords to take up waifes & estray's upon the said moores & / commons saith he know's the mannor of Helaugh & for all the said tyme of his knowledge the midd streame of the river Swale ha's been accounted & reputed the boundery betwixt the said mannors of Grinton & Helaugh saith he has lived as a near neighbour at Fremington all his / tyme & never knew or heard that the Lord Wharton or any of his ancestors ever {heard} wrought any mynes or claimed any royalties in West Grinton or Harkaside or in the moores or commons till about two years agoe /

3 & 9 To the third & ninth interrogatories this deponent saith that the said mannor of Grinton is bounderd on the east by the mannor of Ellerton at Stallerstone Style & on the south by the Duke of Boltons mannors & on the west by Haverdale beck & on the north by the river Swale as aforesaid / saith that the said midd streame of the <said> river Swale is alsoe generally accounted a boundery betwixt the adjacent mannors of Fremington & Grinton Marrick & Grinton Marrick & Ellerton Marrick & Stainton & beleives it is soe downwards /

4 & 17 To the fourth & seaventeenth interrogatories this deponent saith that the hamlett of West Grinton & Harkaside consists of a few stragling houses lyeing at <some> distance from each other saith the Lord Wharton ha's some coppyhold lands in Harkaside the vallue whereof this deponent / know's not & he ha's alsoe coppyhold lands in Whitaside & Fremington the tennants of all which appeare & doe suite & service at his lordshipps court at Helaugh /

10 To the tenth interrogatory this deponent saith that the inhabitants of Grinton Harkaside & Whitaside have used to have their cattle goe by way of a rake upon one anothers commons without any disturbance from each other /

John Hodgson of Fremington miner aged forty nine years sworne & examined saith /

2 11 & 12 To the second eleaventh & twelfth interrogatories this deponent saith that he know's the mannor or reputed mannor of Grinton & the places called West Grinton Harkaside & Whitaside & the moores & commons thereto belonging called Harka hill Harka Moore Grinton Moore Grinton Greetes / Grinton How or Howgroves & ha's soe knowne the said places moores & commons for forty years last past saith that the Swale's & the Hillary's have been the reputed lords of the said mannor dureing his knowledge till Mr Marriott recovered against Swales saith he was a miner & workeman under / the present Sir Solomon at Harka or Harkaside Moore the place in question & continued a workeman there under the said Sir Solomon till the said Mr Marriott turned him out of possession & that the said Sir Solomon had pretty good gettings since which tyme the said Mr Marriott has wrought the / said mynes both on the How & Harka the place in question & ha's had good gettings saith that he this deponent saw a mare in Christopher Naylors hands which was taken up on Grinton commons & brought in as a waife to Swale Hall where she remained above a yeare unclaimed & / became forfeited to Sir Solomon Swale as lord of the said mannor of Grinton saith that he know's the mannor of Helaugh & for the tyme aforesaid the midd streame of the river Swale ha's been the reputed boundery betweene the said mannors of Grinton & Helaugh as it is alsoe that between / the mannors of Fremington & Grinton Marrick & Grinton Marrick & Ellerton saith that he this deponent ha's lived most of his tyme at Fremington over against Grinton & frequently ha's been a miner & travelled through the moores & commons of Grinton West Grinton & Harkaside & / in all his tyme never heard that ever the Lord Wharton nor any of his ancestors ever wrought any myne or mynes for lead or coale in or upon any of the moores waists or commons of West Grinton or Harkaside or in any other place within the said mannor of Grinton /

3 & 9 To the third & ninth interrogatories this deponent saith that the said mannor of Grinton is bounderd on the east by the mannor of Ellerton at Stallerstone Style & on the south by the Duke of Boltons mannors & on the west by Haverdale beck & on the north by the midd streame of the river Swale /

4 & 17 To the fourth & seaventeenth interrogatories this deponent saith that the hamlett of Harkaside consists of a few stragling houses most standing at some distance from one another saith he beleives the Lord Wharton hath some coppyhold lands in Harkaside the particular / vallue whereof this deponent cannott set forth saith he hath alsoe some coppyhold lands in Whitaside & alsoe in Fremington all the tennants whereof appeare (as this deponent beleives) at his lordshipps court at Helaugh /

James Hall or Fremington in the county of Yorke miner aged forty two yeares or thereabouts sworne & examined saith /

2 8 11 & 12 To the second eight eleaventh & twelfth interrogatories this deponent saith that he know's the mannor or reputed mannor of Grinton & alsoe the places called West Grinton Harkaside & Whitaside & the moores & commons thereunto belonging & hath knowne the same for about thirty five / years & dureing all his tyme the said places moores & commons have been generally accounted to lye within & belong to the said mannor of Grinton & that the Swale's & Hillary's & of late Mr Marriott have been the reputed lords of the said mannors saith that Hillarye used to /

[membrane xiij *verso*][126]

worke mynes of coale upon Grinton Moore & Sir Solomon Swale wrought mynes of lead upon Harka or Harkaside Moore & had sometymes good gettings & continued to worke the same till about eleaven years agoe that the said Mr Marriott recovered the same from him & then Mr Marriott / became possessed thereof & wrought mynes there & enjoyed the same till about two yeares agoe that the present Lord Wharton entred upon him saith he ha's heard that the Swale's used to take up waifes & estrays upon the said moores & commons further saith he know's the mannor of Helaugh / & that the river Swale ha's for all his tyme been the parting or division betwixt lord & lord of the said mannors of Grinton & Helaugh as it is betwixt the adjacent mannors of Fremington & Grinton Marrick & Ellerton & soe downwards as this deponent beleives saith that he hath lived at / Fremington & been a near neighbour to the said places moores & commons ever since he was borne & never knew or heard that the present Lord Wharton or any of his ancestors ever wrought any mynes or claimed any royalties in West Grinton & Harkaside or any the said moores or commons till about <two yeares agoe> /

3 & 9 To the third & ninth interrogatories this deponent saith that the said mannor of Grinton is bounderd on the east by the mannor of Ellerton at Stallerstone Style & on the south by the Duke of Boltons mannors & on the west (as he ha's heard) by Haverdale beck /

4 & 17 To the fourth & seaventeenth interrogatories this deponent saith that the hamlett of West Grinton & Harkaside consists of a few houses most of them standing at some distance from one another saith that the Lord Wharton hath some coppyhold lands in Harkside & alsoe in Whitaside & Fremington the tennants of all / which (as this deponent beleives) appeare at his lordshipps court at Helaugh /

7 To the seaventh interrogatory this deponent saith that the dale or vale called Swaledale is about twelve or thirteene myles long saith it is not reputed a mannor of it's selfe but there are four mannors within it Vizt the mannors of Helaugh & Mukar belonging to the Lord Wharton the / mannor of Grinton belonging to Mr Marriott & the mannor of Fremington belonging to Madam Wharton all lyeing within the parish of Grinton & that Grinton (where the church stands) is the cheife towne there /

Thomas Hall of Fremington miner aged thirty four years sworne & examined saith /

2 11 & 12 To the second eleaventh & twelfth interrogatories this deponent saith that he know's the reputed mannor of Grinton & the places now called West Grinton Harkaside & Whitaside & the moores waists & commons called Harka hill Harka Moore Grinton Moore Grinton Greetes Grinton / How or Howgroves & ha's knowne the same for twenty years & upwards & saith that they are all reputed to belong to Grinton mannor & that the Swale's the Hillarys & Mr Marriott of late have been the reputed lords of the said mannor saith that he as a miner & workeman / wrought mynes for lead on Harka the place in question under the present Sir Solomon Swale & had pretty good gettings & continued a miner there untill Mr Marriott came into the possession of the said mines saith that he was a workeman & miner under Mr Marriott for about two years / who had very good gettings saith he saw a sheep brought in as waife to James Naylor a servant of Sir Solomon's which sheep was taken up as waife upon some parte of Grinton Moores & was loosed by Allen Lafeild know's Helaugh mannor saith the river Swale is a / parting or division between the

[126] Parchment, 730mm x 650mm.

said mannors of Grinton & Helaugh saith he ha's been a miner severall years & lived most of his tyme at Fremington & is well aquainted with the moores & commons of West Grinton & Harkaside & never knew or heard till about two years since / that ever the Lord Wharton or any of his ancestors ever wrought any mine or mines upon the said moores or commons of West Grinton or Harkaside or in any other place within the said mannor of Grinton for either lead or coale /

Richard Bowes[127] of Grinton whitesmith aged forty six or thereabouts sworne & examined saith /

2: 11: & 12: To the second eleaventh & twelfth interrogatories this deponent saith that hee knows the manor or reputed manor of Grinton & alsoe the places called West Grinton Harkaside & Whitaside & the moors & commons called Harka Hill Harka Moore Grinton Moore Grinton Greets / Grinton How or How Groves & has soe knowne the same for about thirty years & dureing the said time the said places & moors & commons have been commonly reputed to belong & be parte of the said manor of Grinton & that the Swales & Hillarys have been the reputed lords thereof / saith Hillarys wrought mines upon the How & the Swales wrought mines of lead upon Harka Hill or Harka Moore the place in question & Sir Solomon Swale had good gettings there & continued in possession thereof till Mr Marriott recovered the same / against him after that Mr Marriott sunck severall new shafts & enjoyed the same till about two yeares agoe that the present Lord Wharton enterd upon him saith both Swales & Hillarys used as lords to take upp waifes & estrays within the said moors & commons further / saith hee knows the manor of Helaugh & that the river Swale has been all his time reputed the boundery betwixt the manors of Grinton & Helaugh saith hee has lived from his youth a near neighbour to the said places & never knew or heard that the Lord Wharton / or any of his ancestors ever wrought any mines or claimed any royalties in West Grinton or Harkaside or any the said moors or commons till about two yeares agoe; /

3: & 9: To the third & ninth interrogatories this deponent saith that the said manor of Grinton is bounderd on the east by the manor of Ellerton at Stallerstone Stile & goes upp southwards to White Stone oth Sharrow thence to Snowden Man / where begins the boundaries betwixt the Duke of Boltons manors & the said manor of Grinton & soe proceeds to Brownsett thence to Grinton Currock so to Grinton Greets thence to Gibbon Hill so to High Carle thence to Pickerstin Rigg / thence to Wind Yate Currock thence to Bull head so to Bloody Wall & thence downe Blagill to Haverdale Beck which is the west boundary of the said manor of Grinton saith that the said river Swale is alsoe generally accounted a / division between lord & lord of the adjacent manor of Fremington & Grinton Marrick & Grinton Marrick & Ellerton Marrick & Stainton Marrick & Downham & beleives it is alsoe so downwards; /

7: To the seaventh interrogatory this deponent saith that the dale or vale called Swaledale is not accounted a manor of it selfe but there are four manors within it Vizt the manors of Helaugh & Mukar belonging to the Lord Wharton / the manor of Grinton belonging to Mr Marriott & the manor of Fremington belonging to Madam Wharton all lying in the parish of Grinton & that Grinton (where the church stands) is the chiefe towne therein; /

10: To the tenth interrogatory this deponent saith that the inhabitants of Grinton West Grinton Harkaside & Whitaside used to have their cattle goeing from each others

[127] In May 1708 Richard Bowse, aged 46, gave evidence for Marriott at the trial in London (LI Misc 536, Ward MSS, notebooks vol. 6, f. 112v).

commons by way of a rake without any disturbance from one another / saith that the inhabitants of West Grinton & Harkaside claime a right of getting ling peate & turfe upon their owne commons from the Grinton Moors or commons so farr as the Ox Intack & the White Stone on How / & that they claimed those to be boundary marks for that purpose betwixt the said commons of Harkaside & Grinton; /

John Arundale[128] of Harkaside aged thirty eight years sworne & examined saith /
2 11 & To the second eleaventh & twelfth interrogatorie this deponent saith that he know's
12 the reputed mannor of Grinton & the places called West Grinton Harkaside & Whitaside & the moores & commons thereto belonging called Harka hill Harka / Moore Grinton Moore Grinton Greetes Grinton How or Howgroves & ha's soe knowne the said places moores & commons for twenty years & upwards saith the same are reputed to be parcell of & to belonging to the said mannor of Grinton & that the Swale's & Hillary's / have been the reputed lords of the said mannor & that Sir Solomon Swale wrought mynes of lead at Harka or Harkaside Moore the place in question & saith that he this deponent was a miner under the said Sir Solomon & continued workeing at the mynes in question untill / Mr Marriott came into the possession of the said mynes saith that he ha's been a miner & workeman under Mr Marriott on Harka the place in question & continued working under him there untill the present Lord Wharton entred upon him about two years {agoe} since & till then / the said Sir Solomon Swale & the said Mr Marriott wrought the mynes in question without any interrupcion that ever this deponent knew or heard of saith that he beleives & ha's heard that the Swale's & the Hillary's both took up waifes & estray's on / West Grinton & Harkaside Moore saith that he know's the mannor of Helaugh & dureing the tyme aforesaid the midd streame of the river Swale was reputed to be a {boundery} division between Helaugh & Harkaside & saith that this deponent ha's lived / most of his tyme at Harkaside & that he & his three brothers ha's freehold & coppyhold lands there descended from their father called the intacks for which they pay a joynt rent to the Lord Wharton of eighteen shillings four pence per annum for which they / appeare at Helaugh court but does not know or can he be positive whether or noe the said lands lye in Grinton manor or Helaugh manor but saith the said intack lands were taken in before this deponents time nor did hee ever heare by his ancestors when the same / was taken in saith that hee this deponent & severall others of Harkaside people used as their right to gett stones in the river Swale as farr as the mid streame for keeping upp their banks & that the people on the Helaugh side used to come noe further / for their getting of stones & always either side tooke the said mid streame to be the division between them saith that the Harkaside people claime the sole right of commoning for their cattle & sheep to goe & depasture upon Harkaside Moors & that noe other persons has / right of commoning there as this deponent beleives & that about one year since & severall times before James Lonsdaile of Helaugh asked leave to lead stones from Harkaside Moore of this deponent & his brother saith that hee has lived all his time at Harkaside & / never knew or heard that ever the Lord Wharton nor any of his predecessors ever wrought any mine or mines upon West Grinton of Harkaside Moors till of late as aforesaid; /
4: & To the fourth & seaventeenth interrogatories this deponent saith that hee knows the
17: hamlett of West Grinton & Harkaside saith the same consists of about twenty four

[128] John Arrundell of Harkaside, a miner, swore an affidavit on 5 June 1707 (E207/114/2) to which he signs his name John Arandill. In May 1708 John Arundell, aged 'near 40', gave evidence for Marriott at the trial in London (LI Misc 536, Ward MSS, notebooks vol. 6, f. 113r).

houses lying some few together & others at a distance from each other saith that hee beleives the Lord Wharton[s] / coppy hold lands are not above a fifth or sixth parte of the said hamlett for that as he beleives in the said lords parte there are not above two fire houses of the said twenty four houses saith that the Lord Wharton has alsoe coppy hold lands at Whitaside and / Fremington the tennants whereof appeare at his lordshipps court at Helaugh as alsoe some freeholders of Harkside saith there are alsoe Crown lands Priory lands & some freehold lands in Harkaside that doe not appeare at Helaugh court; /

10: To the tenth interrogatory this deponent saith that the inhabitants of Harkaside Whitaside & Grinton intercommoned with each other cattle on their respective moors & commons without disturbance & that this deponent & the other inhabitants of Harkaside claimed a sole / right upon Harkaside commons for the getting of ling peats & turfe & that the Ox Intack nooke the Gray Stone in the How the Gray stone in the Greets were bounder marks to that purpose saith that hee never heard them called bounder marks for any / other end till about two years since that the present Lord Wharton caused a boundary to be rid on the said moors; /

William Lee[129] of Grinton in the county of Yorke miner aged thirty eight years or thereabouts sworne & examined saith /

2: 8:11: To the second eigth eleaventh & twelfth interrogatories this deponent saith that hee
& 12: knows the manor or reputed manor of Grinton & alsoe knows the places called West Grinton Harkaside & Whitaside & the moors & commons called Harka Hill Harka Moore / Grinton Moore Grinton Greets Grinton How or How Groves & has knowne the same for neare thirty years & that during all the time of this deponents knowledge the said places moors & commons have been reputed to lye within & belong to the manor of Grinton & that / the Swales & Hillarys & of late Mr Marriott have been the reputed lords of the said manor & that both Hillarys & Swales as lords used to take upp waifes & estrays upon the said moors & commons saith hee knows the manor of Helaugh & that during the said time of his / remembrance the river Swale has been commonly reputed to be the ancient boundery betwixt the said manors of Grinton & Helaugh saith hee has lived all his time at Grinton & never heard that the Lord Wharton or any of his ancestors ever wrought any mines / or claimed any royalties in West Grinton or Harkaside or in any the said moors or commons till about two yeares agoe saith the Hillaries used to worke mines upon the How & Sir Solomon Swale wrought mines of lead upon Harka Hill or Harkaside Moor / the place in question & had good gettings there till Mr Marriott recovered the same & got the possession thereof & afterwards Mr Marriott sunck <about> eighteen new shafts there & had very good gettings & enjoyed the same till about two years agoe that the Lord / Wharton enterd upon him; /

3 & 9: To the third & ninth interrogatories this deponent saith that the said manor of Grinton is bounderd on the east by the manor of Ellerton begining at the mid streame of the river Swale & soe goes upp to Stallerstone Stile & soe to the Gray Stone on Sharrow / thence to Snowdon Man (where begins the boundery between the said manors of Grinton & the Duke of Boltons manors) & soe proceeds to Brownsett thence to Redmere Standard thence to Grinton Greets thence to Gibbon Hill thence to high Carle on / which still continues the boundery betwixt the said

[129] William Lee of Grinton, a miner, swore an affidavit on 5 June 1707 (E207/114/2). In May 1708 William Lee, aged 40, gave evidence for Marriott at the trial in London (LI Misc 536, Ward MSS, notebooks vol. 6, f. 113r).

manor of Grinton & the Duke of Boltons manors & thence proceeeds to Pickerstin Rigg thence to Wind Yate Currock thence to Sumerlodge Tarn Currock thence to Bullhead thence to Bloody / Wall & soe down to Haverdale Beck & from thence to the mid streame of the said river Swale which is the west boundery of the said manor of Grinton & soe the said mid streame is the north boundery of the said manor all along down to Stallerstone Stile saith / that the said mid streame of the said river Swale is alsoe the comon boundery of the adjacent manors Vizt the manors of Fremington & Grinton Marrick & Grinton Marrick & Ellerton Marrick & Stainton & beleives it to be soe downwards saith hee / has heard ancient men in his neighbourhood & alsoe his father now dead who was betwixt fifty & sixty years old discourse of the said mid streame of the said river to be a boundary as aforesaid /

4: & To the fourth & seaventeenth interrogatories this deponent saith that hee knows the
17: Hamlet of West Grinton & Harkaside & that the same consists of a few houses some standing at a distance from one another & that the Lord Wharton ha's copyhold lands there as hee / has alsoe at Whitaside & Fremington the tennants of all which (as this deponent has heard) appeare at his lordshipps court at Helaugh & that all the manor of Fremington ly's within & belongs to the constablery of Reeth /

7: To the seaventh interrogatory this deponent saith that the dale or vale called Swaledale is about twelve miles long that hee never heard it called a manor of itselfe but there are four manors within the same Vizt the manors of Helaugh <& Mewkar> belonging to / the Lord Wharton the manor of Grinton belonging to Mr Marriott & the manor of Fremington belonging to Madam Wharton all lying in the parish of Grinton & that the towne of Grinton (where the church stands) is the cheife towne therein /

10: & To the tenth & thirteenth interrogatories this deponent saith that the inhabitants of
13: Grinton Harkaside & Whitaside used to have their cattle going from one anothers commons by way of a rake without interrupcion saith that the inhabitants of Harkaside have claimed / a right of getting ling peats & turfe upon their own commons separate from the commons of Grinton so farr as the marks called the ox Intack & the White Stone on the How & has heard those were division marks for that purpose betwixt the said commons but / never knew or heard they were <any> boundary marks for any other purpsoe till they were pretended soe to be in the rideing of a boundary by the said Lord Whartons agents about two years agoe; /

[membrane xiiij *recto*][130]

George Emerson[131] of Grinton miner aged thirty six years or thereabouts sworne & examined saith /

2: 8: 11: To the second eigth eleaventh & twelfth interrogatories this deponent saith that he
& 12: knows the reputed manor of Grinton & the places called West Grinton Harkaside & Whitaside & the moors & commons called Harka Hill Harka Moor Grinton Moore Grinton Greets Grinton How or / How Groves & has knowne the same for twenty six years & that during all the time of this deponents knowledge the said places moors waists & commons have been reputed to be within & to belong to the said manor of Grinton & that the Swales the Hillarys & of late Mr Marriott / have been the reputed lords of the said manor & that both Hillarys & Swales as lords used to take upp waifes

130 Parchment, 730mm x 650mm. The sheet is age stained but fully legible.

131 In May 1708 George Emerson (no age given) gave evidence for Marriott at the trial in London (LI Misc 536, Ward MSS, notebooks vol. 6, f. 113r).

& estrays upon the said moors & commons saith that the Hillarys used to worke mynes upon the How & Sir Solomon Swale wrought mines of lead on Harka on / Harkaside Moore the place in question & had good gettings there till Mr Marriott recovered the same & got the possession thereof & afterwards Mr Marriott wrought the said mines & sunck about eighteen shafts on Harka & had very good gettings & enjoyed the same / till about two years agoe that the Lord Wharton enterd upon him saith that he is a miner & workman under Mr Marriott which is the reason why hee is able to depose as aforesaid saith hee knows the manor of Helaugh & that dureing the said time of this deponents / remembrance the river Swale ha's been commonly reputed to be the knowne boundary between the said manors of Grinton & Helaugh saith that he ha's lived for the most parte of his time at Grinton & never knew or heard that the Lord Wharton or any of his / ancestors ever wrought any mines or claimed any royalties in West Grinton or Harkaside or any the said moors or waists or commons till about two yeares agoe; /

3: & 9: To the third & ninth interrogatories this deponent saith that the said manor of Grinton is bounderd on the east by the manor of Ellerton begining at the mid streame of the river Swale & soe goes upp to Stallerstone Stile & soe to the Gray Stone on the Sharrow thence to Snowdon Man / (where begins the boundary betwixt the said manor of Grinton & the Duke of Boltons manors) & soe proceeds to Brownsett thence to Redmere Standard thence to Grinton Greets thence to Gibbon Hill thence to High Carle which still continues the boundery betwixt the said / manor of Grinton & the Duke of Boltons manors & thence proceeds to Pickerstone Rigg thence to Wind Yate Currock thence to Sumerlodge Tarn Currock thence to Bullhead thence to Bloody Wall & soe down to Haverdale Beck the west boundary & from thence to the mid streame / of the river Swale which is the north boundary of the said manor all along down to Stallerstone Stile saith that the mid streame of the said river Swale is alsoe the common boundary of the adjacent manors to witt the manor of Fremington & Grinton Marrick & Grinton Marrick / & Ellerton Marrick & Stainton & beleives it to be soe downwards saith hee has heard ancient men in his neighbourhood alsoe his father now dead & who was old discourse of the river Swale being a boundary as aforesaid /

4: & 17: To the fourth & seaventeenth interrogatories this deponent saith that the hamlett of West Grinton & Harkaside consists of a few strागling houses some whereof stand at a distance from each other & that the Lord Wharton has some coppyhold lands there as alsoe some at Whitaside the / tennants of all which appeare at his lordshipps court at Helaugh saith that Fremington manor lies in the constablery of Reeth; /

7: To the seaventh interrogatory saith the Swaledale is about twelve miles in length & ha's within it four manors to witt Helaugh Mukar Fremington & Grinton that the whole dale is in Grinton parish & that the church stands at Grinton /

10: To the tenth interrogatory this deponent saith that the inhabitants of West Grinton Harkaside Whitaside & Grinton has a rake for each others cattle on each others commons without disturbance & that Harkaside claime a sole right in their commons for the getting of peats / turfe & ling /

13: To the thirteenth interrogatory this deponent saith that hee never knew or heard of any boundary being rid between West Grinton Harkaside Whitaside & Grinton & the moors & commons thereof till about two years since that the Lord Wharton caused a boundary to be rid between the said places moors / & commons saith that hee saw the said boundary rid but beleives it not to be a true boundary to the purpose it was rid for that he never heard of any such boundary before; /

Nicholas Tomlin[132] of Midleham in the county of Yorke miner aged thirty eight or thereabouts sworn & examined saith /

2: 8: 11: To the second eigth eleaventh & twelfth interrogatories this deponent saith that he
& 12: knows the manor or reputed manor of Grinton & alsoe knows the places called West Grinton Harkaside & Whitaside & the moors & commons called Harka Moore Grinton Moore Grinton Greets / Grinton How & the How Groves & has knowne the same for about twenty five years & for all the time of his said knowledge the same have been commonly reputed to belong to & be parte of the said manor of Grinton & that the Swales & Hillarys & of late Mr Marriott have been the / reputed lords of the said manor saith that hee has heard by ancient neighbours that Hillarys wrought mines of lead upon Harka Moore & alsoe mines of cole upon Grinton Moore and that hee this deponent knows that Sir Solomon Swale wrought mines of lead upon Harka / Moore the place in question & had good gettings there & continued to worke the same till Mr Marriott recovered the same from him & afterwards the said Mr Marriott sunck eighteen new shafts upon the said Harka Moore & had good gettings there & continued in possession therof / till about two years agoe that the present Lord Wharton enterd upon him saith that the Swales used to take upp waifes & estrays upon the said moors & commons & has heard that Hillarys used alsoe to take upp waifes & estrays there further saith that hee this deponent knows / the manor of Helaugh & that for all the said time of his remembrance the mid streame of the river Swale has been comonly reputed the boundary betwixt the said manors of Grinton & Helaugh saith hee has for twenty years last past been frequently upon the said moors / & commons & for some time wrought in the mines there & never knew or heard that the present Lord Wharton or any of his ancestors ever wrought any mines or claimed any royalties in West Grinton or Harkaside or any of the said moors or commons till about two / years agoe saith hee has heard that the mines upon the said moors & commons were formerly in Mr Wharton of Gilling by lease from the Crown /

3: & 9: To the third & ninth interrogatories this deponent saith that the said manor of Grinton is bounderd on the east by the manor of Ellerton at Stallerstone Stile & on the west by Haverdale Beck & on the south by the Duke of Boltons manors & on the north from the manor of Helaugh / by the river Swale as aforesaid saith that the mid streame of the said river Swale is alsoe commonly reputed to be the boundery betwixt the adjacent manors of Fremington & Grinton Marrick & Grinton Marrick & Ellerton & beleives it may be soe downwards; /

4: & To the fourth & seaventeenth interrogatories this deponent saith that the hamlett
17: of West Grinton & Harkaside consists of some few houses most of them standing stragling from one another saith the Lord Wharton hath some copyhold lands at Harkaside & alsoe at Whitaside / & Fremington all the tennants whereof as this deponent beleives appears at his lordshipps court at Helaugh and further saith that all Fremington lies within the constablery of Reeth /

7: To the seaventh interrogatory this deponent saith that the dale or vale called Swaledale is as this deponent has heard in length about twelve miles that it is not reputed a manor of it selfe but that there are four manors therein Vizt the manors of Helaugh & Mukar / belonging to the Lord Wharton the manor of Grinton belonging to Mr Marriott & the manor of Fremington belonging to Madam Wharton all lying in

[132] Nicholas Tomlin of Midleham, a miner, swore an affidavit on 5 June 1707 (E207/114/2) to which he signed his name with the letter 'N'. In May 1708 Nichloas Tomblin, aged 39, gave evidence for Marriott at the trial in London (LI Misc 536, Ward MSS, notebooks vol. 6, f. 113r).

the parish of Grinton & that Grinton where the church stands is the cheife town therein /

10: & 13 To the tenth & thirteenth interrogatories this deponent saith that the inhabitants of Grinton Harkaside & Whitaside used to have their cattle going upon each others commons without disturbance from one another saith that he has heard of the boundery rid / by the Lord Wharton about two years agoe betwixt the moors & commons of Grinton Harkaside Whitaside & never in all his time heard of a boundary rid there or thereabouts for that purpose; /

Roger Baley of Dycas[133] of miner aged sixty years sworne & examined saith /

2: 11: & 12: To the second eleaventh & twelfth interrogatories this deponent saith that he knows the manor or reputed manor of Grinton & the places called West Grinton Harkaside & Whitaside & the moors & commons thereto belonging called Harka Hill Harka Moore / Grinton Moore Grinton Greets Grinton How or How Groves & has soe kowne the said places moors waists & commons for above thirty five years last past saith that the same are reputed to lye in the said manor of Grinton & that the Swales the Hillarys / & Mr Marriott of late were the reputed lords of the said manor saith that Sir Solomon Swale wrought mines of lead both on the How & Harka Hill the place in question & had very good gettings in both places saith he continued working till Mr Marriott / recovered the said mines since which time the said Mr Marriott wrought the said mines & had good gettings in both places & the reason of such this deponents knowledge is that hee has been a miner under the said Mr Marriott since his being in / possession of the said mines & saith that during this deponents working at the said mines there was never no disturbance till about two years since that the present Lord Wharton enterd upon the said mines in question saith that hee this deponent has / knowne waifes & estrays taken upp on Harkaside Moore & carryed to the servants or agents both for the Swales & the Hillarys saith that he knows the manor of Helaugh & that during the time aforesaid the mid streame of the river Swale has been the reputed / boundary between the said manor of Grinton & Helaugh saith that hee this deponent lived at Gunnerside in Swaledale or nigh hand it for above twenty years & has lived at Harkaside above tenn years & in all that time never knew or heard that / the present Lord Wharton or any of his ancestors did ever worke any mines or claime any royalties in West Grinton or Harkaside moors or in any other place within the said manor of Grinton till about two yeares since as aforesaid; /

10: To the tenth interrogatory saith that the inhabitants of Grinton West Grinton Harkaside & Whitaside have a rake for their cattle thro' each others commons without interruption & that as to the getting of peats turfe & ling the inhabitants of West / Grinton & Harkaside claime a sole right upon their own commons distinct from Whitaside & Grinton & saith the White Stone on the How is a division marke for the divideing the said commons of Harkaside & Grinton for the getting of ling / peate & turfe & for noe other purpose that this deponent knows of /

13: To the thirteenth interrogatory saith that about two years agoe he beleives the present Lord Wharton caused a boundary to be rid of the moors & commons of West Grinton & Harkaside was not at the said riding but saw the same rid saith hee / never heard till then of any boundary being rid of the said moors & commons & beleives the bounder marks which they then agreed to be the bounder marks for the dividing the said moors commons & inclosures of West Grinton & Harka / =side from Whitaside

133 By 'Dycas' is meant 'Dyke House', a farm and farm house in Harkaside.

& Grinton was only marks that divided the commons of West Grinton & Harkaside for the getting of peats turfe & ling as aforesaid & to noe other purpose; /

Leonard Hudson[134] of Fremington in the county of Yorke miner aged thirty years or thereabouts sworne & examined saith; /

2: 8: 11: To the eighth[135] eleaventh & twelfth interrogatories this deponent saith that hee knows
& 12: the manor or reputed manor of Grinton & alsoe the places called West Grinton Harkaside & Whitaside & the moors & commons called Harka hill Harka Moor Grinton Moore Grinton Greets Grinton / How or How Groves & has kowne the same for about twenty years & that dureing all the said time the said places moors & commons have been commonly reputed to belong to & be parte of the said manor of Grinton & that the Swales & Hillarys & of late Mr Marriott have been / the reputed lords of the said manor saith that the present Sir Solomon Swale wrought mines of lead upon Harka or Harkaside Moore & had very good gettings there (this deponent having wrought under him there for some time) and hee the said Sir Solomon continued / in the possession thereof till about eleaven years agoe that the said Mr Marriott recovered the same from him & afterwards the said Mr Marriott sunck about fifteen new shafts there & for some time had pretty good gettings & hee this deponent wrought there for severall / years under the said Mr Marriott who continued in the possession thereof till about two years agoe that the present Lord Wharton enterd upon him saith hee hath known waifes & estrays taken upp both by the said Sir Solomon & the Hillarys & likewise of late by the said Mr Marriott as lords / upon the said moors & commons & perticularly the said Sir Solomon has taken upp severall waifes & estrays upon Harka Moore or Harkaside & this deponents wife told him that her brother carried one or two waife sheep from Harkaside Moore to the agent of Mr Hillary / saith hee knows the manor of Helaugh & that the mid streame of the river Swale has for all his time been commonly reputed & accounted by ancient men in the neighbourhood to be the boundery betwixt the manors of Grinton & Helaugh saith he has lived / all his time except for about two years at Fremington & been a near neighbour to & frequently upon the said places moors & commons & never knew or heard that the present Lord Wharton or any of his ancestors ever wrought any mines or claimed any royalties / in West Grinton or Harkaside or any the said moors & commons till about two yeares agoe /

3: & 9: To the third & ninth interrogatories this deponent saith that the said manor of Grinton is bounderd on the east by the manor of Ellerton beginning at the mid streame of the said river Swale & soe goes upp to Stallerstone Stile thence to the White Stone otha Sharrow thence / to Snowdon Man (where begins the bounder betwixt the said manor of Grinton & the Duke of Boltons manors) & soe proceeds to Brownsett thence to Redmere Standard thence to Grinton Greets thence to Gibbon Hill thence to High Carle thence to Pickerstin Rigg / (soe farr the south bounderies betwixt the said manor of Grinton & the Duke of Boltons manors) & soe proceeds to Tarnsant thence to Wind Yate Currock thence to Sumerlodge tarn Currock thence to Bullhead so to Bloody Wall soe down Blagill & soe down to / Haverdale Beck & thence to the mid streame of the said river Swale which is the west boundary of the said manor of Grinton & the north boundary of the said manor is the mid streame of the said river downe from Haverdale Beck as aforesaid to neare Grinton betwixt the said manors of Grinton / & Helaugh & soe downe to Stallerstone Stile as aforesaid

134 Leonard Hudson of Fremington, a miner, swore an affidavit on 5 June 1707 (E207/114/2) which he signed with his own name. In May 1708 Leonard Hudson, aged 31, gave evidence for Marriott at the trial in London (LI Misc 536, Ward MSS, notebooks vol. 6, f. 113r).

135 The word 'second' omitted in the original.

& this deponent has for all his time knowne the said boundaries & well remembers when he was a boy at school he has been shewn parte of the said boundaries as aforesaid & gone there in procession with other school boys & / parishioners[136] saith that the mid streame of the said river Swale is & has been alsoe generally accounted the boundery betwixt the manors of Fremington & Grinton Marrick & Grinton Marrick & Ellerton & beleives it to be soe downwards saith that this deponent hath / frequently gone a fishing with his father & grandfather who are both dead about twenty years agoe & his this deponents father has been carefull in lying his fishing hooks in the said river Swale betwixt Harkaside & Helaugh & often said hee durst not lye them on / the northside of the said mid streame for feare of the Lord Whartons stewards or agents and his this deponents grandfather alsoe used to observe the mid streame of the said river in his fishing alsoe betwixt the said <severall> manors & this deponent has often heard them say that [the] / mid streame of the said river was always a boundary betwixt the said severall manors as aforesaid Alsoe saith that one Alderidge Swale uncle to the said Sir Solomon used frequently about twelve years agoe to tell this deponent & other miners that was working together of the said boundary / of the said manor of Grinton as before sett forth /

4: & 17: To the fourth & seaventeenth interrogatories this deponent saith that the hamlett of West Grinton & Harkaside consists of about twenty four houses some lying at a distance from one another saith that in West Grinton & Harkaside there are some Crown lands & some Priory lands / & some freehold lands & the Lord Wharton hath alsoe some coppyhold lands there & beleives he has alsoe some little freehold lands but beleives he has not above two or three of the said houses belonging to all his estaite there saith the Lord Wharton hath alsoe coppy / hold land in Whitaside & alsoe in Fremington the tennants of all which (as this deponent has heard & beleives) appeare at his lordshipps court at Helaugh & saith that all Fremington lyes within the constablery of Reeth /

7: To the seaventh interrogatory this deponent saith that the dale or vale called Swaledale is about twelve miles in length that it is not a reputed manor of it selfe but that there are four manors therein Vizt the manors of Helaugh & Mukar belonging to the Lord Wharton / the manor of Grinton belonging to Mr Marriott & the manor of Fremington belonging to Madam Wharton all lying in the parish of Grinton & that Grinton where church stands is the cheife towne therein /

10: & To the tenth & thirteenth interrogatories this deponent saith that the inhabitants of
13 Grinton Harkaside & Whitaside used to have their cattle going from one anothers commons by way of a rake without disturbance from each other saith the inhabitants of Harkaside claime / a right of getting ling peats & turfe upon their own commons separately from Whitaside & Grinton men & that the Harkaside men claime the said right upon their said commons so farr as the marks of Ox Intack & the White Stone in the How & this deponent has been / shewn the said marks by one of the inhabitants of Harkaside to be the division marks betwixt their commons & the commons of Grinton for that purpose saith that hee this deponent saw the pretended boundary rid about two years agoe by the Lord Whartons agents betwixt the / said places moors & commons but never heard of riding any such boundary by the Lord Wharton or any on his behalfe before that time saith he cannot beleive the same to be a boundary for any other intent or purpose but only observed as {a} division / marks betwixt

136 The traditional day for beating the parish bounds was on Holy Thursday (Ascension Day), forty days after Easter.

each others commons for getting ling peat & turfe upon their respective commons separately from each other; /

Robert Waller of Dycas[137] yeoman aged sixty one years or thereabouts sworne & examined saith /

11: & 12 To the eleaventh & twelfth interrogatories this deponent saith he knows the mines in question claimed by the defendant the Lord Wharton as alsoe by the relator Mr Marriott as lessee under the Crowne as this deponent beleives saith the same lye on Harka or Harkaside Moore & were reputed to belong / to the Swales & Hillarys as lords of Grinton who enjoyed the same for forty years last past save for the time Mr Marriott has been in the possession of them & for all the time aforesaid the present Lord Wharton nor any of his ancestors ever were in the possession of or claimed any / right or title to the same or any parte thereof that ever this deponent knew or heard of till about two years since that the present Lord Wharton enterd upon the said mines in question saith {saith} <that> hee knows the places called Harka alias Harkaside & West Grinton & the moors & waists / thereunto adjoining & the places called Harka Hill Harka Moore Grinton Moore Grinton Greets Grinton How or How Groves & has soe knowne the same for forty years last past & that they are reputed to belong to the manor of Grinton saith that hee this deponent never heard of / any person that wrought any mines for lead or cole in any of the said places moors waists & commons but the Swales the Hillarys & Mr Marriott till about two years since that the Lord Wharton enterd upon the said mines in question on Harka on Harkside Moore as aforesaid /

13: To the thirteenth interrogatorie this deponents saith that hee knows the present Lord Wharton caused a boundary to be rid two years agoe between the places of West Grinton & Harkaside saith till then hee never heard that any such boundary was rid and that when this deponent was / a boy & that the inhabitants of Grinton offerred to gett peats & turfe on Harkaside Moore then the said inhabitants of Harkaside rid a boundary between the said moors & commons begining at the Ox Intack for ascertaining their right of their sole getting peats ling & / turfe upon their own moors & further saith that hee beleives the Ox Intack the Gray Stone on the How & the Gray Stone in the Greets were boundarys or markes for that purpose as this deponent has severall times heard; /

Thomas Wood[138] of Grinton in the county of Yorke miner aged thirty seaven years or thereabouts sworne & examined saith /

2: 11: & 12: To the second eleaventh & twelfth interrogatories this deponent saith that hee knows the manor or reputed manor of Grinton & alsoe knows West Grinton Harkaside & Whitaside & the moors & commons called Harka Hill Harka Moore Grinton Moor Grinton Greets Grinton How or Howgroves /

[membrane xiiij *verso*]

& ha's knowne the same for twenty six years & that for all his time the said places moors & commons have been reputed to lye within & to belong to the said manor of Grinton & that the Swales & Hillarys & of late Mr Marriott have been the reputed lords of the said manor & / saith Hillary wrought some mines upon the How & Sir Solomon Swale wrought mines of lead upon Harka Hill or Harkaside Moore the place

137 By 'Dycas' is presumably meant 'Dyke house', a farm and farm house in Harkaside.

138 Thomas Wood of Grinton, a miner, swore an affidavit on 5 June 1707 (E207/114/2) which he signed with the letter 'T'.

in question & had very good gettings & continued to work the same till Mr Marriott recovered the same from him & after / that Mr Marriott wrought the said mines & sunck severall new shafts & continued in possession thereof till about two years agoe that the said Lord Wharton enterd upon him saith that both Hillarys & Swales as lords used to take upp waifes & estrays upon the said moors & commons / as hee beleives & has seen severall waifes & estrays in the custody of their agents saith the hee knows the manor of Helaugh that for all his time the river Swale has been accounted the boundery betwixt the said manors of Grinton & Helaugh and / further saith hee ha's lived all his time (except one yeare) at Grinton & has been very frequently upon the said moors & commons & never knew or heard that the present Lord Wharton or any of his ancestors wrought any mines or claimed any royalties in / West Grinton or Harkaside or any the said moors or commons till about two years agoe /

3: & 9: To the third & ninth interrogatories this deponent saith that the manor of Grinton is reputed to be bounderd on the east by the manor of Ellerton begining at the mid streame of the river Swale & goes upp to Stallerstone Stile & thence to the White Stone oth Sharrow thence / to Snowdon Man (where begins the boundery betwixt the said manor of Grinton & the Duke of Boltons manors) & soe proceeds to Brownsett thence to Redmere Standerd thence to Grinton Greets thence to Gibbon Hill thence to High Carle thence to Pickerstin / Rigg those said marks being the south boundery marks betwixt the said manor of Grinton & the Duke of Boltons manors & soe the boundery of the said manor of Grinton proceeds westwards to Tarnsant thence to Wind Yate Currock thence to Bull / head thence to Bloody Wall & soe down to Haverdale Beck & so downe to the mid streame of the said river Swale which is the west boundery of the said manor of Grinton & the said manor of Grinton is bounderd on the north from the said manor of / Helaugh from Haverdale Beck by the said mid streame of the said river Swale down to neare Grinton & soe to Stallerstone Stile as aforesaid saith that the said mid streame of the river Swale is alsoe commonly reputed the boundery betwixt the / adjacent manors of Fremington & Grinton Marrick & Grinton Marrick & Ellerton Marrick & Downham & beleives it to be soe downwards saith that hee this deponent lived with one Christopher Naylor about twenty years agoe who farmd / the milne at Swale Hall of the Swales & this deponent has heard him severall times say that the said river Swale was a boundery as aforesaid downe to Richmond; /

7: To the seaventh interrogatory this deponent saith that the dale or vale called Swaledale is in length about twelve miles that it is not a reputed manor of its selfe but there are four manors within it Vizt the manors of Helaugh & Mukar belonging to Lord / Wharton the manor of Grinton belonging to Mr Marriott & the manor of Fremington belonging to Madam Wharton all lying in the parish of Grinton & that Grinton (where the church stands) is the cheife towne in the said dale; /

10: & To the tenth & thirteenth interrogatories this deponent saith that the inhabitants of
13: Grinton Harkaside & Whitaside used to have their cattle goeing from one anothers commons by way of a rake without interruption or disturbance from one another / saith that the inhabitants of Harkaside have alwaies claimed a separate right upon their owne commons of getting peats ling & turfe thereon & that neither the Grinton or the Whitaside men should have any such right upon the said commons of Harkaside but / that the marks called the Ox Intack & the White Stone on the How have been all along the divisions or boundery marks betwixt the said moors & commons of Harkaside & Grinton for that purpose of getting ling peats & turfe upon their moors & commons / separately & distinctly from Grinton Moors saith that hee see the agents of the Lord Wharton ride a pretended boundery about two years agoe by the Ox Intack & the White Stone oth How & soe forwards to Grinton Greets saith

he never knew or heard / that the Lord Wharton or any of his ancestors ever rid the said boundery before nor did this deponent ever heare that the said boundery or division marks were for any other purpose but for division marks betwixt the said moors & commons that the said / inhabitants claime for getting their ling peats & turfe separately upon their severall commons; /

Alexander Dunn of Grinton miner aged fifty years sworne & examined saith /

2: 11: To the second eleaventh & twelfth interrogatories this deponent saith that he knows
& 12 the manor or reputed manor of Grinton & the places called West Grinton Harkaside & Whitaside & the moors & commons thereto belonging called Harka Hill Harka Moore / Grinton Moore Grinton Greets Grinton How or How Groves & has soe knowne the said places moors & waists for thirty years last past saith the same are reputed to belong to the manor of Grinton & that the Swales the Hillarys & Mr Marriott of late have been / the reputed lords of the said manor saith that Sir Solomon Swale wrought mines for lead on Harka on Harkaside Moore the place in question & had good gettings & continued working untill Mr Marriott recovered them from him saith that Mr Marriott wrought / the said mines after hee came into possession & had good gettings till the present Lord Wharton about two years since got into possession of the said mines in question saith that hee this deponent lived about thirty years at Grinton & has been both a miner under the present / Sir Solomon Swale as alsoe under Mr Marriott & wrought in the said mines in question for them both without the least disturbance till the present Lord Whartons entry upon the mines in question as aforesaid and for all the time aforesaid never heard that the / present Lord Wharton nor any of his ancestors ever wrought any mynes for either lead or coales upon or in the moores waists or commons of West Grinton or Harkaside or in any other place within the said mannor of Grinton till the entry as aforesaid by the present Lord Wharton / saith that {the} waifes & estrayes have beene taken up upon Grinton Moores both by the Swailes & the Hillaryes & knew a mare taken up in Grinton street & carried as waife to Roger Hillary as lord of Grinton saith that he knows Helaugh mannor & that the river Swaile / is the parteing or division betweene Grinton & Helaugh but saith that till of late he never heard the river Swaile called <a> boundary /

7th To the 7th interrogatory saith that he knowes Swailedale & that is in length about twelve miles & has <in it> foure mannors to witt Helaugh & Mukar belonging the Lord Wharton Fremington to Madam Wharton & Grinton to Mr Marriott the whole dale lyes in the / parish of Grinton & the parish church stands at Grinton /

10th & To the tenth <& thirteenth> interrogatories this Deponant saith that the inhabitants
13th of Grinton West Grinton Harkaside & Whitaside have allways had a rake upon each others commons for theire cattle without disturbance & saith that he has knowne for tenn yeares / last the tennants & the inhabitants of West Grinton & Harkaside claimed a separate right of their owne common for the getting of peets ling & turf & that the White Stone on the How was one of their devision markes for that purpose saith that he never heard of any / boundary markes or stones to divide the moores waists commons & inclosures of Harkaside or West Grinton Whitaside & Grinton till aboute two years since that he <heard> the present Lord Wharton rid a boundary betweene the said places & that they then rid up by the / Elm tree to Jenny Wrens Cow House the White Stone on the How & soe to the Greets never heard of the rideing of the said boundary before & beleives the said markes to be onely for the devideing the said moores & commons; /

James Close of Whitaside in the county of Yorke yeoman aged sixty years or thereabouts sworne & examined saith /

2: 11: & 12 To the second eleaventh & twelfth interrogatories this deponent saith that he knows the manor or reputed manor of Grinton & alsoe knows the places called West Grinton Harkaside & Whitaside & the moors & commons called Harka Hill Harka Moore Grinton Moore Grinton / Greets Grinton How or How Groves & has knowne the same for about fifty years & that for all the time of this deponents knowledge the said places moors & commons have been reputed to belong to & to be parcell of the said manor of Grinton & that the Swales & Hillarys / have been the reputed lords of the said manor saith that the Hillarys wrought West[139] ore upon the How & mines of cole upon Grinton Moore & the Swales wrought mines upon Harka or Harkaside Moore the place in question & Sir Solomon Swale had very good / gettings there till about eleaven years agoe that Mr Marriott recovered the same from him & from that time the said Mr Marriott wrought the said mines & sunck severall new shafts & continued in the possession thereof till about two years agoe that the / present Lord Wharton turned him out of possession saith that Sir Solomon Swale used to take upp waifes & estrays upon the said moores & commons & perticularly upon Harkaside Moore & that hee has seen waifes & estrays in the possession of the / agents of Hillary which this deponent beleives were taken upp upon some of the said moors & commons saith hee has lived all his time at Whitaside except about three years & been a neighbour to & frequently upon the said moors & commons & / never heard that the Lord Wharton or any of his ancestors ever wrought any mines or claimed any royalties in West Grinton or Harkaside or any the said moors & commons till about two years agoe /

7: To the seaventh interrogatory this deponent saith that the dale or vale called Swaledale is about twelve miles long that it is not reputed a manor of it selfe but there are four manors within it Vizt the manors of Helaugh & Mukar belonging to the Lord Wharton the manor / of Grinton belonging to the said Mr Marriott & the manor of Fremington belonging to Madam Wharton all lying in the parish of Grinton & that Grinton (where the church stands) is the cheife towne in the said dale. /

James Raw of Summerlodge <yeoman> aged fifty four years or thereabouts sworne & examined saith /

2: & 10: To the second & tenth interrogatories this deponent saith that hee knows the manor of Grinton & alsoe knows Grinton now called West Grinton Harkaside & Whitaside & has soe knowne the same for thirty years last past saith they are reputed to belong to the said manor of Grinton / & that Sir Solomon Swale & the Hillarys have been the reputed lords thereof saith that Sir Solomon Swale wrought mines on Harka on Harkaside Moore the place in question & had pretty good gettings saith that hee has heard that both the Swales & Hillarys have / taken upp waifes & estrays on Grinton moors & commons never heard the river Swale was or was not called a boundary between Helaugh & Grinton saith that for thirty years last past this deponent never heard that the present Lord Wharton nor any of / his ancestors ever wrought any mines or claimed any royalties in Grinton or Harkaside Moores till about two years since saith that the inhabitants of Harkaside Whitaside & Grinton have a rake for each others cattle on each others moors & commons / saith that the Harkaside people claime a separate right from the Grinton & Whitaside people on their own moors for eating their commons by themselves as farr as Browna Gill; /

Arthur Emerson of Grinton in the county of Yorke miner aged forty years sworne & examined saith /

[139] Presumably a scribal error for 'lead'.

2:11: & 12 To the second eleaventh & twelfth interrogatories this deponent saith that he knows the manor or reputed manor of Grinton & alsoe knows the places called West Grinton Harkaside & Whitaside & alsoe the moors & commons called Harka Hill Harka / Moore Grinton Moore Grinton How {or How} or How Groves & has knowne the same for about thirty two years & for all the said time the said places moors & commons have been reputed to lye within & belong to the said manor of Grinton & that the / Swales & Hillarys & of late Mr Marriott have been the reputed lords of the said manor saith Hillaries wrought old shafts or mines of lead upon the How & the Swales wrought mines of lead upon Harka or Harkaside Moore & had very good gettings / there till about eleaven years agoe that Mr Marriott recovered the same from him & afterwards Mr Marriott wrought there & sunk severall new shafts & continued in possession thereof till about two years agoe that the Lord Wharton enterd upon him saith both / Hillarys & Swales as lords thereof used to take upp waifes & estrays upon the said moores & commons as hee beleives hee this deponent having seen the same in their severall agents possessions saith hee knows the manor of Helaugh & that for all the said / time the river Swale has been the reputed boundary betwixt the said manors of Grinton & Helaugh saith hee hath lived all his time except for about one year at Grinton & never knew or heard that the present Lord Wharton or any of his / ancestors ever wrought any mines or claimed any royalties in West Grinton or Harkaside or any the said moors or commons till about two yeares agoe /

3: & 9: To the third & ninth interrogatories this deponent saith that the said manor of Grinton is comonly accounted to be bounderd on the east by the manor of Ellerton begining at the mid streame of the river Swale & goes up to Stallerstone Stile & thence / to the Gray Stone in the Sharrow thence to Brownsett soe to Snowdon Man & there begins the boundary betwixt the said manor of Grinton & the Duke of Boltons manors & soe proceeds to Redmere Standerd thence to Grinton Greets thence to Gibbon Hill / thence to High Carle thence to Pickerstin Rigg those marks being the south boundarys betwixt the said manor of Grinton & the Duke of Boltons manors & from thence goes westward to Tarnsant thence to Wind Yate Currock thence to Summer / lodge Currock thence to Bullhead thence to Bloody Wall so down Blagill to Haverdale Beck & soe to the {said} mid streame of the river Swale being the west boundary of the said manor of Grinton & from thence downe the said mid streame to Stallerstone Stile / as aforesaid saith the said mid streame of the said river is alsoe the common reputed boundery betwixt the manors of Fremington & Grinton Marrick & Grinton Marrick & Ellerton Marrick & Stainton Marrick & Downham & beleives it is alsoe soe downwards /

10: & 13: To the tenth & thirteenth interrogatories this deponent saith that the inhabitants of Grinton & Harkaside & Whitaside used to have their cattle going from one anothers commons by way of a rake without disturbance from each other saith that hee saw & was present when / the agents of the Lord Wharton about two years agoe rid the pretended boundary betwixt the moors & commons of Grinton West Grinton & Harkaside & never knew or heard of any boundery rid there before that time or that the said marks then rid were for any / other purpose before that time than for getting of ling peats & turfs upon their severall commons separately; /

John Kendall of Grinton yeoman aged forty two years sworne & examined saith /

2: 11: & 12 To the second eleaventh & twelfth interrogatories this deponent saith that hee knows the reputed manor of Grinton & the places called West Grinton Harkaside & Whitaside & the moors waists & commons thereunto belonging & has soe knowne the same for above / thirty years last past saith that the said places moors & waists are reputed to belong & to be within & be parcell of the said manor of Grinton & that the Swailes the Hillarys & of late Mr Marriott have been the reputed lords of the said manor of Grinton saith that / this deponents father was a miner & wrought for lead under the old Sir Solomon Swale as alsoe under the present Sir Solomon on Harka on Harkaside Moore the place in question & this deponent himselfe was a miner & workman under Mr Marriott upon Harka the place in / question & continued working there untill the present Lord Wharton enterd upon the said mines saith that one Blaides & Naylor were two persons that tooke upp waifes & estrays for Sir Solomon Swale on Grinton commons & this deponent has heard that they did frequently take / upp waifes & estrays there for the use of the said Sir Solomon knows the manor of Helaugh & for the time aforesaid the river Swaile has been the reputed boundary between the said manors of Grinton & Helaugh saith that he this deponent has lived at Grinton for thirty years / & been frequently upon the moors & commons & wrought at the {said} mines on the said moors & in all that time never heard that the present Lord Wharton nor any of his ancestors ever wrought any mine or mines upon West Grinton or Harkaside moors or in any other place within the said {manor} / manor of Grinton till about two yeares since that the Lord Wharton enterd upon the mines in question; /

3 & 9th To the third & ninth interrogatories this deponent saith that the manor of Grinton is reputed to be boundered on the north by the mid streame of the river Swale on the east by Ellerton manor at Stallerstone Stile from thence to the Gray Stone in the Sharrow from thence to Snowdon Man whence the / Duke of Boltons manors begins & are the south boundary of the said manor of Grinton & from Snowdon Man to High Brownsitt from thence to Redmere Standerd from thence to Grinton Greets from thence to Gibbon Hill from thence to High Carle Currock & from thence to / Pickerstin Rigg where the said south boundary ends & from thence to Tarnsant & from thence to Wind Yates & from thence to Sumerlodge Currock & from thence to Bullhead & from thence to Bloody Wall & from thence to Blagill Beck & soe to Haverdale Beck which is the west / boundary of the said Grinton manor & soe downe to the mid streame of the said river Swale as aforesaid saith that the said river Swale is alsoe a boundary between the manors of Fremington & Grinton Marrick & Grinton Marrick & Ellerton Marrick {& Downham} / {Stainton} & Downham & soe downwards saith that he has heard his father who was a fisherman <&> has been dead about thirty years & was when he dyed about fifty years old say that the mid streame of the said river Swale was a boundary & has heard severall / other ancient men say the same; /

4: & 17 To the fourth & seaventeenth interrogatories this deponent saith that he knows the hamlett of West Grinton & Harkaside saith the same consists of a few stragling houses some whereof stands together others at a considerable distance from each other saith the Lord Wharton hath / coppyhold lands at Harkaside Fremington & Whitaside the tennants whereof this deponent beleives appeare at his lordshipps court at Helaugh /

7: To the seaventh interrogatory this deponent saith there are four manors in Swaledale & that the whole dale is in Grinton parish & that the church stands at Grinton the cheife towne in the said dale /

10: To the tenth interrogatory this deponent saith that the inhabitants of West Grinton Harkaside & Whitaside have a rake for their cattle on each others commons but as to the getting of peats turfs & ling the Harkaside & West Grinton people claime a sole right in their own commons & that the Ox Intack nooke the Gray / stone on the How & soe upp to the Greets are reputed to be division marks between Harkaside West Grinton & Grinton commons for that purpose /

13. To the thirteenth interrogatory this deponent saith that about two years since the present Lord Wharton caused a boundary to be rid between Harkaside & Grinton

never knew or heard of any boundery rid there before beleives severall of the marks that were observed in that rideing were only marks for the divideing the said moors & commons / as to the getting of ling peats & turfe; /

[membrane xv *recto*][140]

Marke Kendall of Grinton in the county of Yorke miner aged forty seaven years or thereabouts sworne & examined saith /

2: 11: & 12 To the second eleaventh & twelfth interrogatories this deponent saith that hee knows the manor or reputed manor of Grinton & alsoe knows the places called West Grinton Harkaside & Whitaside & the moors & commons called Harka Hill Harka Moore Grinton Moore Grinton Greets Grinton How or the How Groves & has knowne / the same for about forty years saith that for all the said time the said places moors & commons have been reputed to lye within & be parte of the said manor of Grinton & that the Swales & Hillarys & of late Mr Marriott have been the reputed lords of the said manor saith that Hillarys wrought mines [. . .] / How & the Swales wrought mynes of lead upon Harka or Harkaside Moors & Sir Solomon Swale had very good gettings there till about eleaven years agoe that the said Mr Marriott recovered the same from him & after that the said Mr Marriott wrought the said mines & sunck severall [shafts] / & continued in the possession thereof till about two years agoe that the Lord Wharton enterd upon him saith that hee has seen severall waifes & estrays which hee has heard were taken upp both by the said Hillarys & Swales as lords upon the said moors & commons saith he knows the manor of Helaugh [&] / for all the said time <the> mid streame of the river Swale has been the reputed boundary betwixt the said manors of Grinton & Helaugh saith he ha's lived all his time at Grinton & been a neare neighbour to & frequently upon the said moors & commons & never heard that the Lord Wharton or any [of his] / ancestors ever wrought any mines or claimed any royalties in West Grinton or Harkaside or in any the said moors or commons till about two yeares agoe; /

3: & 9 To the third & ninth interrogatories this deponent saith that the said manor of Grinton is comonly reputed to be bounderd on the east by the manor of Ellerton begining at the mid streame of the river Swale & goes upp to Stallerstone Stile & soe to the White Stone on the Sharrow thence to S[nawden] / Man where begins the boundary betwixt the said manor of Grinton & the Duke of Boltons manors & soe proceeds to Brownsett thence to Redmere Standerd thence to Grinton Greets thence to Gibbon Hill thence to Pickerstin Rigg & from thence to Tarnsant thence to Wind [Yate] / thence to Bullhead thence to Bloody Wall & soe down Blagill to Haverdale Beck & soe to the mid streame of the said river Swale saith that the said mid streame of the said river Swale is alsoe generally reputed to be the boundery betwixt the adjacent manors of Fremington & Grinton / Marrick & Grinton Marrick & Ellerton Marrick & Stainton Marrick & Downham & beleives it is soe downwards saith that about thirty six or thirty seaven years agoe this deponent was with his father where he was fishing in the said river Swale a little below Gunnerside [on the] / Helaugh side of the said river & some person then threatned to take his rod from him upon which his said father asked him to let him take out his line & thereupon his father

[140] Parchment, 760mm x 610mm. The sheet is discloured and the right-hand margin is badly torn and decayed. Paper used to carry out repairs has become semi-opaque with age rendering the line endings on both *recto* and *verso* partially illegible.

went thro the said river unto the southside of the said river to Whitaside & then sett him at defiance; /

Martha wife of Luke Robinson[141] of Harkaside aged fourty seaven years sworne & examined saith /

2: 9: & To the second ninth <&> eleaventh interrogatories this deponent saith that shee
11: knows the reputed manor of Grinton & the places called West Grinton Harkaside & Whitaside & the moors & waists thereto belonging saith the same are reputed to lye within the said manor of Grinton & that the Swales the Hillarys / & of late Mr Marriott have been the reputed lords of the said manor saith that shee has heard that the Swales wrought mines for lead on Grinton How & Harka the place in question & once shee saw the workmen at worke on Harka under Sir Solomon Swale she having occasion to goe there to speak / to her husband who was a miner att the said mines has heard that Hillaryes wrought for lead & cole on Grinton Moors & that waifes & estrays were taken upp on Grinton Moors both for the Swales & the Hillarys knows the manor of Helaugh saith that during all this deponents time the mid / streame has been the reputed boundary between the manors of Grinton & Helaugh saith that shee this deponent heard her husbands father William Robinson an old man of eighty six years now dead in his life time say that the mid streame of the river Swale was a boundery between / the said manors of Grinton & Helaugh as farr as from Haverdale Beck to Stallerstone Stile & the reason of her said fathers discourse was that he went to Yorke to be a witness for Sir Solomon Swale there & discoursing of his title brought in the discourse of what was the boundery of the said river Swale as aforesaid /

4: & 17: To the fourth & seaventeenth interrogatories this deponent saith that the hamlett of West Grinton & Harkaside is a hamlett consisting of about twenty five houses some standing together others at a distance from each other saith the Lord Wharton has coppyhold lands at Harkaside & at Fremington as / she has heard the tenants whereof appeare at Helaugh court saith that Fremington is in the constablery of Reeth saith that shee has lived all her time in Harkaside West Grinton & Grinton & that she knows there is in Harkaside crowne lands; lands called Bridlington lands & Free lands / that the tennants doe not appeare at Helaugh court saith that about five or six years agoe Cornelius Freer the Lord Whartons bailiffe bought some stones of this deponents husband which were got by him the said Freer & others on the south side of the mid streame of the river Swale in the river by Harkaside / over against this deponents husbands lands & within his liberty to supporte the banks on the north side of the said river; /

10: To the tenth interrogatory saith that the inhabitants of West Grinton & Harkaside claime a sole right in their owne commons for the getting of peats turfe & ling to the Ox Intack or thereabouts but that the cattle of the inhabitants of Harkaside West Grinton & Whitaside & Grinton have a rake on each others / commons without disturbance saith that when shee this deponent was a girle she remembers that one Blaides who was formerly a servant to Sir Solomon Swale lived at West Grinton & is now dead was making a carle[142] or fence in the said river Swale to secure his grounds there & hee & his / sonns then said they could goe no further to get stones than the mid water of the said river; /

141 In May 1708 Martha Robinson, aged 46, gave evidence for Marriott at the trial in London (LI Misc 536, Ward MSS, notebooks vol. 6, f. 113v).

142 The use of the word 'carle' to signify a fence is unrecorded in either the *OED* or the *EDD*.

Luke Robinson[143] of Harkaside in the county of Yorke mason aged forty nine years or thereabouts sworne & examined saith /

9: To the ninth interrogatory this deponent saith that the mid streame of the river Swale has been for all the time of this deponents remembrance the reputed boundery betwixt the manor of Helaugh & the manor of Grinton hee this deponent having lived most of his time at Harkaside saith that his father when / this deponent was a boy used frequently to fish in the said river Swale & this deponent has often heard his father (who was near ninety years of age when hee dyed & whoe dyed about seaven years agoe) say that the mid streame of the said river Swale was the ancient boundary betwixt the said manors of Grinton & / Helaugh from Haverdale Beck to Stallerstone Stile & this deponent well remembers that about twenty four years agoe when this deponent was about fifteen years of age his father was fishing (this deponent being with him) in the said river at the lowe end of Whitaside on the south side of the mid streame of [the] / river & was then in the said river fishing with a leister and one James Fryer then a bailiffe or agent to the Lord Wharton came to his father & told him that he got a great deale of salmon & that he beleived he got some on their side or the north side of the said river & that he might some time give him [. . .] / & his father then said he would give him a fish or a salmon if he would come downe to Ann Milners: Further saith that about four or five years agoe one Cornelius Fryar then bailiffe or agent to the Lord Wharton bought about a hundred load of stone of this deponent which lay in the said river on the south [side of the] / mid streame of the said river at Harkaside & belonged to this deponent as his ground adjoined thereto; which this deponent & his brother & two or three other persons helpt the said Fryer to take out of the said river and the said Fryer & others with him tooke the same to mend or make upp the banks or bray [on the] / north side of the said river over against Harkaside & he paid this deponent both for the said stones & his labour in getting out the said stones & mending the said bray, and this deponent further saith that he has heard his said father & his uncle one John Blades now dead say that about fifty years agoe there was [. . .] / estrayd steer taken upp on Harkaside Moor & carried to the said John Blades's father who was agent to the Swales & that the said John Blades after he had kept him for above a year fed him & killd him for his own use saith that hee has alsoe heard his father & uncle say that [. . .] / agoe his said uncle tooke upp a mare as agent or steward to Sir Solomon upon some of the moors or commons belonging to the manor of Grinton & that after he had kept her for above a year he sold her; /

John Spence[144] of Midlam miner aged fifty years sworne & examined saith /

2: 11: To the second eleaventh & twelfth interrogatories saith that hee knows the <reputed>
& 12: manor of Grinton & the places called West Grinton Harkaside & Whitaside & the moors & waists & commons thereto belonging & has kowne the same for twenty five years last past saith the said places & moors are reputed to be [parcell of the] / said manor of Grinton & that the Swales the Hillarys & of late Mr Marriott have been the reputed lords of the said manor & the Swales wrought mines of lead on Harka the place in question & that the said Mr Marriott wrought the said mines & had good gettings untill the present Lord Wharton / about two years since got into the possession of the said mines & the reason of such his knowledge is that hee this deponent was a miner under the said Mr Marriott saith that hee hath heard that both

143 In May 1708 Luke Robinson, aged 49, gave evidence for Marriott at the trial in London (LI Misc 536, Ward MSS, notebooks vol. 6, f. 113v).

144 In May 1708 John Spence, aged 50, gave evidence for Marriott at the trial in London (LI Misc 536, Ward MSS, notebooks vol. 6, f. 112r).

the Hillaryes & <the> Swales tooke up waifes & estrays on Grinton & Harkaside Moores saith / hee himselfe released two waife sheep from Roger Hillary saith that he knows the manor of Helaugh & that the mid streame of the river Swale has been the reputed the boundary between the said manors of Grinton & Helaugh saith that for twenty five years last past this deponent [. . .] / knew or heard that the Lord Wharton nor any of his ancestors ever wrought any mine or mines upon Harkaside or West Grinton Moors or in any other place within the said manor of Grinton till about two yeares since; /

3: & 9: To the third & ninth interrogatories this deponent saith that the manor of Grinton is bounderd by the mid streame of the river Swale on the north by Ellerton manor at Stallerstone Stile on the east by the Duke of Boltons manors on the south & as this deponent has heard [by] / Haverdale Beck on the west saith that hee has heard old Mr Benlows who was steward or court keeper for the late Lord Wharton & the present Lord Wharton at his courts of Helaugh & Mukar the deponent travelling with the said Mr Benlows & another gentleman [about] / sixteen or seaventeen years since over the said moors & commons belonging to Harkaside & West Grinton the gentleman asked the said Mr Benlows in whose liberties are wee now & the said Mr Benlows replyed that they were in the liberties of Sir Solomon Swale & Roger Hillary & this deponent / saith they travelled further & came to Scabba Wath & when they came on the other side of Swale the said Mr Benlows said now wee are on the north side of the river Swale & in the libertys of the Lord Wharton /

Parcivall Thwaites of Redhurst in the county of Yorke miner aged fifty eight years or thereabouts sworne & examined saith /

2: 11: & 12: To the second eleaventh & twelfth interrogatories this deponent saith that hee knows the manor or reputed manor of Grinton & alsoe knows the places called West Grinton Harkaside & Whitaside & the moors & commons called Harka Hill Harka Moore Grinton Moore Grinton Greets Grinton / How or the How Groves & has knowne the same for forty five years & that for all the said time of this deponents knowledge the said places moors & commons have been reputed to belong to & be parte of the said manor of Grinton & that the Swales & Hillarys & of late Mr Marriott have / been the reputed lords of the said manor saith that the Swales have wrought mines of lead upon Harka Hill or Harkaside Moore the place in question & Sir Solomon Swale continued to worke the same till Mr Marriott recovered the same & got the possession thereof & that Mr Marriott / afterwards enjoyed the same till the Lord Wharton enterd upon him saith hee hath carried waifes & estrays from the moors of Whitaside to the agents of Sir Solomon Swale & has heard that the Swales used to take upp waifes & estrays upon the said moors & commons saith / he knows the manor of Helaugh & that the river Swale has been for all his time the reputed boundery betwixt the said manor of Helaugh & the said manor of Grinton saith his business has occasioned his going frequently every yeare upon the said moors & commons & never / heard that the Lord Wharton or any of his ancestors ever wrought any mines upon any the said moors or commons till about two yeares agoe; /

3: & 9: To the third & ninth interrogatories this deponent saith that the said manor of Grinton is comonly reputed to be bounderd on the east by the manor of Ellerton at Stallerstone Stile & on the south by the Duke of Boltons manors & on the west by Haverdale Beck & on the north from the mannor / of Helaugh by the river Swale down from Haverdale Beck to near Grinton & so to Stallerstone Stile as aforesaid saith he was borne at Whitaside & brought upp there neare to the said places & his father & ancestors lived thereabouts all their time saith that about thirty years agoe / he was present when one Lomas an inhabitant of Grinton was fishing in the said river Swale upon the south side of the said river at Harkaside & at the said time there was two fishermen fishing upon the north side of the said river over against him & he forct them to goe back & keep within / the mid streame of the said river on the said northside thereof saith that hee has heard ancient men & perticularly one William Spencley his relation who lived at Whitaside & dyed about twenty years agoe & was about seaventy years old say that the said mid streame of the said river / Swale was alwaies a boundary from Haverdale Beck all along down the said river & the same has been commonly reputed & accounted a boundery betwixt the manors of Fremington & Grinton Marrick & Grinton Marrick & Ellerton Marrick & Stainton Marrick & / Downham & the places of Thorpe & Applegarth & beleives it is so down to Richmond bridge; /

Thomas Hillary of Arkingarthdale aged forty yeares sworne & examined saith /

2: To the second interrogatory this deponent saith that he knows the manor or reputed manor of Grinton & knows the places called [West] Grinton Harkaside & Whitaside & has knowne Grinton thirty years & has knowne Harkaside & Whitaside for twenty years saith that the said places are reputed / to be within the manor of Grinton & that the Swales & Hillarys have been the reputed lords thereof saith hee has heard that Sir Solomon Swale wrought mines for lead on Harka the place in question & beleives he did soe for about sixteen years since one Bryan / Birbeck bought wood of this deponent for Sir Solomn Swale to be used at the said mines & this deponent saith that hee hired men to carry the said wood & deliver it at the said mines which hee beleives was accordingly done this deponent paying for the said carriage saith he knowes / Helaugh & that for the time aforesaid the river Swale has been reputed to be a boundary between Helaugh & Grinton /

Thomas Porter[145] of Redhurst in the county of Yorke aged fifty five years or thereabouts sworne & examined saith /

2: & 9: To the second & ninth interrogatories this deponent saith that he knows the manor or reputed manor of Grinton & alsoe knows the places called West Grinton Harkaside & Whitaside & has knowne the same for thirty two years & that for all the said time the same have been commonly reputed to lye within [& to] / belong to the said manor of Grinton & that the Swales & Hillarys & of late Mr Marriott have been the reputed lords of the said manor saith that he knows the manor of Helaugh & that the river Swale has been for all the said time of his knowledge accounted to be the boundary betw[ixt the] / said manors of Grinton & Helaugh & has the greater reason to beleive the same because about thirty two years agoe this deponent was servant to one Mr Wilson who then lived at Helaugh & this deponent having occasion to goe over the said river to fetch coles from Hillarys [. . .] / with one Ralph Garth since dead who was tennant to the Lord Wharton at Riddings & was reputed to have an estaite of his owne in the manor of Helaugh of about thirty pounds per annum and they going over the said river betwixt Helaugh & Harkaside this deponent said it was a pretty dale / Garth then told & shewed this deponent that all that parte on the north side of the said river belong'd to the Lord Wharton[s] manor of Helaugh & that parte on the south side of the said river belonged to the manor of Grinton & to the Swales & Hillarys as lords thereof further saith that [the river] / Swale is alsoe comonly reputed to be the boundery betwixt the manors of Marrick & Ellerton & the places of Thorpe & Aplegarth hee this deponent having lived for some time at all the said places; /

145 In May 1708 Thomas Porter, no age given, gave evidence for Marriott at the trial in London (LI Misc 536, Ward MSS, notebooks vol. 6, f. 113r).

William Musgrave of Reeth Moore aged sixty five years sworne & examined saith /

2: & 9: To the second & ninth interrogatories saith hee knows the manor or reputed manor of Grinton & the places called West Grinton Harkaside & Whitaside & the moors waists & commons thereto belonging & has soe knowne the same for about forty years saith the said places [. . .] / waists were reputed to belong & be parte & parcell of the said manor of Grinton during his knowledge of them & that the Swailes & the Hillarys have been the reputed lords of the said manor saith that the Swailes wrought the mines & got considerable quantitys [. . .] / both on Grinton How & Harka on Harkaside Moor the place in question and that the Hillarys wrought waist ore of lead & cole on Grinton Moors saith hee knows the manor of Helaugh & that he hath frequently heard & beleives the river Swale to be the reputed [boundary] / between the said manors of Grinton & Helaugh & the reason of such his beleife is that the said river Swale is a boundary between the manors of Fremington & Grinton Marrick & Grinton Marrick & Ellerton Marrick & Downham {Downham & Stainton} [. . .] / this deponent for most of his time has lived nigh the moors & waists in question & never in all his time knew or heard that the present Lord Wharton or any of his ancestors ever wrought any mine or mines or claimed any royalties in West Grinton or Harkaside / or in any other place within the said manor of Grinton till about two years since /

John Freer of Ellerton aged thirty three years sworne & examined saith /

2: 9: & 11: To the second ninth & eleaventh interrogatories this deponent saith that hee knows the manor or reputed manor of Grinton & the places called West Grinton Harkaside & Whitaside & the moors & commons thereto belonging & has knowne the same for about eighteen [yeares &] / for all the said time of this deponents knowledge the same have been reputed to belong to & lye within the said manor of Grinton & that the Swales & Hillarys & of late Mr Marriott have been the reputed lords of the said manor saith Sir Solomon Swale wrought mines of [. . .] / upon Harka or Harkaside Moore & had good gettings there till the said Mr Marriott recoverd the same from him & after that the said Mr Marriott wrought the said mines & continued in the possession thereof till about two years agoe that the Lord Wharton enterd upon [him] / saith that hee knows the manor of Helaugh & that hee has heard the river Swale has been reputed the boundary betwixt the said manors of Grinton & Helaugh & this deponent beleives the same for that the said river is comonly accounted to be the boundary be[tween the] / adjacent manors of Fremington & Grinton Marrick & Grinton Marrick & Ellerton Marrick & Stainton Marrick & Downham Marsk & Downham & beleives it to be soe downwards saith hee this deponent has been for about sixteen years a jagger or car[ryer of] / lead & has been very often every year upon the said moors & commons & never knew or heard that the Lord Wharton or any of his ancestors ever wrought any mines upon the said moors or commons till about two years agoe /

[membrane xv *verso*]

John Taylor of Ellerton jagger aged fifty years sworne & examined saith /

9: To the ninth interrogatory this deponent saith that the river Swale is the reputed boundary between the manors of Grinton & Helaugh saith that hee this deponent lived three years at Harkaside & farmed lands there of one Christopher Raper & came away [. . .] / yeare & one halfe agoe and wood comeing frequently downe the said river Swaile against the lands this deponent farmed of the said Christopher Raper hee severall times told this deponent that hee might goe as farr as the mid streame of the said river / & take upp either wood or what other things the streame of the said river should bring there but might goe noe further than the said mid streame it being the boundary between the said manors /

Reynold Jefferson of Marrick smelter aged thirty eight years or thereabouts sworne & examined saith /

3: 9: 11: To the third ninth & eleaventh interrogatories this deponent saith that he knows the manor or reputed manor of Grinton & alsoe knows the places called Harkaside & Whitaside & the moors & commons thereunto belonging & has knowne the [same] / for about fourteen years & that for all the said time the said places moors & commons have been reputed to belong to & lye within the said manor of Grinton & that the Sir Solomn Swale & of late Mr Marriott have been reputed the lords of the said manor saith the said Sir Solomon [wrought] / mines of lead upon Harka or Harkaside Moore the place in question & had very good gettings there till about eleaven years agoe that Mr Marriott recovered the same from him & after that Mr Marriott enjoyed the same & continued in the possession of them till about two [years] / agoe that the Lord Wharton enterd upon him saith he has been concerned in the smelting of great quantites of lead that were brought from the mines both in Sir Solomon & Mr Marriotts time saith that Sir Solomon & Hillarys used to take upp waifes & estrays / upon the said moors & commons as this deponent has heard & this deponent about seaven years agoe loosed one sheep of his owne from one Naylor; saith that hee knows the manor of Helaugh & that the river Swale is comonly reputed to divide the lordshipps [of Helaugh] / & Grinton; & the said river is alsoe comonly accounted to be the boundary betwixt the adjacent manors of Fremington & Grinton Marrick & Grinton Marrick & Downham Marsk & Downham & beleives it is soe downwards & this deponent has [been] / fishing severall times in the said river with other fishermen & they have generally observed in their fishing the mid streame of the said river to be a boundary as aforesaid saith he never knew or heard that the Lord Wharton or any of his ancestors [ever] / wrought any mines upon any the said moors or commons till about two years agoe; /

Roger Staveley of Wensley in the county of Yorke aged fifty one years sworne & examined saith /

2: 9: 11: To the second ninth & eleaventh interrogatories this deponent saith that hee knows the manor or reputed manor of Grinton & alsoe knows the places called Harkaside & Whitaside & the moors & commons thereunto belonging & has knowne the same [for] / thirty years & that for all the said time the said places moors & commons have been comonly reputed to lye within & belong to the said manor of Grinton & that the Hillarys & Swales & of late Mr Marriott have been the reputed lords of the [said manor] / saith Hillarys wrought mines of lead in a shaft upon the How & the Swales wrought mines of lead upon Harka Hill or Harkaside Moore the place in question & enjoyed the same till Mr Marriott about eleaven years agoe recovered [the same from] / him & after that Mr Marriott sunck severall new shafts there & continued in possession thereof till about two years agoe that the Lord Wharton enterd upon him saith hee knows the manor of Helaugh & that hee has heard that the river Swale [has] / been generally accounted the boundary betwixt the said manors of Grinton & Helaugh & has alsoe heard that the said river has been alsoe generally accounted the boundary betwixt the adjacent manors of Mukar & Helaugh Fremington & / Grinton Marrick & Elerton Marsk & Downham & this deponent perticularly remembers a dispute about tenn or more years agoe betwixt the lords of Marsk & Downham about the right to two little islands lying about the midle of the said [river Swale] / betwixt their said manors & they at last agreed to divide them betwixt them saith hee never knew or heard that the Lord Wharton or any of his ancestors wrought any mines or claimed any royalties in Harkaside or the [said moors] / or commons till about two years agoe; /

Ann wife of John Carter of Stainton aged fifty seaven years or thereabouts sworne & examined saith /

2: To the second interrogatory this deponent saith that she knows the manor or reputed manor of Grinton & alsoe knows the places called Harkaside & Whitaside being borne at Whitaside & has knowne the same for neare fifty years & that they have been reputed [to belong] / to the said manor of Grinton & that the Swales & Hillarys were reputed & called lords of the said manor saith shee knows the manor of Helaugh & the river Swale is a parting or division betwixt the said places; /

Richard Addison of Woodale yeoman aged thirty nine years sworne & examined saith /

2: To the second interrogatory saith that hee knows the manor or reputed manor of Grinton & the places called West Grinton Harkaside & Whitaside & the moors waists & commons thereto belonging & has knowne them for thirty years saith they are reputed to belong [to the] / said manor of Grinton & that the Swales & Hillarys were the lords thereof saith that Hillary wrought lead mines on the How & tooke upp waifes & estrays on Grinton Moore & that this deponent himselfe bought waife sheep of him & has of the [Br . . .] / this day knows the manor of Helaugh for the time aforesaid the river Swale has been reputed to be a division between the said two manors of Grinton & Helaugh never heard that the present Lord Wharton nor any of his ancestors ever wrought any mines / for lead or cole or claimed any royalties on Harkaside Moore till about two years since /

Margarett Addison of Woodale widdow aged seaventy nine years sworne & examined saith /

2: To the second interrogatory this deponent saith that shee knows the reputed manor of Grinton & the places called West Grinton Harkaside & Whitaside & the moors waists & commons thereto belonging & has knowne the same for above sixty years saith the said places [moores] / & commons were reputed to <be> parcell of the said manor of Grinton & that the Swales & the Hillarys were the reputed lords of the said manor saith about fifty years agoe her husband William Addison was a miner & wrought under old Sir Solomon Swale at the [mines] / on the How & very well remembers that the her said husband used to tell her that at every pay the said Sir Solomon gave the miners twenty shillings amongst them to drinke knows Helaugh saith that the river Swale has been called a division between / Helaugh & Grinton never heard that the present Lord Wharton nor any of his ancestors ever wrought any mines upon Harkaside Moors till about two years since; /

Christopher Coats of Aglethorpe yeoman aged fifty three years sworne & examined saith /

2: To the second interrogatory this deponent saith that hee knows the manor of Grinton & the places called West Grinton Harkaside & Whitaside & the moors & commons thereto belonging & has soe knowne the same for forty four years & are reputed to [belong] / to the said manor of Grinton & that the Swales & Hillarys were the reputed lords thereof during his knowledge saith that the Swales wrought lead mines & got lead on the How as alsoe on Harkaside Moore saith that this deponents father was a miner [under] / the said Sir Solomon Swale there & this deponent has rid of his Galloway to see his father there saith that Hillary wrought for cole on Grinton Moore & that both the Swales & the Hillarys tooke upp waifes & estrays on the said moores [. . .] / Helaugh saith the river Swale has been a division between Grinton & Helaugh manors as alsoe betwe[e]n the manors of Fremington & Grinton Marrick & Grinton Marrick & Ellerton Marske & Downham & saith the reason of this [deponents] / beleiving the said river to be a division as aforesaid is that in all this deponents time hee never heard that the present Lord Wharton or any of his ancestors ever claimed any royalties or wrought any mines upon the south side of the [said] / river or any of the said moors or commons till about two years since; /

Tobias West[146] clerke master of arts viccar of Grinton in the county of Yorke aged fifty eight years or thereabouts sworne & examined saith /

2 8 11 &12 To the second eight eleaventh & twelfth interrogatories this deponent saith that he ha's been thirty four years viccar of Grinton & that he know's the mannor of Grinton & alsoe know's the places called West Grinton Harkaside & Whitaside & the moores & commons / called Harka hill Harka Moore Grinton Moore Grinton Greetes Grinton How or the Howgroves & ha's knowne the same for about thirty four years & for all the said tyme the said places moores & commons have belonged to & been parte of the said mannor of Grinton & that the Swales & / Hillary's & of late Mr Marriott have been called or reputed lords of the said mannor saith that about twenty four or twenty five years agoe one Ralph Wensley & Richard Harker (now dead) discovered & opened a quarrey of flaggs & gott some stones upon Harka or Harkaside / Moore the place in question & Roger Hillarye hereing thereof seized the same or a great parte thereof as being found & wrought within his mannor or royalties & without his leave & gave parte of them to one Richard Close saith that the said Hillarye wrought mines / of coale upon Grinton Moore & heard that he alsoe gott some lead there saith the Swale's wrought mynes of lead upon Grinton Moore Grinton How Harka or Harkaside Moore & alsoe upon Berks & Grena hills in Whitaside & that for about four or five [years] / together the said mynes upon Harka soe wrought by the present Sir Solomon proved very rich & he & his partners (as this deponent ha's good reason to beleive) gott about thirty thousand pounds worth of oare there within that tyme & the said Sir Solomon continued to worke / there till about ten or eleaven years agoe that Mr Marriott recovered the same from him & after that the said Mr Marriott wrought the said mynes & sunck severall new shafts & this deponent ha's heard that he gott about ten thousand pounds worth of oare there & that / Mr Marriott continued in possession thereof till about two years agoe that the Lord Wharton entred upon him saith he ha's heard that both Hillary's & Swale's as lords used to take up waifes & estray's upon the said moores & commons saith he know's the mannor of Helaugh / & for all the said tyme of this deponents knowledge the midd streame of the river Swale ha's been the antient & knowne boundery betwixt the said mannors of Grinton & Helaugh from Haverdale beck all along downe to over against the foot of Arkle beck (where the / mannor of Helaugh ends) & soe downe to Staller Stone Style saith that neither the Lord Wharton nor any of his ancestors did ever worke any mynes or claime any royalties in West Grinton or Harkaside or any the said moores {&} <or> commons (that ever this deponent / knew or heard off) till about two years agoe further saith that he this deponent ha's heard of & seen some writeings relating to a lease from Queen Elizabeth to Lord Scroop & one Phillips for one & twenty years of all the veins or mynes of lead open or not open in / the feilds or territory's of Grinton Harkaside & Whitaside parcells of the mannor of Grinton part of the late dissolved monasterye of Bridlington and this deponent ha's alsoe heard that there was such a lease of the said mynes in the

[146] Tobias West (*c.*1649–1717), born at Woolley, near Barnsley, was the son of the Rev. Lewis West. Educated at Sedbergh and St John's College, Cambridge, he received his BA in 1668/9 and MA in 1673. Vicar of Grinton from 28 June 1673 to 1717. He died at Grinton and is buried there (Borthwick Institute, clergy card index). In May 1708 Tobias West, aged 58, gave evidence for Marriott at the trial in London (LI Misc 536, Ward MSS, notebooks vol. 6, f. 110r).

said places granted by / James the first to one Humphrey Wharton of Gillingwood esquire but know's not for what terme saith that one James Metcalfe an old man who died about seaven or eight years agoe & was about ninty four years <of> age when he died told this deponent that he was formerly / an apprentice to a blacksmith & that his master & he were employed to worke the smith worke for Mr Wharton's mynes within the mannor of Grinton & that he was sent by his master to the said Mr Wharton with his masters bill who accepted the same and / promised to pay him in iron saith he know's of noe other lessees of the said mynes under the Crowne save the relator Mr Marriott /

3 & 9 To the third & ninth interrogatories this deponent saith that the said mannor of Grinton is bounderd on the east by the mannor of Ellerton begining at the midd streame of the river Swale & soe goes up to Staller Stone Style from thence to the Gray Stone o'th Sharrow thence to [Snowdon] / Man where ends the mannor of Ellerton & begins the boundery betwixt the Duke of Bolton's mannors & the said mannor of Grinton & soe proceeds up Brownsett soe to Ridmer Standerd thence to Grinton Greetes thence to Gibbon hill thence to High Carle thence to Pickerstin [Rigg] / thence to Tarnsant thence to Windyate Currack thence to Somerlodge Tarn or Tarn Currack thence to Bull head thence to Bloody vale or Wall soe downe Blagill to Haverdale beck which is the west boundery of the said mannor of Grinton & from Haverdale beck to the midd streame / of the said river Swale & soe downe the mid streame till over against Stallerstone Style which midd streame of the said river from Haverdale beck all along downe to Staller Stone Style is the north boundery of the said mannor of Grinton & the Duke of Boltons mannors [. . .] / is the south boundery thereof saith that the said midd streame of the said river Swale is alsoe the knowne antient boundery betwixt the adjacent mannors of Fremington & Grinton Marrick & Grinton Marrick & Ellerton Marrick & Stainton Marrick & Downham / Marsk & Downham & ha's alsoe heard it generally accounted the boundery betwixt Thorpe & Applegarth & the mannors of Hudswell & Richmond & Easby & Colbourne & beleives it is alsoe the boundery as far downwards as the said river beares the name of Swale / and this deponent ha's often heard fishermen observed the said midd streame of the said river to be a boundery /

4 & 17 To the fourth & seaventeenth interrogatories this deponent saith that the hamlett of West Grinton & Harkaside consists of about twenty four houses some lyeing stragling at a distance from one another saith there are lands of severall tenures there Vizt crowne lands / free lands & Priory lands which are of a considerable vallue (tho' this deponent know's not the particular vallue thereof) saith the Lord Wharton hath some free lands there for which his tennants Anthony Hird & John Hird pay him a rack rent of six pounds thirteen / shillings & four pence per annum saith he hath alsoe some copyhold lands there in the possession of the four brothers Arundells & Mr Hartley Ralph Wensley <&> Christopher Raper {& Ralph Addyson} the particular vallue whereof this deponent know's not saith that the [. . .] / Mr Hartley & Christopher Raper & Ralph Addyson have alsoe some little free {hold} lands belonging to the said Lord Wharton for which they pay some very small rents & that his tennants of all his said lands both coppyhold <& free> lands appear at his lordshipps court at Helaugh / saith that the said Lord Wharton hath but two of the said twenty four houses belonging to his said estate there & that his said coppyhold lands there are not a fourth parte of the lands belonging to Harkaside & West Grinton tho' by reason of the custome of [. . .] / they are in soe many hands and that all his coppyhold & free lands are far the less parte of the other lands Vizt the said crowne lands the said free lands & the said priory lands in Harkaside & West Grinton the tennants of all which doe not appeare or [doe] / suite & service at his lordshipps court at Helaugh saith

his lordshipp hath alsoe coppyhold lands in Whitaside & Fremington the tennants of which said coppyhold lands appeare (as this deponent beleives) at the court at Helaugh & further saith that all Fremington lyes / within the constabulery of Reeth & yet is a distinct mannor from Helaugh /

[membrane xvi *recto*][147]

6 To the sixth interrogatory this deponent saith that he ha's heard & seen severall old writeings that the said mannor of Grinton formerly belonged to the Priory of Bridlington /

7 To the seaventh interrogatory this deponent saith that the dale or vale called Swaledale is about twelve myles in length that it ha's not been accounted a mannor of its selfe but there are four mannors within the same Vizt the mannors of Helaugh & Mukar belonging to the Lord Wharton (the / said mannor of Mukar as this deponent ha's heard formerly belonging to the Abby of Rivalx) & the mannor of Grinton belonging to Mr Marriott & the mannor of Fremington belonging to Madam Wharton <of Gilling> or her son all lyeing in the parish of Grinton & that Grinton where the church stands & is the cheife towne there /

10 & 13 To the tenth & thirteenth interrogatories this deponent saith that the inhabitants of Grinton Harkaside & Whitaside have used to have their cattle intercommoning or goeing in one anothers commons without interrupcion & alwaies understood that they might turne their cattle into one anothers commons / saith that the inhabitants of Harkaside & West Grinton used to claime a sole right of getting ling peat & turfe upon their owne moores & commons soe far as the Ox intack & the White Stone in the How & Grinton Greetes betwixt them & the moores of Grinton & as farr as Browna Gill betwixt / them & the moores of Whitaside & that the said severall marks were alwaies looked upon to be division marks for that purpose onely saith he hath heard of the boundery ridd by the Lord Whartons agents about two years agoe but never heard of any such rid before saith that he cannot beleive / the same was ever a boundery to divide any mannor but that most of the said marks were onely looked upon & accounted to be the division marks of their commons as aforesaid for getting peat ling & turfe & for no other purpose whatsoever & do's veryly beleive all the said pretended boundery marks lye within & belong to the said mannor of Grinton /

14 To the fourteenth interrogatory this deponent saith that soone after the rideing of the pretended boundery by the Lord Whartons agents one Ralph Wensley told this deponent that he & one Miles Close refused to signe a certificate concerneing the rideing of the said {mannor} boundary which was offerred to them / by some of the said agents because they did not beleive them to be bounderys of any mannor as that certificate imported & one James Naylor alsoe told this deponent that he alsoe had refused to signe the said certificate for the same reason /

Anthony Hudson[148] of St John Chapple in Wardale in the county of Durham miner aged forty two years sworne & examined saith/

2 11 & 12 To the second eleaventh & twelfth interrogatories this deponent saith that he know's the mannor or reputed mannor of Grinton & the places called Grinton Harkaside & Whitaside & the moores & commons thereto belonging & ha's soe knowne the same

147 Parchment, 730mm x 66mm. The bottom right-hand corner of the *recto* and the top right-hand edge of the *verso* are dirt stained and decayed. Paper used for repair has become semi-opaque with age with consequent loss of legibility.

148 In May 1708 Anthony Hudson, aged 42, gave evidence for Marriott at the trial in London (LI Misc 536, Ward MSS, notebooks vol. 6, f. 113r).

for two & thirty years last past saith the same belong to / the mannor of Grinton & are reputed to be parte of the said mannor & saith the reputed lords of the said mannor were the Swale's & Hillary's & of late Mr Marriott saith that Sir Solomon Swale wrought mynes of lead on the How & Harka the place in question & since Mr Marriott both on / How & Harka & that Roger Hillarye wrought mynes for coale on Grinton Moore saith that he ha's heard that the Swale's & the Hillary's tooke up waifes <& estrays> on Grinton & Harkaside Moores but do's not know they did soe of his owne knowledge saith he know's the mannor of Helaugh & that / for the tyme aforesaid the mid streame of the river Swale ha's been the reputed boundery betweene the said mannors of Grinton & Helaugh saith that he this deponent was borne at Fremington & is a freeholder in & lived & was brought up there till about fourteene years since that / he left the said place & since the tyme of his remembrance altho' he lived nigh the place in question he never heard or knew that the present Lord Wharton or any of his ancestors ever claimed any royalties or wrought any mynes for lead or coale upon Harkaside Moores or commons / or within any other place within the said mannor of Grinton till about two years since /

3 & 9 To the third & ninth interrogatories this deponent saith that the mannor of Grinton is bounderd on the north by the mid streame of the river Swale on the east by Ellerton mannor at Stallerstone Style & on the south by the Duke of Boltons mannors & on the west by Haverdale Beck saith the said / mid streame of the river Swale is the boundery between Fremington & Grinton Marrick & Grinton Marrick & Ellerton saith that when this deponent was a boy & learneing to fish with his father & as they were fishing in the river Swale one Roger Hillarye a reputed lord of / Grinton came & discharged this deponents father & himselfe from comeing any further to fish in the said river Swale then the mid streame without his leave & then the said Hillarye declared that to be a boundery between Grinton mannor & Fremington and this deponent / further saith that he ha's severall tymes heard his father (who was sixty three years of age when he dyed) say that the said mid streame of the said river was a boundery as aforesaid & that if any offerred to fish further the said Hillarye might trouble them for that he / claimed as far as the mid streame to be his boundery /

George Miller[149] of Harkaside yeoman aged sixty years & upwards sworne & examined saith /

2 8 11 & 12 To the second eight eleaventh & twelfth interrogatories this deponent saith that he know's the mannor or reputed mannor of Grinton & alsoe the places called West Grinton Harkaside & Whitaside & the moores & commons called Harka hill Harka Moore Grinton Moore Grinton Greetes Grinton How or / Howgroves & ha's knowne the same for near fifty yeares & for all the said tyme the said places moores & commons have been alwaies reputed to belong to & be parte of the said mannor of Grinton that the Swale's & Hillary's & of late Mr Marriott have been the reputed lords of the said mannor saith that / Hillary's gott lead in the waste hillocks upon Grinton How & wrought mynes of coale upon Grinton Moore & old Sir Solomon Swale & Sir Henry Swale & the present Sir Solomon Swale wrought mynes of lead upon Harka or Harkaside Moore the place in question & the present Sir Solomon had very / good gettings there & continued to worke the same till about ten or eleaven years agoe that the said Mr Marriott recovered the same against him & after that the said Mr Marriott wrought the said mynes & continued in the possession thereof till about two years agoe that the present Lord Wharton entred upon him /

[149] In May 1708 George Miller, aged 50 (error for 60?), gave evidence for Marriott at the trial in London (LI Misc 536, Ward MSS, notebooks vol. 6, f. 109v).

saith both Hillary's & Swale's used as lords to take up waifes & estray's upon the said moores & commons saith he know's the mannor of Helaugh & that the mid streame of the river Swale ha's been all his tyme reputed & accounted the boundery betwixt the said mannors of Grinton & Helaugh / saith he was borne at Harkaside & lived there for all his tyme except for about one yeare & an halfe & never knew or heard that the Lord Wharton or any of his ancestors ever wrought any mynes or claimed any royalties in West Grinton or Harkaside or any the said moores or commons / till about two years agoe except that about three or four years agoe his lordshipp sunck a shaft in his owne freehold lands at Harkaside in the occupation of John Hird but beleives gott very little /

3 & 9 To the third & ninth interrogatories this deponent saith that the said mannor of Grinton is bounderd on the east by the mannor of Ellerton begining at the mid streame of the river Swale & goes up {by} <to> Stallerstone Style thence to the White Stone o'th Sharrow thence to Snowden Man (where begins the boundery betwixt the Duke of / Boltons mannors & the said mannor of Grinton[150] & soe proceeds to Brownsett thence to Redmer Standard thence to Grinton Greetes thence to Gibbon Hill thence to High Carle thence to Pickerstin rigg thence to Wind Yate Currock thence to Tarnsant thence to Bulls head thence to Bloody wall thence to / Blagill & soe down to Haverdale beck which is the west boundery of the said mannor of Grinton the said mannor being bounderd on the south by the Duke of Boltons mannors as aforesaid & on the north from Haverdale beck downe the mid streame of the said river Swale & soe all along soe farr as / Stallerstone Style as aforesaid saith that the said mid streame of the said river Swale ha's been alsoe the common reputed boundery betwixt the adjacent mannors of Fremington & Grinton Marrick & Grinton Marrick & Ellerton Marrick & Stainton Marrick & Downham & beleives it to / be soe downwards saith he was well acquainted with one William Robinson (who used to fish in the said river & who was about eighty years of age & died about eight or ten years agoe & this deponent ha's heard him often say that the mid streame of the said river Swale was a boundery as aforesaid /

4 To the fourth interrogatory this deponent saith that the hamlett of Harkaside & West Grinton consists of four & twenty houses most of them standing stragling at some distance from one another saith there are crowne lands freehold lands & Priory lands lyeing in Harkaside & West / Grinton which are of a considerable vallue tho' this deponent cannott particularize the full vallue thereof all the tennants or occupyers whereof doe not appeare or doe any suite or service at the Lord Whartons court at Helaugh saith there are alsoe coppyhold lands & alsoe some freehold / lands belonging to the Lord Wharton lyeing there the rents of the said coppyhold lands as this deponent beleives being about four pounds per annum payable to the Lord Wharton & the rack rents of his said freehold lands being six pounds thirteene shillings four pence <per> annum paid by the two / Hirds <the tenants of which said copyhold lands & freehold lands appeare at his lordshipps court at Helaugh> saith that the Lord Wharton has but two houses in Harkaside & West Grinton belonging to all his said freehold & coppyhold lands & that this deponent beleives the vallue of his said coppyhold lands is not a fourth parte of the other lands in Harkaside & West Grinton /

6 To the sixth interrogatory this deponent saith that he ha's heard that the mannor of Grinton formerly belonged to the Priory of Bridlington /

7 To the seaventh interrogatory this deponent saith that the dale or vale called Swaledale

[150] This bracket is not closed.

is not reputed a mannor of it selfe but that there are four mannors therein Vizt the mannors of Helaugh & Mukar belonging to the Lord Wharton the mannor of Grinton belonging to Mr / Marriott & the mannor of Fremington belonging to Madam Wharton of Gillingwood all lyeing in the parish of Grinton & that Grinton where the church stands is the chiefe towne in the said dale /

10 & 13 To the tenth & thirteenth interrogatories this deponent saith that the inhabitants of Grinton West Grinton Harkaside & Whitaside used to have their cattle goe from one anothers commons without disturbance saith that the inhabitants of Harkaside have all along claimed a sole & separate / right of getting ling peates & turfe upon their owne moores & commons from Grinton & Whitaside & that the moores[151] called the Ox intack & White Stone o'th How & Gray Stone o'th Greetes have been alway's reckoned & accounted the division or boundery marks betwixt their said / moores & commons & the moores & commons of Grinton for that purpose & that the marke called Browna Gill ha's been alway's reckoned the division or boundery marke betwixt the said moores & commons of Harkaside & the moors & commons of Whitaside for that purpose only / of getting ling peat & turfe upon their owne severall commons & never heard that any of the said marks were division or boundery marks for any other purpose saith that he know's the gate hanging at the elme tree betwixt Grinton & Harkaside & that the inhabitants of / Grinton repaire the highway to that gate & that the same ha's been taken notice of by the inhabitants of both places as an old place or marke for them to meet to repaire their highways & that each side used to repaire soe far but never heard it called a boundary / for any other purpose till lately pretended soe upon the rideing of a pretended boundery by the Lord Whartons agents saith he hath heard of a boundery ridd about two years agoe by the Lord Whartons agents betwixt Grinton West Grinton Harkaside & Whaitaside but never heard of any / such boundery ridd before saith he never heard or can beleive that the same was ever claimed before to be the boundery of the mannor of Helaugh but that the said boundery or division marks soe as before described by this deponent betwixt the said moores & commons of Grinton West / =Grinton Harkaside & Whitaside have been alwaies accounted by the neighbourhood to be onely division for their getting of peates ling & turfe separately upon their said severall commons & for noe other purpose /

17 To the seaventeenth interrogatory this deponent saith that there are some lands Vizt about two or three acres lyeing on the north side of the river Swale in the mannor of Helaugh about Scabba Wath which belongs to the constablery of Grinton & pay their taxes to Grinton tho' lyeing in the / mannor of Helaugh /

James Loftus[152] of Bucktin in Craven husbandman aged forty eight years or thereabouts sworne & examined saith /

2 9 11 & 12 To the second ninth eleaventh & twelfth interrogatories this deponent saith that he know's the mannor or reputed mannor of Grinton & alsoe know's the places called West Grinton Harkaside & Whitaside & the moores & commons called Harka hill Harka Moore Grinton Moore Grinton Greetes Grinton How / or the Howgroves & ha's knowne the same for about thirty five yeares & for all that tyme the said places moores & commons have been comonly reputed to belong to & lye within the said mannor of Grinton & that the Swale's & Hillary's & of late Mr Marriott have been reputed lords of the said / mannor saith Hillary's wrought mynes of coale

[151] Probably a scribal error for 'marks'.

[152] In May 1708 James Loftus, no age given, gave evidence for Marriott at the trial in London (LI Misc 536, Ward MSS, notebooks vol. 6, f. 113v).

upon Grinton Moore & the Swale's wrought mynes of lead upon Harka or Harkaside Moore the place in question & this deponent ha's carryed severall quantity's of lead from thence saith Sir Solomon Swale had sometymes very great / gettings there & enjoyed the same till about ten or eleaven years agoe that the Mr Marriott recovered the same from him & after that Mr Marriott wrought the same & sunck severall new shafts & continued in possession thereof till about two years agoe that the present Lord Wharton entred upon him saith / he ha's knowne waifes & estray's frequently taken up by the agents of both the Swale's & Hillary's & beleives they were taken up upon some of the said moores & commons as they claimed to be lords thereof saith he know's the mannor of Helaugh & that the mid streame of the river Swale / ha's for all his tyme been reputed to be the antient boundery betwixt the said mannors of Grinton & Helaugh & this deponent is the more induced to beleive the same because about twenty eight years agoe this deponent lived as servantt for about two years with one James Robinson of Reeth who / was about one hundred & three years old & dyed about twelve years agoe & dureing the tyme of this deponents liveing with him he the said James very frequently told this deponent that the mid streame of the river Swale was alwaies the boundery betwixt the said mannors of Grinton & Helaugh from / Haverdale beck all along downe to neare Stallerstone Style & Reeth where they lived being neare to the said places the said James <often> took occasion to talk thereof and this deponent further saith that one Anthony Fryar an agent of the Lord Whartons told this deponent about three years agoe that the Lord Wharton / could not goe further then the mid streame of the said river Swale for his mannor of Helaugh saith that he ha's frequently travelled the said places moores & commons & never heard or knew that the Lord Wharton or any of his ancestors ever wrought any mynes or claimed any royalties in West / =Grinton or Harkaside till about two years agoe saith the said midd streame of the said river Swale ha's been alsoe all this deponents tyme reputed & accounted the boundery betwixt the adjacent mannors of Fremington & Grinton Marrick & Grinton Marrick & Ellerton Marrick & Downham & has [heard] / it may be so downwards /

Thomas Pratt[153] of Grinton yeoman aged fifty six years sworne & examined saith /

1 To the first interrogatory saith that he know's the relator Mr Marriott but doe's not know the defendants nor did he know the late Lord Wharton in his life tyme /

2 8 11 & 12 To the second eight eleaventh & twelfth interrogatories this deponent saith that he know's the mannor or reputed mannor of Grinton & the places called West Grinton Harkaside & Whitaside & the moores & commons thereto belonging called Harka hill Harka Moore Grinton Moore Grinton Greetes / Grinton How or How groves & ha's soe knowne the said places moores & waists for forty five years or thereabouts saith they are reputed to belong to the said mannor of Grinton & for the tyme aforesaid the Swale's & Hillary's & of late Mr Marriott have been the reputed lords thereof saith that[for] / thirty five years since this deponent was a miner & wrought under old Sir Solomon Swale at his lead mynes upon Harka or Harkaside Moore the place in question for severall years together & saith he & the other miners gott in the

153 Thomas Pratt senior of Grinton, miner, swore an affidavit on on 5 June 1707 (E207/114/2) which he signed 'Thomas pratt'. Thomas Pratt of Grinton, yeoman, aged 55, swore an affidavit on 18 May 1706 (E207/112/1) which he signed with the letter 'T'. Thomas Pratt of Grinton, miner, swore an affidavit on 5 June 1707 (E207/114/2) which he also signed with the letter 'T'. In view of his stated age, this deponent is probably Thomas Pratt junior who was both yeoman and miner. In May 1708 Thomas Pratt, aged 58, gave evidence for Marriott at the trial in London (LI Misc 536, Ward MSS, notebooks vol. 6, f. 110r).

said mynes considerable quantitys of oare and this deponent further [saith] / that this deponents father Thomas Pratt John Robinson Marke Hutchinson William Colling & this deponent wrought the said mynes at their owne charge & smelted the oare there gott at the late Duke of Boltons smelt mills into piggs of lead & then sold the same & out of every fother of [lead] / paid to the said Sir Solomon Swale twenty shillings saith that after the death of old Sir Solomon Swale one Roger Hillarye claimed the said mynes as lord of Grinton & gott into the possession thereof & wrought & gott lead in the waists that lay upon Harka or Harkaside Moore the place / in question & saith that the said Roger Hillarye sunck shafts & wrought for lead on Grinton How but had but small gettings saith the said Hillarye wrought for coale on Grinton Moore neare the Duke of Boltons mannors & gott coale there saith that Christopher Naylor & James Naylor [. . .] / servants or agents to the Swale's tooke up waifes & estray's on Grinton Moores saith that he this deponent was a tennant to the Hillary's for severall years & rented of them thirty six pounds per annum at Grinton & when he was tennant he received severall waifes & estray's that were brought to [. . .] / to Grinton upon account of the said Hillary's being or claimeing to be lords of Grinton mannor saith that he ha's {heard} had waifes brought in by Harkaside people & particularly Thomas Miller & James Hird but upon what moores they took them up they told not [to] / this deponent nor doe's he know saith he know's the mannor of Helaugh & for all his tyme the river Swale ha's been the reputed boundery between the said mannors of Grinton & Helaugh saith that he this deponent ha's lived at Grinton from his childhood & has been frequently upon the / moores & commons & never in all his tyme ever knew or heard that the late Lord Wharton the present Lord Wharton or any of their ancestors ever wrought any mynes or claimed any royalties in or upon West Grinton or Harkaside moores or in any other place within the said mannor [. . . till] / about two years since that the present Lord Wharton entred upon the mynes in question /

3 & 9 To the third & ninth interrogatories this deponent saith that the mannor of Helaugh lyes on the north side of the river Swale & that the mannor of Grinton lyes on the south side thereof & that the mid streame of the river Swale is the reputed boundery between the said mannors & to [. . .] / mannor is bounderd by the said mid streame of the said river Swale on the north & by Ellerton mannor at Stallerstone Style on the east from thence to the White Stone in the Sharrow from thence to Snawden Man (where the Duke of Bolton's mannors begin) from thence to / Brownsett from thence to Ridmer Standard from thence to Grinton Greetes from thence to Gibbon hill from thence to High Carle from thence to Pickerstin rigg & from thence to Tarne Hurrock & from thence to [Wynd Yates] whereabouts the Duke of Boltons mannors end / which mannors are the south boundery of Grinton mannor & from Wynd yates to a place nigh Somerlodge Tarne Hurrock from thence to Bull head from thence to bloody Wall from thence to Blagill from thence to Haverdale beck which is the west boundary of the said mannor of Grinton / & from Haverdale beck to the mid streame of the river Swale the north boundery as aforesaid saith that the said mid streame of the said river Swale is alsoe the boundery between the mannors of Fremington & Grinton parte of Marrick & Grinton & Marrick & Ellerton /

[membrane xvi *verso*]

4: & 17: To the fourth & seaventeenth interrogatories saith the hamlett of West Grinton & Harkaside consists of a few stragling houses some standing together others at a considerable distance from each other & all lye within the manor of Grinton & are reputed to be parcell [. . . saith that] / the Lord Wharton has copyhold lands at Harkaside the tennants whereof appears at his lordshipps court at Helaugh & are within the constablery of Reeth & saith the severall freeholders in West Grinton & Harkaside that appeare at the sheriffs turne & not at Helaugh court [. . .] / tennants of Swale Hall Ralph Wensley George Miller John Blaides Miles Close Robert Waller & now George Simpson having bought his lands & Leonard Galloway saith that he having been constable of Grinton has received money of the aforesaid severall persons & [. . .] / to essoine them at the sheriffs turne which was sometimes kept at Midlam & sometimes at Askrigg saith there are copyhold lands at Whitaside & Fremington that appeare at the Lord Whartons court at Helaugh saith that all Fremington is in the constablery of Reeth [. . .] / alsoe there are lands that lyes on the north side of the river Swale that are in the constablery of Grinton & manor of Helaugh; /

6: To the sixth interrogatory this deponent saith that the manor of Grinton has been reputed formerly to belong to the Priory of Bridlington but cannot sett forth the particular lands in West Grinton & Harkaside belonging to the said Priory but saith that Miles Closes lands [of the] / yearly value of eleaven or twelve pounds & Robert Wallers lands now George Simpsons of about the same value are reputed to belong to the said Priory; /

7: To the seaventh interrogatory this deponent saith that Swaledale is about twelve miles in length & has in it four manors & that the whole dale is in Grinton parish & that the parish church stands at Grinton the cheife towne in the said dale; /

10: & To the tenth & thirteenth interrogatories this deponent saith that the inhabitants
13 of Grinton Whitaside West Grinton & Harkaside have a rake for each others cattle in each others commons without interrupcion & that the same are not separately enjoyed save that the West Grinton & Harkaside / people claime a sole right in their owne commons for getting of peates ling & turfe as far as the Ox intack to the White Stone in the How saith that about two years since the present Lord Wharton caused a boundery to be ridd betweene West Grinton Harkaside & Grinton & as they never rid any [. . .] / pretended boundery one Mr Christopher Plewes came to them & desired to know what was the meaneing of their soe rideing & Mr Smales a sollicitor for the Lord Wharton replyed it was there boundary & they would ride it upon which the said Mr Plewes in the names of Mr John Hillarye & Mr [Roger] / Hillarye both of the City of Yorke discharged them from rydeing any further & told them if they did they should be all indicted upon which Mr Smales replyed they should be all defended & soe they rid it on saith he this deponent never knew or heard of any other [boundery] / ridd betweene the said places save this late boundery /

Matthew Bowes[154] of Ellerton yeoman aged fifty <years> sworne & examined saith /

2 11 & To the second eleaventh & twelfth interrogatories this deponent saith that he know's
12 the mannor or reputed mannor of Grinton & the places called West Grinton Harkaside & Whitaside & the moores & commons thereto belonging & ha's soe knowne the same for forty years past [. . .] / the same are & were dureing this deponents tyme reputed to belong to the said mannor of Grinton & that the Swale's & Hillary's were the reputed lords thereof saith that the Swales wrought & gott lead on the How & Harka the place in question & that Mr Hillary gott lead on the How & coale [on] / Grinton Moores & that he this deponent ha's severall tymes fetched coales from the said Hillary's coale pitts saith that he ha's severall tymes loosed waifes & estray's belonging to this deponents father from Christopher Naylor a servant {to} <of> the

154 In May 1708 Mathew Bowse, aged 50, gave evidence for Marriott at the trial in London (LI Misc 536, Ward MSS, notebooks vol. 6, f. 113r).

{Lord} Hillary's which waifes were taken up on [Grinton] / moores as this deponent was told saith the last year he loosed a waife sheep of his owne from Thomas Pratt a servantt to Mr Marriott which was taken up on Grinton Moore saith that he know's the mannor of Helaugh & for the said forty years last past [the mid streame] / of the river Swale ha's been the reputed boundery betweene the said mannors of Grinton & Helaugh & is alsoe a boundery betweene the mannors of Fremington & Grinton part of the mannor of Marrick & Grinton Marrick & Ellerton Marsk & Downham & the places of / Applegarth & Thorpe & soe downewards saith that he this deponent lives at Ellerton Abbey about two myles from Grinton & ha's been acquainted with the moores & commons of West Grinton for forty years & in all that tyme never knew or heard that the present Lord Wharton [nor] / any of his ancestors ever wrought any mynes or claimed any royalties in or on West Grinton or Harkaside Moores or in any other place within the said mannor of Grinton till about two years since that the present Lord Wharton entred upon the mynes in question /

9 To the ninth interrogatory this deponent saith that some small tyme before this suite was begun he heard one John Parke an old man of about seaventy years of age who lived at Hunton on the south side of the river Swale & now dead say that he was stopped when he / was fishing on the north side of the river by Grinton men for comeing further then the midd streame of the said river Swale & told him that the said midd streame was their boundery between them & those that fished on the north side of the said river /

Robert Browne of Inghead yeoman aged eighty six years sworne & examined saith /

2 & 9 To the second & ninth interrogatories this deponent saith that he know's the mannor or reputed mannor of Grinton & the places called West Grinton Harkaside & Whitaside & the moores & commons thereto belonging & ha's soe knowne the same for sixty years & upwards saith that dureing / his knowledge thereof the said places moores & waists have been reputed to belong & to be parcell of the mannor of Grinton & that the Swale's & the Hillary's have been the reputed lords of the said mannor saith that the Swale's wrought mynes of lead both on the / How & Harka on Harkaside Moore the place in queston & had good gettings till a suite happened between the Swale's & the Hillary's about the said mynes but whether gott the better of the said suite or whether the same was tryed this deponent know's not saith that / he ha's heard that both the Swale's & the Hillary's tooke up waifes & estray's on Grinton Moores saith he know's the mannor of Helaugh & for sixty years last past the tyme of this deponents knowledge thereof the midd streame of the river Swale has been the / reputed boundery between the said mannors of Grinton & Helaugh saith he ha's heard severall antient men say the said midd streame of the said river was a boundery as aforesaid and that he this deponent about fifty six years agoe being fishing with others with netts / in the said river Swale on Harkaside a Harkaside man came & threatned this deponent & his company to take their netts from them & then this deponent & his company went on the other side of the mid streame of the said river & then thought themselves safe from him / saith the he this deponent for all his tyme never heard that the present Lord Wharton nor any of his ancestors ever wrought any mynes or claimed any royalties in West Grinton or Harkaside moores or in any other place in the said mannor of Grinton till of late /

Edmund Jefferson of Catherick in the county of Yorke yeoman aged seaventy three years or thereabouts sworne & examined saith /

2 & 9 To the second & ninth interrogatories this deponent saith that he know's the mannor or reputed mannor of Grinton & alsoe know's the places called West Grinton Harkaside & Whitaside & the moores & comons thereunto belonging & ha's knowne the

same for near sixty / yeares & dureing all the said tyme the said places moores & commons have been accounted & reputed to belong to & be parcell of the said mannor of Grinton & that for all his said tyme the Hillary's have been called & reputed <lords of> the said mannor saith / he know's the mannor of Helaugh & that dureing all the said tyme of his knowledge the mid streame of the river Swale ha's been the reputed antient boundery betwixt the said mannors of Grinton & Helaugh saith that the said / mid streame of the said river is & ha's been alsoe accounted the boundery betwixt the adjacent mannors of Fremington & Grinton Marrick & Grinton Marrick & Ellerton Marrick & Sta<i>nton Marrick & Downham <Marsk & Downham> Thorpe & Aplegarth / & Hudswell & Richmond as far as Richmond bridge saith that he this deponent has very often & for many yeares fished in the said river & the said mid streame of the said river ha's been generally accounted amongst the fishermen to be / a boundery as aforesaid /

William Wynn of Carlton yeoman aged eighty two years or thereabouts sworne & examined saith /

2 9 11 To the second ninth eleaventh & twelfth interrogatories this deponent saith that he
& 12 know's the mannor or reputed mannor of Grinton & the places called Harkaside & Whitaside but know's not West Grinton for that this deponent ha's alwaies thought / there was but one Grinton saith he know's the moores & commons called Harka hill Harka Moore Grinton Moore Grinton Greetes Grinton How & the How groves & ha's knowne the said places called Harkaside & Whitaside & all / the said moores & commons for about sixty years he this deponent being borne at Ellerton near to Grinton & haveing lived for most of his tyme at Ellerton & Whitaside saith that {all} the said places moores & commons have for all the / tyme of this said deponents remembrance been reputed to lye within & belong to the said mannor of Grinton & that the Hillary's & Swale's have been the reputed lords of the said mannor saith Hillary's wrought mynes / of lead upon Grinton Moore & that both Hillary's & Swale's have usually (as this deponents has heard) taken up waifes & estray's upon the said moores & commons saith he know's the mannor of Helaugh & that the / river Swale ha's for all his tyme been the reputed boundery between the said mannors of Grinton & Helaugh as the said river Swale is the common reputed antient boundery betwixt all the severall mannors down to / Richmond bridge /

Thomas Waite
Robert Hilton
Thomas Thomson
Thomas Gyll

[membrane xvij *recto*][155]

Cross interrogatories Exhibited to be administred to the wittnesses to be produced / sworne & examined on the part and behalfe of her Majestys Attorney Generall in a cause depending / in her Majestys Court of Exchequer att Westminster wherein her <said> Majestys Attorney Generall of the relacion of / Reginald Marriott Esq is informant against the right Honorable Thomas Lord Wharton / now Earle of Wharton and the honorable Philip Wharton Esq comonly called Lord Winchendon and / others are the defendants /

155 Parchment, 390mm x 560mm.

1: Imprimis Are you a tennant of East Grinton ? Doe you as such claime right of common or any and what other rights and libertys upon the / moores of Grinton, East Grinton, Herkaside and West Grinton or any and which of them ? Are you any ways interested in the matters in question in this / suite ? Doe you expect to gain or loose by the same in any respect whatsoever ? Have you received or have you any and what promise or hope of / receiving any and what reward increase of wages or other gratification whatsoever if this cause goes either for or against the defendants ? /

2: Doe you know of any and what boundary that was att <any time and when> ridd or perambulated for the manor or pretended mannor of Grinton ? Sett forth / the said boundary and how long you have knowne the same and when by whom and upon what occasion the same was soe ridd or perambulated / Doe you know or have heard of the boundary or extent of Harkaside West Grinton and Whitaside or any and which of them ? Sett forth the / same distinctly to the best of your knowledge or beleife and the names of the said boundarys divideing the same ? /

3: Within what mannor constablery and wapentack do's the vill or township of Reeth and Helaugh or either of them lye or are or is reputed to lye ? / Doe you know of any and what lands and tennements scituate and being upon the south side of the river of Swale that lye within the / same mannor constablery & wapentack or any & which of them ? Sett forth the said lands and tennements and the names thereof and in whose / possession the same now are to the best of your knowledge or beleife. and in perticular the lands and tennements in Harkaside and West Grinton / which lye within the same <mannor> constablery or wapentack <with the said vill of Reeth> or have been reputed to lye within the same during the time of your remembrance ? / Is the river of Swale any boundary to devide the said <mannor> constablery and wapentake from the constablery of Grinton or East Grinton or from / the wapentack of Hang West in the perticular place in question called or knowne by the name of Harkaside ? Declare the truth upon your oath /

4: Doe you know or have heard that the inhabitants of Harkaside have att any time when and how long contributed with the inhabitants / of Reeth to the payment of the summe of seaven pounds two shillings and two pence or any and what other summe for the modus prescript / payment or composicion for the tyth hay and corne of the township of Reeth ? Doe you know or have heard that the inhabitants or / tennants of Harkaside have contributed with the tennants or inhabitants of Reeth in any and what other payments taxes or assessments /

5: Doe you know have heard or doe you beleive that the monastry of Bridlington or any claiming by from or under the said monastry did att any / time when and how long and under what grant or grants and when <& to> and by whom made hold and enjoy all and every the lands and grounds / within the places called Stallerstone and Haverdale Beck ? Did the <said> monastry or any and what person claiming by from or under it ever enjoy / Harkaside and West Grinton wholly and intirely and all or any and which of those lands or grounds within the same which are now held by the / Lord Wharton or his freehold coppyhold or leasehold tennants of the same ? Doe you know or beleive that the said monastry or any claiming by / from or under it had any other interest or estate in Harkaside or West Grinton save a right to some particular tennements or parcells of / ground within the same ? If yea, declare your whole knowledge of the same with the particular grounds or reasons soe induceing you to beleive the / same ? /

6: Have the lords or owners of the pretended mannor of Grinton or East Grinton or any and which of them att any time when and how long had or / taken any fineable rents gressoms allienacion fines, fines upon death or other fines for any and which of the said lands of Harkaside or West / Grinton or the moores or commons thereunto belonging ? Have any of the said lords or owners thereof had any and what escheates or forfeitures / therein or appointed any and what officers to oversee the game or to grant any and and what lycenses touching the same ? Have or had the / said lords or owners aforesaid or any of them and when any and what freehold or coppyhold tennants in Harkaside and West Grinton / that pay any and what rents for any and what freehold or coppyhold tennements within the same ? Declare what you know touching each / of the said particulars ? /

7: Doe you know have heard or beleive that the monastry of Bridlington or any and what person or persons claiming by from or under the said / monastry did att any time before the obtaining the lease in question from the Crowne <&> when and how long work any and what lead mines in / any and what part of the moors of Harkaside or West Grinton ? Was any and what profitt made of any and which of the said lead mines soe / wrought and for how long time and by whom and when ? and what was or were the name or names of the said lead workes soe wrought ? / Did you know any and which of the miners and workmen or persons concerned or imployed about the said workes ? Are they living or dead / and what were or are their names or places of abode ? /

8: Are the coppyhold or customary lands in Harkaside or West Grinton or any and which of them descendible to all or any and which of the sons / of the tennants dyeing seized according to the custome of the mannor of Helaugh or are they descendible according to the custome of the mannor or / pretended mannor of Grinton or East Grinton ? /

9: Doe you know or have heard of any and what priviledges claimed by the tennants or inhabitants of Harkaside or West Grinton of useing or / enjoyeing the common pinfold of or in the towne of Reeth or of useing or enjoying the common feilds moors or pastures of the said towne or any and / what other libertys or customes within the same ? /

/s/ John Pratt
/s/ N: Lechmere

[membrane xviij *recto*][156]

Ex parte defendant

Mr Benjamin Purchase formerly examined on the part and behalfe of the complainants and now produced sworne and examined on the part and behalfe of the defendants deposeth as follows /

[156] Parchment, 620mm x 500mm. The bottom right-hand corner of the *recto* is torn and rubbed with consequent loss of legibility. The top right-hand corner of the *verso* is likewise torn and very fragile. Paper used for repair has become semi-opaque with age and the text here is consequently very difficult to read.

3: To the third interrogatory this deponent saith that he beleives that the townships or vills of Reeth and Helaugh in this interrogatory named doe lye in the mannor of Helaugh constablery of Reeth and wapontake of Gilling West saith that there are two farmes / and severall closes lying in Harkaside which lyes on the south side of the river of Swale and formerly were coppyhold lands as this deponent beleives but whether they still continue soe or that the defendant the present Lord Wharton or his father hath / bought the same of the coppyholders or otherwise altered the tenure this deponent knowes not and that the occupyers of those coppyhold lands in Harkaside allways appeare att the court leat and court baron held for the mannor of Helaugh and doe their / suit and service as he beleives there and that there are other lands lying att Harkaside alsoe belonging to Lord Wharton which are lett to persons inhabiting there and that the said severall persons doe alsoe appear att the said courts at Helaugh for and in respect / of the said lands soe held there but not for their owne lands that they have lying att Harkaside that all or most of the lands in Harkaside are in the constablery of Reeth and wapontake of Gilling West saith that the river Swale is the reputed boundary / betwixt the two mannors of Helaugh and Grinton all along as farr as the same joynes but not betweene the said constablerys and wapentakes. /

Anthony Alderson gentleman formerly sworne and examined on the complainants behalfe and now sworne and examined on the behalfe of the defendants. /

3: To the third interrogatory saith that the vills or townships of Helaugh & Reeth in the interrogatory mencioned lye within the mannor of Helaugh constablery of Reeth and wapontack of Gilling West and saith that severall lands & tennements lying in / Harkaside on the south side of the river Swale in the respective possessions of Richard Close Luke Robinson Robert Robinson & John Hird are in the constablery of Reeth & Wapontack of Gilling West & have pasture gates upon Helaugh Moor or / Reeth Moor & saith that all the rest of the lands in Harkaside except Swalehall which is about one hundred pounds per annum and about thirty pounds per annum of the Priory of Bridlington are alsoe in the said constablery & wapontack but in / the mannor of Grinton /

8: To the eigth interrogatory saith that the coppyhold or customary lands in Harkaside are descendable to all the sons alike of the tennants dying seised according to the custome of the mannor of Helaugh but that there is noe such custome in the mannor / of Grinton that mannor being all freehold. /

Joseph Robinson formerly examined on the plaintiffs behalfe & now sworne & examined on the defendants behalfe deposeth as followes /

3: To the third interrogatory this deponent saith that there are severall inhabitants living att Harkaside that appeare & doe suit & service att the defendants courts att Helaugh but dos not know for what perticular lands /

4: To the fourth interrogatory this deponent saith that he very well remembers that his father being constable of the township of Reeth which lyes on the north side of the river of Swale that the inhabitants of Harkaside which lyes on the southside of the / said river either came to the common watch att Reeth or paid their proporcion for setting on the said watch /

Robert Maltus formerly sworne & examined as a witness on the part & behalfe of the relator Mr Marriott & now sworne and examined on the part of the defendants /

1: To the first interrogatory saith he is a tennant in Grinton or East Grinton to one Mr Blake of London of an house & land of the yearely rent of eight pounds tenn shillings which he holds by lease for one & twenty years & as tennant thereto / claimes right of common on the moores of Grinton or East Grinton under his said landlord /

Thomas Haykins formerly sworne & examined on the relator Mr Marriotts behalfe & now sworne & examined on the behalfe of the defendants the Lord Wharton & others deposeth as followeth /

1: To the first interrogatory saith he is a tennant to Mr Roger Baynes of a cottage in Grinton under the yearely rent of nineteen shillings per annum saith he claimes right of getting turfes upon Grinton Moores as he is a tennant to the said / cottage & hath gott turfes there when he had occasion. /

John Coates formerly sworne & examined on the relator Mr Marriotts behalfe & now sworne & examined on the behalfe of the defendants the Lord Wharton & others deposeth as followeth /

1: To the first interrogatory saith he is a tennant to Jennet Metcalfe spinster of a cottage in Grinton under the yearly rent of thirteen shillings and six pence saith he getts ling & turfs upon Grinton Moors as he is a tennant to the said cottage & / hath gott both ling & turfes there when he had occasion. /

John Garth formerly sworne & examined on the part & behalfe of the relator Mr Marriott now sworne & examined on the part & behalfe of the defendant the Lord Wharton /

2: To the second interrogatory saith he knowes not of any boundary that was ever ridd or perambulated for the mannor of Grinton saith that Browney gill is the reputed boundary between Harkaside & Whitaside. /

3: To the third interrogatory saith that the township of Reeth lys within the mannor of Helaugh within the constablery of Reeth & that the towne of Helaugh lys within the said constablery but knowes not in what wapontacks they doe / lye saith that the Lord Wharton hath coppyhold lands att Harkaside but knowes not the tennants thereof & that he hath alsoe severall other coppyhold lands att Crackpott Satteron & Spring End all on the south side of Swale which said severall / places are within the mannor of Helaugh & constablery of Reeth & saith that the Lord Wharton hath alsoe about thirty days mowing att Whitaside in the possession of Mr Walbanke & Tamar Fryer & Widdow Park and saith that they have / allways called the river Swale a division to divide the mannor of Helaugh from Grinton but never heard it called a boundary /

Ralph Close formerly examined as a witnesse on the complainants behalfe & now sworne & examined on the part of the defendants /

2: To the second interrogatory saith he never knew of any boundary att any time rid or perambulated for the pretended mannor of Grinton /

3: To the third interrogatory saith that he hath heard & beleives the Lord Wharton hath coppyhold lands att Harkaside but cannot sett forth the same perticularly saith he never heard till within these twelve moneths that the river / Swale was reputed a boundary to any place whatsoever. /

James Raw formerly examined on the complainants behalfe & now sworne & examined on the defendant the Lord Whartons behalfe /

3: To the third interrogatory saith that the townships of Reeth & Helaugh are within the mannor of Helaugh & saith that the Lord Wharton hath coppyhold lands att Whitaside Crackpott Satteron & Spring End all lying on the south side / of Swale & are as this deponent beleives within the mannor of Helaugh for that the tennants thereof hold the said lands by coppy of court roll of the said mannor of Helaugh & the said tennants doe suit & service at Helaugh court for the same / and saith he never heard the river Swale reputed a boundary to divide the mannor of Helaugh till this suit begann. /

Tobias West clerk viccar of Grinton formerly sworne & examined on the complainants behalfe & now sworne & examined on the behalfe of the defendants deposeth as followeth /

1: To the first interrogatory saith that as viccar of the parish of Grinton he holds & enjoyes the viccarage house & some glebe lands there but claimes noe common rights upon the moores in the interrogatory mencioned having granted away all his common / right there /

2: To the second interrogatory saith that the extent of Harkaside & West Grinton is from the gate near the elm tree in the highway leading from Grinton or East Grinton to West Grinton on the east to Browney Gill on the west & to the height of the moors / adjoyning upon Wensleydale on the south & to the river Swale on the north /

3: To the third interrogatory saith that the vills & townships of Reeth & Helaugh are within the mannor of Helaugh & constablery of Reeth & wapontack of Gilling West saith that there are some lands [lying] att Harkaside on the south side of the / river of Swale part of them the Earle of Whartons own lands part of them coppyhold land & part freehold lands the tennants whereof he beleives appear att the court holden for the mannor of Helaugh saith he beleives the same lys in the / constablery of Reeth & wapontack of Gilling West but doth not know whether the same lye in the mannor of Helaugh or not & for the perticulars of the said lands & the tennants thereof this deponent [referreth] himselfe to his deposition made [in] / answer to the fourth & seaventeenth interrogatories administered to him on the complainants behalfe and saith that the river Swale is a boundary in the particular place in question called by the name of Harkaside betwixt the mannor of Helaugh / & the mannor of Grinton but knowes not that the said river is a boundary there to the said constablery or wapontake or what is the boundary there to the <said> constablery & wapentake /

5: To the fifth interrogatory saith he doth not know whether or noe the monastry of Bridlington or any claiming under the said monastry did ever enjoy any of those lands or grounds in Harkaside which are now held by the Lord Wharton or his freehold / coppyhold or leasehold tennants of the same. /

6: To the sixth interrogatory saith that he cannot say that in his time the lords or owners of the mannor of Grinton have taken any fineable rents gressomes allienacion fines fines upon death or other fines for any of [their] lands in Harkaside / or West Grinton or the moores or commons thereto belonging. /

8: To the eigth interrogatory saith that the coppyhold or customary lands in Harkaside & West Grinton are descendible upon the death of the tennant dyeing seized to all the sons of such tennant alike according to the custome of the mannor of / Helaugh but dos not know whether the said lands are in the mannor of Helaugh or noe. /

George Miller of Harkaside in Swaledale formerly examined on the complainants behalfe & now sworne & examined on the defendant the Lord Whartons behalfe /

2: To the second interrogatory saith he knowes there are some boundary markes divideing Harkaside from Grinton or East Grinton (vizt) the Ox Intack nook the White Stone on the How & the Gray Stone on the Greets & saith that the markes are / for distinguishing the commons as to getting ling & peates & turfes & that Browney Gill is the boundary betwixt Harkaside & Whitaside downe as farr as Scabba Yate for the purpose aforesaid /

3: To the third interrogatory saith that the townships of Reeth & Helaugh are within the mannor of Helaugh constablery of Reeth & wapontake of Gilling West saith there are coppyhold lands lying att Harkaside in the possession of William / Arundell Mr Marmaduke Hartley Ralph Wensley & Christopher Raper & that the tennants thereof appeare & doe suit & service at the Lord Whartons courts att Helaugh in respect of those lands & that there are freehold lands lying / alsoe att Harkaside in Anthony Hird John Hird & Ralph Addison possession which said lands are within the constablery of Reeth and that the river Swale is a boundary [to divide] the mannors of Grinton & Helaugh but / knowes not that it is a boundary to divide the said constablery of Reeth & the wapontake in the place called Harkaside but as above. /

[membrane xviij *verso*]

[5:] To the fifth interrogatory saith that there are lands in Harkside which are reputed to belong formerly to the Priory of Bridlington in the possession of Miles Close of about twelve pounds per annum alsoe other lands there in [George Milners] possession of / about six pounds per annum alsoe lands in George Simpsons possession of eleaven pounds or twelve pounds per annum & beleives some of the lands in Leonard Galloways possession about four pounds per annum [but knowes noe other priory lands] / there saith he never heard that the freehold coppyhold or leasehold lands lying att Harkaside & now belonging to the defendant the Lord Wharton were ever held or enjoyed by the Prior of Bridlington or any claiming by from or / under the said Prior /

[6:] To the sixth interrogatory saith he never knew or heard that the lords or owners of the pretended mannor of Grinton or any of them ever had or took att any time any fineable rents gressomes allienacion fines fines upon death or / any other fines for any lands att Harkaside & West Grinton or the moores & commons thereto belonging or had any escheats or forfeitures paid them there & never heard of any free rents demanded of the freehold or coppyhold / tennants of Harkaside & West Grinton by the lords or owners of the said pretended mannor of Grinton or East Grinton but saith there is a pepper corne reserved by his deed to be paid for his lands att Harkaside & West Grinton to the / cheif lord thereof yearly /

[8:] To the eighth interrogatory saith that the coppyhold or customary lands in Harkaside & West Grinton att death of the tennants thereof are descendible to all the sons alike according to the custome within the said mannor of Helaugh & / that theres noe such custome within the mannor of Grinton or East Grinton. /

/s/ Thomas Waite
/s/ Robert Hilton
/s/ Thomas Thomson
/s/ Thomas Gyll[157]

[157] The signatures of the four commissioners are followed by the *liberatur*, or certificate of delivery to the court (see p. 20 footnote 48).

Thomas Lord Wharton v. Attorney General and Reginald Marriott

Pleadings[158]

E112/915 EBOR 100 – document 1[159]

[membrane i][160]

To the right honourable Sidney Lord Godolphin lord high treasurer of England Henry Boyle esquire chancellor and under treasurer of her majesty's Court of Exchequer at Westminster Sir Edward / Ward knight lord cheife baron of the same court and to the rest of the barons there. /

Humbly complaineing sheweth unto your honours your orator Thomas Lord Wharton Baron Wharton <& Philip Wharton son & heir apparent of the said Thomas Wharton an infant by the said Thomas Wharton his next friend & guardian> (debtors farmers & accomptants to her majestie as by the records of this honorable court & otherwise doth & may appear) That Philip late Lord Wharton your <said> orators father <& grandfather> being in his life time for sixty yeares & upwards & to the time of his death was / seized of and in the dominion lordshipp & mannor of Swaledale & of Helaugh alias Helaw alias Ryeth in Swaledale with its rights members & appurtenances in the parish of Swaledale or Grinton in Swaledale in the county of York within the antient boundaries whereof there being diverse vills townshipps or hamletts with diverse moors wasts and / commons thereunto belonging & divided or distinguished from each other & in particular the vills townshipps or hamletts commonly called or knowne by the names of East Grinton West Grinton & Herkeside your orator sheweth that the said Philip <late Lord Wharton> did during all the time aforesaid quietly & peaceably hold & enjoy diverse demesne lands rents services / royalties priviledges & other profitts in Herkeside & West Grinton aforesaid & the moors wasts & commons thereunto belonging or therewith used occupied or enjoyed And your orator further sheweth that the tenants of Herkeside & West Grinton aforesaid as well freehold & coppyhold as leasehold tenants have for / yeares last past & for time whereof the memory of man is not to the contrary paid their rents & done accustomed suit & service at the severall & respective courts of your orators ancestors & those under whome they respectively claime in Swaledale aforesaid And your orators sheweth that your orators ancestors [and those] / under whome they

158 In TNA, the pleadings are contained in portfolio E112/915 EBOR 100.

159 Bill of complaint. The bill is endorsed in the top left-hand corner with the legal term and regnal year, Easter 5Anne [1706], but is otherwise undated. The last membrane has been signed by the chief baron, Sir Edward Ward. The bill was amended by order of the court on 26 June 1706 when Philip Wharton's name was added as a party (E161/65, f. 79v). A note to this effect, signed by [Thomas] Harris, Lord Wharton's sworn clerk, appears in the left-hand margin of the first membrane. As a result of this amendment it became necessary to insert Philip's name at several points in the bill and alter various nouns, pronouns and verbs accordingly. Other lengthy interlineations on the fourth membrane may have been inserted at the same time as they do not appear to have been separately authorised by the court.

160 Parchment, 810mm x 760mm. The lower right-hand edge is torn and rubbed.

claime as aforesaid [&] their tenants agents and assigns have time beyond memory constantly held used and enjoyed the said wasts and commons of Herkesyde & West Grinton not onely by feeding and depasturing their cattle thereon & by digging and taking of peates & turves & by pulling of ling but / alsoe by workeing the mines & quaries digging & takeing the stones thereof & by a constant & uninterrupted enjoyment of free chase and free warren within & throughout the same and have done severall other acts from time to time as often as there was occasion to manifest & prove their respective right & title to the said wasts & / the soyle & the royalties thereof And your orators sheweth that whenever your orators ancestors or those whose estate they have in the premisses <have> thought fitt to use or make any profitt of the soil or quarryes of the said wasts or to grant licence to inclose or cutt downe wood within the same that noe person or persons whatsoever / have disturbed or pretended to interrupt or disturbe your orators & their tenants or those under whome they respectively clayme <in> the peaceable enjoyment & perception of the profitts of the said wasts & commons but more particularly of the wasts & commons of Herkesyde & West Grinton or either of them And your orator[s][161] / sheweth that they haveing of late under the antient right & title of their ancestors & those under whome they claime as aforesaid sunk for lead oar in a certaine part of the wasts & commons scituate & being neare to the lead works called How Groves now wrought by Reginald Marriott esquire (one of the / defendants hereafter named) yett within the knowne or reputed boundaries of Herkesyde & West Grinton or the one of them or of the wasts & commons thereunto belonging or scituate & being within the wasts & commons of right belonging & appertaineing to your orators & their tenants in Swaledale aforesaid [And] / your orators well hopes that your orators their servants agents miners & workmen should have proceeded & continued to worke the said mines within the precincts territories & limitts of the wasts moores & commons belonging to your orators & their tenants and within the liberties thereof without interrupcion / or disturbance of him the said Reginald Marriott or any other person or persons whatsoever But now soe it is may it please your lordshipps that the said Reginald Marriott haveing applyed himselfe to Sir Edward Northy knight her majesties Attorney Generall to aid and assist him in his pretence to your / orators mines under colour of her majesties right in the said mines & he the said Marriott combineing & confederateing himselfe together with George Tussingham of London gentleman (one of the clerks of her majesties Court of Exchequer aforesaid) and to and with Toby West clerke parson or vicar of the / parish church of Grinton or Swaledale aforesaid)[162] and to & with Sir Solomon Swale barronett (one of your orators freehold tenants in West Grinton) Humphrey Bellamy Andrew Hopegood & Edward Shaller[163] (whoe claimed some estate or interest by mortgage or otherwise from the said Sir Solomon in his / freehold tenement in West Grinton aforesaid) and to and with --------- Hillary of the city of Yorke (the pretended lord or owner of the pretended mannor of Grinton alias East Grinton) & to and with other of the tenants of certaine small peices or parcells of land in Herkeside & West Grinton / the one of them and to and with diverse other persons to your orators unknowne (whose names when discovered your orators praye may be inserted herein & they made parties hereunto with apt words & clauses to charge them) severall of which said confederates have of late entered into <a> partnershipp combinacion [or] / engage-

161 Letter missing in the original.

162 This bracket is not opened.

163 Humphrey Bellamy, Andrew Hopegood and Edward Shaller were Sir Solomon Swales's mortgagees.

ment by bond covenant articles or otherwise how to deprive your orators of the benefitt & advantage of their said works and of the mines & royalties of the wasts mores & commons aforesaid wherein the orators has sunk their said shaft grove or pitt for the getting of lead oar He the said Reginald / Marriott with intent to assert his private purposes & to enrich himselfe out of the mines & royalties of your orators estate under the pretence of defending & asserting the rights & possessions of the Crowne did of late years whilest he was an officer under the Crowne & had the opportunity of inspecting the lease / made and the surveys then of the Crownes possession and under colour of examineing the same for the benefitt & advantage of the Crowne by some wayes or meanes unknowne to your orators did obtaine a lease from the late King William the third in the name of the said George Tushingham (amongst other things) / of the mines & veines of lead within the territories & feilds of Grinton < Whitaside> & Harkaside as being parcell of the lordshipp or mannor of Grinton or of the possessions of the late monastery of Bridlington in the county ofYorke to hold him the said George Tushingham for the terme of one & thirty years [from] / the makeing thereof under the rent of twenty shillings and one full tenth part of the clear yearly profitts of the said mines which said lease soe obtained by the said Reginald Marriott & the said George Tussingham or the one of them he the said Marriott (as your orator has reason to believe[164] was / encouraged to take upon some prospect of defeating the then pretended title of the said Solomon Swale he the said Sir Solomon pretending the freehold tenement in West Grinton soe held of your orators to be a mannor & he the said Sir Solomon haveing in right of such said pretended mannor by a ficticious title / and which after several tryalls has been adjudged & found against <him> claymed not onely the mines in West Grinton & Herkesyde now wrought by your orators but alsoe all & every other the mines in Grinton alias East Grinton & Whitaside soe leased or pretended to be leased as aforesaid And he the said Reginald / Marriott being soe encouraged either from the weakness & extravangance of Sir Solomons said claime or for that your orators deeds with the antient evidences & writeings relateing to your orators estate & possessions in Swaledale aforesaid were at that time & have ever since continued till within these three / monthes last past or thereabouts out of your orators custody or knowledge he the said Reginald Marriott under the pretence and colour of the said lease hath not onely wrought the lead mines or works called by the name of the How Groves the said lying as pretended within that part of the wasts or commons [. . .] / Ho alias How which belong as is likewise pretended to the vill or towne of East Grinton or [which] is enjoyed by the tenants thereof But he the said Marriott (still presumeing your orators evidences to be all of them lost or not knoweing that your orators had lately found out & recovered <severall of> the said antient evidences [. . .] / their estate hath of late attempted to passe & transgresse the knowne boundaryes divideing betwixt the wasts & commons belonging to the said vill or towne of East Grinton & those belonging to West Grinton & Herkesyde & hath attempted to follow the vein of the ------- mine or workes called the How Gr[oves] / not onely through the precinct or territoryes of the wasts of East Grinton but into the place where your orators are now workeing within the precincts & territoryes of West Grinton & Herkeside & threatene to follow & worke the said mines aswell through your orators said wasts belonging [to] / West Grinton & Herkeside as other their wasts in Swaledale aforesaid & to interrupt & disturbe your orators & their workemen agents & servants in workeing the said mine though your orators are in the actual possession thereof

164 This bracket is not closed.

And to colour such his illegal proceedings the said confederates [or] / some of them doe some times pretend that the mannor of Grinton (as the same is mencioned in the lease soe made to the said George Tussingham) was antiently vizt at or about the time of the conquest reputed to be a reall mannor & that the same is accordingly found to be described by the antient record of Doomesday Booke & to contain [. . .] / league in length & halfe a league in breadth And they doe further pretend that afterwards to witt in or about the reigne of King Henry the first and Walter de Gant & Matilda his wife (the daughter of Stephen Earle of Britany & Richmond being seized of the said mannor or reputed mannor of Grinton aforesaid he did grant [. . .] / with the church of Grinton or Swaledale aforesaid to the priory of monastery of Bridlington in Yorkshire And the said confederates doe likewise pretend that the said mannor of Grinton haveing come to the Crowne upon the attainder of the prior of Bridlington or by the dissolucion of the said monastery That [. . .] / did afterwards about the one and fortieth yeare of her reign grant the said mannor or reputed mannor of Grinton to one Fitch & Wiseman[165] (under whome the confederate Hillary or the said Marriott by some conveyance lately made by him the said Hillary pretended to have some right or title) But yett they pretend [that] / the said pretended mannors to Fitch & Wiseman haveing an excepcion of the lead mines therein that the said lead mines have ever since the dissolucion of the said monastery as well before as since the said grant to the said Fitch & Wiseman been separately enjoyed by the Crowne & their lessees & farmers And [. . .] / at other times pretend that the said pretended mannor of Grinton & the territoryes & feilds thereunto belonging are & imimerorially have been a large tract or peice of ground containeing severall demeanses inclosures lands grounds wasts & moors of a very great extent all lying together & butted & bounded by severall other mannors [. . .] / wasts [x x x][166] & other boundaries but that noe part of any other mannor lands or wasts doe lye intermixed with the said pretended mannor wast & moores of Grinton aforesaid or in any wise divide or breike in [. . .] / And it is likewise pretended that the said mannor of Grinton is separated & bounded northwards by the river Swale & soe by the severall mannors northward from the said river And that the severall places called or knowne by the names of Harka Harkahill Harkeside & West Grinton & the wasts & moors / thereunto belonging aswell as the moores & wasts called Grinton How, Whitaside & other places are all of them part & parcell of the said pretended mannor of Grinton & comprehended within the boundaries & precincts thereof which said boundaries of the said pretended mannor of Grinton aforesaid are pretended to [. . .] / particularly from the mannor or lordship of Ellerton or the bounder marke there called Stallerstone Stile on the east to a certaine rivulet called Haverdale beck on the west and from the said river of Swale on the north to & beyond a certaine place called the High Carle on the south adjoyning to Wensleydale / the lands or estate therein belonging to his grace the Duke of Bolton And that a part of the orators mannors or lands doe lye within the said pretended boundaries or extend southward beyond the river of Swale aforesaid Whereas the said confederates doe very well know or cann sett forth & [. . .] / it appears that there is noe such mannor of Grinton now remaineing nor of any such dimensions or extent as the same is described or pretended or pretended to be described in Doomsday aforesaid And your orators shew that the dimensions or extent of the said pretended

165 In November 1599 the manor of Grinton was granted by the Crown to Francis Fitch and Richard Wiseman, citizens of London, for the sum of £1463 9s 2d (LMA CLA/044/01/112).

166 Half a line has been erased.

mannor Grinton as mencioned / in the aforesaid survey called Doomsday Booke is much different from the other pretended boundary which is now claimed by the said confederates & as pretended to extend from the aforesaid lordshipp or mannor of Ellerton or the bounder marke there called Stallerstone stile on the [east & by] / the aforesaid rivulett called Haverdale beck on the west & from the said river of Swale on the north to & beyond the High Carle on the south And in case any such mannor there now be of such dimensions & extent as are mencioned & described in the said survey of Doomsday [&] / different from the said pretended mannor of Grinton now claymed by the said confederates or some of them your orators hath reason to beleive that they your orators are best intituled to the said old mannor & the royalties thereof For the said confederates but particularly the said Reginald Marriott [&] / Sir Solomon Swale aswell by their inspeccion of antient records evidences & writeings & of copies or abstracts taken thereof or otherwise by intimacion of such as have beene imployed by them respectively to search the records & evidences relateing to the title or boundaryes of the lordshipp of / Swaledale & of the said pretended mannor of Grinton doe very well know or are satisfyed & beleive or have been informed that the antient mannor of Grinton as described & mencioned in the said survey of Doomsday Book haveing in the time of William <the> Conqueror come to the Crowne together [with] / the severall other mannor towns vills hamletts & lands in Swaledale aforesaid the said King did by his charter grant all those his towns & lands in Yorkshire scituate & being near or about the castle or towne of Richmond in the said county which thence afterwards [. . .] / the name or title of the honour of Richmond or Richmondshire unto his nephew Allan (the then Earle of Britany) & his heires including therein the lordship of Swaledale as part & parcell of the said honour And they are likewise well satisfied & cannot deny but that Stephen Earle of Britany & Richmond / being afterwards (vizt) about the reign of King Henry the first seized of the said honour of <Richmond or> Richmondshire (includeing Swaledale as parcell thereof) did thereupon grant in frank marriage with his daughter Matilda to Walter de Gant the whole & intire dominion lordshipp or territory of Swaledale / included therein all & every the antient mannors vills & hamletts in Swaledale aforesaid particularly the mannor of Grinton as mencioned & described in Doomsday booke the royaltyes of all which mannors or reputed mannors vills or hamletts in Swaledale aforesaid became [. . .] & / the said Walter de Gant & his said wife or the one of them in right of their dominion & lordshipp of Swaledale And the said confederates doe very well know & are satisfied & beleive or have been informed & your orators doubte not but it will appear by the grants made to the said monastery or by [. . .] / coppyes thereof now in the custody of the said confederates or which they or some of them have seen or can or may come by that the said Walter de Gant & his said wife granted onely the vill of Grinton alias East Grinton or some tenements therein to the said monastery & that what right & [. . .] / monastery had or ever enjoyed either in any of the lands of West Grinton or Herkaside (in case they had ever any right therein) or in Whitaside Cogden & severall other places within the said pretended boundaryes was granted to the said monastery by the lord of Swaledale by diverse part [. . .] / subsequent to the grant of Grinton aforesaid & were not enjoyed by the said monastery under their original grant of Grinton And the said confederates doe likewise know that if the said monastery of Bridlington had ever any right or interest in the lands of West Grinton or [. . .] / same was onely in certaine particular peices or parcells of land scituate & being within the respective precincts & territories thereof & under the said particular grants so made as aforesaid And that the said monastery was never seized of all the said lands of West Grinton [. . .] / Herkeside or either of them but on the contrary that the lords of Swaledale

& of Helaugh aforesaid did during the continuance of the said monastery constantly enjoy all or the greatest part of the said lands or rents issueing thereout And yett the said confederates findeing by some records coppies [. . .] / evidences or other writeings that the severall lands & possessions in Swaledale aforesaid held & enjoyed by the monastery of Bridlington are therein mencioned to be all of them scituate lying or being betwixt the said lordshipp of Ellerton or the bounder marke there called Staller Stone [Stile on the east] / and the said Haverdale Beck on the west & betwixt the river Swale on the north & the said High Carle or the bounders of Wensleydale on the south they the said confederates under colour thereof insist & pretend that the aforesaid places are the proper boundaryes of the said pre[. . .] / Grinton Whereas the said confederates very well know that the said Stallerstone Stile is one of the knowne & antient bounder markes of your orators lordshipp of Swaledale aforesaid divideing the same from the lordshipp of Elerton & that the said Stallerstone Stile & the said [. . .] / being the most remarkeable & notoriously knowne in those parts are onely mencioned to describe the scituacion of the said severall possessions which then belonged to the said monastery or to shew that none of their said possessions in Swaledale aforesaid extended further than the [. . .] / your orators beleive{s} the said monastery lands did all of them lye within the same Yet your orators sheweth that the said monastery did never enjoy all the possessions lands & tenements scituate & being betwixt the said pretended boundary For your orators insist & the said confederates [cannot] / deny but that the lords of Swaledale & their tenants in Herkeside & West Grinton have time beyond memory & dureing the whole continuance of the said monastery & all a long hitherto without interrupcion or disturbance constantly & peaceably held & enjoyed diverse possessions [&] / other priviledges within the same pretended boundaryes And that the said monastery never enjoyed any possessions lands or liberties within the said pretended boundaryes but what they held under some grant or grants from the lord of Swaledale or some [. . .] / the said confederates deny but that your orators are now well intituled to all such possessions rights <and> roylatyes within the said pretended boundaryes & the dominion lordshipp & mannor of Swaledale as have not yett been granted away by the lords of Swaledale or some of them [. . .] / doe like wise know that the river of Swale was never a bounder to the lordshipp dominion or mannor of Swaledale as is some times pretended by the said confederates nor doe they find by any records or antient evidence that the said river was ever accounted the boundary [. . .] / on the contrary the said confederates must owne & cannot deny but that the lordshipp dominion or mannor of Swaledale as the same was granted to the said Walter de Gant did antiently extend it selfe to all the lands vills & hamletts within the said dale both on the north & south side of [the] / river Swale & particularly to the vill of Grinton alias East Grinton aswell as to Herkeside & West Grinton aforesaid And that the antient boundary of the lordshipp of Swaledale did more particularly extend it selfe from the aforesaid bounder marke called Stallerstone Stile [which] lyse on the / midle or bottome of the said dale near unto & on the southside of the river of Swale)[167] to a certaine antient & knowne bounder marke standard or hurrock of stones lying at the west end of the said lordshipp called Hollow Mill Crosse alias Hollow Miln [. . .] / of the river of Swale where the same bounder marke divides not onely betwixt the said lordshipp of Swaledale & the next adjoyning lordshipp or mannor there but betwixt the parish of Swaledale or Grinton & the parish next adjoyning thereto as alsoe divides the county of Yorke from the / county of Westmorland & is soe taken

[167] This bracket is not opened.

notice of in severall antient printed mapps discripcions & boundaries of the said countyes And your orators [sheweth] that the river of Swale which runneth through the middle or bottom of the dale or lordshipp of Swaledale being by nature separate & divided from the other rivers [which] / runn through the adjacent dales (vizt) Wensleydale on the southside & Arkildale or Arkilgarth Dale on the north side thereof by certaine mountaines & very high hills which for many miles together runn all along on the north & south side of the river Swale and that the said river Swale being [. . .] / supplyed by certaine springs rivuletts [&] becks which descend from the aforesaid hills & mountaines on each side of the same river & runn into the same through certaine gills or lesser vales your orators sheweth that the said dale or lordshipp of Swaledale [on each side of the river Swale] / does naturally extend to the heads of every of the said gills springs rivuletts or becks & soe includes therein all such grounds from whence the raine water had any discent into the said dale or river aforesaid And the said dale or lordshipp does not onely naturally or visibly take in all / the said grounds gills & rivuletts but it more particularly extends further to the very sumitt or height of the said hills & mountaines whence the rain water as the same falls divides & deals part into one dale & part into the other And soe your orators shew that the said boundary of the said / lordshipp on the north & south sides of the said dale proceeds by the sumitt or height & ridge of the hills as the heaven water deals or divides it selfe And your orators shew that the said boundary of the lordshipp of Swaledale vizt from Stallerstone Stile [to] Hollow Mill Crosse as heaven water deales / on each side of the said dale [. . .] not onely the said vill of Grinton but the severall other places comprized within the said pretended boundaries of the said pretended mannor of Grinton And your orators shew & the said confederates know it to be true that [. . .] boundary from Stallerstone Stile [to] / Hollow Mill Crosse Holow Miln or [ ][168] min as heaven water deals on each side of the said dale was not onely the reputed boundary of the lordshipp of Swaledale before the time that the said Walter de Gant & his wife made the said grant unto the monastery of Bridlington but / that a long time after (to witt) in the reigne of King Edward the First one Gilbert de Gant (then lord of Swaledale) as appears by the records of the justice in eyre for the county of Yorke and which the confederates or some of them [have seen] & perused appeared and claimed / before the said justice free chase & free warren in Swaledale aforesaid within the same boundaryes vizt from Stallerstone Stile to Holow Min *sicut* heaven=water *radit ex utroqe parte* And that the said Gilbert de Gant did not onely soe claime [the] liberties aforesaid but obtained a / verdict & judgement for the same upon a writt of *quo warranto* whereby it was found that the said Gilbert had no way inlarged his said liberties And the said confederates doe very well know that the said vill of Grinton with other the vills on the southside of the river / of Swale may be in some parts bounded by the said river from the other vills on the northside thereof yett the said river was never any boundary to the dominion lordshipp or mannor of Swaledale or of your orators mannor of Helaw or the liberties and priviledges thereof / as the said confederates endeavour to insinuate And your orators further shew{eth} that the said confederates doe some times admitt that the said monastery of Bridlington had not originally the mannor of Grinton or the royalties thereof granted to them but that / the [x x x][169] estate of the said monastery in Grinton aforesaid was originally granted to the said monastery by the name of a vill or village or by the name of Grinton without any further descripcion

168 Damage to the parchment has caused the loss of a single word.
169 One word has been erased.

or addition thereto And they some times admitt or cannot deny but that / the said monastery held their lands in Harkeside Whitaside or other places within the pretended bounder aforesaid by grants subsequent to the originall grant of <Grinton> to the said monastery But yett they pretend that the said lands being by their tenure in /

[membrane ii][170]

frank almoign discharged & exempt from suit of court & service or services the priors of Bridlington had good right (as is alledged) in all their said lands soe granted off & exempt as aforesaid to hold courts & enjoy the rights & priviledges of a mannor And did likewise pretend that by custome useage [& long enjoyment] / ever since the time of the said severall & respective grants or before the said grants that the said vill of Grinton with the said other particular peices or parcells of ground enjoyed therewith by the said priory within the said pretended boundaries have assumed the name & reputacion of a mannor And that the priors of Bridlington / amongst their owne tenants have all a long exercised a distinct & separate jurisdiccion & have been noe way accountable to any of the lords of Swaledale or done suit or service at any of their courts And the said confederates at other times pretend that the lords of Swaledale or some or one of them some time after / the aforesaid grants soe made to the said priory by Walter de Gant or his said wife did by some other subsequent grant or grants to the said priory grant or confirme their estate of Grinton or other their lands & possessions within the said pretended boundaries by the name of a mannor or lordshipp / or by words equivalent to conclude the claime & antient rights of the lords of Swaledale Whereas the confederates have reason to beleive that noe such grant or confirmacions were ever really made and that they well know that there are noe licenses or pardons of alienacion to be found upon record to [warrant] / the makeing of any such grants or to inable the monastery to take any interest or estate thereby nor can they sett forth any writts of *ad quod dampnum* that were thereupon awarded as in the like cases nor are the same grants or pretended charters of confirmacion mencioned or comprized within any of the *reguli* / charters of *inspeximus* or confirmacion made to the said monastery tho' the same or most of them are all remaineing upon record in her majesties record office in the Tower & all or most of them have been inspected & perused by the said confederates or their agents And the said confederates doe further insist that / upon the attainder of the prior of Bridlington about the nine & twentieth year of the reigne of <King> Henry the Eighth some inquisicion was taken of the severall possessions of the said priory or monastery of Bridlington whereby it appears (as is pretended) that the said Grinton alias East Grinton which formerly belonged to / the aforesaid monastery is mencioned by the name of a mannor though the confederates know that the originall inquisicion upon the attainder of the said prior is not now extant or if the same be extant the said confederates doe conceal & will not discover to your orators in whose custody or power the same now is or in / what office which your orators hope that said confederates shall be obliged to sett forth And the said confederates doe further insist or pretend that in some other survey or surveys <that> have been taken of the lands & possessions of the said priory either upon the dissolucion thereof or since the same came to the Crowne as alsoe / in all rentalls particulars & leases made from the Crowne thereof & in all the accounts of the rents & profitts & likewise in the afore-

170 Parchment, 820mm x 735mm. The lower right-hand edge is rubbed and decayed.

said grant to Fitch & Wiseman[171] the said Grinton (as pretended) is named or described to be a mannor And it is further alleadged that the lands thereat & formerly enjoyed within / the said pretended boundary & particularly such peices & parcells of land as the said monastery enjoyed within the precincts & teritoryes of West Grinton & Herkeside are mencioned to be held as members or appurtenances of such said pretended mannor Whereas the confederates refuse to discover & will / not sett forth what records grants leases surveys rentalls particulars <and> accounts they have or know of or in whose custody or power the same are or be whereby it may appear that Grinton is either named a mannor or that West Grinton & Herkeside are mencioned as members of the same And your orators charge that / there are no such records evidences or writeings wherein the same is called a mannor but that on the contrary there are severall antient & other records & diverse inquisicions post mortem taken <up> on the deaths of the respective lords of the lordshipp of Swaledale or mannor of Helaugh which are now remaineing / in her majesties record office in the Tower & amongst the records of this court & in other offices which the confederates can discover & sett forth in which said records the estate of Grinton belonging to the said priory is mencioned & described by the name of a vill or village & soe found to be held of the lords of / Swaledale or Helaugh aforesaid or as <part> parcell or member of their lordshipp or estate in Swaledale aforesaid And the said confederates can likewise sett forth that there are severall other records evidences or writeings which prove or evidence the said Grinton alias East Grinton (as occupyed or enjoyed by the said priory / or as enjoyed by the Crowne or some claymeing under it) to be noe reall mannor & that there have been diverse disputes suites & causes about the reality of the said mannor & particularly about the time of King James the First at which time one Henry Simpson (then vicar of Grinton) pretending right to a mannor / and to severall royalties & priviledges therein as lord of the pretended mannor of Grinton which the said Simpson soe claimed as is pretended under the said grant from Queen Elizabeth to the aforesaid Fitch & Wiseman the confederates know of have heard or been informed that the said Simpson was / opposed therein by the tenants of Grinton whoe defended themselves in a certaine suit which long depended in this honorable court wherein diverse witnesses were examined & deposicions taken certified & published in order to prove that the said Grinton was noe mannor of which said proceedings the / confederates or some of them have court coppies or other transcripts or knowe of the same but by some means or other unknowne to your orators (but well known to some of the confederates) the bill answeres & deposicions & the originall proceedings in the said cause are taken off the proper files & / destroyed or concealed to your orators prejudice & the decree made in the said cause is likewise concealed & nothing to be found thereof but interlocutory orders in the said cause And the said confederates or some of them doe know have <been> informed or have reason to beleive that a decree or dismission was obtained / against the said Simpson in the said suites or one of them and that he the said Simpson was prohibited from holding courts at Grinton to which he pretended to have a right & that he was likewise prohibited from inclosing any parts of the moors wasts or commons to which he pretended / title in right of such pretended mannor And diverse other suites have been commenced by or against the confederate Hillary or his ancestors or others which the said confederates or some of them conceal whereby it would appear that the said Grinton is no reall mannor & by which (if the same were to be discovered & brought to light) your / orators

171 See p. 165 footnote 175.

right in the said premisses would more plainly & fully appeare And your orators further shew that in case the said Grinton alias East Grinton was a reall mannor or that the said monastery was well intituled to the royaltyes thereof yett the said confederates very well know or have heard are satisfyed or beleive that the / said monastery of Bridlington never had or claimed to have any other right either in the moore or wast called Grinton moore or in the wasts or moors of Herkeyside & West Grinton either in right of such said mannor or in right of other their particular tenements or inclosed lands [or][172] have <a> right of common upon the said / moors wasts or commons or some other small inconsiderable libertyes cheifly for the benefitt & advantage of their tenants And that they never interrupted your orators ancestors or those whose estate your orators have ever claimed of or in the aforesaid premisses in their use or enjoyment of their rights royaltyes or other priviledges / in the severall moors wasts or commons of East Grinton West Grinton or Herkeside aforesaid or any of them And your orators shew that the said confederates have not onely found out or discovered or know of diverse evidences or writeings which demonstrate prove or shew what rights or priviledges the lords of Swaledale / have heretofore enjoyed or claimed to enjoy in West Grinton & Herkeside & the moors wasts or commons thereof but alsoe in the moors wasts or commons of East Grinton aforesaid & in severall other places within the said pretended boundaryes of the said pretended mannor of Grinton And the said confederates some or one of them / doe in perticular know of a certaine grant made or mencioned to be made by Robert de Gant unto one Ralph the son of Godman & his heires whereby the said Robert de Gant grants twenty acres & a halfe of land in Swaledale on the west part of the vill of Grinton & thereby limitts & setts out what common the aforesaid / tenant & his heires should hold & enjoy with his <said> tenement upon the commons of Grinton aforesaid And the said confederates doe alsoe know of other grants fines or conveyances whereby the said monastery held or enjoyed the aforesaid tenement or land aforesaid with the common of pasture therewith granted by the aforesaid [Robert] / de Gant And the said confederates have read seene or been informed & doe beleive & cannot deny but that Robert de Gant was son of the aforesaid Walter de Gant And the said confederates doe likewise know have seene <read> or been informed of one other grant made or mencioned to be made by Gilbert de Gant to the said [monastery] / entitled a grant *de curia dilatanda* whereby the said Gilbert de Gant grants that the prior & convent of Bridlington aforesaid might inlarge their court or inclosure or severalty of Grinton towards the west & near to the church yard by five score & six foot in length & twenty foot in breadth [that] / they might have liberty there to build one house And the said confederates some or one of them doe likewise know have seene read or beene informed of one other deed or grant whereby the prior & convent of Bridlington doe grant or confirm to Gilbert de Gant & his heirs that he or they might hold [. . .] / enjoy their free parke in Whitaside & processe thereupon in his the said Gilbert de Gants court of Helagh And the said confederates some or one of them doe likewise know have seen <read> or been informed of one other deed grant or confirmacion whereby the said prior <or> convent consent or [. . .] / that Gilbert de Gant <& his heirs> should hold or enjoy their wood in Swaledale or within the lands there which were held or enjoyed by the said monastery as alsoe their free parke there with the attachments of vert & vension with all wafts escapes & the liberty of takeing [. . .]mends & other liberties [And] / the said confederates or some or one of them either have the said deeds grants or confirmacions in their

172 Word omitted in the original.

power or custody or know where the originalls of some or one of them are or be or otherwise have seen perused or been informed of some coppies or entryes thereof & yett will not discover in what / custody the same now are or where your orators or their agents <may> peruse the same & such other deeds evidences <or> writeings as demonstrate or shew your orators right <or the right> of the lords of Swaledale aforesaid of in & to the moors wasts or commons of East Grinton West Grinton & Herkeside aforesaid & other the commons [aforesaid] / claymed by the said confederates And the said confederates deny & will not owne that the Lords of Swaledale have ever enjoyed the moors wasts & commons aforesaid or any of them by grazeing or feeding their deer thereon tho' the said confederates or some of them doe very <well> know or have read heard or [been] / informed that the deer belonging to the lords of Swaledale or enjoyed by them within their lordshipp or forest of Swaledale have time beyond memory grazed depastured & fedd without interrupcion or disturbance aswell upon the moors wasts & commons of East Grinton aforesaid as other the wasts & <commons> before sett / forth and that the said lords of Swaledale have from time to time since the originall grant of Grinton granted license to inclose severall parts of the said wasts And the said confederates cann sett forth diverse grants which have been obtained by the aforesaid monastery of & from the lords of Swaledale / whereby the said monastery has beene authorized & impowered to inclose some part of the aforesaid commons or otherwise to hold or enjoy the same And your orators shew that tho' the deer belonging to the lordshipp or forest of Swaledale did alwayes graze and depasture upon all & every the aforesaid moors / wasts & commons yett the monastery of Bridlington nor any other clayming by from or under them had ever any deer upon the said moors or the liberty of grazeing chaseing or depasturing deer upon the said moors wasts & commons or any of them or any other right therein save a common of turbary & common of pasture / for their cattle to graze with the deere belonging to the lords of Swaledale And the said confederates or some or one of them cann likewise sett forth diverse records evidences & writeings whereby it appears what right & title the lords <of Swaledale> or what pretences of right or title the monastery of Bridlington had to the moors wasts or / commons of the severall places within the said pretended boundaryes of the said pretended mannor of Grinton or to some of them or to the soil or mines thereof which yett they conceal & will not discover And your orators further shew that the monastery or priory of Bridlington in case they had any / right to any <such> mannor as is pretended yett never had nor claimed to have title to the soil or mines of any the moors wasts or commons of East Grinton West Grinton & Herkeside aforesaid nor to any other mines within the pretended boundaryes aforesaid save <within> such of the inclosed grounds severaltyes or feilds / or within the territories or precincts of such said severaltyes as were held or enjoyed by the said monastery distinct from the moors & wasts aforesaid or which the said monastery had liberty or license to inclose by vertue of any grant or grants from the lords of Swaledale or some or one of them And the said / confederates have seen perused or beene informed of diverse antient grants & licences which the said monastery have obtained from time to time from the lords of Swaledale or some of them whereby the said monastery had been impowered to inclose some particular parts of the said moors without such / lycense the said monastery had noe authority to inclose the same by vertue or pretence of the originall grant of Grinton to the said monastery & your orators shew that diverse parts of the said moors have been accordingly inclosed by vertue of such subsequent grants & licences & not otherwise yett the said / confederates will not discover where the said grants or licences or the enrollments or entryes thereof are now remaineing or how the confederates know the same which your orators hope they shall be obliged to sett forth & discover And your orators shew that the confederates pretended / mines of Grinton Herkeside & Whitaside or some of them (as parcells of the possessions of the said monastery of Bridlington) were not onely leased or demised to Henry Lord Scroop & Arthur Phillips by Queen Elizabeth[173] & after to Humphrey Wharton esquire by King Charles the First[174] but there have / been severall other leases thereof granted by the Crowne or of some part of the said mines (as parcell of the possessions of the said monastery[175] And the confederates pretend or insist that such said leases extend to & take in the mines aswell of the moors wasts or commons by [express name] as of / the inclosed grounds feilds or territoryes of the severall places therein named & yett they the confederates at other times own or they must now confesse it to be true that the mines of the said moors wasts or commons are not expressed in such said leases [. . .] / insist that <the> said mines of the aforesaid moors wasts or commons are not expressed in such leases that yett the Crowne or their lessees or some clayming under them have from time to time wrought the mines within the moors wasts or commons of East Grinton West Grinton Herkeside [&] / Whitaside & other the moors within the confederates pretended boundaryes as being part or parcell of the estate or possessions of the monastery of Bridlington or of their said pretended mannor of Grinton Whereas the confederates doe very well know have heard beleive & have been informed or can otherwise make / it appear that neither the Lord Scroop & Phillips nor the said Humphrey Wharton under their said severall & respective leases <or> any other person or persons claimeing under them or under the estate of the said monastery did ever worke or pretend to worke any mines within any of the moors wasts or / commons before sett forth And the said confederates or some of them have seen or perused the books of account of the said Humphrey Wharton relateing to the lead works by which <it> is pretended that it does appear that twenty shillings per annum rent has been for some time answered to the Crowne for / tho' the said Humphrey Whartons <mines> within East Grinton yett it noe way appeares nor cann the said confederates prove that he the said Humphrey Wharton or his agents or assignes ever wrought or pretended to worke under the said lease within any of the commons of East Grinton aforesaid or / within any of the grounds of Herkeside & West Grinton or either of them And your orators have beene informed that it does not appeare by the books of accounts of the said Humphrey Wharton or those clayming under him & which the said confederates have inspected or perused & ought to dis[cover &] / sett forth that any mines were wrought under the said lease soe taken by the said Humphrey Wharton save within the estate of Fremington which was likewise demised to him by the said lease And yett the said confederates doe sometimes pretend that some lease or leases were [. . .] / him the said Humphrey Wharton or some clayming under him to one Broderick[176] or others of some mines within the pretended bound<aryes> which have been accordingly wrought under such lease or leases as is pretended & that there are still the appearances or markes of some [. . .] / works within the moors or wasts aforesaid but the said confederates will not discover their knowledge or informacion by whome the same have been at any time wrought or

173 In February 1583 the mines of Grinton were leased by the Crown to Henry Lord Scrope and Arthur Phillips for a term of twenty-one years – see TNA LR1/184.

174 After the death of Arthur Phillips, in 1597, the Grinton mines were leased by the Crown to Humphrey Wharton of Gillingwood (d.1635), Receiver General of Land Revenues for the Archdeaconry of Richmond and the Counties of Durham and Northumberland.

175 This bracket is not closed.

176 The identity of 'Broderick' has not yet been established.

attempted to be wrought or under what leases right or authority And the said confederates have gott into their / custody {or} <and> doe know of have seen perused or been informed of severall other books of account & diverse records surveys rentalls evidences writeings accounts & other memorandums whereby it appears what places were really wrought under such leases as were at any time granted / by the crown & within what places the agents of the Crowne ever wrought the mines within the premisses soe demised & for what mines in particular the rent to the Crowne was soe answered <or> whereby it appears that the rent answered or in charge to the Crowne for the mines / of the premisses or some part thereof was answered charged mencioned or taken notice of by the name of *novus redditus* or as a new rent & <though> they well know where the said books evidences writeings or accounts are now kept yett they conceal & refuse to discover the [. . .] evidences / or proofes they the said confederates have or can make touching the premisses And the said confederates doe give out in speeches that either the monastery of Bridlington or the Crowne since the dissolucion of the said monastery & their lessees farmers agents or [. . .] / as occasion required with<out> interrupcion or disturbance wrought very great & considerable quantities of lead oar in the severall moors wasts & commons of East Grinton West Grinton & Herkeside aforesaid & made very <great> advantages & profitt thereof & pretend they have done severall other [. . .] / right & title to the soil & mines of the aforesaid wasts as parcell of the estate of the aforesaid monastery or of the said pretended mannor of Grinton whereas the said confederates doe know have heard or verily beleive that if the said monastery of Bridlington or the Crowne or any of their former lessees or agents did ever worke any of the said [. . .] / moors wasts or commons of East Grinton West Grinton or Herkeside aforesaid that the profitts thereof made were inconsiderable or that the same were soe wrought during the minority of your orators ancestors or during their absence in forreigne parts or whilst they were distant from the premisses or in case any tryalls [. . .] were made for [the] / findeing of oar within the said moores wasts or commons without the consent of your orators ancestors or those under whome <they> claymed as aforesaid the said tryalls were connived at by them as is usuall till the veins were found or became profittable but soe soone as the same have become rich or advantageous the confederates know that no person or persons [. . .] / ever attempted or presumed to worke out the said mines but that the greatest part of all the said mines have remained unwrought till the said confederates attempted the same And the said confederates likewise know that though the orators ancestors & those under whome they claime <as> aforesaid have been constantly imployed [in] / workeing the mines in diverse parts of their lordshipp of Swaledale & Helaugh aforesaid which proveing successfull & very rich have for some time prevented their makeing tryalls within the said moors wasts or commons of East Grinton West Grinton or Herkeside aforesaid yett neither the crown <nor> any other person or persons clayming [. . .] / lease made from the crown have for these fifty years last past & upwards pretended to worke or make any tryalls for the findeing of oar within any of the said moors wasts or commons And the said confederates have reason to beleive & can sett forth that what estate right or pretence of right the crown ever had in any of the said mines was [. . .] / since by some one or more patent or patents granted from the Crowne & that the rents reserved upon the said leases made to the said <Lord> Scroop & Phillips or Humphrey Wharton or any other for & in respect of the mines to them respectively demised [x x x] [177]or pretended to be demised were therefore for [. . .]

[177] One or more words have been erased.

years of the / respective terms in such said leases [never] accounted for answered or in charge to the crown yett they refuse to discover such <said> grants to whome <the> same were made & when or in what years the said rents were answered to the Crowne or not answered for the said mines during the continuance of such said leases or any of them And your orators shew that if any / lease have been at any time granted from the Crowne of the mines within any of the said moors wasts or commons the same were soe granted in right of your orators mannor of Helaw or some part thereof which was then in the Crowne by reason of the attainder of Sir Francis Bygod[178] But the said confederates with intent to conceal what right the Crowne had in the moors wasts & commons of / East Grinton West Grinton & Herkeside aforesaid & other the moors wasts or commons within the aforesaid <extended> boundaryes to the lead mines therein have gott into their or one of their custodyes or possessions the inquisicion or survey taken of the estate of the said Sir Francis Bygod upon his attainder as aforesaid or doe know of the same or have some coppy thereof & yett / will not discover in what office or in whose custody the same or any entry <thereof is> now remaineing or where they beleive or have been informed the same now is which your orators hope the said confederates shall be obliged to doe And the said confederates or some or one of them haveing seen or perused the ministers accounts of the lordshipp of Helaw whilst the same / remained in the Crowne by reason of the attainder of the said Sir Francis Bygod as aforesaid but in particular one account thereof taken or entered in the augumentation office in or about the fourth or fifth year of King Edward the Sixth or otherwise relateing to the said time And the said confederates findeing by the same account of the lordshipp of Helaw as parcell of / the possessions of the said Sir Francis Bigote that an account is therein rendered by the office or minister therein mencioned of fifty shillings of the issues & profitts of the lead oar of Grinton & Helawe in Swaledale as detained in the hands of one Humphrey Coolwiche[179] therein named for one years rent thereof ending at Michaelmas in the fourth / year of the said king And that notice is taken in the same account of a lease or demise thereof made by indenture to the said Humphrey Coolewicke at fifty shillings per annum And the said confederates or some or one of them and in particular the said Reginald Marriott or some on / his behalfe haveing not onely perused or seene the said last mencioned account of the estate of Sir Francis Bigote [x x x][180] /

[membrane iii][181]

But other accounts therein mencioned & referred to as also the said lease to the said Humphrey Coolewicke or some entry of copy thereof they the said confederates refuse to communicate or discover the same nor will confess or owne whether the said lease of the lead ure in Grinton aforesaid be soe made or mencioned to be made by the said Sir Francis Bigote himselfe or by the Crowne as / parcell of the estate of the said Sir Francis And your orators shew that the said confederates have gott into their custody know of or endeavour to conceale severall antient evidences court rolls

178 Sir Francis Bigod of Settrington (1507–37) held one moiety of Helaugh manor. He was executed in 1537 for his part in the Pilgrimage of Grace after which his estates were confiscated by the Crown (*VCH*, p. 241).

179 In 1550 the Crown leased of a portion of the ore raised from the Grinton mines to Humphrey Coolwiche or Cholwick (Tyson, Spensley and White 1995, p. 12).

180 Half a line of text has been erased.

181 Parchment, 805mm x 780mm.

deeds grants & other writings which formerly belonged to Sir Francis Bigote or his estate in Swaledale aforesaid the same haveing come to the Crowne or to some other persons to the crowns use about the / twenty nynth yeare of King Henry the Eighth when the said Sir Francis became attaint & have also in their custody know of or endeavour to conceale the severall accounts taken or certified by the officers or ministers of the crowne relating to the estate of the said Sir Francis betwixt the time of the said attainder & till the estate of the said Sir Francis was granted from the / Crowne for your orators shew that there are noe ministers accounts thereof to be found in the augmentacion office during all the time aforesaid save the said single account before sett forth the'other annual accounts of the like nature & of those times are still remaining in the said office And the said confederates are not onely unwilling & refuse to discover what ministers / accounts particulars or surveys have been at any time taken or certifyed of the said Sir Francis Bigotes estate or of the estate of your orators & the rents issues & profitts of the severall mines thereunto belonging & in what office or where the same may be found but they also conceale what leases have been at any time granted by the said Sir Francis or the / Crowne relateing to any of the said mines By all which or at least some of them it will appeare when the same are produced to this honourable court)[182] that the mines of Grinton or of the moors wasts or commons therewith occupied or enjoyed & the soyle & royalties of the said moores have been held & enjoyed as part or parcell of your orators estate in Swaledale aforesaid And / your orators further shew that some of the confederates doe sometimes pretend that there is noe such place as your orators claymes by the name of West Grinton & doe deny that the vill or village of Grinton soe enjoyed by the priory of Bridlington was ever distinguished called or knowne by the name of East Grinton whereas the said confederates find by severall writings that they have / inspected that the said East Grinton & West Grinton have been both antiently & of late distinguished both in name tenure & other respects And that East Grinton & West Grinton by those names are not only particularly named & distinguished by the books of account of the said Humphrey Wharton herein before mencioned but also in & by diverse leases & grants from the Crowne / & in & by diverse perticulars of the said estate of Sir Francis Bigote now remaining upon record in this honourable court & in & by diverse other records evidences & writings which the said confederates know of but will not discover as they ought to doe & as your orators pray they may doe And the said confederates doe at other times pretend that neither the said West Grinton or Herkeside is any vill or township / or hamlett or that there are any houses or tenements therein sufficient to make a vill or hamlett whereas the confederates very well know that there are severall houses thereunto belonging & that the nearest of them to the vill of East Grinton stand at a considerable distance from thence & that there have been antiently more houses thereunto belonging & severall old house steads still visible & the ruines / or marks of such said houses still remaining And the confederates very well know or have heard & are convinced that the lords of Swaledale or those under whome your orators clayme have had or now have good right & title to the mines & royaltyes of the moors wasts or commons of Herkeside & West Grinton either by those names or by any other name or names soever by which the said vills / hamletts or places have been distinguished reputed or called And the said confederates doe at other times pretend that neither the said West Grinton or Herkeside nor either of them is any distinct vill or hamlett but that the same is part & parcell of the vill or Reeth or of the reputed mannor of Reeth aforesaid your

[182] This bracket is not opened.

orators / insist that he your orators ha<ve> good right & title thereunto & to <the> mines & royalties of the moores wasts & commons thereunto belonging for the said confederates doe very well know or cannot deny but that the said vill of Reeth with its severall rights members & appurtenances belongs to your orators & is part & parcell of their lordship of Swaledale & Helaugh aforesaid And your orators shew that the courts held for / the lordship of Swaledale haveing been some times kept at the vill of Helaugh & at other times kept at the vill of Reeth within the said lordship [&][183] that there are severall records grants writings & antient evidences which the said confederates know of but refuse to discover wherein the said lordship of Swaledale is sometimes called by the name of the lordship or mannor of Helaugh & at other times / called by the name of the lordship or mannor of Reeth to all which said lordship mannor vills hamletts or places of Swaledale Helaugh Reeth Herkeside West Grinton & every of them as now held or enjoyed by your orators & by whatsoever other names or boundaryes the same now are or may hereafter appeare to have been formerly knowne or distinguished your orators now claymes to have right & title as also / to the severall moores wasts & commons thereunto belonging & to the mines & royalties within the same And your orators shew that the said confederates doe at other times pretend that Herkeside & West Grinton & the moores wasts or commons held used or enjoyed by the tenants thereof are noe part or parcell of your orators said lordships mannors & vills in Swaledale aforesaid or any of them Whereas your orators shew & doe{s} insist / upon itt & the said confederates know it to be true that tho the same be noe part of such said lordships mannors or vills that yet your orators {is} <are> well intituled to the same And your orators further shew that the said confederates doe at other times owne or they cannot deny or shew to the contrary but that the said East Grinton West Grinton & Herkeside aforesaid & the moores wasts or commons thereunto belonging / are separated & divided by known boundaries which yet the confederates refuse to discover nor will they set forth what they really clayme to be the boundaries thereof nor of the vill or pretended mannor of Grinton or East Grinton nor what deeds grants evidences records or other writings they have or know of or intend to make use of to prove the same or what they have or know of whereby the contrary may any way / appeare And some of the confederates of late presumed to draw up or cause to be drawne some paper or writings containing the pretended boundaries of East Grinton West Grinton & Herkeside aforesaid or some of them & have prevailed on & influenced severall persons to subscribe or acknowledge the same or to make affidavits of the truth of such pretended boundaryes with intent as your orators beleive / to engage or prevaile upon such persons to become witnesses against your orators on the behalfe of the said confederates or some of them And your orators shew that the said confederates with further intent to give some colour to the pretended bounder of the said pretended mannor of Grinton & to obscure your orators title to the moores & wasts of Herkeside & West Grinton distinct from the commons of East Grinton aforesaid they doe / at other times give out that there is noe boundary dividing betwixt the same & that there was never a bounder ridding till your orators caused one to be lately made for the distinguishing of the commons aforesaid Whereas the confederates doe very well know have heard or beleive that the same commons are sufficiently distinguished by their severall distinct & proper names & that the severall names of East / Grinton West Grinton & Herkeside aforesaid as also Grinton Moore Grinton Ho Herka or Herka Hill & that the names of Hughs Ditch How Dike or Hodic & of the White

183 Word omitted in the original.

Stone in the How & the Gray Stone in the Greets & the name of Stamburghaues are all of them remarkeable & notoriously knowne in the said neighbourhood And the said confederates doe likewise know that the commons of East / Grinton are sufficiently distinguished from those of West Grinton & Herkeside aforesaid by the separate enjoyment of the commons thereof by the severall tenants of each of the said respective vills & that the tenants of East Grinton never enjoyed nor do now clayme to have or enjoy any right of common upon the moores or wasts of West Grinton & Herkeside but that the tenants of the said vills have always / distinguished their said commons And in case the tenants of East Grinton have at any time pulled ling digged turfes or gott stones or grazed their cattle upon any of the moores wasts or commons of Herkeside or West Grinton the same had been by the leave lycence or consent of your orators or their tenants to whome they have paid mony or made other satisfaccion or acknowledgement for their soe doeing & all the / said confederates very well know that the said commons have been time beyond memory & all along hitherto soe well distinguished & knowne aforesaid that there was never yet occasion till the said confederates have given itt themselves to ascertaine the said boundaryes by rideing thereof And your orators shew that the said confederates or some or one of them when your orators lately caused the boundaries / to be ridd betwixt the said moores wasts or commons did by themselves or their agents attempt to discourage & prevent severall persons in appeareing thereat or shewing the same & did prevaile upon severall other persons to conceale their knowledge of the said bounder & did then endeavour to draw your orators agents into mistakes in rideing the same & they doe since give out & pretend that your orators agents did / mistake the said boundary in some part thereof but refuse to discover in what part thereof the same is mistaken tho they know & have reason to beleive that the said boundary was truely ridd in that part thereof neare to the lead works now in question And your orators hope that in case their agents or those employed to ride the said boundary in your orators absence have by mistake or by misinformacion of the said / confederates or any employed on their behalfe been misguided in rideing any part of the said boundary (which the said confederates endeavour as aforesaid to conceale) that yet the same shall not prejudice your orators right or the rights of the tenants of the said vills And the confederates know that what question or dispute may happen to arise in the manner of rideing the said boundary or / any part thereof that the same are noe way materiall to the right of the mines in the place now wrought by your orators And the confederates cannot deny but that the place so wrought by your orators is within & part & parcell of the moores or wasts belonging to West Grinton & Herkeside aforesaid or the one of them And your orators shew that the said confederates not withstanding their / pretences aforesaid at other times owne that there are certaine metes & boundaries dividing the commons of East Grinton from those of West Grinton & Herkeside aforesaid & doe sometimes admitt or cannot deny but that the boundary of West Grinton & Herkeside & of the moors wasts & commons thereunto belonging does proceed & extend itselfe & does devide the same from East Grinton & the other / adjoyning lordships or lands in manner & forme following (vizt) from the mid streame of the river Swale neare the east corner of a close called Holme from thence cross the land to the east corner of Low Close leaving the elme tree upon the west & Cuddy Spout on the east up a ridge to Benson cowhouse or Ginny Wren cowhouse soe up by the east wall of Ox Intack to the corner & from thence by direct line / to the White Stone neare unto the west end of the How & so by direct line from the said stone to the place where the Gray Stone at the east end of the Greets formerly layd & was lately removed from thence towards the south where an antient standard or hurrock stood & was also lately removed where the

Duke of Bolton adjoynes to Herkasyde for the mannor of Bolton (the former places being / the bounders between West Grinton & Herkeside & East Grinton) and soe westwards as heaven water deales to a place called Gibbon Hill & in the like manner westwards to an hurrock or standard called Gibbon Hill standard or hurrock from which place westward comes in the bounder of Whitasyde & from the said standard or hurrock northwards downe a gill called Brownah Gill to a place called / Horse Pasture corner and soe downe the said gill to a yate or gate called Scabba Gate & soe to the river Swale which last mencioned bounder divides Herkasyde & Whitaside And the said confederates doe at other times pretend that the boundarey of West Grinton & Herkeside aforesaid & the commons thereof were in the time of Walter de Gant ascertained by him to proceed & extend in the other manner following vizt / from the phile[184] or middle of the river Swale towards the South extending it selfe by the hollow of a certaine ditch called Hughs Ditch meeted & bounded between the lands of the said Walter de Gant of East Grinton & the lands of West Grinton unto the head of a certaine ditch abutting upon the moore & from the head of the aforesaid ditch towards the west by the wall about the land of West / Grinton which was called Gaytys Croft unto the gate of the moore And beyond the said gate by the wall about the close called Bowans unto the land of Rigghouses & soe by the meets & bounds to the land of Harkey houses & so between his the said Walter de Gants land of Harkey houses & the lands of West Grinton to the bounds betwixt the land of Reeth & of the said West Grinton which is called West / =syke Riddinges & by the meets & bounds extending towards the east end of a certaine ditch under Kepick house and from the east end of the said ditch turning lineally towards the north between the lands of Reeth & the lands of the said mannor unto the phile or middle of the water unto the aforesaid bounds towards Hughs Ditch aforesaid And the said confederates doe at other times pretend that the boundary of the / moores wasts or commons of or belonging to West Grinton & Herkeside does otherwise proceed & extend it selfe in manner following vizt from the aforesaid great ditch called Hughs Ditch Howdike or Hodic to the head or riseing of a certaine spring or fountaine lying on the west of Hodic aforesaid & so from the said spring or fountaine unto a certaine place called Stambarghause / & from thence downe towards the dale or valley of Swale unto certaine lands which anticntly belonged to one William the son of Alured & so back againe by the boundary of the said lands to Hodic aforesaid And your orators shew that by such severall <&> different boundaries the said confederates would endeavour to puzle & confound the true & just boundary of the said commons & would conceale within which of them / it is that your orators have now wrought & sunk {his} <their> said shaft pit or grove & they will neither discover within which of them the same lyes nor wherein they clayme the same to be tho' they the confederates or some of them know of or have seen perused or been informed of diverse records & antient proceedings upon diverse writts of *curia claudenda* (which have been brought / either by or against the lords of Swaledale or their freehold tenants or Herkeside or West Grinton or some of them or by or against the priors of Bridlington or the freehold tenants of East Grinton aforesaid) as also proceedings upon diverse writts *de rationabilibus divisio* or upon other antient writts

184 The word commonly used to describe the middle of a river or stream regarded as a boundary was 'thread'. The word 'phile' in this sense is not recorded in the *OED* but is clearly an attempted angicisation of the Latin *filum* (thread) as used in the phrase *filum aquae*. There is some discussion of the phrase at, for example, the entry for *filum aquae* in Wharton's *Law Lexicon*, or at *Ad medium filum viae* (or *aquae*) in Jowett's *Dictionary of English Law*. This information has been kindly supplied by John Simpson, Chief Editor, *Oxford English Dictionary*.

or speciall commissions for setting out or ascertaining the / boundaries aforesaid or of the said severall vills of the wasts moores or commons thereof whereby the separate boundaries thereof not onely appeare but the antient right of the lords of Swaledale to the soyle & mines of the said moores & to diverse royalties & antient priviledges reserved to them within the same And your orators shew that the said confederates doe at other times / endeavour to misrepresent your orators case & would obscure & confound your orators right & title to the said lands of Herkeside & West Grinton by pretending that the lands of Herkeside & West Grinton aforesaid answer at severall other courts besides your orators & to that purpose the confederates alleadge that part of the said lands of Herkeside & West Grinton had antiently / answered & does still answere & owe suit & service at the court held for the honour of Richmond & that some other part of the same lands had antiently answered & does still answer at the court held for the mannor of Arkildale & that other part thereof had antiently answered & does still pay answer & owe suit & service at the court held for the / pretended mannor of Grinton aforesaid Whereas the confederates doe very well know or have heard are satisfyed or beleive that there are noe such lands in Herkeside & West Grinton that owe suit & service at the said severall & respective courts of Richmond Arkildale & Grinton aforesaid or if there be that the same are onely small peices or parcells of ground scituate & being / at the east end of Herkeside aforesaid & which the confederates doe some times owne were all of them formerly the lands or tenements of the said monastery tho' antiently & originally part or parcell of your orators mannors (& in some respect still held thereof) And the confederates doe likewise know or have been informed that the said severall peices or parcells of ground in / Herkeside aforesaid soe belonging to the said monastery haveing come to the Crowne by the dissolucion of the said monastery & that the tenants thereof being thereby obliged to pay & answere their rents to the Crowne (as the rest of the tenants of East Grinton did) that the said tenants of the Crowne did thereupon (or by other meanes which the said / confederates ought to set forth[185] answere & pay such said rents at the respective courts held for the Crowne within the honour of Richmond aforesaid (within which honour the said lands of Herkeside with the rest of Swaledale antiently lay) And so the said tenants of such parcells paid the rents at the severall places where the same courts were respectively held / (to witt) sometimes at Richmond & at other times at Arkildale & other places within the said honour And your orators shew that after the dissolucion of the said monastery the Crowne haveing made severall grants as well of part of such lands in Herkeside which formerly belonged to the said monastry as of the rents & services reserved upon such / respective grants & leases thereof And that the Crowne haveing likewise made other grants of the honour of Richmond & mannor of Arkindale to the citty of London who have again granted of & divided the same the tenure of such places or parcells of land which soe formerly belonged to the said monastery have by such meanes become confused & the tenants thereof / in the payment of their rents have answered severally as aforesaid And the confederates takeing advantage of the obscurity & confusion of their owne title contrive to conceale & refuse to discover what perticular lands in Herkeside or West Grinton did formerly belong to the said monastry or out of what lands the said monastery had rents / issueing & in whose possession the said lands now are & to what courts the said lands or the tenants thereof respectively answere or pay their rents or owe their respective suite & service or to what court they ought to answere or for what reasons they are soe obliged And

[185] This bracket is not closed.

the confederates pretend that the rents issueing & payable to your orators out of the lands in Herkesyde aforesaid / are but small & petty rents Whereas the confederates cannot deny but that the freehold & copyhold rents issueing & payable to your orators are all of them very antient rents & that the lands out of which the same are issueing are of very great & considerable value & much more in quantity than the other lands which formerly belonged to the said monastery And the said confederates / likewise know that your orators demeasne lands now in the possession of his leasehold tenants in Herkesyde aforesaid are of considerable value to the said tenants & of great extent tho your orator thinks fitt for the ease of his tenants to continue the same at the antient rents which have been formerly paid for the same And the confederates cannot deny but / that most of the lands in Herkysyde & West Grinton doe still pay their rents & answer at your orators severall courts And they doe some times owne that all the lands whereof the tenants soe answere at your orators said courts have been time beyond memory & still are reputed to lye within the vill or constablery of Reeth in Swaledale / aforesaid belonging to your orators & were never reputed to lye within the constablery of Grinton or of East Grinton And the said confederates doe sometimes pretend that the lands of Herkesyde & West Grinton or the one of them have been antiently distinguished & still are distinguished or knowne by the names of the old lands & the new / lands but at other times they deny that there is such distinction in the said lands nor will they discover or set forth what part thereof is called new lands or what part thereof is called old lands which your orators praye{s} they may sett forth And your orators further shew that the said confederates with intent to blemish your orators title doe / at other times give out that your orators have made a forceable entry upon the said mines whereof your orators {is} <are> now in possession And that your orators till very lately had never possession of the said commons or any of the profitts of the soyle of the said wasts Whereas the confederates cannot deny but that your orators & their ancestors & those who / state <they> have & clayme have time beyond memory been in the actuall & peaceable possession of the demeane lands & inclosed grounds of Herkesyde aforesaid & that the tenants of Herkesyde & West Grinton (as well for & in respect of the said commons as of their other rights therein whereof they have time beyond memory had & still have / the quiet enjoyment) have all along paid their rents & done their respective suits & services to your orators & their ancestors & those under whome they respectively clayme in manner as aforesaid And the confederates cannot deny but that the said moores wasts & commons of Herkesyde & West Grinton as also the said demeanes & inclosed grounds / whereof your orators have such actuall & peaceable possession receipt of rents suit services & other royalties as herein before are mencioned & set forth are parcell of your orators mannor of Swaledale or your orators estate within the same And the said confederates know that your orators ancestors have granted licences from time to time to take in & improve severall / parts of the said commons which have been accordingly inclosed & enjoyed under the same from time to time & that the tenants of such improvements have paid rents to your orators & their ancestors for the same & that your orators & their ancestors have granted lycences to the coppyholders of Herkesyde for severall other purposes And tho / your orators have such cleare right & undoubted title to the soyle mines & other royalties of the moores & wasts of Herkesyde & West Grinton yet your orators were soe cautious to prevent any wrong in extending the libertie of your orators said mannor in Herkesyde in West Grinton beyond their antient & knowne bounds that your orators / before they began to sinke at the place now wrought by them for that the same lyes neare the lymitts and confynes of the said moores & wasts of Herkesyde & West Grinton did therefore on or about the eleventh day of September last past give order to ten of their principall agents there & did thereby require them / to ride the said boundary of Herkesyde & West Grinton takeing to their assistance such of the most antient & knowing inhabitants as could give them informacion thereof and did thereby require them to give due notice to the neighbouring & adjacent lords of the time & certaine place of meeting in order to ride / the same And your orators shew that in pursuance of such said authority & command publick notice was given upon Fryday the one & twentieth day of September last at the market cross in Reeth about one of the clock of the said day as also on the Sunday following in the parish church yard of Grinton (after divine service / & sermon ended) by your orators bayliffe of the rideing the said bounds by your orators agents or commissioners upon Wednesday the six and twentieth day of September last meeting att ten of the clock of the same day at a certaine elme tree in the lane upon the east side of Swale Hall which is very well / knowne to the said neighbourhood And your orators shew that your orators agents did accordingly then and there meet together with severall of the tenants and inhabitants as well of East Grinton as of West Grinton & Herkesyde & of other of the said adjoyneing lordshipps And your orators shew / that the said boundary was thereupon ridd according to the best informacion & accounts that your orators agents could at that time gett and the said boundary being drawne up in writeing as the same was then ridd was signed and subscribed by above sixty persons that were then present as the true and reall / boundary of Herkesyde & West Grinton and of the moores wasts and commons thereof or therewith used occupied or enjoyed as separate & distinct from East Grinton and the commons thereof and from other the adjacent lands and lordshipps And your orators shew that tho they <are> advised that {he is} <they are> well intituled / as well to the soyle & mines of the commons of East Grinton as to those of Herkesyde & West Grinton which was soe ridd in manner aforesaid and tho' the said Marriott continues to worke and makes very great profitt and advantage of the said mines yet your orators have never offered to interupt his possession the / same being soe lately gained nor would stopp his workeing till the same be directed by this or some other court And yet the confederates notwithstanding the legall methods soe taken by your orators endeavours to impose on the neighbourhood of Grinton and on severall others and attempts to amuse them with / false stories and intimacions about your orators said proceedings and of the danger they are in as is alleadged of being disturbed by your orators and their agents in the peaceable enjoyment of their estates Whereas the proceedings of the confederates themselves doe manifestly tend to subvert and destroy / the antient rights and undoubted priviledges of all the tenants of Herkesyde & West Grinton and to remove and overthrow the antient land markes deviding betwixt them and the tenants of East Grinton And your orators further shew unto your honours that the other confederates / Solomon Swale Humphrey Bellamy {and} Andrew Hopegood <& Edward Shaller> set up severall claymes to the said moores wasts and commons of East Grinton West Grinton and Herkesyde aforesaid much different from and inconsistent with the other claymes and pretences before set forth And to give colour / to such clayme and pretence of the said Sir Solomon your orators shew that some of the said Sir Solomons ancestors or those under whome he <or those claimeing under him> now enjoy his freehold tenement or lands in West Grinton (which are really held of your orators estate in Swaledale aforesaid and which owest / suite and service to your orators courts and is charged or chargeable to your orators with the payment of an antient annuall free rent of four pence per annum) deriveing their pedigree and title from one Alured (the supposed nephew and principall chamberlaine of the aforesaid Walter / de Gant have and sett upp certaine pretended grants made or pretended to be made by the said Walter de Gant unto him the said Allured of the lands of West Grinton by the name of the mannor of West Grinton by which said grants or some or one of them the confederates / pretend that the boundary of the demeasnes and inclosed lands of such said pretended mannor are not onely sett forth and perticularly described and mencioned to be divided from the lands of East Grinton by the aforesaid ditch called Hughs Ditch alias Hodicke / or Hodic (and which is expressed in such said grant by the Latin phrase of *Fossa Hugains*) but the moores wasts or commons pretended to be assigned by such said grant to the said pretended manor of West Grinton are thereby alsoe set forth and described / by perticular meets and boundaries And as the same are thereby sett forth comprehend therein not onely the commons of East Grinton aforesaid but the commons of West Grinton and Herkesyde and also Whitaside and the severall other wasts from / Stollerstone Style to the aforesaid rivulett called Haverdale Beck (which the said Marriott and others of the confederates suggest as aforesaid to be the boundary of their said other pretended mannor of Grinton alias East Grinton which soe formerly belonged to the priory of / Bridlington as is pretended) and under such said grant or grants the said Sir Solomon Swale claymes to have his coate of armes as then belonging to the said pretended lordship or lands of West Grinton and as thereby granted by Walter de Gant to the said / Allured & his heirs And by the said grants he pretends to clayme likewise the liberties of free chase with diverse priviledges as thereby granted by speciall words not only within the aforesaid boundaries but within all other the liberties of your orators estate in Helaugh Reeth and Swaledale aforesaid /

[membrane iv][186]

And by the said deeds the said Solomon claimes to hold the said mannor of West Grinton & the liberties aforesaid by rendering therefore to the said Walter de Gant & his heires once in a life a leash of grayhounds for all manner of secular services exactions & demands which by mans heart might be imagined spoke with tongue or could be expressed And he the said Sir / Solomon Swale does likewise claime to be cheife forester of your orators forest of Swaledale aforesaid & to have some other grant or grants of the said office of cheife forester made by some of the Gants or those under whome your orators claime But your orators shew that notwithstanding the said pretended grants neither the said Sir Solomon nor / any of his ancestors nor those whose estate he has or claimes ever held any power within such pretended mannor of West Grinton aforesaid nor are there any freehold or coppyhold tenants within the same that now hold or ever held of the same mannor or ever did suit or service thereat Nor does it appear by any of the records of the justices / in eyre for the county of Yorke that any of the ancestors of the said Sir Solomon ever appeared before the said justices to claime any of the royalties or other liberties pretended to be granted by the said deeds Nor cann the said Sir Solomon produce any allowance in eyre of any of his said liberties Nor are there any antient records wherein the / said pretended deeds are pleaded or allowed or wherein the <said> Sir Solomons estate in West Grinton aforesaid is named or described to be a mannor nor any inquisicion *post mortem* taken upon the deaths of any of the ancestors of the said Sir Solomon whereby it may appear that they or any of them ever dyed seized of any such estate but / on the contrary there are severall records & antient evidences which the said confederates or some of them know have seen or perused wherein the said Sir Solomons estate is otherwise described & wherein the same is taken notice of

186 Parchment, 810mm x 545mm.

onely as a particular freehold tenement held of your orators lordshipp aforesaid or under the payment of the / aforesaid free rent of four pence per annum And your orators shew{eth} that the said pretended grants have been fraudulently contrived by some under whome the said Sir Solomon now claimes & are much different in style & form <from> the deeds of that age in which they are pretended to have been made Nor is it probable that any coat of armes / could be then granted or did then belong to such estate in right of or appurtenant thereto And your orators shew{eth} that the said deeds are wholey inconsistent with severall other deeds really made by the said Walter de Gant & his decendants of diverse lands within the said pretended bounders & which your orators & {his} <their> ancestors constant enjoyment of diverse other / possessions within the same as parcell of the estate of him the said Walter (to which your orators are now intituled) And the said pretended grants are much different in expression & forme from the severall other deeds still extant which were really made by the said Walter But your orators shew that some under whome the said Sir Solomon now claymes (haveing an / opertunity about the time of the dissolucion of Bridlington monastry when not onely the lands of the said monastery but also part of your orators possessions were in the Crowne to sett upp such pretended title) with intent to create some claime & pretence as well to the possessions of the said priory as to some of your orators lands & possessions have since that / time in some of their private settlements stiled or called their said estate some times by the name of the mannor of West Grinton & their capitall messuage by the name of Swale Hall Whereas the said messuage of late years was onely a house of mean appearance & the estate thereto belonging according to its present rent & improved value but inconsiderable / And the said Sir Solomons claime & pretended title has of late after severall tryalls been sett aside & noe credit given to the aforesaid deeds And your orators shew{eth} that the ancestors of the said Solomon Swale or those under whome the said Solomon claymes the said freehold tenement in West Grinton aforesaid being tenants to the lords of Swaledale aforesaid And the freehold tenants of the / aforesaid lordshipp & mannor of Swaledale haveing or pretending to have or claime a right to the mines within their inclosed <freehold> tenements & the severall tenants of your orators lordshipp & mannor <& pretending> to have by the custome of minery within the said lordshipp a right to search for winne & worke all such mines as they could find within the moores & wasts of the said lordshipp for the benefit & / advantage of the lords thereof And the tenants of your orators & their ancestors haveing or pretending to have a right for particular purposes to digg winne & worke the stone quarryes lying & being within the moors & wasts of the said lordshipp It may be true that those under whome the said Sir Solomon soe claimes might under colour or pretence of such right as / aforesaid worke the mines lyeing & being within the inclosed grounds & tenements or might make some tryalls upon the moors & wasts belonging to your orators as the other tenants might have done But your orators shew that none of the lead works ever wrought by the said Sir Solomon or his ancestors did ever exceed or countervail the charge of / working And your orators hope{s} that whether such mines were soe wrought under colour or pretence of the tenants libertyes or of the pretended grants before sett forth the same shall not prejudice your orators right the weaknesse of such pretended title & the means of supporting it being now detected And your orators shew{eth} that the said / other confederate Toby West (the present parson or vicar of the church of Grinton) tho' he very well knows the boundaryes of his parish & that the boundaryes thereof are the very same with those of the lordshipp of Swaledale aforesaid belonging to your orators is likewise acquainted with the division & extent of the severall towns / vills & hamletts within his said parish & with the proper & antient names thereof he being intituled in right of his said church to diverse tenements & lands with right of common & other rights within the same Yett he being a partner with the said Marriott in the said mines & haveing received severall presents guifts or g[ratuities] / from the said Marriott for the services done by him in the severall suites wherein he the said Marriott has been heretofore engaged And haveing some expectacion of further reward for further services he the said Toby West conceals his knowledge of the said boundaryes & refuses to discover what bookes writeings & / evidences he has in his custody which either belonged to the priory of Bridlington or the rights of his parish & the possessions of his church within the said severall vills & hamletts or what right of common or other right he claimes to have within the same & by what boundaryes the same are divided or are pretended to be divided & / he the said Toby West has drawne upp or been present at & a witness to the drawing of severall affidavits testimonialls or accounts of the boundaryes of the said East Grinton West Grinton & Herkeside or some of them & has discoursed with severall persons about the same & gott them to subscribe such papers or writeings or knowes of their soe [doing] / & yett will not discover the contents thereof to your orators nor what use is designed to be made thereof or what persons have subscribed or promised to be witnesses to prove the same And the said confederates Marriott or some to his knowledge or by his order have prevailed upon severall persons to become partners with him in / his lead works & to assist & contribute to the charge & expence of carrying on or defending suits against your orators And he the said Marriott has applyed himselfe lately to severall persons & had consultacion with severall of them in order to invalidate your orators title & to support the unjust pretence of him the said Marriott And he has made / proposalls to some of the said persons to become partners & to convey some estate <or> interest to them or some of them of & in the mines within the lands & wasts whereof your orators are now in possession or other the mines to which the said Marriott claymes title And the said person or persons whose names the said Marriott will not disclose are / agreed to come into the said partnershipp And your orators have reason to beleive that some other person or persons were interested or concerned in obtaining the aforesaid lease of the said mines from the said late king & that such person or persons or some others beside the said Marriott are still interested in the same And that he the said Marriott by some private / agreement is onely to have part of the profitts thereof in consideracion of his undertakeing the management thereof & of all suits to be comenced about the same And yett the said Marriott endeavours to conceal & will not discover the title & interest of such other persons & how farr he himselfe is perticularly interested And he threatens to comence actions & suites in the name of / himselfe or of some other persons against your orators {his} <their> agents & servants for workeing the mines within the premisses And has actually exhibited an informacion in this court in the name of her majesties Attorney Generall wherein he the said mister Attorney Generall does aid & assist him the said Marriott & threatens to disturbe your orators & their agents in the peaceable possession of the / said mines & will not discover the severall records evidences & writeings which he has found perused or seene relating to your orators right & title but endeavours to suppresse & conceale the same Nor will he discover what he insists to be the boundaryes of the lordshipp of Swaledale & mannor of Helaugh & of East Grinton West Grinton & Herkeside / aforesaid & the severall & respective wasts & commons thereof And tho' the confederates & every of them cann sett forth the <said> severall boundaryes & have or know of diverse records bookes manuscripts surveys evidences deeds memorandums coppies of originalls or other writeings declareing & setting forth / the said boundaryes or some of them or the rights or claymes of the monastery of Bridlington or of the Crowne or of the lords of Swaledale & Helaugh aforesaid Yett the confederates endeavour to suppresse & conceal the same & refuse to discover in whose possession or custody the same now are And by / such & the like contrivances he the said Marriott has wrought your orators mines under the colour & pretence of the said lease which yett noe way includes any of the mines within the moors wasts & commons aforesaid or any of them Nor could the same passe by the said lease tho' the same were mencioned or expressed / therein And he the said Marriott & other the confederates have wrought lead oar out of your orators mines to the value of one hundred thousand pounds & upwards & have for some years raised above the yearly summe of five thousand pounds or some other great & considerable summes [. . .] / clear profitt as your orators are informed it appears by the said Marriotts owne accounts of the tenths thereof reserved to the Crowne by the said lease tho' your orators doubts not but to charge the said Marriott with much greater summs had & received And he the said Marriott does also absolutely refuse to come to any [. . .] / with your orators touching the same or to distinguish what profitts he has made out of the mines of Herkaside & West Grinton separate & distinct from those of East Grinton nor will produce the bookes of accounts of the profitts made of the said mines nor suffer your orators or their agents to [. . .] / the same or take coppies thereof Nor will he consent to bring the same into this honourable <court> to be lodged there for your orators security till such times as this court has determined the right of the said mines but in truth the said Marriott has burnt destroyed or otherwise suppressed severall of his books of / account & other writeings relateing to the workeing of the said mines & the profitt & advantage made thereof All <which> doeings of the said confederates are contrary to right equity & good conscience & tend to the distrucion of your orators rights in the said premisses In tender consideracion / whereof & for as much as your orators witnesses whoe could prove their title to & enjoyment of the said wasts moors & commons aforesaid & the lead mines therein are either dead or gone beyond the seas or into places remote And for that some of them whoe are still liveing and very aged or / infirme soe that your orators {is} <are> like to loose the benefitt & advantage of their testimony in the premisses unlesse the same be very soon taken & preserved on record in this honourable court And for as much as the confederate Marriott has had the comand custody & ordering of the severall records books / & other evidences relateing to the premisses which were formerly kept in the auditors office & did some time agoe (as your orators {is} <are> informed) presume & take upon him without the order or direccion of this honourable court to take into his private custody out of the severall treasuries & record offices of this honourable / court & elsewhere the severall records & most materiall evidence relateing to the premisses & under the colour & pretence of carrying the same to the assizes at Yorke for assisting the Crownes right to the said premisses And under that pretence did carry great numbers of the said publick records / into Yorkshire And for as much as diverse of the records belonging to the said office & relateing to the premisses are now wanting or otherwise mislaid or concealed from your orators but more perticularly those before sett forth As alsoe the originall grants of the [. . .] / monastery of Bridlington which came to the Crowne upon the dissolucion of the said monastery with such court rolls as at any time belonged to the pretended mannor of Grinton aforesaid & severall surveys leases particulars boundaryes & other evidences relateing thereto for all which / records he the said confederate Marriott is accomptable to your orators and all other persons concerned therein And your orators hope that this honourable court will oblige the said Marriott & all other persons concerned with him in the ordering & disposall of the said records to / give a full & perfect [account] thereof But for as much as your orators have noe way to compell the confederates to discover the said premisses or what they know relateing thereunto save by the ayd & assistance of your honours in this honourable court where they <being> severally obliged to appear must or ought upon / their severall corporall oathes to sett forth & discover the truth of all & singular the premisses To the end therefore they may soe doe & may true & perfect answers make to all & every the premisses as fully as if the same were herein severally repeated & inter[. . . is] / the prayer of this bill & in perticular may severally set forth or answer to their respective answeres schedules or abstracts of all such grants or coppyes of grants which <have> at any time been made to the monastery of Bridlington of any lands or possessions in East Grinton / West Grinton & Herkeside aforesaid or elsewhere within Swaledale <And that they may set forth whether it any way appears to the said confederates or any of them of or from whom the pretended mannor of Grinton or any lands in Grinton East Grinton West Grinton Herkeside & Swaledale aforesaid or any of them was or were held by the said monastery or by the said Earle of Richmond or by the said Walter de Gant his heirs or descendants & under what rents or services <what other grant or grants> the same was or were soe held & how the same appears & whether they or any of them have seen perused or know of any deeds or evidences wherein are described or / mencioned what estate lands or tenements in the places aforesaid or any of them were granted or enjoyed by the said Earle or by the said Walter de Gant his heires or descendants or the said monastery & in whose custody the same [severally] were & now are> & that they may give an account [what] registers cartularyes & other writeings they have or know of relateing to the possessions of the said monastery & where the same are & what inquisicions surveys or records / or any of them know or have been informed of or have been seen or perused relateing to any of the premisses now in question in whose office or in whose custody the same now are <And in particular whether he the said Reginald Marriott or other the confederates ever had in his or their custody or ever see the inquisicion or survey taken of the estate & lands of the said Sir Francis Bigott upon or after his attainder & in whose custody the same then was & now is & what person by name ever showed the same <either to him or any other person on his behalfe> to any of the confederates & whether the same be now amongst the records of this court & in what particular office> & whether the said Reginald Marriott & his agents have not searched the severall offices of record [for] / records & evidences relateing to the premisses & what he or they have found therein & what surveys grants or records any of them have found seene or knowne of touching or concerning your orators estate in Swaledale aforesaid & the boundaryes thereof or any part of the same / premisses now in question And that they may discover <whether any & what records were carryed to Yorke assizes or perused there upon the tryall with Sir Solomon Swale or any other tryall relateing to the premisses & how & by whom the same were sent & by whom & by whose order the same were delivered out of court or out of the respective offices to which the same belonged> to what person or persons the severall records carryed to York by the said confederates or which have beene in their custody were delivered & what office & in whose custody the same are now lodged & in whose hands the same / <were all along kept whilest the same were out of court & when the same were taken out & how long they were kept out of court & whether any of the same were altered or lost And that the said Marriott> may annex to his answer a particular of the heads & titles of such records as were taken out of court & which he has had or made use of about the premisses & where the same are now respectively lodged & that they may set forth the said affidavits paper or writeings subscribed as aforesaid & what estate right or title they or any of them have or clayme either / to the wasts & commons of <Grinton alias East Grinton

or to the wasts & commons of> Herkeside & West Grinton & may sett forth the respective boundaryes thereof & what evidences they have to prove the same & what right or title they or any of them know or have reason to beleive your orators {has} <have> of & in the said wasts moores or commons or any / of them & that they may sett forth what right or title any other person or persons have or claime in & to the said premisses & in whome the interest <or> terme of the said lease under which the said Marriott claimes is now vested & who are interested or have parts or shares with him therein & whether he / claimes the said old mannor as mencioned or described in Doomsday Booke or the said other mannor as otherwise bounded or the mines of them or either of them & whether he claimes any right to the said mines by & under the title lately purchased from the confederate <Bellamy> & that he may discover what right / he claimes to the said mines under the severall titles sett upp by him And to the end that a commission may be awarded to perpetuate the testimony of your orators witnesses & that an injunction may be granted to quiet your orators in <their> possession of the said mines & of the moores & wasts wherein / they have sunk the aforesaid shaft pitt or grove And to stay the confederates proceedings at law & that the said Sir Solomon Swale Humphrey Bellamy & Andrew Hopegood <& Edward Shaller> & the tenants in Herkeside & West Grinton who owe suite & service at your orators courts may be obliged to performe the / same And that the said Marriott & other the confederates pray account with your orators touching the profitts of the said mines And that the bookes of accounts & other writeings relateing to the profitts of the said mines may be lodged & brought into this <court And that the evidences & particularly the said [. . .] > And that your orators may be releived in all & singular the / premisses according to equity & good conscience May it please your honours to grant unto your orators her majesties most gracious writt or writts of *subpena* under the seale of this court to be directed to the said Reginald Marriott George Tushingham Toby West Sir Solomon Swale Humphrey Bellamy / Andrew Hopegood Ed Shaller & Hillary & other the confederates when discovered thereby commanding them & every of them at a certaine day & under a certaine paine therein to be limited personally to be & appear before your honours in this honourable court then & there to answer the [. . .] / upon their severall corporall oathes & further to stand to and abide such further order & decree as your honours shall think fitt to make in the premisses And your orators as in duty bound shall ever pray etc

/s/ N Lechmere

E112/915 EBOR 100 – document 2[187]

[membrane i][188]

The several answers of Reginald Marriott esquire one of the defendants to the bill of complaint of the right honourable Thomas Lord Wharton complainant

This defendant saving & referring to himself now and at all times hereafter all benefit and advantage of exception to the manifold untruths incertainties & imperfections in the complainants said bill of complaint contained for answer thereunto or

[187] Answer of Reginald Marriott.

[188] Parchment, 735mm x 655mm. The lower left-hand corner is torn and dust stained. The final membrane is endorsed at the foot with the note of swearing, the *jurat*, dated 24 May 1706, and the order for process, the *fiat*, is signed by the chief baron, Sir Edward Ward.

unto so much thereof / as this defendant by his council is advised, is material for him this defendant to make answer unto he answereth & with that he doth not know or believe that Philip late Lord Wharton the complainants father was at any time seised of or in the supposed dominion lord- / ship or mannor of Swaledale but believes he was seised of the mannor or reputed mannor of Helaugh in the bill named & that the same do's lye in Swaledale but whether any or what part thereof doe extend into the parishes of Swaledale or Grinton, or what are the / boundarys thereof this defendant doth not know And saith it may be true that there are within the boundarys of Swaledale & Helaugh severall moors wastes townships vills or hamletts And this defendant saith he doth not know that there is any such vill or town calld East / Grinton but believes and doubts not to prove that there are within the mannor of Grinton severall places calld West Grinton & Harkaside & that the same do lye within & are part of the mannor of Grinton And saith he doth not know that the complainants father did at any time hold / or enjoy any demesne lands rents services royalties priviledges or profits in Harkaside or West Grinton aforesaid or the moors wasts or commons or any of them or any part of them thereunto belonging or therewith used save only that this defendant hath heard that the complainants father had / & that the complainant hath and was & is seised of some freehold lands lyeing in Harkaside and West Grinton or one of them or of some small free rents charged upon or issueing out of some lands in Harkaside West Grinton and Reeth which in all amounted to a sum not exceeding ten shillings per annum in Ph[ilip][189] & / Mary's time when the mannor of Helaw [was gr]anted[190] to Thomas Lord Wharton And saith he doth not know that any freehold leasehold or copyhold tenants in Harkaside or West Grinton at any time paid their rents or have done suite or service at the courts of the complainant or his ancestors in Swaledale / And he saith he doth not know or hath ever [heard][191] or beleives that the complainant or his father had any court in Swaledale except the courts for the mannor of Helaugh, and beleives the complainant and his said late father might order and direct such tenants as they had out of the said mannor of any lands / or tenements to repaire to the said mannor court of Helaugh and there for convenience to pay their rents, but whether they owed of right any suite or service there this defendant doth not know, And this defendant saith he is inclined to beleive that the complainant or his ancestors or their tenants agents or assigns / have sometimes used or enjoyed a tenant right upon part of the wasts or commons of Harkaside or West Grinton by feeding of cattle digging of peate or turfe or by pulling of ling but denyes they did ever worke the mines or quarrys or did any other acts manifesting the ownership of the soyle or had any right or / title to the soyle or ground or any thing under ground or ever pretended to meddle therewith Nor does this defendant know that the complainant or his ancestors ever had any free chase or free warren thereon or within or throughout the same or any part thereof And saith when any of the complainants or his ancestors / or their tenants cattle were fedd thereon it was either by sufferance or as they had or pretended some right of common only and not as owners of the soyle And saith he does not know or beleive that the complainant or his ancestors or those whose estate he pretends to clayme did ever use or make any profit of the / soyle or quarryes of the said wasts or grant any lycences to inclose or cutt downe wood, And this defendant saith he knows it to be true that the complainant or his agents have within

189 Name left incomplete in the original.
190 There is some minor loss of text due to a small hole in the parchment.
191 As above.

these 9 months sunck for lead oar in the commons or wasts of Grinton neere the lead workes of this defendant called Grinton Moore / or the How Groves but this defendant doth insist that the same is done in wrong & prejudice to this defendant and without any lawfull right or title in the complainant And that the complainant hath no right or title to the wasts or commons soyle or ground in or near Harkaside or West Grinton to digg for lead / oare there And this defendant saith and insisteth upon it that Harkaside & West Grinton and the moores wasts and commons belonging to or about the same and particularly the moores wasts commons and places now claymed or wherein the complainant pretends to clayme the mines or digg for / lead oare are and imemorially have been part of the mannor of Grinton in the county of Yorke and doe belong to the said mannor And that the said mannor of Grinton does comprehend the townshipps or vills of Harkaside and West Grinton and the moores wasts and commons belonging to / the same and severall other vills or townships and that the complainant hath not nor his ancestors had any right or title to the same or to the soyle or ground thereof or the mines therein or any of them And saith he doth beleive & hopes to prove that the said mannor of Grinton is / bounded on the east by Cogden & Stallerston Stile and the mannor of Ellerton on the south by the High Carr or Carle or Close and several other places in a direct line on the west by the Bloody vale & Somer Lodge and the beck or rivelett runing along there called Haverdale Beck / and on the north by the river Swale And that all the moores wasts and commons lying & being within the boundaryes aforesaid are part & parcell of the said mannor & doe belong to the same And that the said mannor of Grinton doth comprehend the town of Grinton Harkaside Whitaside / West Grinton and the places called Cogden & Somerlodge with severall other vills and places and all the wasts moores & commons belonging to them and the moores wasts and commons called Grinton How or <Moore> Harkahill & Whitahill And the complainant hath noe right or / title to the same or the soyle thereof or any mines therein, but this defendant saith he doth not beleive that the said mannor of Grinton doth lye within that tract of land called Swale Dale which this defendant doth beleive and hath been informed is a vast tract of land and does / comprehend many mannors or estates to which this defendant supposeth the complainant neither hath nor claymeth any title And this defendant doth deny that the mannor of <Grinton or any part of the mannor of> Swale Dale, if any such manor of Swaledale there be, which this defendant doth not know that there is And this defendant doth deny / that hee hath entered into any partnership combination or engagement to deprive the complainant of any right interest or title that he hath or to any benefit or profitt to be made thereby of or from the said mines moores or wast, but this defendant saith he doth verily beleive the complainant hath noe right / or title to the same but is a meere intruder & trespasser therein And this defendant doth utterly deny that he ever intended to inrich himselfe out of any mines royalties or estate of the complainants But doth confess that while he acted as one of the auditors of this court for an other / devision and not for the county or division of where the mines in question doe lye he did find out and discover that the lead mines within the said mannor feild and territoryes of Grinton aforesaid did belong to and were part of the possessions of the Crowne and that the same / were and had been concealed for severall yeares from the Crowne and thereupon this defendant did make a discovery thereof very fairly and by very lawfull wayes and meanes and by such did obteyne a lease in the name of George Tushingham a defendant in the bill named his trustee / from the late King William the third of all the mines and veines of lead aswell open as to be found ariseing and being within the territoryes and feilds of Grinton Whitaside & Harkaside aforesaid parcell of the lordshipp or mannor of Grinton aforesaid and sometime a parcell of the / possessions of the late monastery of Bridlington and the free liberty to digg for the same And which premisses were formerly demised to Humphrey Wharton esquire by letters patents of King Charles the first to hold to this defendants said trustee for the terms and under the rent & covenants / in the said bill mencioned as by the said lease inrolled in this court & to which this defendant for more certainty herein referreth himselfe more at large it doth and may appeare, And this defendant saith that when he obteyned the said lease he apprehended and was advised that Sir Solomon Swale in / the bill named had no right or title to the mines in question but that the same were in the crown But this defendant doth not remember or beleive that the said Sir Solomon pretended that he held any lands from or under the complainant or his ancestors but pretended the wastes where the said mines are did belong / to him as lord of the pretended mannor of Swaledale or West Grinton or by some such title and saith true it is that upon two several trials at law the right of the said mines was found in the crown and against the said Sir Solomon Swale And this defendant hath since then untill very lately accordingly enjoyed the same / And this defendant saith he doth not know that the said Sir Solomon held or had any freehold tenements in West Grinton held of the complainant but pretended to claim a mannor in Swaledale & that the said mines were within such his mannor & thereby claimed the mines in Grinton & Harkaside that are now wrought or / pretended to by the complainant and also all other mines so leased to this defendant altho' as appeared without any just title And this defendant saith he neither was encouraged by nor did know that the complainants deeds evidences or writings relating to his estates & possessions in Swaledale were out of the complainants / possession or that the same have <soe> continued till lately as in the bill is suggested Nor do's this defendant know or beleive that the complainant hath or ever had any deeds evidences or writings that make him out any rights or title to all or any of the mines in question And saith he this defendant hath not only wrought / the mines called Howgrove or Grinton moor which lies in Grinton but such other mines in Grinton and Harkaside as he thought fit And saith he do's not know that there is or ever hath been any such vill as East Grinton unlesse the same hath been so called in contradistinction to a small number of houses / calld West Grinton but saith Grinton is the general name & comprehends that place which the complainant calls West Grinton & also Harkaside And saith he doth not know or ever heard of any wastes or commons belonging to West Grinton exclusive of the mannor of Grinton or so called or distinguisht nor do's he know the / boundary thereof but beleives West Grinton & also Harkaside & all the wastes & commons to them belonging to be part of the mannor of Grinton and heretofore also parcell of the possessions of the monastery of Bridlington And saith he doth not know nor can set forth the boundarys of the wastes & commons / between West Grinton & Harkaside if any such there be but beleives the names in the bill for that purpose specified are only used for a description of the places and that the same and all the wastes & commons are within and do belong to and are part of the mannor of Grinton And saith from the res- / -pective trials with the said Sir Solomon Swale and the last whereof was about eight years ago this defendant hath been (untill lately interrupted by the complainant) in the possession of & worked the mines both at West Grinton Harkaside & How Grove or Grinton moor & intends to continue his possession and work the same / as he shall find for his interest and conveniency within and thro' any moores wastes or places within the said mannor of Grinton and hopes it is lawfull for him to interrupt the complainant in case the complainant shall continue or attempt to digg oar within the said mannor of Grinton and particulerly in or upon / Harka waste or the place where the complainant now pretends to dig for oar or sink shafts the complainant as this defendant beleives having no right or title thereunto, nor was ever in possession thereof until lately as aforesaid by his entry upon this defendant to the knowledge or belief of this defendant And saith the place or / mines now pretended to be in working by the complainant were in the possession of the said Sir Solomon Swale when this defendant recovered the same in ejectment upon a trial as aforesaid And this defendant saith it may be true that the said mannor of Grinton was a mannor & so reputed about the time of the Conquest / but how the same is described in Doomsday book or whether any and what grant was made thereof in Henry the first time by Sir Walter de Gant and his wife to the priory of Bridlington this defendant refers to the Doomsday book and grant when produced And saith he does not know nor can set forth / whether the wife of Walter de Gant was the daughter of Stephen Earl of Richmond & Britany But this defendant doth believe that the said mannor of Grinton did come to the crown upon the attainder of the prior of Bridlington or by the dissolucion of the said monastry and that Queen Elizabeth did afterwards / about the time in the bill alledged make such grant thereof to Fitch & Wiseman under whom the said Hillary in the bill named and this defendant claim the said mannor as in the bill alledged And that in the said grant there was and is an excepcion of the leadmines and that the said mines have been separately / leased and granted by the crown both before and ever since as by the said grants & leases to which the said defendant for more certainty referreth himself when produced may appear And this defendant saith the said mannor of Grinton and the territories and fields thereunto belonging are and immemorially have been a large / tract or piece of ground conteining severall demesnes inclosures lands grounds wastes and moors lyeing mostly together and butted and bounded by severall other mannors lands wastes and other boundaries as before mencioned But this defendant saith he doth admit and believe it to be true that some small fields / of land may or do ly intermixed within the said mannor wastes and moors of Grinton and do break into the same, but whether any such lands be part of any other mannor or how the same are butted or bounded this defendant knoweth not But this defendant doth not know or believe that any moors wastes / or commons within the boundarys of Grinton are part of or do belong to any other mannor And this defendant doth deny that he doth know or believe or can make it appear that there is no such mannor of Grinton And saith he can not distinguish the boundarys of the said mannor of Grinton in Doomsday book / from the boundary now claimed by the owners of the said mannor or the extent thereof as described in and by the said bill of complaint And this defendant doth deny that by inspection of records evidences or writings or any of them or by copys or abstracts of them or any of them or by any informacion or otherwise / he do's know or is satisfy'd or believes that the said mannor of Grinton as described or mencioned in Doomsday book or otherwise together with the severall other mannors townes hamlets or lands in Swaledale were granted to Allan Earl of Britany as in the bill is pretended but / referreth himself to such grants if any such there be And saith he doth not know nor can set forth when the town or castle of Richmond took the name of the honor of Richmondshire nor do's this defendant know that Stephen Earl of Britany was about the time in the bill seised / of the said honour of Richmond or Richmondshire or that the same did include therein Swaledale as parcell thereof or that he made such grant to Walter de Gant of the dominion lordshipp or territory of Swaledale as in the bill is pretended or that such grant included therein the said mannor / of Grinton or that the royaltys thereof became united or vested in the said Walter de Gant & his wife or one of them as in the bill is pretended in right of the lordshipp of Swaledale And this defendant doth deny that he doth know or is satisfy'd or believes or hath been informed by the grant made to the / said monastry of Bridlington or by the entries or copies thereof now or at any time in this defendants custody or which he doth know hath seen or can come by or otherwise that the said Walter de Gant & his wife granted only the vill of Grinton alias East Grinton as in the bill is alledged or some / tenements therein to the said monastery or that what right or interest the said monastry had or enjoy'd in any of the lands in West Grinton Harkaside Whitaside or Cogden was granted to the said monastry by the lords of Swaledale by particuler grant subsequent to the grant of Grinton or were / not enjoy'd by the said monastery under their original grant or grants And this defendant doth deny that he knows that the said monastery had only a right in certain particular peices or parcells of lands in West Grinton or Harkaside or the precincts or territories thereof but saith & insisteth / upon it that the said monastry had the whole mannor of Grinton and the lands moors and wastes therein and that West Grinton & Harkaside and all the moors and wasts thereunto belonging was parcell thereof And this defendant do's deny that to his knowledge or belief the lords or pretended / lords of Swale Dale or of Helaugh during the continuance of the said monastry did enjoy all or any part of the lands or rents issueing out of West Grinton or Harkaside other than the free rents above mencioned not exceeding ten shillings per annum or that any lands or rents by them enjoyed / (if any such they had) was or were any part of the lordshipps of Swaledale or Helaugh And this defendant denys that he do's know that Stallerstone Stile is one of the bounder marks of the lordshipp of Swaledale or divides the same from the lordshipp of Ellerton or that the complainant hath any such lordshipp / or manor as Swaledale And this defendant saith he doth not know or believe that the lord[192] of Swaledale or their tenants in Harkaside or West Grinton if any such they had did time beyond memory or during the continuance of the said monastery or since enjoy without interrupcion / any possessions lands royalties or <priviledges> within the said boundarys of the mannor of Grinton <but believes the monastry enjoyed the possessions lands & libertyes within the said boundaryes of the mannor of Grinton> but from or under whom they held the same this defendant knoweth not And this defendant doth deny that the complainant is intituled to the mannor of Grinton or to the possessions right or royaltys within the / said boundaryes of the said mannor but what possessions right or royaltys the complainant is intituled unto within the mannor or pretended mannor of Swaledale this defendant knoweth not And saith he supposeth the river of Swale to be a bounder to the lordshipp or mannor of Swaledale if any such / there be And this defendant saith he doth not know or believe that the lordshipp dominion or mannor of Swaledale as granted to Walter de Gant if any such there be did extend to all the lands vills or hamlets within the said dale both on the north & south sides of the river Swale or to / the vills of Grinton Harkaside or West Grinton or any of them Nor do's this defendant know or can set forth the boundarys or extent of the same nor whether the ancient bounder of the lordshipp of Swaledale if any such there be doth extend itself from Stallerstone Stile on the south side of the / river Swale to Hollow mill in the bill named near the head of the river Swale or that any bounder mark there devides between Swaledale and the adjoyning mannor or between the parish of Swaledale or Grinton and the parish next adjoyning thereto nor do's this defendant know / whether the same divides the county of York from the county of Westmorland or that it is so taken notice of in printed maps descriptions or boundarys of the said countys, but saith & insisteth upon it that the mannor and parish of Grinton are distinct & separate from the mannor & parish of Swaledale / if any such there be And this defendant saith he believes the river of

[192] The word 'lord' is written in the singular.

Swale may run thro the bottom of the dale & is separate from the other rivers running thro' th'adjacent dales vizt Wensleydale & Arkilgarthdale by mountains & high hills on the north & south sides of the said river Swale and believes / the said river may be fed by springs and becks which descend from the said hills or mountains on each side of the said river & run into the same but this defendant doth deny that the said dale or lordshipp of Swaledale (if any such there be) <does> extend to the heads of the said rivulets gills or becks or include all such ground from / whence the rain water has any descent into the said dale or river as by the bill is suggested or that the said lordshipp extends to the sumit or heigth of the said mountains whence the rainwater falls divides or deals part into one dale & part into th'other or that the boundary of the pretended lordshipp on the north or south / of the said dale proceeds by the sumit or heigth or ridge of the hills as the heaven water deals or divides it self as in the bill is suggested But this defendant doth confess that from Stallerstone Stile to Hollow mill cross as heav'n water deals on each side of the said dale do's take in the mannor of Grinton & the several places / [compris'd] in the same or the boundary thereof And this defendant saith he doth not know or believe that from Stallerstone stile to Hollow mill cross was at any time the reputed boundary of the lordshipp of Swaledale nor do' this defendant know what appears by the records of the justices in eyre for the county of Yorke about the lands / of Gilbert de Gant or the boundarys of Swaledale but referreth himself to the records themselves when produced Nor do's this defendant know whether <the said Gilbert> obtein'd any such verdict or judgement as in the bill is suggested And this defendant saith he believes the said river of Swale was & is the ancient bounder of the complainants mannor of Helaugh / And this defendant saith he believes & doubts not to prove that the said monastry was seised in fee of the said mannor of Grinton and had all the royalties thereof & held courts & did all other acts of ownership as lords of the mannor and that the same came to the crown and that the said monastry had the same originally so granted / to them and neither did nor could ever pretend to keep courts and enjoy the rights and priviledges of a mannor as being exempt from secular service as in the bill is pretended in the lands they held or possessed And saith he believes and doubts not to prove that the said mannor of Grinton was & is a real mannor and the / prior of Bridlington did all along [. . .] a distinct & separate [jurisdiction] [. . .] or tenants there & were no ways accountable to any of the lords or pretended lords of Swaledale or did suit or service at any of their courts And this defendant saith he hopes to prove that either originally or by [some] subsequent [ancient] / grants the prior of Bridlington [. . .] the said mannor of Grinton & all the rights & royaltys belonging to the same and that in case the said Walter de Gant or the complainant or any of his ancestors have or ever had any such lordshipp dominion mannor of Swaledale [comprehending] such boundarys as in the bill is pretended / or by the boundaries [. . .] mannor of Grinton [. . .] ever this defendant doth not know believe or admit Yet the said mannor of Grinton may subsist or be a mannor within a mannor or lordshipp as this defendant is advised without any inconsistency therein And this defendant saith he doth not know what / grants were made to the prior of Bridlington nor whether there be any & what lycences or pardons of alienacions to be found [. . .] warrant the making of such grants or to enable the monastry to take any interest or estate thereby Nor do's this defendant know whether there did issue out any writts / of *ad quod dampnum* nor what grants [or] charters of continuation mencioned in any *reguli* charters of *inspeximus* or confirmation made to the monastry or how many of 'em remain upon record in the record office all the said matters not being necessary to be made out at this distance of time as this defendant insists / And this defendant saith [. . .] he has made search in the record office in the Tower & in severall other record offices in this court & elsewhere for what was or can be found or appears therein [. . .] concerning the matters in question or the contents of the records there found or <per>used by this defendant he referreth the complainant to the records themselves if he /

[membrane ii][193]

shall think fitt to search for or peruse the same and saith he hath taken copies of such records as he thought fitt and proper and hopes it was just and lawfull for him so to doe and the complainant if he so pleaseth this defendant beleives may at his own charges have copyes of the same from the proper offices in / that behalfe And this defendant saith he doth not know or beleive that it appeares by any records or entryes which this defendant or his agents have seen or perused that the complainant or his ancestors had or hath any right or title to the said mannor of Grinton or to the wasts or moores aforesaid / or any mines in or under the same or to the soyle or mines of Harkaside or West Grinton aforesaid or that the same are or were part of the mannor of Helaugh or pretended mannor lordshipp or dominion of Swaledale And this defendant saith he beleives it to be true that an inquisition was / found & taken upon the attainder of the said prior of Bridlington and that the same remaines in the office of augmentacion to which this defendant refers the complainant for the contents thereof and beleives the surveys taken of the possessions of the said monastery either upon the dissolucion or / since the same came to the Crowne and the particulars and leases made <by the Crowne and the accounts of the rents & profitts of the grant made> to Fitch & Wiseman are upon record in the proper offices for that purpose or entered or inrolled in the same and in the custody of the proper officers viz. the auditor of Yorkshire's office <the augmentacion & treasury remembrancer> / unto which this defendant referreth the complainant for the contents of the same And saith and humbly insisteth upon it that he is not obliged to set out or discover the records by him inspected or perused in the respective offices, but being known and publick offices and records the complainant / as this defendant is advised ought to resort to the same for his satisfaction therein and not enforce this defendant to set out particularise or discover the same or the contents heads or titles thereof or which of them this defendant intendeth to make use of in his defence against the complainants pretences / and saith he this defendant hath not ever had in his power or custody any records entryes particulars rentals leases surveyes grants or charters concerning the premises but what are upon record in the respective offices and where this defendant did see peruse or take copyes of the same or / such of them as he thought proper And this defendant saith that for what antient or other records or inquisicions *post mortem* taken upon the death of the lords or pretended lords of Swaledale or Helaugh which are in the Tower or amongst the records of this court and in other offices of / record this defendant referreth himselfe to the same when produced but does not know or beleive that in any of them or in any records evidences or writings whatsoever the estate of the said monastery of Bridlington is mencioned or discribed by the name of a vill or village or found or / appeares to be held of the lords of Swaledale or Helaugh or as part or member of their lordshipp or estate in Swaledale as in the bill is suggested or that Grinton was noe reall mannor, nor does this defendant know whether there have been any or what disputes about the reallity / of the said mannor And sayes he hath bin informed & beleives that vicar Sympson in the bill named did clayme Grinton as a mannor but never heard of any suites were begun between him and the tennants

193 Parchment, 745mm x 670mm.

concerning certaine priviledges claymed by the tennants or concerning the / reallity of the mannor nor of the substance of the bills or answers in such suites or any of them if any such there were or what witnesses were examined or depositions taken or published in that cause this defendant knoweth not nor can set forth And this defendant denyes that he or any / person on his behalfe or for his use have or ever had any court copies or other copie or coppies or transcripts of the same or any of them or does know what is become of them or any of them (if any such there ever were[194] or that the bills answers deposicions decrees or proceedings in the said cause or / any of them were taken off the file or files or destroyed or concealed Nor does this defendant know or has been informed or hath reason to beleive or can set forth whether any decree or dismission was obteyned against the said Sympson in the said supposed suites or any of them or that he the said / Sympson was prohibited from holding courts at Grinton or that he was prohibited from incloseing any part of the moores & wasts altho it is probable that the tenants might and did oppose any inclosures of the moores and wasts in regard the same would be prejudiciall to their / rights of common And this defendant saith he doth not know or hath heard that any suites have been commenced against the said Hillary or his ancestors or any of them or any other whereby it did appeare that Grinton was no mannor or whereby any right of the complainant to the premisses did or would appear / And this defendant doth deny that he do's know or hath heard or is satisfyd or beleives that the monastery of Bridlington had not nor claimed any other right than of common upon Grinton moor or the wasts or moors of Harkaside & West Grinton or some libertys for the benefit of their tenants but on the / contrary he beleives that the monastery had and claimed the soil of the same and all the mines therein or under the same & all other royaltys as lords of the said mannor & owners of the same and of the wastes and moors aforesaid and that neither the complainant or any of his ancestors or those under / whom he claims did ever use or enjoy or attempt to use (untill now lately as aforesaid) any rights royaltys or privileges in the moors wastes or commons of Grinton West Grinton or Harkaside or any of them or pretended to be owners of the soil thereof or to be intituled to any mines therein / as this defendant believes And this defendant doth deny that he doth know of hath found out or discovered any evidences or writings which do make out prove or shew that the lords of Swaledale have enjoyed or claimed to enjoy any rights or privileges in West Grinton or Harkaside or the moors wastes or commons / thereof or in any moors wastes or commons of Grinton or in any place or places within the said boundarys of the said mannor of Grinton and this defendant saith he doth not know any thing of the pretended grant in the bill alledged to be made by Robert de Gant unto Ralph the son of Goodman in the bill mencioned / or the contents of the same or of any other grants fines or conveyances whereby the said monastery held the supposed tenement or land & common pretended to be granted to the said Ralph the son of Goodman by the said Robert de Gant And saith he knoweth not nor can set forth whether Robert de Gant was the son of / Walter de Gant but is a stranger to the pedigree of that family And saith he doth not know hath seen read heard or been informed to the best of his remembrance of the pretended grant made or mencioned to be made by Gilbert de Gant to the said monastery intituled a grant *de curia dilatanda* or the / contents of the same or of the other pretended deed or grant whereby 'tis pretended that the said prior & co[n]vent[195] did grant or confirm to Gilbert de Gant to have free park in Whitaside & process in his court of Helaugh or of the contents thereof or of

194 This bracket is not closed.

195 The letter 'n' is omitted in the original.

the other pretended deed grant or confir- / -macion made to the said Gilbert de Gant by the said prior & co[n]vent[196] of their wood in Swaledale & other particulers in the bill mencioned or of the contents of the same And this defendant doth deny that he or any person by his order or for his use hath or ever had any of the said deeds grants or confirmacions / in his or their custody or power or know where the originals of them or any of them were are or be nor hath seen perused or been informed of any copys or entrys thereof or of any of them And this defendant saith that he doth not know that the lords or pretended lords of Swaledale or any of them have or ever / had any right to the said mannor of Grinton or the wastes moors or commons thereof or any of them or ever enjoyed the same or any of them or grazed or fed any deer thereon or hath ever read heard or been so informed And saith he doth not know nor can set forth what grants were obtained by the / said monastery of or from the lords of Swaledale if any such there ever were or any of them nor the contents of any of them And saith he cannot set forth whether the deer belonging to the lordshipp or forest of Swaledale (if any such place there be) did graze or depasture upon the said moors / wastes or commons or any of them but beleives the said monastery of Bridlington and those claiming under them did claim and enjoy all the said moors wastes & commons & soil of & mines in or under the same & not common of turbary or pasture as in the said bill of complaint is suggested And this defendant saith / he hath not nor ever had in his power or custody nor doth know of or can set forth any records copys of records deeds evidences or writings whereby it appears that the lords or pretended lords of Swaledale or any of them had any right or title in or to the said moors or wastes of Grinton Harkaside / or West Grinton or any of them or the soil thereof or mines therein or any of them but beleives & hopes to prove that there are severall records in the Tower & in this court in the respective offices that make out the title of the said monastery to the same & to the soil thereof & the mines and / royaltys therein And this defendant doth deny that he hath seen perused or been informed of any grants or licenses obtained from the lords of Swaledale or any of them or any inrollments or entrys whereby the monastery was empowered to inclose any part of the said moors Nor doth this / defendant know or believe that any part of the said moors were inclosed by any license from the lords or pretended lords of Swaledale or any other person except by license of the said monastery or those claiming under them And this defendant saith he hopes to prove by the records of this court that the / mines of Grinton Harkaside & Whitaside (as parcell of the possessions of the said late monastery) were leased or demised by Queen Elizabeth to Henry Lord Scroop & Arthur Philips and afterwards by King Charles the First to Humphrey Wharton as appears by the records of this court And / that such respective leases do extend to and take in the mines as this defendant is advised of the moors wastes & commons & the inclosed fields grounds & territories of Grinton as by the said leases which for more certainty this defendant referreth himself may appear But believes it to be true / that the moors wastes & commons are not in such leases particulerly named but only the mines within the territories & fields of Grinton Harkaside & Whitaside which this defendant doth insist & is advised do's effectually comprehend & pass the mines in the moors wastes & commons as if particulerly / named and that accordingly they have been wrought & enjoy'd under such grants & leases and that the Lord Scroop Philips & Wharton did possess & enjoy and as they thought fit did work the mines in the moors & wastes of Grinton Whitaside & Harkaside & West Grinton when they / found or discovered the same And this defendant saith that he doth not know or hath heard or

196 As above.

believes or hath been informed of the contrary or can make the contrary to appear as in the bill is suggested And this defendant saith that true it is he hath seen an account book of the said Humphrey / Wharton and therein is entred as paid a rent of twenty shillings per annum to the crown for the mines in Grinton (& not in East Grinton to the best of this defendants remembrance[197] and believes the same refers to the said lease and according to this defendants opinion it appears by the said / book that the said Mr Wharton did work the mines in Grinton and verily believes the said mines were in the commons of Grinton or some of them and that the said Mr Wharton had a right and claimed all the mines as well in the said wastes moors & commons as in the fields or inclosures And this defendant / hopes to prove he wrought the same when he found it for his benefit & as occasion required but do's not remember that the particuler places are specified in the said book of accounts and says he believes the mines in Fremington were also leased to the said Mr Wharton and that it appears by the / said book of accounts that he wrought the same And this defendant believes it very probable that the said Mr Wharton did make some <under> lease or leases of the mines in the moors or wastes of Grinton & within the boundarys of the said mannor but whether to one Brodrick in the bill mencioned he / knows not <or> he or his agents or assignes wrought the same under such lease but saith that there are still appearances or marks of old works within the said moors or wastes where such were made & carry'd on But this defendant of his own knowledge doth not know or hath otherwise been / informed to his remembrance than as aforesaid by whom the same were wrought but believes the same was done under and by virtue of the crown lease aforesaid And this defendant doth deny that he hath or ever had in his custody or do's know of hath seen perused or been inform'd of any / book or books of account records surveys rentals evidences writings accounts or memorandums whereby it appears what particuler places by name <were wrought under the leases granted by the Crowne or within what particuler places by name> the agents of the crown wrought the said mines or for what mines in particuler the rent to the crown was answer'd or whereby it / appears that the rent answer'd or in charge to the crown for the said mines or any part of them was answerd charged mencioned or taken notice of by the name of *novus redditus* or as a new rent or where any such books evidences writings or accounts are or were kept except only / the said book of accounts of the said Mr Wharton which was produced by old Mr Smales at the bar of this court in the trial against Sir Solomon Swale and which is now in the custody of his son Mr Smale as he believes And wherein this defendant do's not believe any of the particulers aforesaid / do appear except as aforesaid and except such accounts records surveys rentalls evidences & memorandums as are upon record & remain in this honourable court to which for the particulers & contents of them this defendant referreth the complainant And this defendant saith that he doth believe that / the said monastery and the crown since the dissolucion thereof and their lessees farmers agents & assigns have from time to time as occasion required without any interruption or disturbance wrought considerable quantitys of lead ore in the several moors wastes & commons / of Grinton West Grinton & Harkaside and made considerable advantage and profit thereof and have allalong done severall other acts to prove their right & title to the soil & mines of the said wastes as part of the estate of the said mannor of Grinton And doth deny that / he doth know hath heard or believes that the same were so wrought used or done during the minority of the complainants ancestors or their absence in foreign parts or while they were distant from the premisses And saith he believes that trials for ore were made

197 This bracket is not closed.

from time to time / without any leave consent or interrupcion of the complainants ancestors or those under whom he claims or any of them and that when a rich or profitable vein or mine was found out or discovered the same was quietly [x x x][198] wrought out by the lessees or grantees of the crown and no claim / as this defendant knows or believes made by or any interrupcion given by the complainants ancestors or any of them And this defendant saith he doth not know whether the complainants ancestors or those under whom they claim had any mines in any part of their pretended lordshipp of Swaledale Yet he / believes the complainant & his ancestors had & did work some mine or mines in the mannor of Helaugh but do's not believe the same prevented the complainant or his ancestors in making trials for mines elsewhere Nor do's this defendant believe they had any right or title to make trials in / any of the moors wastes or territorys of Grinton Harkaside or West Grinton And this defendant saith he do's believe that the said mines were concealed from the crown from the restoracion or civil wars till discovered by this defendant but that Sir Solomon Swale & some others under his pretences / did enter upon & work the same and that the complainant or his ancestors never had any possession of or worked any mines in or upon the manor of Grinton or the wastes or moors aforesaid or any of them And this defendant doth deny that he doth believe or can set forth that the said / mine[s][199] were or are granted away by or from the crown except the lease so made to this defendant's trustee as aforesaid And says he believes the rents reserved upon the said lease made by the crown to the said Lord Scroop & Philips & Humphrey Wharton were constantly accounted for / answer'd or in charge to the crown but herein & the times for which they were answer'd accounted for or stood in charge to the crown this defendant referreth himself to the records of this court And this defendant doth deny that any leases made by the crown of the said mines were made / or granted in right or as part of the mannor of Helaugh or any part thereof And this <defendant> doth deny that he hath or ever had in his power <or custody> the inquisicion or survey taken of the estate of the said Bygod upon his attainder or any copy thereof or doth know where the same is or are or / were or any entry thereof or in whose custody except amongst the records of this court if any such there be And this defendant saith that in the presence of the proper officers he hath seen & perused the ministers accounts of the lordshipp of Helaugh in the augumentation office / in the bill mencioned for the contents thereof he refers to the same And saith that he did observe therein some notice taken of a lease to one Coalwiche in the bill mencioned and did request the proper officer in that behalf to shew him this defendant the said lease and was / told by him the said officer that an agent of the complainant had been searching there some days before and that he the said officer could not find or discover any such lease or the entry or inrollment thereof And saith he doth not remember whether in the said accounts the said supposed / lease be mencioned to be made by the Bigots or by the crown as part of the estate And this defendant doth deny that he or any other person by his order or for his use to his knowledge hath or ever had in his custody or doth know of or endeavour to conceal any evidences court rolls / deeds grants or writings which ever did belong or concern the said Sir Francis Bigod or his estate in Swaledale or any other accounts taken or certifyd by any officer or minister of the crown or any other person relating to the estate of the said Sir Francis Bigod at or / for any time whatsoever except what remain in the auditor's office and augmentation office which this defendant hath seen as aforesaid and this defendant saith he doth not

198 One or more words erased.

199 The word 'mine' is written in the singular.

know nor can set forth what minister's accounts particulers or surveys have at any time been taken or certifyd / of the estate of Sir Francis Bigod or of the complainant or the rents issues or profits of any mines thereunto belonging or in what office or where to be found nor what lease or leases have been granted by the said Bigot or the crown except only what appears in the said augmentation office or / auditor's office before mencioned relating hereunto to which this defendant referreth himself And this defendant saith he do's not know or believe that by any account particulers surveys leases court rolls or otherwise it doth appear that the mines of Grinton or the moors wastes or commons / therewith occuy'd or belonging or the soil or royaltys of the same [or][200] any of them have been held or enjoy'd <as> any part of the complainants estate in Swaledale And this defendant saith he doth not know or believe there is or ever was any such vill or place as East Grinton or so known or / distinguish't but confesseth he hath heard & believes that a few houses lying west of Grinton are called by the name of West Grinton but conceives the same is so done by way of distinction only and saith & insisteth upon it that Grinton is the general name and that the same is a / mannor and that the name of Grinton do's comprehend the whole as well Whitaside Harkaside & West Grinton as all the moors & wastes to them belonging and that they are all comprehended in & part of the mannor of Grinton And this defendant denys that by any writing / or writings that he hath seen that either in name or tenure or any other respects Grinton hath been or is or was distinguished by the names of East Grinton & West Grinton or that there is or ever was any such vill or name as East Grinton or any other such distinction / either in the books of accounts of the said Humphrey Wharton or in any leases or grants from the crown or in any particulers of Bigot's estate remaining on record in this court or in any other records writings or evidences whatsoever to the remembrance or belief of this defendant / and this defendant doth confess that there are severall houses & tenements standing in the places called West Grinton & Harkaside but whether the same be in law vills townships or hamlets this defendant knoweth not but submits to the court & saith that he computes that the place call'd / West Grinton is distant from Grinton about half a mile and that Harkaside is distant from Grinton about a mile and saith he cannot set forth whether antiently there hath been any or how many more houses in those places nor what ruines or marks thereof do appear And / saith he doth not know or hath heard or is convinced that either the lords of Swaledale or those under whom the complainant claims ever had or have any right or title to the mines or royaltys of the moors wastes or commons of Harkaside or West Grinton or any of them by those or / any other name or names whatsoever And saith he doth not know or believe that West Grinton or Harkaside is or ever was part of the vill of Reeth or of the mannor or reputed mannor of Reeth nor do's this defendant know whether the said vill or mannor or reputed mannor / of Reeth be or ever was any part of the lordshipps or pretended lordshipps of Swaledale or Helaugh and saith he doth not know whether any court for the pretended lordshipp of Swaledale have been sometimes kept at Helaugh or sometimes at Reeth or that it <so> appears by any records grants / writings or evidences whatsoever or that it appears by any of them or otherwise that the pretended lordshipp of Swaledale was or is sometime call'd by the name of the lordshipp or mannor of Helaugh and sometime by the name of the lordshipp or mannor of Reeth nor do's this defendant know what / right or title the complainant hath or claimeth to the same by the names aforesaid or by any other names whatsoever And this defendant saith he

200 The word 'or' is omitted in the original.

doth not know or believe that there is or ever was any <such> mannor lordshipp or dominion called Swaledale as by the bill is pretended & wherever this defendant in this his / answer hath called the same by those names it is done in answer to the terms in the complainant's bill only and not any ways admitting that there is or ever was any such lordshipp mannor or dominion & so this defendant explains himself & prays to be understood And this defendant saith he doth not / know or believe that the complainant hath any right or title to the places calld West Grinton or Harkaside or any of the moors or wastes thereof or thereunto belonging or in Grinton aforesaid or to the mines in them or any of them either as part or parcell of any mannor or mannors or / otherwise howsoever And this defendant saith he doth not know that there are any settled or known boundarys of West Grinton & Harkaside or the moors & wastes to them belonging other than what the inhabitants upon account of their common rights have as marks of distinction / or boundarys among themselves nor can he set forth how the same are separated and divided but insists that the whole are within and part of the mannor of Grinton and which is bounded as herein before described And saith such deeds grants evidences records & writings /

[membrane iii][201]

as he doth know of he dos intend to make use of for the defence of his interest & the title of the crown as he shall think fit or be advised which he hopes is lawfull for him to do without particulerizing the same to the complainant and humbly insists that he ought not to set them forth / or the contents of them or discover his evidences or what he intends to make use of to the complainant And this defendant doth admit that he or his order have drawn up a paper containing the boundarys of the said mannor of Grinton and which boundarys are as herein before described & set forth / and that severall ancient persons have published the same as the true boundarys of the said mannor as he hopes it was lawfull for him to do & believes & admits that as severall of them did swear the same at the former trials with Sir Solomon Swale so they have again made / affidavits of the truth thereof which was done in order to apply to this court to quiet this defendants possession & not with any intent to preengage or prevail upon such persons to be witnesses against the truth or for any other unjust purpose but 'tis probable the said persons may be witnesses / against the complainant in case he continue his pretences And this defendant saith he do's not know there's any bounder dividing the moors or wastes of Harkaside or West Grinton distinct from the commons of Grinton except for the purposes before mencioned nor did this defendant hear that / there was ever any bounder riding made or pretended to be made or ridd by the complainant or any of his ancestors or their orders separating or distinguishing the same untill that lately made or rode by the complainants order and which this defendant conceives to be an untrue bounder & utterly / mistaken besides this defendant doth believe & hath been informed that the complainant directed the same to be made and rode as the pretended bounder of the mannor of Helaugh only and believes the said boundary was made or set on foot on purpose to create an evidence & to disturb this / defendant in his this defendants said possession and mines And this defendant saith true it is that the commons moors and wastes belonging to the said mannor of Grinton are distinguished by the severall names of Grinton moor Grinton how Harka hill & Whitamoor or hill but do's not know that / those of West Grinton are any ways

201 Parchment, 735mm x 570mm.

distinguished from Grinton except as aforesaid and says the names of Hugh's ditch Howdike or Hughdick Whitstones in the how Graystone in the Greets & Stamburgham in the bill named may be known places but do's not know or remember / to what they perticularly relate other[wise][202] than as he believes some of them may be boundarys to the Duke of Bolton's estate and saith he do's not know how the commons of Grinton are known or distinguished from the commons of West Grinton or Harkaside by any separate enjoyment / of the said commons by the tenants there [Nor][203] do's this defendant know whether there be any separate enjoyment or not or whether all the tenants in Grinton do comon all over the said moors without distincion but believes the tenants of Grinton do claim & have enjoyed common / all over the said moors & lordshipp of Grinton and also liberty of turbary and pulling of ling as well on the moors & wastes of Harkaside & West Grinton as elsewhere within the said mannor of Grinton without any leave or license from the complainant or his tenants or without paying / or making any satisfaction or acknowledgement for the same And this defendant doth deny that he by himself or his agents did attempt to discourage or prevent any persons to appear or conceal their knowledge of the complainants bounders of his mannor of Helaugh or to draw them into / any mistakes but believes his this defendants agents did oppose or discourage persons to take in the mannor or wastes of Grinton into their riding or boundarys and believes they the said complainant & his agents did mistake their bounders when they took in the mannor moors or wastes of / Grinton & therein & in that particuler this defendant doth insist & believe the said bounder was not truly ridd & as to the rest of the said bounder this defendant conceives himself not concerned therein And this defendant saith & insisteth upon it that the places now working by the complainant or / his agents for lead ore are in the mannor of Grinton and that the complainant or his ancestors till very lately to wit 'till within nine months last had never any possession thereof but that the same have ever since the said trials with the said Sir Solomon Swale been in this defendants / possession and do of right belong to him this defendant and not to the complainant but saith he believes the said places are part of the moor called Harkaside & <part> of the moor called Grinton how or moor and insists that both the moors as well as [the][204] vill called Harkaside & the place called / West Grinton are within & parcell of the mannor of Grinton And this defendant doth deny that the boundary of West Grinton & Harkaside & of the moors wastes & commons thereunto belonging do's proceed & divide the same from Grinton in such manner as in the bill is mencioned / to the belief of this defendant but do's believe some of the descriptions do divide & bound the said mannor of Grinton from the Duke of Bolton's mannor and that the rest of the descriptions are imaginary and no true bounders altho' it may be true for ought this / defendant doth know that there are such places & known by such names as in the bill of complaint <are> mencioned but this defendant doth deny that he or any person by his order or privity ever removed the Gray Stone or hurrock in the bill named or any boundary stone / whatsoever and do's deny to his belief the places in the bill for that purpose mencioned are or were the boundarys between Grinton & West Grinton or Harkaside or Whitaside or any of them And this defendant saith he doth not know or believe that the / boundarys of West Grinton or Herkaside & the commons thereof were in the time of Walter de Gant ascertained by him to proceed & extend in the

202 The word 'otherwise' is not fully legible on account of some minor damage to the parchment.
203 As above.
204 Word omitted in the original.

manner in the bill alledged nor whether then there were or yet are such places or names as in the bill for that purpose / is alledged or that the moors wastes or commons of or belonging to West Grinton or Herkaside do proceed or extend themselves from the great ditch called Hugh's ditch in such manner as therein is mencioned but saith all the said moors & wastes are part of & do belong to the mannor / of Grinton & which mannor is bounded as herein before is mencioned and that all the other pretended boundarys & separations between the vills places or commons of West Grinton Harkaside & Whitaside are no boundarys of any separate mannors or interests but all comprehended / in and part of the mannor of Grinton & no part of the complainants mannor or mannors and this defendant denys that he doth know of or hath seen possessed or been informed of any records or proceedings upon any writ or writs of *curia claudenda* brought by or against the pretended / lords of Swaledale or any their tenants or by or against the priors of Bridlington or any their tenants or any proceedings upon any writ or writs *de rationalibus divisis* or upon any writ or comission for the ascertaining the boundarys of the said vills places wastes / moors or commons or otherwise whereby the separate boundarys or any right of the lords or pretended lords of Swaledale to the soil or mines of the said moores or to any royalties or privileges therein do's appear And this defendant saith he do's not know that any of the lands / at Harkaside or West Grinton or the tenants thereof do or ever did ow suit or service or answer either at the complainants court or att any courts held for the honour of Richmond or mannor of Arkildale but believes they all ow suit and service and ought and do & anciently / did answer & appear at the said mannor court of Grinton And this defendant denys that he do's know or believe that any pieces or parcells of ground at the east end of Harkaside were ever part of the complainants mannor or held of the same or ever paid their rents at the courts held for / the crown within or for the honour of Richmond nor do's this defendant know the extent of the honour of Richmond nor whether the lands of Harkaside or Swaledale lye within the compass or extent of the same And this defendant saith it may be true that upon the dis- / -solucion of the said monastry the crown might make severall grants as well of part of the lands at Harkaside as of the rents & services belonging to the same and likewise that the crown might make severall grants of the honour or parts of the honour of Richmond and / mannor of Arkildale to the citty of London and that they afterwards may have granted off and divided the same but for more certainty herein this defendant referreth to the severall grants when produced if any such there be But whether any or what confusion hath / thereby or otherwise been made in the tenure of such lands this defendant knoweth not And says he believes all the lands in Harkaside & West Grinton did formerly belong to the said monastry and that the said monastry had rents and profits out of the same And that the said / lands & tenements thereof did answer and pay their rents suit and service to the said mannor court of Grinton And saith he believes that the rents payable to the complainant out of any lands in Harkaside do not exceed eight or nine pounds per annum and that the lands out of / which they are issueing are but a very small part of the lands in Harkaside And says he does not know that any of the complainants rents in Harkaside are copyhold rents but believes the rents be ancient rents And says all the rents the complainant hath in Harkaside by / grant from the crown are not of above the yearly value of ten shillings as he believes And saith he doth not know or believe that any the lands in Harkaside or West Grinton or the tenants thereof do pay their rents or answer at the complainants courts except the complainants tenants / there so do for their conveniency or by the complainants order or direction and not of right but whether any particuler lands in Harkaside claimed by or belonging to the complainant do pay suit or service at any of the complainants courts this defendant conceives & is advised is no ways material / to this defendants right and title under the crown to the said lead mines And says he doth not know what lands in Harkaside do lye in the constablery of Grinton nor what within the constablery of Reeth And saith he doth not know nor can set forth whether the lands in Harka- / -side or West Grinton be distinguished by the names of the old lands & new lands nor whether there be any such lands or names And saith he doth believe the complainant hath made a forcible entry on the said mines and this defendants possession thereof And that he the said complainant / never had possession of any of the said commons or of the profits of the soil of the said wastes or of the said mines untill lately as aforesaid And says the complainant and his ancestors have as this defendant believes have only been in possession of the particuler lands tenements & rents above mencioned / to belong to him and not of all the inclosed grounds of Harkaside And saith he doth not know or believe that the said complainants lands in Harkaside or rents in West Grinton if such there be or the moors wastes or commons aforesaid are parcell of the mannor of Swaledale if any / such there be or of the complainants estate within the same or that the complainants ancestors <have> granted any license to take in or improve any part of the wastes or commons or that the same was accordingly done as in the bill is pretended Nor do's this defendant know that / the complainant hath any copyholders in Harkaside And this defendant saith he doth not know that the complainant gave any such orders to his agents to ride the boundarys of Harkaside & West Grinton as in the bill is pretended this defendant being informed that the complainants / orders were to ride the bounders of his manor of Helaugh Nor do's this defendant know whether such notices were given of riding the said boundaries and at such times and places as in the bill is alledged But believes it to be true that some pretended boundary was ridd and / drawn up in writing and signed by several persons the complainant having in answer to this defendants bill in this court set forth the same but this defendant believes and doubts not to prove that there are very great and many mistakes therein and altho' the complainant / by his bill pretends title to the mines now in working by the defendant Yet in his said answer in this court he do's not pretend to the same as this defendant is advised And says if the said boundarys so ridd by the complainants agents and the complainants pretences to the lordshipp of / Swaledale and it's extent should be <true or> insisted upon by the complainant it would take in and disturb the possessions of great numbers of people as this defendant believes and therefore the respective persons have just reason to be alarmed thereat But this defendant doth deny / that he endeavourd to impose on the neighbourhood of Grinton or others to amuse them with false storys or insinuacions about the complainants proceedings as in the bill is alledged And this defendant denys that his this defendants proceedings do tend to subvert / or destroy the rights or privileges of the tenants of Harkaside or West Grinton or to remove or overthrow any land marks whatsoever And this defendant saith that as to the pretences and claims of the said Sir Solomon Swale and other <the> charges in the bill / relating thereunto this defendant at present conceives himself not concern'd therein And this defendant doth deny that Toby West in the bill named was or is a partner in the said mines or any of them with this defendant and says he hopes it was & is lawful for him this defendant to make what presents / and give what gratuities or guifts to the said Mr West he this defendant thought fit And this defendant saith he believes the said Mr West in the bill named to be a very honest person and that he hath & will do this defendant any acts of friendship he can in a lawfull way and doth not know or believe that the said / Mr West dos any other ways intermeddle or concern himself in the dispute between the complainant and the defendant and believes the said Mr West was present at the drawing up of some accounts testimonialls or affidavits of the boundarys of Grinton mannor in order to have the same made / use of in this court to quiet this defendant's possession in the premisses & that for that purpose he did discourse [with][205] severall ancient & knowing persons in that behalf & believes the contents thereof to be the boundarys of the mannor of Grinton as herein before described and the possession of / this defendant of the mines in question And this defendant doth deny that he or any other to his knowledge or by his order hath prevailed with any person or persons to become partners in the <said> lead works or to assist or contribute to the charge or expences of carrying on or defending the suits / with the complainant but confesseth he hath apply'd himself to several persons and had consultacion with his council & others to maintain his own right and the right of the crown and to invalidate the complainant's (as this defendant conceives and is advised) unjust pretences but denys he / hath made any proposals to any persons <to become partners> or to convey any estate or interest to them of or in the mines in question or to which this defendant claims title or which the complainant is or pretends to be in possession and saith he conceives the complainant no ways concerned to know whether any / person or who was concerned with this defendant in obtaining the said lease And saith he this defendant hath the legal estate & interest in the said lease & mines & is solely concern'd in the same And saith true it is this defendant hath declared that he will bring and do's intend to bring / as he shall be advised actions against the complainants agents touching the premisses And that this defendant hath exhibited such informacion in this court in Mr Attorny Generalls name as in the bill is alledged and says he has neither found nor seen any record evidences or writings / that make out to his beleife any title to the complainant to all or any the mines in question or the lands grounds moores & wastes aforesaid or any of them And this defendant doth utterly deny that he or any person by his order hath suppressed concealed or imbesselled any records any wayes relateing to the complainants title or some / of the matters or premisses in question And this defendant denyes he hath wrought any mines of the complainants and conceives and is advised that the complainant is not concerned to know what lead oare this defendant hath raised or the value thereof or the profitt made by the same, the rather for that the complainant in his answer to the said bill or / informacion in this court does not pretend any title to the mines now working by this defendant and sayes true it's he this defendant doth refuse to account with the complainant or to produce or shew the complainant his booke of account or to distinguish what profitts he hath made out of the mines of Harkaside or West Grinton separately / distinct from the other mines and does with submission refuse to bring his bookes of account into this court at the complainants instance or for his benefitt this defendant haveing duely accounted to the Crowne according to the covenants of his lease and is advised the complainant hath no pretence to have any account / or discovery herein before he have affirmed his right and title at law or in this court which this defendant believes the complainant will never be able to do And this defendant doth deny that he this defendant hath burnt distroyed or suppressed any of his bookes of account or any part of them or any other writeings / writeings[206] relateing to the workeing the said mines or the profitt or advantages made thereof or thereby And this defendant doth deny that he hath had the comand custody or ordering of any records bookes entryes or evidence relateing to the premisses which were formerly kept in / the a[udi]tors[207] office or any other office of records or did take the same or any of them into his custody out of the Treasuryes record offices of this court or elsewhere as in the bill is untruely suggested

205 Word omitted in the original.
206 The word 'writeings' is repeated in the original.
207 A small hole in the parchment has resulted in some minor loss of text.

Nor doth this defendant believe that any records belonging to the said / offi[ce][208] or relateing to the premisses are wanting mislade or concealed which were in being or found before or made use off at the said former tryall with the said Sir Solomon Swale or any of them or which were in being when this defendant first discovered the said mines to be / concealed from the Crowne and sayth he never had nor saw the originall grants of the monastry of Bridlington or the court rolls belonging to the said mannor of Grinton or the surveys leases particulars boundarys or other evidences relateing thereto other than what this defendant / did see [or peruse in][209] the Tower and the office of this court and where all those which this defendant did see or peruse yet remaine as he verily believes And this defendant doth utterly deny that he hath taken out of this court or any other office or imbezelled suppressed secreted or concealed any / records bookes [. . .][210] grants leases rolls or other evidences concerning the said lead mines or boundaryes or any the matters in question or any wayes relateing to this complainant or his estate And this defendant sayth he hath and claymeth the right & title to the said mannor of Grinton and / to the moores wasts & commons thereunto belonging and to the soyle thereof by purchase from John Hillary Mary Hillary & Samuel Hillary & the mines therein by and under the lease from the Crowne herein before mencioned and believes the complainant hath noe right or / title to any the moores wasts & commons or mines aforesaid And saith he this defendant hath & claymeth both the said mannor moores wasts commons & mines howsoever the same are mencioned or described and by all lawfull wayes & meanes which the law alloweth And this defendant / saith he observeth by and in the complainants bill a very great incertainty in his pretences sometimes setting up one right or title and sometimes another sometimes pretending the premisses to be within the mannor of Helaugh sometimes Reeth sometimes in Swale Dale, and sometimes / the mannor lordshipp or dominion of Swale Dale and sometimes as this defendant apprehends the complainant pretends West Grinton or Harkaside to be a mannor of itself, by all which very greate incertaintyes it appears that the complainant is at a loss to find out what title to pretend to the premises as this defendant is advised / And this defendant doth deny all confederacy and combinacion in the bill charged without that that any other matter or thing in the complainants said bill of complaint contained materiall or effectual for this defendant to make answer unto and not herein have answered or avoided traversed or / denyed is true to the knowledge of this defendant all which matters and things by him herein alledged this defendant is ready and willing to averr maintaine and prove as this honourable court shall direct and humbly prays to be hence dismissed with his reasonable costs and / charges in this behalfe most wrongfully sustained. /

/s/ Sam Dodd[211]

E112/915 EBOR 100 – document 3[212]

208 As above.

209 As above.

210 As above.

211 See p. 7, footnote 22.

212 Parchment, 260mm x 130mm. Commission of *dedimus potestatem* written in Latin on parchment giving authority to Thomas Pulleine Esq., Roger Baynes gent., Thomas Dickonson gent., Peter Rawsthorn gent., Mathew Smales gent. and Anthony Hammond gent. to take the answers of Toby West. The commission was issued on 12 June 1706 and is signed by the Queen's Remembrancer, [Henry] Ayloffe. On the *verso*, the memorandum of execution is signed by Thomas Pulleine and Roger Baynes. Thomas Pulleine, or Pulleyn, was Marriott's father-in-law. Roger Baynes was Marriott's solicitor and presumably a relative of the Jacob Baynes who served the court injunction on Lord

E112/915 EBOR 100 – document 4[213]

[membrane i][214]

The answer & disclaimer of Toby West clerk viccar & impropriator of Grinton in the county of Yorke one of the defendants to the bill of complaint of the / right honourable Thomas Lord Wharton & his son /

This defendant saveing to himselfe now & at all tymes hereafter all benefitt & advantage of excepcion to the manifold incertainties & imperfecions in the complainants said bill of complaint for answer thereunto or to soe much thereof as this defendant by his councell is advised is / anywayes materiall for him this defendant to make answer unto He answereth & saith that he doth not know of what dominions lordshipps mannors or lands Philip late Lord Wharton the complainants father in the bill named was seized or possessed nor of what the complainants or either of them / were or are seized or possessed of interested in or intituled unto And this defendant saith he hath heard & beleives that in Swaledale there are or be four mannors or lordshipps or reputed mannors or lordshipps vizt the mannor of Grinton the mannor of Helaugh the mannor of / Muker & the mannor of Fremington But this defendant doth not know or beleive that there is any such mannor or lordshipp as the mannor or lordshipp of Swaledale And this defendant saith he beleives that all Swaledale is within the parish of Grinton & that he this defendant claimeth & / beleives he hath right to all the tythes in the said parish of Grinton but saith he hath noe lands or tenements in any other of the townes towneshipps vills or hamletts in the bill named except the towne of Grinton onely where the church & viccaridge house doe stand & the gleab / lands lye Neither does this defendant know the divisions or extents of the other townes or vills of the said parish or any of them or the divisions or extents of the moores or wasts of them or any of them except as herein after mencioned And this defendant saith he doth not know the / boundaryes of the parish of Grinton And saith the said parish is soe very long & large being twelve miles & upwards in length & the moores & wasts of soe vast an extent & compass that this defendant doth not beleive any one person doth know or can set forth the boundaryes of / the said parish But this defendant doth very well know the boundaryes & extent of the mannor or lordshipp of Grinton haveing lived there upwards of thirty three yeares & does beleive the boundaryes & extents thereof to be according to two papers the coppyes whereof are subscribed or <hereunder annext> / & have been soe reputed for these thirty three yeares last past ever since he this defendant has been viccar of Grinton aforesaid And this defendant saith that the myne or mynes which Mr Marriott by his agents is now workeing for lead ore as likewise the shafts or pitts in / Harkaside & Grinton Moore where the Lord Whartons agents doe now sink are parcell of & within the bounders of the said mannor or lordshipp of Grinton according to the best of his knowledge & as he beleives And this defendant doth deny that he hath or ever had any parte share / or interest in all or any the mynes in

Wharton's miners. After Marriott won the case he was appointed barmaster for the Grinton mines (Tyson, Spensley and White 1995, p. 20). Mathew Smales was Lord Wharton's solicitor.

213 Answer and disclaimer of Toby West, vicar of Grinton, accompanied by two schedules. Toby West's answer was given at Christopher Bowes' inn at Wensley, in Wensleydale. The document is endorsed at the foot with the *jurat* signed by Thomas Pulleine and Roger Baynes and dated 6 November 1706. On the *verso* the note of delivery, or *deliberatur*, dated 26 November 1706, is signed by baron J[ohn] Smith. The messenger who delivered the answer to the court is named as John Aislabie.

214 Parchment, 750mm x 565mm.

controversion or in the bill named or any benefitt profitt or advantage out of the same or any of them (except this defendant as viccar onely & in that right or as impropriator be entituled to any tythe thereof which this defendant / however doth not know that he is) And this defendant saith that Mr Marriott one other of the defendants in the bill named hath been a kinde friend to this defendant since this defendant was acquainted with him and this defendant hath often owned & acknowledged his kinde / favours & respects to this defendant & this defendant hath been & still is ready & willing to doe him the said Mr Marriott any just & lawfull service that lyeth in this defendants power and he the said Mr Marriott never desired any other of this defendant And this defendant saith he is & for thirty three / yeares past hath been vicar of the parish church & vicaridge of Grinton & is alsoe impropriator of the same And this defendant doth deny that he knowes the boundaries of his said parish or of the lordshipp of Swaledale (if any such lordshipp there be) or that he this defendant is acquainted / with or doth know the division or extent of the severall townes vills or hamletts within the said parish or any of them or any other names then those by which they are commonly called (except as hereinafter mencioned) nor is this defendant in right of the said church or otherwise / intituled to any lands or tenements in the said parish vills or hamletts or any of them (except as aforesaid) But beleives he may in right of his gleab be intituled to common upon the moores & wasts in the said parish or some parte thereof And this defendant denyes that he is or ever / was a partner with the said Mr Marriott in the said mynes in the bill mencioned or any of them or in any other mynes whatsoever & saith he conceives & is advised the complainants are not concerned to know nor this defendant obliged to set forth whether the said Mr Marriott ever made or / gave this defendant any or what presents gifts or gratuities but denyes that he this defendant ever took or that the said Mr Marriott ever gave <any> such for any unjust or unlawfull intent or purpose whatsoever or to conceale this defendants knowledge of the bounders of the said parish or places as in the bill / is suggested & this defendant doth deny that he hath or ever had any bookes writeings or evidences in his custody or power which belonged to or concerned the priory of Bridlington in the bill named or the rights of this defendants said parish of Grinton or the possessions of the church of Grinton / within the severall vills or hamletts in the bill mencioned or elsewhere or any of them except onely what deeds concerne the impropriation of the said church nor does this defendant claime any other right therein then as aforesaid And saith he hath noe publick writeings in / his custody or power or any writeings which any wayes concernes the complainants or either of them or their estate or pretences or the mynes or matters in question or any of them or the <said> parish of Grinton (except as aforesaid) & the common register of births marryages & buryalls And this / defendant doth deny that he has influenced or does endeavour to influence or prevaile upon any of the tennants of Harkaside or West Grinton to withdraw their suite or service from the complainants <court> or to conceale the rents which they ought to pay to the complainants or to disowne / the complainant the Lord Wharton as their lord as in the bill is suggested Neither doth this defendant know what suite or service or rent or rents are due or payable from the tennants of Harkaside or West Grinton or any of them unto the Lord Wharton but hath heard that some of / the tennants in Harkaside doe pay some rent to the Lord Wharton but by what right or tytle this defendant knoweth not And this defendant doth deny that he hath drawne up or been present at or a witness to the drawing of any affidavits testimonialls or deskriptions / of the boundaries of Grinton West Grinton or Harkaside in the bill named or any of them (except as hereinafter mencioned) or has discoursed with any persons about the same or gott any person to subscribe such paper writeing or knowes of their soe doeing / (except as hereinafter mencioned) But this defendant doth acknowledge that he did discourse with severall of the antient inhabitants of the parish of Grinton touching the bounders of the mannor of Grinton & that this defendant & the said inhabitants did subscribe two papers / purporteing to be certificates of the said bounders a true coppy of which papers or certificates together with the names of the persons that soe subscribed the same & who declared that they verily beleive the bounds of the said mannor to be as the same are in the said paper described is / hereunto subscribed <or annexed> But neither this defendant nor any other of the said persons whose names are subscribed to the said papers or certificates did to this defendants knowledge or beleife make any affidavits thereof or swear to the same And this defendant saith that he doth not know or remember / the names of any other person or persons that promissed to be witnesses to prove the same And saith he beleives since the controversie in question both he this defendant & many of the persons have in their common usual discourse talked about the same & the boundaryes of / the premisses but beleives there are severall other persons that can & will testifye the same when required And this defendant doth deny that he was or is any wayes concerned in interrest in the mynes in question or any of them or any profitt out of them / or any of them (except tythe be due to him as aforesaid) And this defendant doth disclaime all right & tytle to the mynes in question or in the bill mencioned & to all profitt out of the same (except for tythes as aforesaid) And this defendant doth deny all / confederacy & combinacion in the bill charged and humbly prays to be dismist with his reasonable costs & charges in this behalfe susteined

/s/ Christopher Driffeild[215]

The schedules or papers to which the answer referrs[216]

These are to certifye whome it may concerne that wee whose names are under written / doe know Grinton & have heard & doe beleive that places hereafter mencioned vizt on the east / Stollerston steele Whitestone in Sharrow on the south Snowden Man Brownseat Greets / Gibon Hill Highcarle Pigott santi (alias Pigott stonerig) Windyates Torre santi on the west / Bloody Wall Haverdale beck on the north the river Swale are bounders of the said mannor of Grinton / and that Cogden Harkaside Whitaside & Somer Lodge lyeing & being within the bounds above said / are partes & parcells of the above said mannor of Grinton & have alwayes been soe reputed Toby West / vicar of Grinton George Milner Ralph Wensley James Hird Giles Metcalfe John Garth Anthony Close / Christopher Raper Robert Waller <his marke> Anthony Raper /

These are to certifye whome it may concerne that wee whose names are under written doe know & have seen the / lead myne or veine of lead ore now wrought by

215 Christopher Driffeild (*c.*1665–1724), lawyer. Son of Christopher Driffeild of Gray's Inn. Admitted to Christ's College, Cambridge, 1681. Admitted to Gray's Inn 21 June 1682. Described as 'of Ripon' in 1697 and 1702. A letter from Thomas Dickonson, dated 5 July 1706, is addressed to 'Chr. Driffeild Esq, Recorder of Rippon'. *Register of Admissions to Gray's Inn, 1521–1889*, ed. Joseph Foster, privately printed, 1889; *Alumni Cantabrigiensis*; BnL, Special Collections, Marrick MSS, 5/4/85.

216 The schedules are written side by side at the foot of the parchment membrane. Above each one a longitudinal cut, about 100 and 150 mm in length, has been made in the parchment. The intended purpose of these cuts is not apparent.

Mr Reginald Marriott & that it was always called or comonly / knowne by the name of Grinton moor & is parte of the moor or common belonging to Grinton & was alwayes soe reputed / and that all the ground from the said myne southward as far as heaven water deales or the bounds betwixt Wensleydale / & Swaledale belong to the said common of Grinton Toby West Vicar of Grinton George Milner Ralph Wensley Miles Close James / Hird Giles Metcalfe Anthony Close Christopher Raper Robert Waller <his marke> Anthony Raper /

E112/915 EBOR 100 – document 5[217]

[membrane i][218]

The answer of Reginald Marriott esquire one of the defendants to the amended bill of complaint of Thomas Lord / Wharton & Philip Wharton esquire a minor son & heir apparent of the said Lord Wharton by guardian complainants

This defendant saving as in his former answer he hath saved, for answer to the amendments or amended bill Saith that it no way appears to this defendant / to his knowledge or understanding nor to any other of the defendants to his knowledge or belief of or from whom the manor of Grinton or any lands in / Grinton East Grinton West Grinton (if any such places there be) Harkaside or Swaledale in the bill named or any of them was or were held / by the monastery of Bridlington or by the Earl of Richmond in the bill named or Walter de Gant in the bill also named his heirs or descendants / if any of them ever had any interest therein or whether under any or what rent or rents or services or under what title grant or grants the same was / or were held by the said monastry Nor hath this defendant seen perused or knows of any deeds or evidences wherein is or are described or mencioned that any or / what estate lands or tenements in the place aforesaid [or any of them] was or were granted to or enjoy'd by the said Earl of Richmond by the said Walter de Gant his / heirs or descendants or by the said monastry (other than as to the said monastry in his said former answer is mentioned) And this defendant doth deny that he or / any other of the defendants to his knowledge ever had in his or their custody or did ever see the inquisicon or survey taken of the estate or lands of Francis Bygot / in the bill named or any part thereof upon or after his attainder if any such there be or ever was nor was any such ever shewn to him or any person / on his behalf or any other of the defendants to his knowledge or belief Nor doth this defendant know or can set forth when the same was or is if any such / there be And this defendant saith he believes there were some records or copys of them carry'd to York assises & produced there to make out the title of / the Crown upon the trial of Sir Solomon Swale but this defendant doth not now remember the particulers of them but believes the same were sent by & in the / custody of the proper officer or person by him employ'd for that purpose But this defendant doth deny that he this defendant had the custody or power of such records or any of / them or was intrusted with the same of did take them or any of them out of this court or out of any offices to which they or any of them did belong & saith the same / and severall other records in this court were afterward upon the trial at bar in this court with the said Sir Solomon Swale produced and made use of at the said / trial at bar when & where the proper officers or persons by them employ'd

217 The second answer of Reginald Marriott. The document is endorsed at the foot with the note of swearing, the *jurat*, dated 19 June 1707 and signed by the chief baron, Sir Edward Ward.

218 Parchment, 390mm x 340mm.

for that purpose attended with them as this defendant remembers & believes And saith that all / the records so produced at either of the said trials were carry'd back & returned into the respective offices to which they belong'd as the defendant verily believes And this / defendant doth verily believe & knows nothing to the contrary that all the records memorials & entrys belonging to any officer or offices or in their or any of their custody / which were used at either of the said trials or taken out of court were presently after & as soon as possible returned into court & have ever since been & still are in the / respective offices & were returned to the same without any alteration or loss whatsoever But whether they or any of them were deliver'd out of court by any order / this defendant cannot set forth nor when they or any of them were taken out of court nor how long they or any of them were kept out of court But believes it was no / longer than was necessary to make out the title of the Crown upon the said controversy & that they were always kept by the proper officer or person by him intrusted for that / purpose with the same altho' at this distance of time this defendant doth not remember the names of such persons And saith he doth not know that there was or ever were any / other trials concerning the premisses than those with Sir Solomon Swale as aforesaid And this defendant doth deny that he did take out of court or ever had in his custody or / that there was by any officer ever delivered him out of court any records whatsoever concerning the said premisses or matters in question or any part thereof And / saith that all the records by him used are now as he verily believes lodg'd in the respective offices & are in the custody of the respective officers to whom they / belong and have been so ever since the same were so used at either of the said trials And this defendant denyeth as in his former answer hee hath / denyed and traverseth as he hath therein traversed /

/s/ Sam Dodd

E112/915 EBOR 100 – document 6[219]

[membrane i][220]

The answer & disclaimer of George Tuchingham gentleman one of the defendants to the bill of complaint of the right honourable Thomas Earl Wharton / and Philip Wharton esquire complainants /

This defendant saving to himself now and at all times hereafter all benefit and advantage of excepcion to the manifold incertaintyes and imperfections in the complainants said bill of complaint / conteined for answer thereunto or to so much thereof as this defendant by his councill is advised is any ways material for him this defendant to make answer unto he answereth & saith, That he doth / not know whether Philip late Lord Wharton the complainants father in the bill mentioned was ever seised or possess'd nor whether the complainants or either of them were or are seised as aforesaid of the dominion / lordshipp or manor in the bill named, or any of them, nor do's this defendant know or can set forth the boundaryes or extents of them or any of them, nor whether the same or any of them comprehend any or what vills / towns townshipps or hamlets or any & what moors wasts or comons nor how the same or any of them are divided or distinguish'd butted or bounded

219 Answer and disclaimer of George Tuchingham.

220 Parchment, 590mm x 280mm. The document is endorsed at the foot with the *jurat*, or note of swearing, dated 11 June 1707 and signed by baron J[ohn] Smith.

And this defendant saith he doth not know nor can set forth of his / own knowledge whether the complainant the Lord Wharton hath sunk for oar in any wasts or commons or places whatsoever And this defendant saith he believes it to be true that Reginald Marriott in the bill / named did obtain a lease in this defendants name from the Crown of severall mines & veins of lead within the territories & fields of Grinton Whitaside & Harkaside for the term of thirty one years at such / rents as in the bill is alledged as by the said lease (inrolled as this defendant believes in this court) and to which in all things the defendant for more certainty referreth himself more at large it doth & may appear / And this defendant saith he was no ways concerned in interest in the said lease but only as a trustee for the said Mr Marriott And saith his this defendant's name was used in the said lease at the request of the said / Mr Marriott & only as a trustee for him, And saith that he this defendant accordingly by deed dated on or about the nineteenth day of August in the year 1696 granted and assign'd over the / said lease unto the said Reginald Marriott for the residue of the term then to come in the same And this defendant doth deny that he is or ever was concern'd or interested in his own right of or in the said lease / or mines or premisses thereby granted or demis'd or of any share or profit to be made by or arising out of the same or any part thereof And this defendant doth dissclaim in his own right all estate / right title or interest of or unto the said lease mines & premisses & all benefit profit or advantage arising thereout or thereby or any ways relating thereunto, never having had any interest therein / but only in trust for the said Marriott as aforesaid & which this defendant hath long since & long before the complainants bill exhibited or any process taken out assigned over to the said Mr Marriott And this defendant / doth deny that he hath or ever had to his knowledge remembrance or belief in his possession power or custody any records deeds evidences or writings whatsoever any ways relating to or concerning the complainants / or either of them, or their estate, or estates, or any ways relating to or concerning the mines or premisses in the bill mencioned or the manor of Grinton or priory of Bridlington or any other the mannors / lands hereditaments or premisses in the bill mentioned And this defendant saith he is a stranger to the rest of the matter & things in the bill charged And this defendant doth deny all confederacy & combinacion / in the bill charged and having disclaimed all right & title in or to the premisses & being no ways concerned in interest therein and having no knowledge of any other of the matters or things in the bill / charged or otherwise than as being a clerk in this court for the said Mr Marriott & by the bills answers affidavits filed in this court & the proceedings in the causes now lately depending in this court this defendant / humbly hopes he shall be hence dismist with his reasonable costs & charges in this behalf susteined. /

/s/ Sam Dodd[221]

221 For Sir Samuel Dodd see p. 7, footnote 22 above and profile in Appendix 3.

Thomas Lord Wharton v. Attorney General and Reginald Marriott

Depositions[222]

E134/6Anne/Mich38 – document 1[223]

E134/6Anne/Mich 38 – document 2[224]

[membrane i *recto*][225]

Interrogatories exhibited to be administered to the wittnesses to be produced sworne and examined on the part and behalfe of the right honorable Thomas Lord Wharton / now Earle of Wharton and the honorable Philip Wharton esquire commonly called Lord Winchindon son and heir apparent of the said Earle of Wharton in a cause depending in her majesty's Court of / Exchequer att Westminster wherein the said Lord Wharton now Earle of Wharton and the said Philip an infant by the said Earle his guardian are complainants against her majesty's Attorney / Generall and Reginald Marriott esquire and other defendants. /

Imprimis Doe you know the partys complainants and defendants in this cause or any and which of them ? Did you know the right honorable Philip late Lord Wharton and when did he dye ? /

2. Doe you know the dale called Swaledale ? within what honor shire or wapentake does the same lye ? What are the extent boundarys or reputed boundarys of the said dale ? Are there any and what parishes / forrests mannors constablerys vills villages and hamletts within the said dale ? Who has been reputed the cheif lord or lords of the said dale called Swaledale ? [x x x][226] Doe you know of any and what court leets that have / been att any time held within the said dale or lordshipp of Swaledale ? When where by whom and how often have the said courts been respectively held ? Doe you know have heard or beleive that any of the lord / or lords of the lordship of Swaledale or mannor of Helaugh have att any time had or claimed to have or enjoy any and what forests chaces warrens or royalties of hunting hawking fishing or fowling / or any and what other royalties franchises or [x x x][227] priviledges in or throughout the aforesaid dale or lordship of Swaledale ? Declare what you know have heard or beleive touching the same with the / grounds and reasons of such your knowledge or beleife ? /

3. Doe you know the mannor of Helaugh ? Has the lordship [x x x][228] of Swaledale

222 In TNA, the following depositions are catalogued as E134/6Anne/Mich38.

223 Parchment, 260mm x 155mm. Commission of *dedimus potestatem*. The commissioners were Thomas Waite Esq, Robert Hilton gent., Thomas Thomson gent., and Thomas Gyll gent.

224 The bundle consists of seventeen parchment membranes tied together with string at the top left-hand corner. The arrangement of the documents within the bundle is similar to that of those contained in E134/6Anne/Mich44 – document 2 (see p. 20, footnote 48).

225 Parchment, 575mm x 700mm.

226 A whole line of text has been erased.

227 One word has been erased.

228 One or two words have been erased.

and Helaugh been att any time and how long enjoyed by the same or by any and what different lord or lords ? How long hath the / said lordship [x x x][229] been called by the said name or names of Swaledale or Helaugh or by any and what other name or names and what was the occasion of such said names as you have heard or doe beleive ? /

4. Doe you know the vill and constablery of Reeth ? Within what lordship or mannor doe the said vill and constablery lye or are or is reputed to lye ? Does the said lordship of Swaledale mannor of Helaugh and the said / vill and constablery of Reeth or any and which of them onely extend to the northside of the river Swale or doe the same extend to the south side alsoe of the said river and how farr does the same extend southwards / from the said river ? Has the said river been reputed a boundary to limitt or divide the said lordship of Swaledale mannor of Helaugh and the said vill and constablery of Reeth or any and which of them ? Sett / forth what you know have heard or beleive touching the severall particulars in this interrogatory together with the grounds and reasons of your beleife ? /

5. Doe you know Harkaside and West Grinton and the moores and commons thereunto belonging ? Are the same divided by any and what boundary <from each other or are they one and the same place> ? Doe you know Grinton and East Grinton and the moores thereof ? Doe you / know or have heard of any boundary dividing the said Grinton or East Grinton from the said Harkaside and West Grinton ? Does the said boundary onely divide the moores or commons of Harkaside or West Grinton from / those of East Grinton or do's the same boundary extend throughout or divide likewise the inclosed lands of each respectively ? Doe you know the high way or road leading from East Grinton to West Grinton and / Harkaside ? Doe you know the gate in the said road or high way there and the elme tree growing [x x x][230] near to the said gate and how long have you knowne the same ? How farr and to what place doe the tennants / or inhabitants of East Grinton repair keep or maintain the said way ? Is the said place to which the tennants or inhabitants of East Grinton repair the said way reputed one of the boundermarkes divideing the said / East Grinton from Harkaside or West Grinton or is there any and what other boundermarke divideing the same in the said high way sett forth the whole boundary soe divideing the said Grinton or East Grinton / from Harkaside or West Grinton with the names and descriptions of each particular boundermarke and the reasons induceing you to beleive the same to be true and remarkable boundermarkes ? Sett forth the / boundary of Harkaside and West Grinton as the same proceeds westwards or divides the same together with the said dale called Swaledale from the other adjoyning dale called Wensleydale ? and sett forth / the boundary as the same proceeds thence and divides the said Harkaside and West Grinton from the hamlett or place called Whitaside ? /

6. Doe you know the lead mines now or lately wrought by the defendant Marriott called the How Groves ? Doe you know the lead mines or workes lately opened by the complainant the Lord Wharton or his agents near / adjoyning thereunto ? Doe the lead workes soe belonging to the said Lord Wharton and opened by him lye within the moores and commons of Harkaside and West Grinton and the boundary thereof or doe the same lye within the / moores and commons of East Grinton or the boundary thereof as you beleive and how near doe the said severall workes lye to each other and the boundary divideing the same or to any and what part of the said boundary / to the best of your judgement ? When did the complainant the said Earle

229 One or two words have been erased.

230 One word has been erased.

of Wharton open the said workes near to the said boundary ? Was the said boundary [x x x][231] ridd by the said Earle of Wharton or his order <before the opening of the said workes> ? When and / by whom was the same soe ridd or agreed unto and what notice was given of the rideing thereof ? Was <or were> the complainant the Lord Wharton or his tenannts of Harkaside and West Grinton in the actuall possession of the said / moores of Harkaside and West Grinton att the time of rideing of the said boundary and opening the said lead works within the same ? and for how long before were the said Lord Wharton and his ancestors or other the / lords or owners of the lordship of Swaledale mannor of Helaugh and vill of Reeth and their severall and respective tennants of Harkaside and West Grinton or other their tennants within the said lordshipp in the / actuall possession or enjoyment of the said moores of Harkaside and West Grinton or any and what part thereof and what enjoyment had they thereof ? Doe you know of any and what enjoyment that the tennants of / East Grinton have att any time had or claimed upon the moores of Harkaside or West Grinton or of any and what disturbance or interruption they or any of them have had therein or of any satisfaction or acknowledgement / that the said tennants of East Grinton have paid for their use or enjoyment of the said moores to whom when and upon what occasions ? /

7. Has the said Harkaside or West Grinton and the moores thereof been att any time when and how long reputed part and parcell of the lordship of Swaledale mannor of Helaugh and of the said vill or township of Reeth or / any and which of them ? Or has the same been reputed part and parcell of the pretended mannor of East Grinton ? Doe you know of any and what lands commonly called demesne or leasehold lands lying or being in Harkaside or / West Grinton now belonging to the complainant the Lord Wharton ? Of what extent or vallue are the said demesne lands and how and by whom and under what rents fines and services have the same been hitherto held ? Doe / you know of any and what freehold tennants in Harkaside or West Grinton that hold any and what freehold lands or tenements therein of the said complainant the Lord Wharton and by any and what rents suite and service / when where <or at what courts> and to whom have the same been hitherto paid done and performed ? Doe you know of any and what customary or coppyhold tennants in Harkaside and West Grinton that hold any and what customary / or coppyhold lands or tenements therein of the said complainant the Lord Wharton and by any and what rents fines harriotts and services ? Are the said customary or coppyhold lands now held of the complainant the said Lord / Wharton held by coppy of court roll of the said complainant's mannor of Helaugh or by copy of court roll of or from any and what other mannor or mannors ? /
8. Doe you know of any and what lands or inclosures adjoyneing to the said moores of Harkaside and West Grinton or commonly called by the name of the intacke or by any other names now belonging to the said complainant / the Lord Wharton or any of the tennants ? How long have the same been taken or inclosed from the said moores and by what rents and of whom are the same held ? /
9. Of what extent are the lands in Harkaside and West Grinton now chargeable or antiently chargeable with the said rents fines and services to the best of your knowledge or beleife ? Are the demesne lands in Harkaside and / West Grinton and the rents fines customes and services now payable to the said complainant the Lord Wharton out of the freehold and coppyhold lands within the same of much greater

231 One word has been erased.

or lesser vallue then other his demesne lands / or freehold coppyhold or customary rents in other the hamletts within the mannor of Helaugh or vill of Reeth ? /

10. Are the coppyhold or customary lands in Harkaside or West Grinton descendible to all or any and which of the sons of the tennants dyeing seized according to the customs of the mannor of Helaugh or are they descendible according to / the custome of the said mannor or pretended mannor of Grinton or East Grinton ? /

11. Doe you know or have heard of any or what modus prescript rent or composicion for tyths <of hay & corne> payable by the vill or township of Reeth ? Does the hamlett of Harkaside [x x x][232] as being part and parcell of the said vill and / township of Reeth contribute any and what part and proporcion to the said modus prescript rent or composicion ? and for how long have you knowne the [same be] paid [x x x] / [x x x][233] /

12. Doe you know the mill in West Grinton or Harkaside now belonging to the owner of Swale Hall by whom and when was the same built and by whose lycense and authority and upon what termes and condicions as you have heard or doe / believe ? Have the owners of Swale Hall been att any time and how long called to the courts held for the mannor of Helaugh as freehold tennants within the same or have they appeared att any other court or courts or have att any / time held any court or courts in Harkaside or West Grinton or had any freehold or coppyhold tenements or any free or customary rents within the same ? /

13. Doe you know the pinfold belonging to the vill or township of Reeth ? Where are the cattle or goods belonging to Harkaside and West Grinton usually impounded ? /

14. Doe you know or have heard that any and which of the complainants tennants of the mannor of Helaugh and vill of Reeth have att any time inter commoned with the tennants of Harkaside and West Grinton and what / ways laynes passages have you knowne betwixt the aforesaid commons ? /

[membrane i *verso*]

15. Doe you know or have heard of any and what suites betwixt the lords and tennants of the mannor of Helaugh and township of Reeth wherein the tennants of Harkaside and West Grinton or any of them / have joyned or contributed for maintaining the custome of the said mannor in respect of the said lands in Harkaside and West Grinton ? Are the lands and rents in Harkaside and West Grinton now / belonging to the complainant the Lord Wharton distinguishable by the same termes and names as the other lands and rents within the complainants said mannor or by any and what other name or names and / how long and upon what occasion have the lands or rents been soe distinguished as you have heard or doe beleive ? Doe you know of any and what assessments taxes or other payments wherein the / tennants of Harkaside [x x x][234] doe usually joyne or contribute with the tennants of Helaugh or Reeth ? Declare your knowledge to each particular in this / interrogatory and the severall grounds and reasons induceing you to believe that the said Harkaside and West Grinton are part and parcell of the complainants lordship of Swaledale mannor of Helaugh and / towne and constablery of Reeth or any of them ? and have you knowne <the> inhabitants of Harkay syde keep watch and ward at Reeth ? /

16. Was the complainant the Lord Wharton in the actuall possession and enjoyment of

232 One or two words have been erased.
233 About a third of a line of text has been erased.
234 About a third of a line of text has been erased.

the lead mines of and within the lordship of Swaledale or mannor of Helaugh att the time of the commencement of the suites / depending betwixt the complainants and the defendants and for how long and before or since that time has the said complainant the Lord Wharton or any of his ancestors been in the actuall possession or enjoyment of any and / which lead mines within the lordship and mannor aforesaid ? Has any and what profitt been made thereof and how long ? Have any and what lead mines been wrought att any time within the moores of / Harkaside and West Grinton before the time of obtaining the lease in question from the crowne ? If yea by whom when where and for how long time were the same soe wrought were the same soe wrought or claimed to / be wrought by pretence or colour of any title thereto by from or under any and what lord or lords of the lordship of Swaledale or mannor of Helaugh or were the same soe wrought by from or under the monastry / of Bridlington or the right and title of the said monastry now claimed by the defendant Marriott[235] [x x x][236] / Doe you know or have heard of any custome of minery att any time observed within the lordship of Swaledale or mannor of Helaugh and what liberty or priviledges had the tennants of the said mannor or / mannors by vertue of such custome to search sink or digg for mines ? Did the defendant Marriott att any time and when enter upon and worke any and what mines in the moores of Harkaside and West / Grinton ? Did he soe enter or worke the same before or after the <decease of the late Lord Wharton or before or after the> entry of the complainant the now Lord Wharton upon the mines of his lordship of Swaledale or mannor of Helaugh ? Who were in possession of the / said mines of the lordship of Swaledale and mannor of Helaugh att the time of the defendant Marriotts said entry and what estate had they therein ? How long did the defendant Marriott after his said entry / continue to work any mines within the moores of Harkaside and West Grinton ? Did he after such his working discontinue his working thereof ? When were the same discontinued and for what reason and how long ? / Declare what you know have heard or beleive touching the severall particulars in this interrogatory with the grounds and reasons of such your beleifs ? /

17. Doe you know Stallerstone Stile Haverdale Beck the river of Swale and High Carle were claimed by the defendant Marriott to be the boundarys of the mannor or pretended mannor of Grinton ? Have you knowne the said / pretended boundary ever ridd or perambulated and when and by whom and upon what occasion ? What reasons have you to beleive that the said places are boundarys or not boundarys to the said mannor or / pretended mannor of Grinton ? /

18. Doe you know or have heard that any and which of the lord or lords of the lordship of Swaledale or mannor of Helaugh have lived att any and what distance from the said mannor and for how long time ? / Doe you know or have heard of any right customes or usages within the said Harkaside or West Grinton that have been att any time discontinued by reason of such distance or of the absence of such lord or lords ? /

19. Does the plott or mapp now to you produced contain a true account or descripcion to the best of your judgement beleife or informacion of any and which of <the> places divisions and boundarys therein described or / represented ? Sett forth what you know of the said plott or mapp and of the care taken in settleing the same ? /

20. Were any and which of the paper and parchment writings now to you produced sealed and delivered in your presence and when where and by whom ? Are the

[235] The question mark has been omitted in the original.
[236] Two thirds of a line of text has been erased.

subscribing witnesses thereto or any and which of them / dead or living ? Are the names of the respective partys or witnesses thereto the proper handwritings of the respective partys as they import ? When where and by whom and in whose possession were the severall / writings now to you produced first found and how long have the same been in the possession of such person or persons ? What other account can you give of the writings now to you produced or any and which / of them ? Or of the person or persons by whom the same were drawne or written ? Are the paper or parchment coppys or writings now to you produced true coppys of the originalls thereof ? Where are the said originalls / or lately were and when and where ? Did you examine the same ? Declare what you know have heard or beleive touching the same with <the> grounds and reasons of such your knowledge or beleifs ? /

21. Doe you know of any and what actions or suites that have been att any time and when commenced or prosecuted by any and which tennants of Harkaside and West Grinton att the court or courts / held for the mannor of Helaugh or any and what suite and service that have been done att any of the said courts by the said tennants of Harkaside and West Grinton or any for or on their behalfe ? /

22. What other matter or thing you know have heard or beleive materiall to be deposed on the complainants behalfe declare the same with the grounds induceing you to depose herein ?

/s/ John Pratt[237]
/s/ N: Lechmere[238]

[membrane ii – *recto*][239]

Ex parte quer[ens]

Depositions of witnesses produced sworne and examined as well on the part and behalfe of the right honourable Thomas Lord Wharton now Earle of Wharton and the honourable Philip Wharton Esq now / commonly called Lord Winchendon an infant by the said Thomas Lord Wharton his father and guardian complainants as on the part and behalfe of her majesty's Attorney Generall and / Reginald Marriott esquire defendants att the house of Christopher Bowes of Wensley in the county of York inholder the twenty ninth day of August in the sixth year of the raigne of our soveraigne / lady Queene Anne over Great Brittain etc anno domini 1707 by vertue of her majesty's comission issued out of her majesty's Court of Exchequer directed to Thomas Thomson gentleman Thomas Gill gentleman / Thomas Wait esquire and Robert Hilton <gentleman> and taken before all the said comissioners as followeth : -- /

Richard Close[240] of Richmond in the county of Yorke yeoman aged seaventy yeares sworne and examined on the complainants behalfe deposeth as followeth. /

1: To the first interrogatory saith he knowes not the complainants but knowes the defendant Mr Marriott and did know Philip late Lord Wharton deceased who dyed about tenn yeares agoe as this deponent hath heard /

237 Sir John Pratt (1657–1725), lawyer and politician. See profile in Appendix 3.

238 Sir Nicholas Lechmere (1675–1727), lawyer and politician. See profile in Appendix 3.

239 Parchment, 650mm x 720mm.

240 Richard Close of Richmond, yeoman, swore an affidavit on 1 June 1706 (E207/112/2). In May 1708 Richard Close, aged 71, gave evidence for Lord Wharton at the trial in London (LI Misc 536, Ward MSS, notebooks vol. 6, f. 116r).

2: To the second interrogatory he saith he knowes the dale called Swaledale and saith the same doth lye in the honour of Richmond and in the wapentake of Gilling west and Hangwest but most part of it in Gilling west saith that the said dale is bounded / on the east by Stollerstone Stile and on the west by Hollow Mill Crosse on the north by Arkingarthdale and on the south by Wensleydale saith there is but one parish in Swaledale and three constablerys to witt Grinton Reeth and Mewkar and there is in it / the mannor of Helaugh and lordship of Mewkar and none else where any courts hath been kept in this deponents remembrance it alsoe containes the severall vills villages and hamletts hereafter mencioned that is to say Fremington Castell Heges / Sturfitt Hall Racroft Raw Reeth Helaugh Parke Hall Kearton Blaides Feetham Low Raw Smarbarr Winteringe Pott Ing Lodge Green Gunnerside Ivelett Calvett house Cracky Pott Hall East stonedale West stonedale Ravenseat Birkdale Keld / Thwaite Mewkarr Rash High Oxnop Low Oxnop Satteron Spring End Crack Pott High Whitay Low Whitay Harkaside West Grinton and East Grinton. Saith the Lord Wharton and his predecessors have been reputed cheif lords of the said dale called / Swaledale except East Grinton and Whitaside on the south side and Fremington on the north side thereof further saith that court leets have been kept during all the time of this deponents remembrance sometimes once a year sometimes twice by / the Lord Wharton and his ancestors att Helaugh and this deponent saith he hath heard and beleives that the said Lord Wharton and his ancestors as lords of Swaledale or the mannor of Helaugh and noe others have all along claimed and enjoyed / the royaltys of hunting hawking fishing and fowling and the deer that happened to be therein in all the said dale except East Grinton Whitaside and Fremington aforesaid and this he is the better enabled to depose for that he this deponent hath / been one of the jury att the said courts when the jury hath presented persons for fishing in the river Swale without lycense from the said lords of Helaugh /

3: To the third interrogatory saith he knowes the mannor of Helaugh and hath knowne the same for threescore yeares being borne in Harkaside and lived there fifty seaven yeares of his time and saith the said lordship of Swaledale and Helaugh have been enjoyed by the / family of Lord Wharton all this deponents time and by noe other and been called all his time by the name of Swaledale or Helaugh and by noe other name that this deponent knows or has heard of. /

4: To the fourth interrogatory saith he knowes the vill and constablery of Reeth and hath knowne the same for sixty yeares past saith the said vill and constablery doth lye and all his time hath been reputed to lye within the lordship or mannor of Helaugh / aforesaid saith the said mannor of Helaugh and the said vill and constablery of Reeth doth extend from Arkingarthdale common on the north quite crosse the river Swale to Wensleydale common on the south and this deponent saith the river Swale in this deponents memory / was never reputed a boundary that this deponent knows of to limitt the said lordship of Swaledale or mannor of Helaugh or constablery of Reeth or any part thereof. /

5: To the fifth interrogatory this deponent saith he knowes Harkaside and West Grinton and the moores and commons thereunto belonging and hath knowe the same for sixty yeares past saith they are not divided from each other by any boundary but have been all / along accounted as one place. saith he knowes in like manner Grinton or East Grinton and the moores thereof and hath knowne the same from the time aforesaid and alsoe the boundary which devides the said Grinton or East Grinton from Harkaside and West Grinton / and <saith> the same doth not onely divide the moores or commons of Harkaside or West Grinton from those of East Grinton but alsoe divides the inclosed grounds of each respectively further saith he alsoe knowes the high way or road leading from East Grinton to West Grinton / and Harkaside and alsoe the gate in the said road and the elme tree growing ngear the said gate and hath knowne the same for sixty yeares saith the tennants or inhabitants of East Grinton have all the time of this deponents remembrance as occasion required / maintained and kept in repair the said road or high way from East Grinton to the said gate and noe farther saith the said gate is one of the boundermarkes which divide the said East Grinton from Harkaside and West Grinton and this deponent saith that the bounders / which divide the said East Grinton from West Grinton and Harkaside (as well the inclosed grounds as the wasts and moores thereof) are as followes and have all the time of this deponents remembrance been accounted to be the true boundarys thereof that is to say from / the river Swale crosse the east end of the Holme and crosse the street on the east side of the aforesaid gate on the west of Cuddy Spout into a close called Low Close and soe along the fence upon a ridge of ground from the Low Close to another close called the / Middle Close along the said ridge to a close called Bensons Close wherein is a cowhouse called Benson's cowhouse standing att the head of the close from Benson cowhouse along Ox Intack wall to Ox Intack corner from Ox Intack / corner in a direct line upon the moore crosse Hughs Ditch to a boundermarke called the White Stone in the How and from the White Stone in the How in a direct line to the Gray Stone in the Greets and soe to the top of the hill where heaven water deales or / divides it selfe which said boundary is the true boundary betwixt East Grinton and Harkaside or West Grinton as this deponent verily beleives and this deponent saith that the said Harkaside and West Grinton moores and waists are divided from the lord Duke of / Boltons mannor or liberty of Wensley dale from the top of the hill beyond the said Gray Stone in the Greets all along westwards as heaven water deales to Gibbon Hill from Gibbon Hill to a standard on the west end of the hill called Gibbon Hill standard and / soe to Browney Gyll head and from Browney Gyll head the said Harkaside and West Grinton is divided from Whitaside by the gyll which runs down to the Horse pasture corner to a gate called Scabbey gate and from / Scabbey gate to the river Swale /

6: To the sixth interrogatory this deponent saith he knowes the lead mines called the How Groves now wrought by the defendant Marriott and alsoe the lead mines or workes lately opend by the complainant the Lord Wharton or his agents near the said Mr Marriotts leadworkes / saith the leadworkes soe belonging -- -- -- -- to the Lord Wharton lye within the moores and commons of Harkaside and West Grinton and within the boundarys thereof soe as aforesaid described by this deponent in his deposition to the next preceding / interrogatory and not on the moores or commons of East Grinton or the boundary thereof but saith both the said lead workes are about an equall distance from the said boundary and in this deponent's judgement about one hundred yards distance from each / other and saith that some time in this next month of September one thousand seaven hundred and seaven it will be {it will be} two yeares since the complainant the Lord Wharton caused the said lead workes to be opened saith that before the opening the said / boundary was ridd by the order of the complainant now Earle of Wharton and the said rideing was in September one thousand seaven hundred and five and the same rideing was made by about forty or fifty persons whereof this deponent was one and Miles Close of / East Grinton another but what other person had notice of the said rideing this deponent knowes not and saith the said rideing was made in the same manner this deponent hath before described the boundarys thereof and not disagreed to by any there present that this / deponent knowes of save one Naylor tennant to Swale Hall and this deponent saith that the said Lord Wharton as reputed lord of the mannor of Helaugh and his tennants of his lands att Harkaside and West Grinton and the other freeholders and tennants there / was or were in the actuall possession of the said moores of Harkaside and West Grinton att the time of the rideing that said boundary and of opening the lead workes within the same and that the said Lord Wharton and his ancestors as lords of the said mannor of / Helaugh and their tennants and other the freeholders and tennants in West Grinton and Harkaside had been in actuall possession of the said moores and waists before that rideing ever since this deponent can remember by their being lords of the mannor and their / tennants useing thereof as this deponent verily beleives nor ever were disturbed in their possession thereof that this deponent <ever> heard and this deponent saith that the tennants of East Grinton have not att any time that this deponent knowes of enjoyed or claimed the / use of the said Harkaside and West Grinton moores saving a rake for their cattle without interrupcion for this deponent saith that once about twenty yeares since one Richard Harker then of East Grinton having pulld ling on the moores within the aforesaid boundary / of West Grinton and Harkaside (this deponent then living att Harkaside) he and the other inhabitants of Harkaside would not suffer the said Harker to lead it away to East Grinton but this deponent and the other inhabitants of Harkaside led it to Harkaside and used it / there and att another time not long before or after the said Harker was getting of stones within Harkaside boundary but this deponent opposed him therein he having noe right there soe to doe whereupon he desisted. /

7: To the seaventh interrogatory this deponent saith that Harkaside West Grinton and the moores thereof have all along (as farr as this deponent remembers) been reputed parcell of the lordship of Swaledale and of the mannor of Helaugh and township and / constablery of Reeth but was never that this deponent knoweth or heard of reputed part or parcell of the pretended mannor of East Grinton saith that he knowes of about thirteen or fourteen acres of demesne lands lying in Harkaside or West Grinton which / belonged to the complainant the Lord Wharton that John Hird and Anthony Hird are his tennants to the same under the yearly rent of five markes a peice saith that there are severall freehold tenements in Harkaside and West Grinton this deponent being formerly / owner of one of them worth about three pounds per annum the same being in the possession of one Mr Hartley for which there is and allways has been nine pence per annum free rent paid to the lord of the mannor of Helaugh alsoe lands lying in / Swalehall now worth about ninety pounds per annum for which four pence per annum free rent <as this deponent has been informed & beleives> is paid likewise to the said lord Ralph Addisons land worth about three pounds per annum for which two pence per annum free rent is alsoe paid to the said lord / and saith the said freeholders and their tennants allways during this deponents time appeared att the said Helaugh court and did suite and service for their said lands saith there are severall coppy hold lands and tenements in Harkaside aforesaid held by coppy / of court roll of the mannor of Helaugh to witt one tenement in Harkaside in the possession of Marmaduke Hartley the yearly coppyhold rent whereof to the lord is one pound one shilling and four pence and the fine thereof to the lord upon death or alienacion is or / should be thirteen pounds eleaven shillings to the best of this deponents remembrance <part> of it being they call new lands and part old lands another coppyhold tenement there in possession of Christopher Raper yearly coppyhold rent five shillings two pence att the / least the fine on alienacion or death either a fifteen penny fine or twenty penny fine but whether of them this deponent knowes not certainly another coppyhold tenement in West Grinton in possession of Ralph Wensley yearly coppyhold rent to the lord five shillings / fine on death or alienacion the like as the former another in Harkaside in possession of William Arrundale and his three brethren coppyhold rent or fine this deponent remembers not the owners of all which coppyhold lands (All the while this deponent lived at /

Harkaside[241] appeared and did suite and service att Healaugh court and alsoe paid the lord five pence each tenement att each court for green hows or gressoms which as this deponent beleives is a liberty of cutting wood in the lords woods. /

8: To the eighth interrogatory saith there are three inclosures adjoining to Harkaside and West Grinton moores commonly called High New Intack Low New Intack and West Intack saith they were inclosed from the said moores before this deponents time and are all in / the possession of William Arrundell and his three brothers for which they pay the lord of the mannor two shillings per annum coppyhold rent and are coppyhold lands holden by coppy of court roll of the mannor of Helaugh and pay a twenty / penny fine on death or alienacion. /

10: To the tenth interrogatory saith the coppyhold lands in Harkaside and West Grinton are descendible to all the sons alike according to the custome of the mannor of Helaugh there being no such custome in East Grinton or the pretended mannor thereof that ever this / deponent heard of. /

11 To the eleventh interrogatory saith he knowes that during all his remembrance there hath been payable by the township of Reeth a certain modus prescript rent or composicion tyth for corne or hay amounting to seaven pounds two shillings and two pence / or near thereabouts and saith that the hamlett of Harkaside contributed to the same but how much this deponent is not certain but remembers that the land which he had there paid six shillings towards it. /

12: To the twelfth interrogatory saith the owners of Swalehall and lands [x x x][242] thereunto belonging by themselves or tennants or stewards have used to be called att Helaugh court and appeared and did suite and service att Helaugh court as did other / freehold tennants in Harkaside during all this deponents time he used to appear att Helaugh court and not att any other courts that this deponent knowes or has heard of and saith the said owners of Swalehall never held any courts in Harkaside or West / Grinton nor had any freehold or coppyhold tennants or any free or customary rents within the same. /

13: To the thirteeenth interrogatory saith that he knowes the pinfold belonging the townshipp of Reeth and saith that the tresspassing cattle or such like goods belonging to Harkaside and West Grinton used to be impounded att Reeth pinfold being in that constablery and / the Whitaside goods used to be impounded att East Grinton pinfold. /

15: To the fifteeenth interrogatory saith that about two or three and twenty yeares agoe the late Lord Philip Wharton being then lord of the mannor of Helaugh did comence a suite with the coppyhold tennants in the said mannor about making the enhansed rents fineable as / the rents of his old lands were in the defence of which suite this deponent and other coppyholders of Harkaside did joyne in charge with the coppyholders of Helaugh and Reeth and other places within the said mannor and it cost this deponent five pounds for his share of that / charge besides five pounds tenn shillings composicion to my Lord Wharton for exempting his lands from being fineable and saith that all the Lord Whartons lands in the mannor of Helaugh as well those in Harkaside as those in Reeth and Helaugh are usually distinguished / all those places by the name of old lands and new lands and saith he hath heard and beleives the said distinction was occasioned by the Whartons purchaseing part thereof of the Molineuxs further saith that for fifty or sixty yeares that this deponent lived att Harkaside / when there happened to be any royal ayds or land taxes the inhabitants

241 This bracket is not closed.

242 Two words have been erased.

of Harkaside were taxed their proporcions to the said assessment bill with Reeth and Helaugh and saith that the inhabitants of Harkaside have all along used to keep watch and ward / att Reeth as the inhabitants of Reeth did or paid for hireing others for watching for them and this deponent whilst he lived att Harkaside hath severall times himselfe kept watch att Reeth by house row which said circumstances and others soe as aforesaid sett forth / by this deponent are inducements to this deponent to beleive that Harkaside and West Grinton are part and parcell of the complainants lordship of Swaledale and mannor of Helaugh and constablery of Reeth. /

21: To the twenty first interrogatory saith that above thirty yeares agoe this deponents father Marke Close being both a coppyholder and freeholder and inhabitant of Harkaside was sued in the court held for the mannor of Helaugh by one James Milner a freeholder in Harkaside / near forty shillings and the cause was tryd in that court and the verdict against the defendant and about thirty yeares agoe one Robert Robinson a freeholder of crown land and inhabitant in Harkaside sued this deponent (who then lived att and was a freeholder and coppyholder there) in the court /

[membrane ii *verso*]

holden for the said mannor for a tree which came to tryall in the said court and the verdict found for this deponent. /

John Milner[243] of Harkaside in the county of York yeoman aged seaventy three yeares and upwards sworne and examined deposeth as followes /

1: To the first interrogatory this deponent saith that he knowes the present Lord Wharton and defendant Mr Marriott and knewe Philip late Lord Wharton who dyed about eleaven yeares ago. /

2: To the second interrogatory saith that he knowes the dale called Swaledale and that the same lys within the honor or liberty of Richmond mannor of Helaugh and wapentacks of Gilling West and Hang West and that the extent and boundaryes therof begins att / Stollerton Stile on the east and reaches to Hollow Mill Crosse on the west and from Arkilgarthdale on the north cross the river Swale to Wensleydale on the south saith there is onley one parish named Grinton one forrest three constablerys (vizt.) Grinton Reeth and / Mewkar that the towne of Grinton and Whitaside and part of West Grinton all on the south side of Swale lye in the constablery of Grinton and wapentack of Hang West and the townes and hamletts following in the constablery of Reeth (vizt.) Fremington Reeth / Sturfitt Hall Rowcroft Castle Heggs Raw Helaugh Helaugh Park or Park Hall Kirton Fetham Low Row Blaides Smarbarr Lodge Green Winterings all these last mencioned places lye on the north side of the river Swale Harkaside Crack Pott Spring End Satteran / and some other odd houses lying on the south side of the river Swale and the townes and hamletts following lye in the constablery of Mewkar (vizt.) Gunnerside Ivlett House Crack Pott Hall Frith East Stonedale and West Stonedale Cold Kennell Birkdale all / these last menciond places lye on the north side of the said river of Swale Keld Thorns Angram Heugh Head Thwaite Mukarr Ralph High Oxnopp Low Oxnopp all these places <lye> on the south side of the said river of Swale and that the present Lord Wharton / and his ancestors have all along been reputed lords of the said dale called

243 John Milner of Harkaside or West Grinton, yeoman, swore three affidavits on 17 May 1706 (E207/112/1), 1 June 1706 (E207/112/12), and 18 October 1707 (E207/115/1). In May 1708 John Milner, aged 74, gave evidence for Lord Wharton at the trial in London (LI Misc 536, Ward MSS, notebooks vol. 6, f. 116r).

Swaledale and that the Lord Wharton and none else keeps court leets att Helaugh for the mannor of Helaugh and att Mewkarr for the mannor of Mewkarr sometimes once and sometimes / twice in a year saith that the Lord Wharton and his ancestors during all the time of this deponents remembrance have and hath all the libertys of hunting hawking fishing and fowling within the said dale called Swaledale and that this deponent was att the killing of a / stagg for the old Lord Whartons use on the moores in Swaledale and that one Fryer is now gamekeeper in Swaledale for the present Lord Wharton and that none comes there that he knowes of to hawk hunt fish or fowle without asking leave of the said lord or his / gamekeeper /

3: To the third interrogatory saith he knowes the mannor of Helaugh and that the lordship of Swaledale and Helaugh during all the time of this deponents remembrance was held and enjoyed by the Lord Wharton and his ancestors hath heard that part of the said mannor / or lordship did belong to one Molineux but that the same was purchased by the late Lord Wharton many yeares agoe and that the said lordship has all along during his time been called by the name or names of Swaledale or Helaugh and by noe other name or names that he / knowes of. /

4: To the fourth interrogatory saith he knowes the aforesaid vill and constablery of Reeth and that the same lyes within the mannor of Helaugh saith that the lordship of Swaledale mannor of Helaugh & constablery of Reeth lyeth on both sides the river of Swale and / that the same extends southward as farr as the boundary of Wensleydale and the river of Swale was never accounted a boundary to limitt or divide the said lordship of Swaledale mannor of Helaugh & constablery of Reeth or any of them during all his time till of late / by some that this suite was begunn. /

5: To the fifth interrogatory saith he knowes Harkaside and West Grinton and the moores and commons thereto belonging and hath knowne the same all his time being borne att Harkaside and having lived there and in Swaledale ever since he was borne saith the same are not / divided by any boundary from each other but have been all along accounted as one and the same saith he knowes Grinton or East Grinton and the moores thereof and saith there are certain knowne boundarys which divide Grinton or East Grinton and the moores and / inclosed grounds thereto belonging from West Grinton and Harkaside and the inclosed grounds and moores belonging the same throughout from the river Swale northwards to the limitts of Wensleydale southwards saith he knowes the highway or road leading from / East Grinton to West Grinton and Harkaside and the gate in the said high way and the elme tree growing near it and hath knowne the same all his time saith the inhabitants of East Grinton have all along repaired the said road from East Grinton near the said gate / when occasion required but noe further and hath seen them repairing the same saith the place to which the East Grinton men repair the said way is reputed one of the boundermarkes beween East Grinton and West Grinton or Harkaside and saith there are severall / other antient knowne boundermarkes divideing the said townes and inclosed grounds and moores thereof as followes (to witt[244] from the river Swale crosse the east end of the Holme to the said road or highway and soe crosse the said road a little east of the gate / and elme tree and not farr from Cuddy Spout soe up a ridge or bank of ground with a fence upon it to Benson cow house by some called Jennywren cowhouse from Benson cowhouse to the east corner of the Ox Intack from the east corner of the Ox Intack in / a direct line upon the moore crosse Hughes Ditch up to the White Stone in the How and from thence in a direct line up to a place called the Gray Stone in the Greets and from thence to where heaven

[244] This bracket is not closed.

water deales between Wensleydale and Swaledale and this deponent saith / that the first time this deponent had reason to take notice of the said boundary was when this deponent was a young lad att school that one Richard Hutchinson who then was about fourscore yeares of age and had land in West Grinton showed this deponent the said boundary marke soe farr / adviseing him to take perticular notice thereof for that this deponent might have occasion to know the same when he became a man and this deponent further saith from where heavenwater deales above the Gray Stone in the Greets the said West Grinton and Harkaside Moores are reputed / to be bounded from Wensleydale as heavenwater deales westwards to a currock near the west end of Gibbon hill from thence to Browney gill head and from Browney gill head the said moores of Harkaside and West Grinton are bounded from Whitaside by the said Browney gill <down> by the / Horse pasture yate and soe to Scabbey yate and from Scabby yate to the river Swale. /

6: To the sixth interrogatory saith that he knowes the lead mines now wrought by the defendant Marriott called How groves knowes alsoe the lead mines and workes lately opened by the complainant the Lord Wharton near adjoyning saith that the same lye within the boundary of the / moores and commons of Harkaside and West Grinton and not in the moores and commons of East Grinton or the boundarys thereof as this deponent verily beleives and that the said severall workes lye near one hundred yards distant from each other and near the boundary dividing the /same. That in September one thousand seaven hundred and five the Lord Wharton opend the said workes and the boundary above sett forth betwixt West Grinton and Harkaside and East Grinton was ridd before the opening of the said workes and hath heard that publick notice / of the rideing thereof was given in Grinton church yard that the said boundary was ridd by some of the Lord Whartons agents and by near one hundred menn besides whereof this deponent was one and that the said Lord Wharton and his ancestors as lords of the mannor of / Helaugh and {other} their tennants and other the freeholders and tennants in West Grinton and Harkaside had been in the actuall possession of the said moores and wasts before that rideing ever since this deponent can remember by their being lords of the mannor and their tennants / useing thereof as this deponent verily beleives nor ever were disturbed in the possession thereof that this deponent ever heard and saith that the freeholders and tennants in East Grinton never had any enjoyment (that this deponent heard of) upon the moores of Harkaside and West Grinton / (save their cattle going from one common to another by wind and aire) but were as often as they heard disturbed when they came to pull ling and digg for peates and grave turves and saith that Mr West the present viccar of Grinton did ask leave of some of the inhabitants of Harkaside / to digg peates upon Harkaside Moor and remembers that one Richard Harker then a freeholder in East Grinton about twenty eight yeares agoe having pulld <some> ling upon Harkaside Moore this deponent and some other inhabitants of Harkaside having notice thereof / went to fetch the same away and that whilest they were att the place the same Harker came and the said Mr West along with him and desired that he might lead the said ling away but they refused to let him have it and some of the <said> Harkaside men lead it / to Harkaside and gave it to a poor man <there> to cover his house with and saith that neither he nor any other of themselves were troubled for soe doing and hath heard and beleives that one Roger Lee an inhabitant in East Grinton paid a shilling to Ralph Wensley then living / att West Grinton for liberty of getting peates upon Harkaside Moor and hath likewise heard and beleives that one George Emerson and some others inhabitants in East Grinton having graved turves upon Harkaside Moor had the same lead away by Harkaside / men without any disturbance that he heard of from the Grinton men save threatening words and further saith that the said

George Emerson having puld ling and graved turves this summer upon Harkaside Moor this deponent and some other inhabitants of Harkaside / together with the Whartons bayliffe went and burnd the same upon the said place and has not been put to any trouble for soe doeing. /

7: To the seaventh interrogatory saith that during all the time of this deponents memory Harkaside and all the moores and part of the grounds in West Grinton in Ralph Wensleys possession have been reputed part and parcell of the lordship of Swaledale mannor of / Helaugh and the constablery of Reeth and saith that the owners or tennants of Swaledale and the lands belonging it doe suite and service att Helaugh court and saith there are two farmes in Harkaside in the occupacion of John Hird and Anthony Hird which are / the complainant the Lord Whartons owne lands for which the said Hirds each of them pay him five markes per annum antient rent and that there are severall other freehold tennants in West Grinton and Harkaside which pay the Lord Wharton severall yearly free / rents hereafter menciond to witt the owners of Swalehall and lands belonging it four pence Anthony Hird one penny Ralph Addison five pence and Mr Marmaduke Hartley and Christopher Raper but their two free rents this deponent knowes not and saith there are severall coppyhold / tennants in Harkaside which hold their lands by coppy of court roll of the mannor of Helaugh (vizt.) William Arundell and his three brothers Christopher Raper Marmaduke Hartley and as this deponent thinks Ralph Wensley but whether Wensley be a freeholder or coppyholder / this deponent is not certain neither <doth > this deponent know the perticular coppyhold rents and fines upon death or alienacion the said coppyholders pay to the lord but saith all the said lands both of my Lord Whartons <owne> and the freehold and coppyhold alsoe have during / all this deponents time been reputed to lye in the mannor of Helaugh and the owners and occupyers thereof have all along during this deponents time appeared att Helaugh mannor court and done suite and service there for these lands. /

8: To the eigth interrogatory saith there is a parcell of ground called the New Intack <now> in the occupacion of William Arundell and his three brethren which was inclosed from Harkaside common within this deponents memory but whether they pay any rent for the same to the complainant / this deponent knowes not /

10: To the tenth interrogatory saith that the coppyhold lands in Harkaside are descendable to all the brethren or sons alike according to the custome {of the court} of the mannor of Helaugh and this deponent doth not know that there is any such custome in any other mannor in Swaledale save in / the complainants said mannor. /

11: To the eleventh interrogatory saith that he knowes that there is a modus prescript or composicion for tythes of corne and hay paid by the vill or township of Reeth to Mr West as present impropriator of the parish of Grinton and that some of the inhabitants of Harkaside joynes / paying part of the said modus but neither remembers what the whole summe is that is paid nor what summe is paid by Harkaside men in part but saith that whilest he was owner of lands in Harkaside he paid one shilling and tenn pence as his proporcion towards the said modus / for his said lands there but saith the priory lands in West Grinton and the lands att Swalehall doe not contribute to the modus. /

12: To the twelfth interrogatory saith that he hath often heard the owners of Swalehall called att the Lord Whartons courts held att Helaugh as freehold tennants within the same mannor and doth not know that they have appeared att any other court or courts neither have / they during all his time held any court or courts att West Grinton and Harkaside or hath any freehold or coppyhold tennants or any free or customary rents within the same /

13: To the thirteenth interrogatory saith that he knowes the pinfold belonging to the township of Reeth and that the same stands in or near the towne of Reeth and that the inhabitants of Harkaside usually impound their cattle in the same and this deponent saith that he once impounded / cattle himselfe in the said pinfold whilest he lived att Harkaside /

15: To the fifteenth interrogatory saith that he hath heard that there was a suite betwixt the old Lord Wharton and his customary tennants within the mannor of Helaugh and that the Harkaside tennants joyned with the other tennants in maintaining the said suite but this deponent / being then a freeholder was not concernd therein saith that the lands and rents in Harkaside and West Grinton belonging to the Lord Wharton are distinguished by the names of old lands and new lands as the rest of his lordships lands are in the other parts of the said mannor of Helaugh / and the reason of such distinguishment as this deponent hath heard and believes is that part thereof was purchased by the old Lord Wharton of one Molineux and <further> saith that all the inhabitants of Harkaside and part of West Grinton pays their land taxes and constable assessments / to the towne of Reeth saith that he hath been upon the watch att Reeth whilest he lived att Harkaside and hath knowne severall other inhabitants of Harkaside doe the same. /

16: To the sixteenth interrogatory this deponent saith that the Lord Wharton and his ancestors during all the time of this deponents remembrance were in the actuall possession of severall lead mines within the mannor of Helaugh and lordship of Swaledale and that the present Lord Wharton / is now in the actuall possession thereof and hath made great profitt thereof by working the same saith that he remembers old Sir Solomon Swale working lead mines upon Harka and young Sir Solomon Swale but none else till now of late that <the> defendant Marriott wrought some there / hath heard old Sir Solomon Swale and young Sir Solomon say that they derived their title to the said mines from the Gaunts saith that the defendant Marriott gott possession of the mines att Harka about tenn yeares agoe and wrought the same for some time beleives the said entry / was after the old Lord Whartons death and before the present lord entred thereupon for <that> he hath heard that all his mines in Swaledale were in trustees for some time from old Lord Wharton and att the time of the defendant Marriotts entry and that the present Lord Wharton / entred not till afterwards saith that Mr Marriott wrought the mines on Harka for some short time but discontinued working thereof by reason the other mines att How proved better as the deponent thinks /

21: To the twenty first interrogatory saith that he remembers there was a suite brought and tryed in Helaugh court above twenty yeares agoe betwixt William Robinson and Marke Close both of Harkaside. /

Anthony Raper[245] of Whitaside in the county of Yorke yeoman aged sixty two yeares and thereabouts sworne and examined /

1: To the first interrogatory saith that he knowes the plaintiff Thomas Lord Wharton and hath seen <the> defendant Marriott, but he knew not Philip late Lord Wharton or when he dyed. /

2: To the second interrogatory saith he knowes the dale called Swaledale and that the same lys within the honor or liberty of Richmond and wapentacks of Hang West and Gilling West that the reputed boundarys of the same dale are from Stollerston Stile on the east / to Hollow Mill Crosse on the west and from Arkingarthdale on the north cross the river Swale to Wensleydale on the south that there is onely one

245 In May 1708 Anthony Raper, aged 60, gave evidence for Lord Wharton at the trial in London (LI Misc 536, Ward MSS, notebooks vol. 6, f. 116v).

parish within the said dale namely Grinton that there are onely two mannors within the said dale belonging to the Lord Wharton / (viz[t].) the mannors of Helaugh and Mewkar that he knowes of that the present Lord Wharton is the reputed owner of the said dale and hath heard that his father the late Lord Wharton was reputed owner in his life time hath heard that there are court leets held / once every yeare by the Lord Wharton or his agents both att Helaugh and Mewkarr and that when deer came into Swaledale they were reputed to be in the Lord Whartons liberty and when in Arkingarthdale that they were then within the Kings libertys never heard / that any other mannor any royalties or priviledges in Swaledale save the Lord Wharton and hath heard that he did claime the libertys thereof /

3: To the third interrogatory saith that he knowes the mannor of Helaugh and that the lordship of Swaledale and mannor of Helaugh have been during all the time of his remembrance enjoyed by Lord Wharton / aforesaid were never called or reputed by any other name or names save Swaledale and Helaugh. /

4: To the fourth interrogatory saith that he knowes the towne and constablery of Reeth and that the same lys within the mannor of Helaugh saith that the lordship of Swaledale mannor of Helaugh and constablery of Reeth extends and lyes on both sides the said river of Swale / both on the north and south sides thereof and that the same southwards extends as farr as the boundary next Wensleydale and that the river of Swale was never reputed a boundary during all his time that he ever heard of or can remember till of late that this suite / begunn to limitt and divide the lordship of Swaledale mannor of Helaugh and constablery of Reeth or any of them for that he was borne att East Grinton and hath lived there and att Harkaside and Whitaside during all his time saith that there was severall places besides / Harkaside and West Grinton (the places in question) on the south side of Swale which are within the mannor of Helaugh (viz[t].) Crack Pott Spring End Satteron and thirty one acres of coppyhold lands in Whitaside all which doe suite and service att Helaugh mannor court /

5: To the fifth interrogatory saith he knowes Harkaside and West Grinton and the moores and commons thereunto belonging and hath knowne the same for upwards of forty yeares that the same are not divided by any boundary one from the other but have been always reputed / during his time one and the same place that he knowes East Grinton and the moores thereunto belonging he hath heard that there is a boundary divideing East Grinton from Harkaside and West Grinton and that the said boundary divides as well the inclosed grounds as / the moores and commons throughout betwixt the said East Grinton and West Grinton and Harkaside that he knowes the highway or road leading from East Grinton to West Grinton and Harkaside and alsoe the said gate in the said road and the elme tree growing near / the same and hath knowne the same for forty yeares and upwards that the inhabitants of East Grinton usually repair and amend the said high way as farr as the said elme tree that the reputed boundary dividing East Grinton from West Grinton and Harkaside / begins att the river Swale crosse the Holme End and soe crosse the said high way betwixt the elme tree and Cuddy Spout and soe up a ridge or bank of earth to Benson cow house from thence along the wall on the moor by Ox Intack side to the east corner of Ox / Intack wall from thence crossing Hughes Ditch to the White Stone in the How in a right line from thence to the Gray Stone in the Greets in a direct line from thence to where heavenwater deales betwixt Wensleydale and Swaledale from thence westwards to / Gibbon hill currock as heavenwater deales betwixt Wensleydale and Harkaside from thence to the head of Browney Gill where the boundary betwixt Harkaside and Whitaside begins and soe downe Scabby Yate and soe to the river Swale and / saith that he never ridd the said boundary but that the same hath allways been reputed the boundary during all the time of his remembrance and hath heard antient men in the neighbourhood near fifty yeares agoe to witt Thomas Waller and /

[membrane iii *recto*][246]

Christopher Hird both of Harkaside both dead often say that the said boundary above sett forth was the antient boundary of Harkaside he haveing discourse with the said old menn concerning the East Grinton men comeing to gett ling turves and stone upon Harkaside Moore. /

6: To the sixth interrogatory saith that he knowes the lead mines now wrought by the defendant Marriott called How Groves and knowes the lead mines or workes lately opened and claimed by the Lord Wharton near adjoyneing thereto and saith that the said lead mines or workes lately opened by the / said Lord Wharton lye within the moores and commons of Harkaside and West Grinton and the boundary thereof and not within the moores and commons of East Grinton and boundary thereof saith that the Lord Whartons workes lye about sixty yards west / of the line of the boundary and the workes of Mr Marriott lye about sixty yards eastwards of the said boundary line hath heard that the Lord Wharton or his agents about two yeares agoe ridd the said boundary betwixt East Grinton and West Grinton / and Harkaside and hath alsoe heard that publick notice of the rideing thereof was given and this deponent saith that the said Lord Wharton as reputed lord of the mannor of Helaugh and his tennants of his lands att Harkaside and West Grinton and / the other freeholders and tennants there was or were in the possession of the said moores of Harkaside and West Grinton att the time of rideing the said boundary and of opening the said lead workes within the same, and that the said Lord Wharton and / his ancestors as lords of the said mannor of Helaugh and their tennants and other the freeholders and tennants of West Grinton and Harkaside had been in the possession of the said moores and waists before that rideing ever since this deponent can remember by their being lords of / the mannor and their tennants and other the freeholders and tennants in West Grinton and Harkaside useing thereof as this deponent verily beleives nor ever were disturbed in the possession thereof that this deponent ever heard and this deponent further saith that he remembers about twenty nine / yeares agoe he remembers one Richard Harker then an inhabitant in East Grinton did sett a man to pull ling upon Harkaside Moore as the said Harker owned in this deponents presence and this deponent living then att Harkaside he and some others of Harkaside men went and / took the said ling away and gave it to a poor man of Harkaside to thatch his house with and further hath heard about fifteen or twenty yeares agoe and verily beleives the same to be true that some other men of East Grinton getting stones upon the Harkaside Moore were disturbed in the / same by the men of Harkaside. /

7: To the seaventh interrogatory saith that Harkaside and West Grinton and the moores thereof have been during all the time of this deponents remembrance reputed part and parcell of the lordship of Swaledale mannor of Helaugh and that two closes of Ralph Wensley in West / Grinton and all Harkaside except some lands reputed formerly to belong to the Priory of Bridlington lye within the constablery of Reeth and that the same were never reputed part and parcell of the pretended mannor of East Grinton that this deponent ever heard of saith he knowes / the lands att Harkaside called leasehold lands belonging to Lord Wharton and in the possession of

246 Parchment, 650mm x 730mm.

Anthony Hird and John Hird and that the same is six pounds thirteen shillings and four pence per annum and saith that he hath seen the said Hirds appear att Helaugh / court for the said lands saith that Christopher Raper Ralph Wensley Ralph Addison and Mr Hartley as freeholders or coppyholders in Harkaside appears att the said Lord Whartons courts att Helaugh. /

10: To the tenth interrogatory saith that the coppyhold lands in Harkaside or West Grinton discend to all the sons alike att the death of any the said tennants according to the custome within the mannor of Helaugh. /

11: 11:To the eleventh interrogatory saith that he hath heard and beleives that there is a modus or prescript rent paid for tythes forth and out of the township of Reeth to the present viccar of Grinton impropriator and that a part of the said modus is paid by some of the inhabitants of Harkaside / but what the said prescript rent is or what part Harkaside pays of it this deponent cannot sett forth. /

13: To the thirteenth interrogatory saith that he knowes the pinfold belonging to the township of Reeth saith that the same stands on the north side of Reeth near the said towne and that the goods belonging to Harkaside and West Grinton are usually impounded in the said pinfold / and that he hath seen some Harkaside cattle impounded therein for tresspasses done in Harkaside. /

15: To the fifteenth interrogatory saith that he hath heard there was a suite depending betwixt the late Lord Wharton and his customary or coppyhold tennants within the mannor of Helaugh and that the customary or coppyhold tennants in Harkaside paid their proporcions towards / carrying on the said suite with those of the rest of the mannor of Helaugh and saith that the customary and coppyhold lands in Harkaside are distinguished by the same names of old lands and new lands as the rest of the customary and coppyhold lands within the mannor of Helaugh / are distinguished and that the old lands pays twenty penny fine and the new lands fifteen penny fine to the lord of the said mannor saith he knowes that the lands in Harkaside belonging to the constablery of / Reeth and that when he lived att Harkaside he did watch att Reeth as the rest of Harkaside men did that belonged to the said constablery. /

21: To the twenty first interrogatory saith that he remembers above twenty yeares agoe an accion was tried in Helaugh court betwixt one Marke Close and William Robinson about a tree the one a coppyholder and the other a freeholder in Harkaside. /

Christopher Raper[247] of Harkaside in the county of Yorke husbandman aged seaventy yeares or thereabouts sworne and examined. /

1: To the first interrogatory saith he knowes the complainant Thomas Lord Wharton and the defendant Mr Marriott and did knowe Philip late <Lord> Wharton who dyed tenn or eleaven yeares agoe. /

2: To the second interrogatory saith he knowes Swaledale and saith it lyes in the honour of Richmond {shire} and in the wapentacks of Gilling West and Hangwest and extends in length twelve miles from Stollerston Stile to Hollow Mill Crosse and in breadth from Wensleydale on the south side to / Arkingarthdale on the north thereof and that the Lord Wharton and his ancestors have been all this deponents time reputed cheif lords of Swaledale and saith the Lord Wharton and his ancestors have yearly kept court baron att Helaugh and Mewkarr and / knowes of noe other courts kept in Swaledale and saith he beleives none but the said lords and their ancestors have claimed the royalties of fishing fowling and hunting within the said dale and hath heard that they appointed gamekeepers there and this deponent saith / he hath

247 In May 1708 Christopher Raper, aged 70, gave evidence for Lord Wharton at the trial in London (LI Misc 536, Ward MSS, notebooks vol. 6, f. 116v).

knowne that deer have some time been taken in the said dale and that the said lords bayliffes have had the same brought to them for the said lords use. /

3: To the third interrogatory saith the complainant and his ancestors have all this deponents memory been owners and enjoyed the mannor of Helaugh in the interrogatory menciond and lordship of Swaledale and noe other that this deponent ever heard of saith he hath heard the Molineuxs / were owners of some lands there but were purchased of by the said late Lord Wharton and this deponent hath knowne the <said> lordship of Swaledale and mannor of Helaugh for fifty five yeares and upwards and never knew the same to goe by any other name. /

4: & 7 To the fourth <and seveanth> interrogatory saith he hath knowne the towne and constablery of Reeth for fifty five yeares past and saith the same lys within the mannor of Helaugh saith the said mannor of Helaugh and constablery of Reeth extends in some parts thereof as well on the south / side of the river Swale as on the north side thereof saith that Harkaside and West Grinton which lye on the east of Whitaside and on the south side of the river Swale are all (save two or three houses and the lands belonging them in West Grinton and some crowne lands in / Harkaside[248] in Helaugh mannor and that Crack Pott Satteron and Spring End which lye west of Whitaside and on the south side of the river Swale are alsoe in the said mannor of Helaugh as this deponent verily beleives and the reason of such his beleife is this deponent hath / done suite and service att Helaugh mannor court for forty yeares by times and hath observed the inhabitants of the places aforemenciond (except as before excepted) to appear att the said court and doe their suite and service there and saith he beleives all / Harkaside lyes in the wapentack of Gilling West and constablery of Reeth for they pay their land taxes with Reeth in the said wapentack. /

5: To the fifth interrogatory saith he knowes Harkaside and West Grinton and the moores thereto belonging and alsoe Grinton and East Grinton and the moores thereof and hath knowne them for fifty five yeares past saith there are noe boundarys dividing West Grinton / from Harkaside but there are certain knowne boundarys betwixt East Grinton and West Grinton and Harkaside as alsoe betwixt Harkaside and Whitaside which divide both the moores and inclosed grounds of East Grinton from the moores and inclosed grounds of West / Grinton and Harkaside and Harkaside from Whitaside and hath knowne the said dividing boundarys as alsoe the high way or road and gate and elme tree in the interrogatory menciond for fifty <five> yeares as aforesaid saith the inhabitants of East Grinton / repair the said highway as farr as the said boundary betwixt East Grinton and West Grinton and noe farther saith the said boundary begins att the river Swale and crosses the Holme to the said road or highway and crosse the same betwixt Cuddy Spout / and the gate or elme tree up the ridge to Benson cowhouse and from Benson cowhouse along the moor by the Ox Intack wall to the east nook or corner of the said Ox Intack wall and from the east corner of Ox Intack wall over the moore crosse Hughs Ditch to the White Stone in / the How and soe to the Gray Stone in the Greets in a direct line and to the top of the Greets where heaven water deales and there begins the boundary betwixt Swaledale and Wenselydale (to witt[249] from the top of the Greets as heaven water deales to Gibbon Hill and from thence / to Browney gill head and there begins the boundary betwixt Harkaside and Whitaside and proceeds down Browney gill to Scabby Yate and soe to the river Swale. /

6: To the sixth interrogatory saith he knowes the lead mines now or lately wrought by

248 This bracket is not closed.

249 This bracket is not closed.

the defendant as alsoe the lead mines or workes opened lately by the complainant saith they lye att about sixty yards distant from each other that those of the complainant lye within the boundarys on Harkaside / moor on that side claimed by the complainant and this depoenent saith he hath heard the aforesaid boundarys were about two yeares agoe ride by the complainant or his agents but this deponent saith he was not att the rideing saith he beleives the Lord Wharton had a right to the moores / of Harkaside but made noe claim to em that this deponent heard of till that rideing. /

11: To the eleventh interrogatory saith that there is a modus or tyth rent for hay and corne of seaven pounds two shillings and two pence paid by the township or constablery of Reeth to which Harkaside contributes but how much this deponent cannot certainly tell. /

12: To the twelfth interrogatory saith the owners of Swalehall hath all the time of this deponents remembrance been called att Helaugh <mannor> court as other freeholders within the said mannor have been and never appeared att any other court nor ever kept any court att Harkaside or / West Grinton or had any free or customary rents within the same that <ever> this deponent heard of. /

13: To the thirteenth interrogatory saith that the cattle or goods belonging Harkaside have usually been impounded in Reeth pinfold and noe other and this deponent hath severall times impounded goods out <of> Harkaside there and hath had cattle of his owne impounded there for tresspasses done / in Harkaside. /

15: To the fifteenth interrogatory saith that about twenty yeares agoe there was a suit betwixt the lord of the mannor of Helaugh and the customary or coppyhold tennants of the said mannor about fines for those they calld enhansed rents towards the charge of defending or prosecuteing / whereof the customay or coppyhold tennants in Harkaside contributed with the rest of such tennants in other parts of the said mannor and saith the inhabitants of Harkaside have all this deponents time in their course kept watch and ward att Reeth with those of Reeth or paid others for doeing / for them and soe hath this deponent often done himselfe. /

Ralph Wensley[250] of West Grinton in the county of Yorke yeoman aged fifty six yeares and upwards sworne and examined. /

1: To the first interrogatory this deponent saith that he knowes the present Lord Wharton and the defendant Mr Marriott and knew Philip late Lord Wharton but remembers not when he dyed. /

2: To the second interrogatory saith that he knowes the dale called Swaledale and that the same lys within the wapentacks of Gilling West and Hang west saith that the boundary or extent of the said dale is from Stollerston Stile on the east to Hollow Mill Crosse on the west and from / Arkingarthdale on the north crosse the river Swale to the height of the hill as heaven water deales adjoyneing on Wensleydale on the south that there is onely one parish within the said dale called sometime Grinton parish and sometime Swaledale parish that he knowes the mannors of / Helaugh and Mewkarr that lys therein and noe more that he has heard of till of late that Grinton and Fremington which are called mannors since this suite began and not before that this deponent heard of that there are three constablerys therein and noe more (vizt.) Grinton Reeth and / Mewkarr that the Lord Wharton and his ancestors have all along during his time been reputed cheife lords of the said dale and hath once every year kept court leets att Helaugh and Mekarr for the said mannors and this deponent hath been a juror att Helaugh court beleives / that the Lord Wharton and his ances-

250 In May 1708 Ralph Wensley, aged 56, gave evidence for Lord Wharton at the trial in London (LI Misc 536, Ward MSS, notebooks vol. 6, f. 116v).

tors have the royalties of hunting hawking fishing and fowling throughout the whole dale hath heard that the present Lord Wharton appoints game keepers within the said dale never heard that they disturbed any from fowling on Harkaside / moore. /

4 & 7: To the fourth and seaventh interrogatorys saith he knowes the towne and constablery of Reeth that all the towne and greatest part of the said constablery lyes within the mannor of Helaugh that most part of Harkaside and two closes in West Grinton belonging to this deponent / (All which lye on the south side of the river of Swale[251] and alsoe Crack Pott Satteron and Spring End which lye alsoe on the south side of the said river are all in the constablery of Reeth and this deponent as coppyholder for the two closes in West Grinton and all the / coppyholders and tennants of the Lord Whartons owne lands and diverse freeholders in Harkaside and coppyholders or customary tennants att Satteron and Spring End doe all appear and doe suite and service att Helaugh court and this deponent beleives / that they are all within the said mannor because they appear there saith he hath heard old Sir Solomon Swale and the present Sir Solomon Swale as owners of Swale hall called att court att Helaugh saith there are divers lands att West Grinton and Harkaside / called Kings lands and Priory lands that doe not appear att the said court knowes not or ever heard the river Swale being a boundary for the mannor of Helaugh or constablery of Reeth till this suite was begun saith that he is owner of two closes in West Grinton / which are coppyhold lands held of the mannor of Helaugh for which att the time of the purchaseing thereof he paid a fine to the Lord Wharton of three pounds fifteen shillings and payes a rent of five shillings per annum for the said closes to the Lord Wharton / as lord of the mannor of Helaugh. /

5: To the fifth interrogatory saith he knowes Harkaside and West Grinton and the moores and commons thereunto belonging and that the same are not divided by any boundary but were allways accounted one and the same place saith he knowes Grinton or East Grinton and that / there is a boundary which divides the same from West Grinton and Harkaside as well on the moores as within the inclosures saith he knowes the high way leading from East Grinton to West Grinton and Harkaside and the gate and elmetree growing near the same and hath / knowne the same about forty yeares that the inhabitants of East Grinton repaire the said highway to the said elmetree which elme tree is about five or six yards beyond the said boundary and that the said boundary betwixt East Grinton and West Grinton and / Harkaside begins att the river Swale and crosse the Holme and the highway a little to the east of the elme tree and west of Cuddy Spout and soe up the ridge to Benson cowhouse from thence up the side of Ox Intack wall upon the moor to the east nook or corner of the said intack wall / and soe to the White Stone in the How from thence to the Gray Stone att the east end of the Greets from thence to the top of the Greets to where heaven water deales adjoyneing upon Wensleydale and soe along as heaven water deales to Gibbon hill from thence to Browney gill / which lyes betwixt Harkaside and Whitaside and soe downe Browney gill to Scabby Yate and soe to the river of Swale. /

6: To the sixth interrogatory saith he knowes the lead mines now or lately wrought by the defendant Marriott called How Groves and alsoe the lead mine lately opened by the Lord Wharton near adjoyneing saith that the lead mines soe opened by the Lord Wharton lye within the moores / and commons of Harkaside and West Grinton and not within the moores and commons of East Grinton and this deponent saith that the said Lord Wharton as reputed lord of the mannor of Helaugh and his tennants of his lands in Harkaside and West Grinton and the other freeholders and tennants / there

251 This bracket is not closed.

was or were in the possession of the said moores of Harkaside and West Grinton at the time when the boundary was ridd which is now near two yeares – by the Lord Whartons agents and others (whereof this deponent was one) and of opening the said lead workes within the the same and / that the Lord Wharton and his ancestors as lords of the said mannor of Helaugh and their tennants and other the freeholders and tennants in West Grinton and Harkaside has been in the possession of the said moores and waists before that riding ever since this deponent can remember by / being lords of the mannor and their tennants and other the freeholders and tennants in West Grinton and Harkaside useing thereof as this deponent verily beleives saith that one Richard Harker of East Grinton <who> was halfe brother to this deponent and then lived in the same house with him att / East Grinton about twenty four yeares agoe pulld about twenty threave of ling upon Harkaside Moor and that the men of Harkaside and West Grinton went and took the said ling away and that Richard Harker gave the Harkaside and West Grinton men noe disturbance for the / doeing that this deponent ever heard of that about three yeares since one Roger Lee an inhabitant in East Grinton having digged or graved peates on Harkaside Moor this deponent and other inhabitants of West Grinton hearing thereof mett att this deponents house and agreed to take / the same away and the said Lee hearing thereof came to this deponent and some others there and own'd himselfe a tresspasser and paid them twelve pence for the damage and promissed to come there noe more. /

[membrane iii *verso*]

10: To the tenth interrogatory saith that all the coppyholder or customary lands in Harkaside or West Grinton are att the death of the tennants thereof descendable to all the sons alike according to the custome of the mannor of Helaugh and never heard of any such custome within the / pretended mannor of East Grinton. /

11: To the eleaventh interrogatory saith he knowes that severall of the Harkaside men contributes with the men of Reeth in paying a modus of seaven pounds two shillings and two pence to Mr West now impropriator for tythes but knowes not what the particular / summe is that the Harkaside men payes. /

20: To the twentyth interrogatory saith that the indenture now produced by this deponent att the time of his examination was sealed and delivered by William Barwick and Mary his wife as their act and deed to this deponent. /

William Waller of Helaugh in the county of the citty of Yorke gentleman aged about sixty yeares sworne and examined. /

1: To the first interrogatory saith he knowes the complainant and the defendant Mr Marriott and did know Philip late Lord Wharton who dyed in February one thousand six hundred ninety five. /

3: To the third interrogatory saith he knowes the mannor of Helaugh in the interrogatory menciond saith the lordship of Swaledale and Helaugh have been all the time of this deponents knowledge thereof {been} enjoyed by the present Lord Wharton and his late father the Lord Philip Wharton and / by noe other and hath been called by the name or names of Swaledale and Helaugh and noe other name or names that this deponent knowes or heard of. /

4: To the fourth interrogatory saith he knowes the vill and constablery of Reeth and saith the same lyes within the mannor of Helaugh and that the said mannor and constablery extends on the south side of the river Swale as farr as the utmost boundary of Swaledale / southwards to the confines of Wensleydale saith the said river Swale was never reputed a boundary to the said mannor or constablery or either of them that this deponent knowes or heard of till now. /

7 & 10: To the seventh and tenth interrogatorys saith that the complainant the Lord Wharton hath severall freehold lands of his owne in Harkaside farmed by the two Hirds att twenty nobles by the year and that there are a good manycoppyholders there which hold of the said mannor / of Helaugh by coppy of court roll which pay diverse coppyhold rents to the lord of the said mannor and att death or alienacion the like fines as other coppyholders of the said mannor doe (to witt) if the same be of the denominacion of old lands a twenty penny fine that is twenty shillings / for every twelve pence coppyhold rent and soe proporcionably but if new lands onely a fifteen penny fine and the said coppyholders in Harkaside alsoe pay the said lord of the mannor of Helaugh a yearely payment of two pence a coppyholder which is called a greenhugh which is / an acknowledgement for leave from the lord to cutt wood in the lords wood and gett peates or turves and ling on the commons for fewell. /

20: To the twentyth interrogatory this deponent saith that severall writings and written bookes now produced and showne to this deponent att the time of his examination and hereafter perticularly menciond were all (save two of them first hereafter menciond) found by this deponent in / Mr John Gunters[252] house att the time of his death (he dying about Martinmass 1685) which said Mr Gunter was for thirty yeares or thereabouts before his death agent for the late Lord <Philip> Wharton in manageing his lordships estate and this deponent beleives the said Mr Gunter / <had formerly> received the said bookes and writings from one Mrs Higginson sister to one Mr Branthwaite[253] long since dead who was formerly agent for the said late Lord Wharton for that this deponent found with the said bookes a memorandum of the said Mr Gunters owne hand writeing as this / deponent beleives (being well acquainted therewith) mencioning such the said Mr Gunters receiving the same from the said Mrs Higginson (that is to say) one antient paper book marked Lib. A. intituled in the seventh folio thereof Extent & *Annual[is] Valor Manerior[um]* the title ending / Anno Dom 1558 one other antient paper book marked Lib. B. intituled a Book of Fines Assessed by Sir Robert Stappleton knight beginning Anno Domini 1578 with an Abstract of Fines before the said title which said two bookes are the two bookes which to the best of this / deponents remembrance this deponent had from Wharton Hall in Westmoreland a house of the complainants about seaven yeares agoe another paper book marked Lib: C. intitled the Gressome Booke betweene the First of May 1595 and the Seaventh of December 1597 another / marked Lib. D. intituled Fines and Gressomes etc Anno 1624 another marked Lib E. being an Account of Fines and Gressomes beginning the Tenth of October 1626 another marked Lib. F. being a Book of Admittances beginning <the> twentyth <of > Aprill 1634 another marked Lib G. / being an Account of Swaledale Admittances beginning Anno 1655. Another marked Lib H. of the same sort beginning Anno 1634 another marked Lib. J. conteining Admittances to Helaugh antient and New lands beginning Anno 1663 another marked Lib: K. conteining Admittances / in Mewkarr mannor Anno 1664 another marked Lib L. containing Admittances in Swaledale beginning Anno 1664 another marked Lib M. being Admittances in Mewkarr mannor beginning Anno 1664 another marked Lib N. conteining Admittances to Helaugh New and / antient Land and Mewkarr Anno 1664 another marked marked Lib. O. of Admittances in Helay mannor <Anno> 1665 another marked Lib. P of Admittances to Helaugh New

252 John Gunter was an Independent minister who had been evicted from the rectory of Bedale at the Restoration. In 1661 he succeeded Edmund Braithwaite as steward of the Whartons' Swaledale estates – see Fieldhouse and Jennings 1978, p. 126.

253 Edmund Braithwaite was a young man brought up by the Whartons in the South. After the Civil War he was made steward of the Whartons' Swaledale estates – see Fieldhouse and Jennings 1978, p. 126.

Land Anno 1665 another marked Lib. Q. conteining Admittances in Healaugh and Mewkarr mannors <Anno> 1666 another / marked Lib. R. conteining the like pro Anno 1667 another marked Lib. S. conteining the like pro Anno 1660 another marked Lib. T. conteining the like pro Anno 1669 another marked Lib. U. conteining the like pro Anno 1671 another marked Lib. W. conteining the like pro Anno 1672 / another marked Lib X. containing the like pro Anno 1672 another marked Lib. Y. containing the like pro Anno 1673 another marked Lib. Z. containing the like pro Anno 1679 five paper writings numbered severally 1.2.3.4.5. being court Rolls or Call Bookes for Healey mannor / court for the yeares 1664 . 1665 . 1666 . 1667 & 1668 a rentall book numberd 6. for Cumberland Westmorland and Swaledale for Pent 1664 and 11th of November 1664 a book purporting to be an Extract of Admittances for Helaugh New Land beginning Anno / 1634 and ending Anno 1678. One paper booke marked Lib. A.2. conteining Mewkarr <& Helaugh> antient Admittances in Anno 1677 one paper book marked Lib. B2. containing Swaledale Admittances in Anno 1677 one paper book marked C. 2. conteining Helaugh new & old lands Admitt[ances] in / Anno 1677 one paper booke marked Lib. D.2. conteining Healey antient and Healey new Admittances in Anno 1678 one paper book marked Lib. E.2. containing Healey New Healey antient and Mewkarr Admittances in Anno 1680 one paper booke marked Lib. F.2. containing Book of Admittances / for Swaledale October Twenty third 1682. One paper booke marked Lib. G.2. conteining Admittances in Swaledale October 1684 one paper booke marked Lib. H 2. conteining Admittances in Swaledale October 1686 one paper book marked Lib. J. 2. being a Book of Admittances in / Swaledale beginning October Seaventeenth 1688 with the entryes therein till the year 1690 all which said bookes (except the two first mencioned) were found by this deponent in the said Mr Gunters house as aforesaid and the same together with the first <2> mencioned bookes have / been in this deponents custody ever since he being concerned as an agent under the said Lord Wharton and saith the rest of the last mencioned book containing the Admittances in Swaledale from the year 1688 exclusive to the year 1699 inclusive and the severall / bookes and papers hereafter menciond are the bookes and papers which have been kept and the entrys therein relating to the Lord Whartons mannors in Swaledale (vizt.) one paper book marked Lib. K. 2. conteining Swaledale Admittances Book for Aprill 1701 one paper / booke marked Lib. L.2 containing Swaledale Admittance Book for Aprill 1700 and for 1702 one paper book marked Lib. M.2. conteining Swaledale Admittance Book Aprill 1703 and 1704 and this deponent hath had the custody thereof as agent for the said Lord Wharton / from the time they have respectively been made and entred as the same purport. /

Miles Close of Harkaside in the county of Yorke yeoman aged fifty six yeares and upwards sworne and examined. /

1: To the first interrogatory saith he knowes the Lord Wharton and the defendant Marriott. /

4. & 7: To the fourth and seaventh interrogatorys saith that he knowes the towne and constablery of Reeth that the same lyes within the mannor of Helaugh saith that the greatest part of Harkaside and West Grinton lyes within the lordship of Swaledale mannor of Helaugh / and towne and constablery of Reeth saith he neither knowes or beleives that the river Swale is a boundary to divide the same or any of them. /

5: To the fifth interrogatory saith he knowes Harkaside and West Grinton and the moores and commons thereunto belonging and that there is noe boundary to divide the same but that they are one and the same place saith he knowes Grinton or East Grinton and the moores / thereof doth know the boundary divideing the said East Grinton from Harkaside or West Grinton and that the said boundary divides as well the <said> moores as the inclosures that the boundary begins att the river or midstreame of the said river of Swale and soe crosse the Holme and / goes <by> near the elme tree and soe up the ridge to Benson cowhouse from thence along the wall of Ox Intack upon the moor to the south east corner of the said intack and from thence in a line crosse Hughs Ditch to a White Stone or Gray Stone in the How from thence to the / Gray Stone near the east end in the Greets from thence to where heaven water deales adjoyning upon Wensleydale and soe along the top as heavenwater deales to Gibbon hill from thence to a standard or currock att the west end of Gibbon hill and soe to the Browney gill head / northwards where the boundary betwixt Whitaside and Harkaside begins and soe downe Browney gill to Whitaside pasture corner and from thence downe to Scabby Yate and soe to the river Swale and this deponent saith that above thirty yeares ago Marke Close and / Thomas Waller both then antient men of Harkaside and since dead told this deponent of this boundary above sett forth and acquainted him with the sevrall remarkable boundermarkes above menciond. /

6: To the sixth interrogatory saith he knowes the lead mines or groves called How Groves now wrought by the defendant Marriott and alsoe the lead mines or workes lately opened by the Lord Wharton near adjoyning and saith that the said workes lately opened by the / Lord Wharton <ly within the moores & commons of Harkaside & West Grinton & not> within the moores and commons of East Grinton saith that about two yeares agoe the Lord Wharton or his agents opened the said workes and that the said Lord Wharton or his agents ridd the boundarys above sett forth before the opening of the said / mines and that this deponent was one that ridd the said boundary att that time and saith that publick notice was given thereof that the inhabitants of Harkaside and West Grinton hath the enjoyment of the moores and wasts belonging thereto by pulling of ling for / thatching their houses and graving peates and turves there and their cattle grasing thereupon saith that he knowes of severall disturbances given to East Grinton men by this deponent and others when they came to pull ling and grave peats upon Harkaside / Moor and that when they had pulld ling he and others carryed the same away and would not suffer them to doe the like when they knew thereof. /

10: To the tenth interrogatory saith that when any coppyhold tennants in Harkaside dye the lands descend to all the sons alike as the rest of the coppyhold lands within the mannor of Helaugh does and that there is noe such custome within the pretended / mannor of East Grinton. /

Christopher Harker of East Grinton in Swaledale in the county of Yorke yeoman aged forty two yeares sworne and examined. /

1: To the first interrogatory saith he hath seen the complainant the Lord Wharton and knowes the defendant Marriott. /

4: To the fourth interrogatory saith he knowes the towne and constablery of Reeth saith the same is reputed to lye within the wapentakes of Gilling West and Hang West that the Lord Wharton and his ancestors during all the time of this deponents remembrance have been reputed cheif lords of the said dale /

5: To the fifth interrogatory saith he hath knowne Harkaside and West Grinton and the moores and commons thereunto adjoyneing ever since he can remember saith he knowes the boundary which is reputed to divide the moores and commons and inclosures of Harkaside and West Grinton / from the inclosures and moores or commons of East Grinton and the same as hath been formerly related to this deponent and which this deponent beleives to be the said boundary is as followes from the river Swale crosse the end of the Holme to near the elme tree in the highway / from thence up a ridge which goes through this deponents grounds to Benson cowhouse and soe to Ox Intack nooke and soe to the Whitestone in the How and soe to the Gray Stone in the Greets and soe to where heaven water deales betwixt Wensleydale and Swaledale saith he / knowes the highway leading from East Grinton to West Grinton and that East Grinton repaires the same as farr as the said elme tree and noe further and this deponent being an inhabitant of East Grinton hath holpen to repair the same himselfe soe farr /

6: To the sixth interrogatory saith that about twenty two yeares agoe this deponent hired a man to pull twenty threaves of ling for him on Harkaside Moores and saith the inhabitants of West Grinton or Harkaside ledd away and disposed of the same as soon as it was pulld / saith this deponent and the rest of East Grinton Men had some thoughts of sueing them for soe doeing but did not comence any suit not knowing whether East Grinton had a right to pull ling on the said moore or noe /

Robert Robinson[254] of Harkaside in the county of Yorke mason aged sixty nine yeares and upwards sworne and examined. /

1: To the first interrogatory saith that he knowes the Lord Wharton and the defendant Marriott and knew the late Lord Wharton who dyed about tenn yeares agoe /

2: To the second interrogatory saith he knowes the dale called Swaledale and that the same lys within the wapentakes of Gilling West and Hang West that the extent is near twelve miles (vizt.) from Stollerstone Stile to Hollow Mill Crosse in the west and from where heaven / water deales joyneing upon Arkingarthdale on the north crosse the <river> Swale to where heaven water deales adjoyning upon Wensleydale on the south that the Lord Wharton and his ancestors during all the time of this deponents remembrance have been reputed cheife lords of the said dale / that there have been every year court leets held att Healaugh and Mukarr for the said mannor by the said Lord Wharton or his agents during all the time aforesaid knowes that the Lord Wharton as lord of Swaledale has the royaltys of hunting fishing and fowling within the said dale / and appoints game keepers there hath heard that deer have been killd within the said dale for the Lord Whartons use /

4: To the fourth interrogatory saith he knowes the towne and constablery of Reeth and that the same lys within the mannor of Helaugh saith that the lordship of Swaledale mannor of Helaugh and the said towne and constablery of Reeth lys as well on the southside as the northside / of the river of Swale and that the same extends southwards as farr as to where heaven water deales adjoyneing upon Wensleydale saith that he never heard the river of Swale during all his time reputed a boundary to divide the lordship of Swaledale and mannor of Helaugh and / the vill and constablery of Reeth or any of them till after the suit was begunn. /

5: To the fifth interrogatory saith he knowes Harkaside and West Grinton and the moores and commons thereunto belonging and that they are not divided by any boundary but are and have been allways accounted during his time one and the same place saith he knowes Grinton or / East Grinton and the moores thereof saith that there is a boundary which divides as well the moores as the inclosures betwixt East Grinton and Harkaside and West Grinton saith he knowes the highway <leading> from East Grinton to Harkaside and West Grinton and the gate in the / said way and the elme tree near thereto and hath knowne the same for fifty yeares and hath heard that East Grinton repaires the said highway to about the said gate and elme tree saith that the boundary betwixt East Grinton and Harkaside or West Grinton /

254 Robert Robinson of Harkaside swore an affidavit on 17 May 1706 ((E207/112/1). In May 1707 he was described by Jacob Baynes as Lord Wharton's 'chief workman or mason' (E207/112/1). In May 1708 Robert Robinson, aged 70, gave evidence for Lord Wharton at the trial in London (LI Misc 536, Ward MSS, notebooks vol. 6, f. 117r).

and the moores and commons thereunto belonging begins att the river Swale crosse the Holme to near the gate and elme tree in the road and soe up the ridge to Ginny Wren cowhouse and soe by the side of Ox Intack wall on the moore to the east corner of the said / intack and soe crosse Hughes Ditch to the White Stone in the How and soe to the Gray Stone on the Greets and soe to where heaven water deales adjoyning upon Wensleydale and soe as heaven water deales westwards to Gibbon hill and to Gibbon hill standard and to / the top of Browney gill where comes in the boundary betwixt Whitaside and Harkaside and downe Browney gill to Scabby Yate and soe to the river of Swale hath knowne this to be a boundary for thirty yeares and upwards and hath heard his father say long / since that it was the antient boundary betwixt the said places. /

6: To the sixth interrogatory saith he knowes the lead mines now or lately wrought by the defendant Marriott called How Groves <and that he alsoe knowes the lead mines or works lately opened by the Lord Wharton thereto near adjoyning> and that the said workes claimed and opened by the Lord Wharton lye within the moores and commons of Harkaside and not within / the moores and wasts of East Grinton and that the said mines lye about eighty yards distant from each other and that the said boundary lys betwixt them saith the said leadworke was opened by the Lord Wharton a day or two after the said boundary / was ridd and that there was a great number of people together <as> were this deponent and his two sons att the rideing near two yeares agoe saith that this deponent was not present att any publick notice of such rideing but it was the common report in the neighbourhood / that such publick notice was given of it in Grinton church yard after divine service the Sunday next before the rideing and in Reeth markett on the markett day before and the said rideing was according to the boundary before sett forth and this deponent saith that the said / Lord Wharton and his tennants and other inhabitants of Harkaside and West Grinton att the time of the said riding of the said boundary and the said Lords ancestors and their tennants and other the inhabitants there during all this deponents knowledge had / been in possession of the moores of Harkaside and West Grinton aforesaid comprised or included within the said boundarys soe as aforesaid described by this deponent by using the said moores and commons in eating the herbage with their horses cattle and sheep getting / of ling and turfes and digging of stones as they had occasion and saith he beleives <& all along hath beleived> that the said Lord Wharton and his ancestors had right to digg for and gett lead or coale upon any part of the said moores in any place within the said boundaryes and / might lawfully have done soe if they pleased saith that the tennants of East Grinton have not had any enjoyment of the said moores during his time but when any of them have pulld ling or graved any turfes then they have not been suffered by the West Grinton /

[membrane iv *recto*][255]

and Harkaside tennants to carry them away saith that if East Grinton cattle or sheep stray of their owne moores upon Harkaside or West Grinton moores the Harkaside men doe not impound them as / tresspassers but may turne them back if they please as in like manner the East Grinton men may doe with the Harkaside and West Grinton cattle in case of their straying upon East Grinton moors /

7: To the seaventh interrogatory this deponent saith that during all his time Harkaside and West Grinton and the moores thereof have been reputed part and parcell of the

[255] Parchment, 430mm x 700mm.

mannor of Helaugh and / constablery of Reeth for that this deponent being a mason was employed by James Loansdale and John Loansdale both of Helaugh to gett and digg stones upon Harkaside Moor to build each of them a house with / att Helaugh aforesaid (the said John Loansdale having a peice of an island by the river side in Harkaside but saith that James Loansdale hath no lands in Harkaside that {ever} this deponent <knew or ever> heard of) which this deponent did accordingly without any / interruption or asking leave of any person whatsoever and that the said stones were carryed to Helaugh for the purpose aforesaid and this deponent did build one of the said houses with part thereof and hath heard and beleived that ling was / pulld and gott upon Harkaside Moor to thatch houses with att Helaugh and that Harkaside or West Grinton was never reputed part and parcell of East Grinton saith that there are lands lying in Harkaside which they call my lords owne lands / farmed now by John Hird and Anthony Hird two brothers att six pounds thirteen shillings and four pence yearly rent saith that there are severall freehold tennants in Harkaside which appear att Helaugh court and beleives they pay free / rents to the Lord Wharton for their said lands lying att Harkaside (vizt.) Ralph Addison Anthony Hird Christopher Raper Ralph Wensley but knowes not what free rents they pay and saith that there are severall coppyhold tennants in Harkaside / which hold by coppy of court roll of and from the mannor of Helaugh (vizt.) William Arundell John Arundell George Arundell James Arundell Mr Hartley Christopher Raper and Ralph Wensley and saith that the said coppyhold tennants appear / att Helaugh court and doe suite and service there as the rest of the coppyhold tennants within the said mannor doe. /

10. To the tenth interrogatory saith that when any of the coppyhold tennants in Harkaside dye seized their coppyhold lands goe equally amongst their sons according to the custome of the mannor of Helaugh. /

13. To the thirteenth interrogatory saith he knowes the pinfold belonging the township of Reeth that the same stands att Reeth town end near the said towne and <that> the inhabitants of Harkaside allways impound their cattle either att Reeth or Helaugh. /

15. To the fifteenth interrogatory saith that there was a suite betwixt the late Lord Wharton and his coppyhold or customary tennants within the mannor of Helaugh and beleives the tennants of Harkaside joyned with the other tennants within the mannor of / Helaugh and paid their proporcions towards carrying on the said suit and further saith that the inhabitants of Harkaside <pay of their land tax & [window] money to the town of Reeth & saith that the inhabitants of Harkaside> that are within the constablery of Reeth did the duty of watch and ward att Reeth and this deponent hath been upon the watch / himselfe there /

16: To the sixteenth interrogatory saith that he knowes that the present Lord Wharton and his father for forty yeares last past have been in the actuall possession and wrought severall lead mines within the lordship of Swaledale or mannor of Helaugh to / witt)[256] at the Old Gang and att Lounythwaite and that the present Lord Wharton workes the same now and that great profitts have been made thereof of late yeares saith that he knowes <that> there was lead mines wrought in Harkaside Moor before the / obtaining the lease from the Crowne under which <the> defendant Marriott claimes by John Richardson and John Coates who wrought severall yeares att Harka and the How under old Sir Solomon Swale who claimed his title to the said mines as this deponent / hath heard from the Gaunts but saith that the gettings there were very mean. /

21: To the twenty first interrogatory saith that about twenty yeares agoe he then and yet

[256] This bracket is not closed.

living att Harkaside brought an action in Helaugh court against Richard Close then alsoe of Harkaside about a tree and tryd the same in the said court. /

Leonard Galloway[257] of Harkaside husbandman aged about fifty yeares sworne and examined /

1: To the first interrogatory saith he knowes the complainant the Lord Wharton and the defendant Mr Marriott. /

5: To the fifth interrogatory saith he hath knowne Harkaside and West Grinton and the moores and commons thereto belonging for about forty yeares past saith they are accounted one and the same place and are not distinguished or divided one from another by any / bounds saith he alsoe knowes Grinton or East Grinton and the moores thereof and hath knowne the boundarys divideing East Grinton from West Grinton and Harkaside for about forty yeares saith the said boundarys divide both the moores and inclosures / of East Grinton from the moores and inclosed grounds of West Grinton and Harkaside knowes and hath knowne for the time aforesaid the high way or road leading from East Grinton to West Grinton and the gate in the said road and the elme tree / growing near it saith East Grinton during all this deponents remembrance hath repaired the said road to within about twenty yards of the same elme tree and that the utmost place whither they repair the same is one of the said bounder / markes saith the said boundary begins att the river Swale crosses the low end of Holme to the said highway and crosse the highway up the ridge to Benson cowhouse or Jinny Wren cowhouse from thence up the east side of Ox Intack / wall to the east corner of the wall from thence over the moore crossing Hughs Ditch in a straight line to the White Stone in the How from thence in a like line to the Gray Stone in the Greets and soe up as farr as heaven water deales / where the boundary betwixt Harkaside and West Grinton and Wensleydale begins and runns westwards as heaven water deales to Gibbon hill saith that Browney gill divides Harkaside from Whitaside from the upper end thereof downe to Horse / pasture corner and soe to the river Swale. /

10: To the tenth interrogatory saith the coppyhold or customary lands in Harkaside and West Grinton are descendible to all the sons alike on death of the ancestor according to the custome of the mannor of Helaugh. /

22: To the twenty second interrogatory saith that about three yeares agoe William Taylor parish clerk to Mr West the viccar of Grinton came to this deponent and told this deponent Mr West desired to speak with him to whom this deponent came / where Mr West read a letter to this deponent which he said he had from Mr Marriott inviting this deponent to goe and dine with Mr Marriott at Bolton accordingly this deponent went and there att his comeing away the said Mr Marriott desired this deponent / to perswade the neighbourhood of West Grinton and Harkaside to acknowledge that they had done East Grinton men wrong in leading some turves away which the {said} East Grinton men had graved upon West Grinton and Harkaside Moors for / he said it would strengthen his title. /

William Arundale of Harkaside and the county of Yorke yeoman aged thirty two yeares or thereabouts sworne and examined. /

1: To the first interrogatory this deponent saith that he knowes the complainant the Lord Wharton and the defendant Mr Marriott. /

2: To the second interrogatory saith he knowes Swaledale and that the extent of the said dale is about twelve miles and reaches from Stollerstone Stile on the east to Hollow

257 Leonard Galloway of Harkaside, yeoman, swore affidavits on 17 May 1706 (E207/112/1) and on 1 June 1706 (E207/112/2).

Mill on the west and from where heaven water deales adjoyning upon / Arkingarthdale on the north to where heaven water deales adjoyning upon Wensleydale on the south and this deponent saith that he hath paid twice an acknowledgement to the Lord Whartons steward att Helaugh court for priviledge of fishing with / the angle rod on both sides of the river and had not paid the said acknowledgement till he was called upon by the lords bayliffe for payment of the same. /

4: To the fourth interrogatory saith that he knowes the towne and constablery of Reeth and the same lys within the mannor of Helaugh and saith that the lordship of Swaledale mannor of Helaugh and constablery of Reeth extends / as well to the south side of the river of Swale <as on the north side thereof and saith that the river of Swale was never reputed a boundary> during all his time to divide the lordship of Swaledale mannor of Helaugh and the said vill or constablery of Reeth or any of them /

7: To the seaventh interrogatory saith that the greatest part of Harkaside and West Grinton and the moores thereunto belonging are reputed part and parcell of the lordship of Swaledale mannor of Helaugh and constablery of Reeth and / the remaining part lye within the constablery of Grinton saith he knowes that there are lands lying att Harkaside that are called the lords owne lands now farmed by John Hird and Anthony Hird for which they pay to the lord yearly / six pounds thirteen shillings and four pence saith that there arethree freehold tennants within Harkaside that appear att Helaugh court saith that this deponent and his three brothers and Christopher Raper Ralph Wensley and / Mr Hartley are coppyholders within Harkaside and West Grinton and doe appear att Helaugh court saith that the coppyholders or customary tennants aforesaid payes a fine att death or alienacion to the Lord Wharton and that / the same is for old Land a twenty penny fine that is for every twelve pence per annum rent twenty shillings fine and for new Land<s> fifteen penny fine which is for every one shilling per annum fifteen shillings fine <and> this / deponent saith that there are some of the coppyhold lands that doe not pay a fine which is called the inhansed rent and saith that he and his brothers payes eighteen shillings four <pence> per annum rent for the coppyhold lands to the Lord / Wharton and that five shillings six [pence] per annum part thereof being called inhansed rent payes noe fine to the lord. /

8: To the eighth interrogatory this deponent saith that part of the coppyhold lands lying att Harkaside belonging to this deponent and his brothers <are> called intacks which are held of the Lord Wharton and verily beleives that part of the eighteen /shillings four pence rent paid to the lord are for the said intacks but knowes not what part thereof are paid for the same nor knowes not when the same have been inclosed the same having been long before his time and saith that the / Lord Wharton has a right of cutting and taking wood for his mills and mines of the coppyhold lands in Harkaside and that the lords bayliffe has cutt wood upon this deponents grounds there and carryed the same away for the said use. /

10: To the tenth interrogatory saith that when any of the coppyhold tennants in Harkaside dyes seized the lands descends to all the sons alike according to the custome within the mannor of Helaugh and that there is noe such custome within the / pretended mannor of Grinton. /

13: To the thirteenth interrogatory saith that there is a pinfold att Reeth but this deponent saith that when he has any cattle tresspasses to him he impounds the same att Helaugh pound and saith he beleives the Harkaside men that lyes in the / constablery of Reeth has a priviledge to impound att Reeth pound if they think fitt. /

15: To the fifteenth interrogatory saith that there was a suite betwixt the old Lord Wharton and his coppyhold tennants within the mannor of Helaugh touching the inhanced rents and to make these pay a fine as the other rents doe / as this deponent hath heard and beleives in which suite this deponents father was a party and saith that he very well remembers that his father paid money to one Anthony Fryer for his part of the charge of that suite /

John Blaides of West Grinton yeoman aged fifty five yeares sworne and examined /

1: To the first interrogatory saith he knowes the present Lord Wharton <and of> the complainant and the defendant Mr Marriott /

5: To the fifth interrogatory saith he knowes Harkaside and West Grinton and the moores and commons belonging to them and that the same are not divided one from the other by any boundary and alsoe knows Grinton or East Grinton and / the moores thereto belonging saith he knowes there are certain boundarys dividing the said East Grinton from <the said> Harkaside and West Grinton saith the said boundarys doe not onely divide the moores or commons of Harkaside or West Grinton / from those of East Grinton but the same boundarys extend throughout and divide the inclosed grounds of each respectively saith he knowes the said boundarys doe begin at the river Swale from thence crosse the east end of the holme up / to the laine or road that leads from East Grinton to West Grinton and Harkaside and crosses the said laine or road east of the gate in the road and east of the elme tree growing near the said gate and soe up the ridge to Jinny Wrenn / cowhouse or Benson cowhouse from Jinny Wrenn cowhouse up the moor on the east side of Ox Intack wall to the corner of the said wall and from thence in a direct line up the moor to Hughes Ditch and crosse that ditch still in a / direct line to the White Stone in the How and from thence in a direct line to the Gray Stone in the Greets and soe up in a line to where heaven water deales betwixt Wensleydale and Swaledale and from thence the / boundary betwixt Harkaside and West Grinton moores and Wensleydale proceeds westwards as heavenwater deales to Gibbon hill and from Gibbon hill to Browney gill head and from thence the boundary betwixt Harkaside and / Whitaside is the said Browney gill as the same runs downe to horse pasture nook and soe to Scabby Yate and soe to the river Swale and saith that the said boundarys ever since he can remember {and} have been accounted the true / and antient boundarys betwixt the said places and he hath been told by his father and grandfather whilst this deponent was young that the <said> boundarys as before described by this deponent were antient boundaries as aforesaid /

10: To the tenth interrogatory saith the coppyhold or customary lands in West Grinton and Harkaside when the ancestors owners die seized thereof doe descend to all the sons of such ancestors alike and not to the eldest onely which is the custome / of the mannor of Helaugh whereof the Lord Wharton is lord and saith there is noe such custome for any descendible lands in Grinton that this deponent ever knew or heard of. /

Giles Metcalfe[258] of Whita in Whitaside in the county of Yorke yeoman aged sixty yeares and upwards sworne and examined deposeth as followes. /

1: To the first interrogatory this deponent saith that he knowes the Lord Wharton and the defendant Marriott and did knowe the old Lord Wharton who dyed about eleven yeares since. /

2: To the second interrogatory saith he knowes Swaledale that it lyes in Gilling West and Hang West wapentacks and that the reputed boundary thereof is from Stollerston Stile

[258] In May 1708 Giles Metcalfe, aged 60, gave evidence for Lord Wharton at the trial in London (LI Misc 536, Ward MSS, notebooks vol. 6, f. 117r).

to Hollow Mill Crosse and from Arkingarthdale on the north to Wensleydale on / the south as heaven water deales that there is two mannors (vizt.) Helaugh and Mukarr in the said dale both belonging to the Lord Wharton where he keeps courts once a year and never knew any other courts kept in Swaledale that there [are] three / constablerys (vizt.) Grinton Reeth and Mukarr and saith that the Lord Wharton and his ancestors have during all his time been reputed cheif lords of the said dale and none other that this deponent heard of and claimed all the royaltys / belonging to them as lords of Swaledale. /

4: To the fourth interrogatory saith that he knowes the towne and constablery of Reeth that the same lys within the mannor of Helaugh saith that the lordship of Swaledale mannor of Helaugh and constablery of Reeth lys on each side / the river Swale and saith that the said river of Swale was never reputed a boundary during all his time to limitt and divide the same or any part thereof. /

5: To the fifth interrogatory saith he knowes Harkaside and West Grinton and the moores and commons thereunto belonging and that the same is not divided by any boundary but was allways accounted one and the same place saith he knowes / the towne of Grinton or East Grinton and the moores belonging thereto saith that theres a reputed boundary which divides the moores and inclosed grounds betwixt Harkaside or West Grinton and East Grinton saith he knowes the road leading / from East Grinton to West Grinton and the gate standing in the same and the elme tree growing near thereto saith that he hath knowne the same above forty yeares saith he hath heard that East Grinton men repaired the said highway to the / said gate that the said boundary dividing East Grinton from West Grinton and Harkaside begins att the river Swale crosses the said high way a little east of the said elme tree and soe up along a bank or ridge to Benson cowhouse or Jinny Wrenn / cowhouse and from thence along Ox Intack wall upon the moor to the east corner of the said wall and soe crosse Hughs Ditch to the White Stone in the How and from thence to the Gray Stone in the Greetes and soe to where heaven water deales /

[membrane iv *verso*]

adjoyneing on Wensleydale and soe along westwards as heaven water deales to Gibbon hill and soe along to the top of Browney gyll where begins the boundary betwixt Harkaside and Whitaside and soe down Browney gyll to / Scabby Yate and soe to the river Swale and saith that he hath heard old men above forty yeares agoe say that this boundary above menciond was the boundary betwixt East Grinton and Harkaside and Harkaside and / Whitaside. /

6: To the sixth interrogatory saith that he knowes the lead mines now wrought by <the> defendant Marriott called How Groves and alsoe knowes the lead mines or workes lately opened and claimed by the Lord Wharton saith that the / said mines soe claimed and opened by the Lord Wharton lye within the mannor and commons of Harkaside and not within the moores and commons of East Grinton and that the said lead mines or workes lye about eighty yards / from each other and that the said boundary lyes betwixt the said mines saith that the said lead workes were opened by the Lord Wharton or his agents about two yeares agoe and that the boundary above sett forth was / ridd by the Lord Whartons agents and severall others (whereof this deponent was one) before the opening <of> the said mines and that publick notice of rideing the said boundary was made by the Lord Whartons bayliffe in the open markett / att Reeth upon markett day next before riding thereof in the hearing of this deponent and this deponent further saith that the said Lord Wharton and his tennants and other inhabitants of Harkaside and West Grinton att the / time of the said rideing of the said boundarys and the said lords ancestors and their tennants and other the inhabitants there during all this deponents knowledge have been in the possession of the moores of Harkaside and West / Grinton aforesaid comprized or included within the said boundaryes soe as aforesaid described by this deponent by using the said commons or moores in eating the herbage with their horses cattle and sheep getting of ling and turves / and digging of stones as they had occasion and saith he beleives and all about hath beleived that the said Lord Wharton and his ancestors had right to digg for and gett lead or coales upon any part of the said moores / in any place within the said boundaryes and might lawfully have done soe if they pleased saith that he hath heard that when East Grinton men puld ling or graved peates or gott turves upon Harkaside Moore that / the same were taken from them by the Harkaside men remembers that one Richard Harker a Grinton man complained to this deponent of the Harkaside men taking away some ling that he had pulled upon Harkaside / moor and saith he knowes and very well remembers that one James Lonesdale of Helaugh about three or four yeares since gott stones or had stones gotten upon Harkaside Moor to build himselfe an house <with at Helaugh> as he verily / beleives because he hath seen the said Loanesdale or his workmen lead the said stones to the place where the said house was built and <saith the house was built with the said stones> beleives the said Loanesdale asked noe leave of any for soe doing or had any / disturbance for doeing thereof that this deponent heard of. /

7: To the seaventh interrogatory saith that Harkaside and West Grinton and the moores thereof have been during all his time reputed part and parcell of the lordship of Swaledale mannor of Helaugh and township of Reeth and was / never reputed part or parcell of the pretended mannor of Grinton that this deponent knowes or heard of saith he knowes that there are Harkaside lands lying att Harkaside belonging to the Lord Wharton called freehold lands farm'd / by the Hirds but knowes not the yearely rent alsoe saith that there are severall freeholders in Harkaside that appear att the Lord Whartons courts att Helaugh and beleives they pay him free rents but knowes not what they / pay saith that there are severall coppyhold or customary tennants in Harkaside <that> have old lands and new lands there which hold of the Lord Whartons mannor of Helaugh but what coppyhold rents or fines they pay for the same / to the said lord this deponent knowes not and saith that there are some freeholders in Harkaside and West Grinton that doe not appear att Helaugh court but the greatest part of them appeares there. /

8: To the eigth interrogatory saith that there are parcells of ground in Harkaside which adjoyne upon the moor called intacks which this deponent beleives have been inclosed of the moores but were inclosed before his time and beleives the / occupyers thereof payes a rent for them to the Lord Wharton but what rent in particular is paid for the same this deponent knowes not. /

10: To the tenth interrogatory saith the coppyhold lands in Harkaside att the death of any tennant descends to all the sons alike according to the custome of the mannor of Helaugh and that there is noe such custome within Grinton / that this deponent knowes or ever heard of /

13: To the thirteenth interrogatory saith that he knowes the pinfold att Reeth and that the same belongs to Harkaside as well as Reeth /

14: To the fourteenth interrogatory saith that he dos not know that the complainants tennants of the mannor of Helaugh and vill of Reeth have intercommoned with the Harkaside men upon their moor but hath heard and beleives that / they have a right soe to doe and one reason is for that there is a place upon Harkaside Moor called Reeth Peat Moor and hath heard long since that Reeth men formerly gott peates and turves there and saith that there is a / laine that lead from Helaugh to Harkaside which is called Neatherd Lonning and beleives the reason of itt being called soe and hath heard old men say soe was because they used formerly to drive their cattle from one / common to another. /

John Harker of Oxnop in Swaledale aged seaventy three yeares or upwards sworne and examined. /

6 & 21: To the sixth and twenty first interrogatorys saith that the Lord Whartons ancestors and their tennants in Harkaside have had enjoyment of the moores of Harkaside during most of this deponents remembrance by their horses cattle and sheep / feeding thereon and their getting ling turves or peates and stone there and saith that above seaventeen yeares agoe one Thomas Wilkinson and his son masons both dead being working att this deponents house they told this deponent that / they had two horses taken up as wafes or strayes upon Harkaside Moor and carryed to my Lord Whartons bayliffe one James Freer by name and that they had given twelve pence a peice to Freer for loosing or redeeming them / and saith that about twenty four yeares agoe this deponent was one of the jury att Helaugh court when and where the jury presented one Richard Harker of East Grinton for digging up and getting flaggs on Harkaside common / without right and fined him twenty Shillings for soe doing. /

George Alderson of Sturfitt hall in the county of Yorke yeoman aged sixty four yeares and upwards sworne and examined /

1: To the first interrogatory saith that he knowes the Lord Wharton and hath seen Mr Marriott. /

2: To the second interrogatory saith he knowes the dale called Swaledale and that the Lord Wharton and his ancestors have allways during the time of this deponents remembrance been reputed cheif lords of the said dale and saith that / the said lords have allways claimed and enjoyed the privledges of hawking hunting fishing and fowling within the same. /

4: To the fourth interrogatory saith he knowes the towne and constablery of Reeth and that the same lys within the mannor of Helaugh that the said lordship of Swale mannor of Helaugh and constablery of Reeth lyes on the / south side of the river Swale as well as on the north side thereof and that the river Swale was never reputed a boundary that he heard of to divide the lordship of Swaledale mannor of Helaugh and constablery of Reeth / or any of them. /

7: To the seventh interrogatory saith that the most part of Harkaside lyes in the constablery of Reeth and wapentack of Gilling West and the township of Grinton and Whitaside in the wapentack of Hang West /

14: To the fourteenth interrogatory saith that one Christopher Raper of Harkaside now living told this deponent about tenn or twelve yeares agoe that <he had> a right of common upon Reeth Moor upon account of his lands att Harkaside and told / this deponent then that he this deponent might put his horse to depasture upon Reeth common upon his account if he thought fitt and he would justify him in it. /

15: To the fifteenth interrogatory saith that the inhabitants of Harkaside that lyes within the constablery of Reeth are joyned with the inhabitants of Reeth and pay the land tax with them and this deponent hath been twice / a collector of the said land tax there and is thereby the better enabled to depose herein. /

James Robinson[259] of Oxnop in Swaledale yeoman aged about fifty five yeares sworne and examined. /

2: To the second interrogatory saith he knowes the dale called Swaledale and saith

[259] In May 1708 James Robinson, aged 'above 40', gave evidence for Lord Wharton at the trial in London (LI Misc 536, Ward MSS, notebooks vol. 6, f. 117r).

it lyeth in the honor of Richmond and Richmondshire and in the wapentacks of Gilling West and Hang West saith it extends {it selfe} from / Stollerston Stile [on the] east to Hollow Mill Crosse on the west and from the height of the moors north to Arkingarthdale to the height of the moores south to Wensleydale saith tis all in the parish of Grinton and there are two mannors / therein (viz[t].) Mukarr and Helaugh <at> both which courts are kept by the Lord Wharton and noe courts else in Swaledale and but three constablerys to witt Reeth Mukarr and Grinton saith that the said Lord Wharton kept / court leet constantly att Helaugh once every year and att Mukarr sometimes once in a year and sometimes but once in two yeares saith the Lord Wharton and his ancestors have all this deponents memory for forty yeares and above / claimed and enjoyed the royalties of hunting fishing and fowling throughout the said dale or lordship of Swaledale and have all along from time to time appointed gamekeepers therein for that purpose who have used to hinder all / people (save such as had license from the said lords) to kill any game therein. /

3: To the third interrogatory saith he knowes the mannor of Helaugh and saith the lordship of Swaledale and Helaugh have all this deponents time been enjoyed by the Lord Wharton and his ancestors as cheif lords thereof and noe / other hath heard that the Molineuxs had formerly some part thereof which was purchased by the Lord Whartons ancestors saith the said lordship hath not been called by any other name then the same now is in all this / deponents time. /

4: To the fourth interrogatory saith he knowes the vill and constablery of Reeth saith the said vill & constablery lyes within the lordship and mannor of Helaugh saith the said lordship of Swaledale mannor of Helaugh / vill and constablery of Reeth extend both to the north side and south side of the river Swale saith that Harkaside Crack Pott Satteron and Spring End doe lye on the south side of the river Swale and saith Satteron is within / the mannor of Helaugh and Crack pott and part of Spring End and most of Harkaside within the mannor of Helaugh and constablery of Reeth saith he never heard that the river Swale was accounted a boundary to / the mannor of Helaugh and constablery of Reeth or either of them nor dos this deponent beleive the said river is a boundary thereto because the mannor of Helaugh and constablery of Reeth lye on both sides the said river /

7: To the seaventh interrogatory saith that during all this deponents knowledge of Harkaside (which is above forty yeares past) most part of the same hath been reputed part and parcell of the mannor of Helaugh and township and / constablery of Reeth and the severall tennants of the Lord Whartons owne lands and the coppyholdes and severall freeholders in Harkaside appear and doe suite and service att the Lord Whartons mannor court of Helaugh / aforesaid but the particular rents they pay the said lord of the mannor or their respective and different termes they hold their land by this deponent cannot perticularise saith that upon death or alienacion of a coppyholder / there is paid to the lord of the said mannor a twenty penny fine if it be the old lands and a fifteen penny fine if new lands. /

10: To the tenth interrogatory saith that when any coppyholder of lands in Harkaside dye seized of any coppyhold lands the said lands descends to all his sons alike be they never soe many according to the custome of the mannor / of Helaugh (there being noe such custome in Grinton or East Grinton that this deponent ever heard of[260]. /

11: To the eleaventh interrogatory saith there is and has been as long as this deponent can remember a modus or prescript rent for tyth of hay and corne amounting in all to seaven pounds two shillings and two pence per annum / paid by the vill and

[260] This bracket is not closed.

constablery of Reeth to the impropriator of the parish of Grinton and saith this deponent hath severall times about thirty yeares agoe collected the same and further saith that Harkaside as part of the / said vill or constablery of Reeth pays a part thereof proporcionable to their lands but cannot recollect himselfe how much the same amounts to but remembers Richard Closes lands paid six shillings per annum and / Christopher Rapers lands eight shillings or thereabouts towards the same. /

14: To the fourteenth interrogatory saith that he hath heard long since that his grandfather and Mathew Metcalfe who were Reeth men used to gett peats in a place in Harkaside Moor called Reeth Peat Moor saith / he beleives they had some land att Harkaside but whether they gott those peats in right of their lands in Harkaside or as Reeth men or intercommoners with the tennants of Harkaside this deponent knowes not. /

15. To the fifteenth interrogatory saith that twenty yeares agoe or upwards there was a suite betwixt the late Lord Wharton and the tennants of the mannor of Helaugh relating to the customes of the said mannor / and hath heard severall who were both coppyholders and freeholders in Harkaside say they contributed with the rest of the tennants in the said mannor towards the charge of maintaining or defending that suit / further saith that the inhabitants of Harkaside which live <with> in the constablery of Reeth have all along paid their land tax and constable assessments and kept watch and ward together with Reeth. /

James Naylor[261] of West Grinton in Swaledale aged thirty yeares or thereabouts sworne and examined. /

1: To the first interrogatory saith that he hath seen the present Lord Wharton and knowes the defendant Marriott but did not know the old Lord Wharton /

2: To the second interrogatory saith that he knowes the vill and constablery of Reeth and that the same lys within the mannor of Helaugh saith that the lordship of Swaledale mannor of Helaugh and constablery / of Reeth <ly> on both sides the river of Swale and that the river of Swale was never reputed a boundary to divide the said lordship of Swaledale mannor of Helaugh or constablery of Reeth or any of them that / ever he heard of. /

[membrane v *recto*][262]

5: To the fifth interrogatory saith that he knowes Harkaside and West Grinton and the moores thereunto belonging saith that there is noe boundary to divide the same but they have been allways during all his time accounted one and the / same place saith he knowes East Grinton and the moores thereof and that there is a reputed boundary to divide the same from West Grinton and Harkaside and that the said boundary divides as well the moores as the / inclosed grounds there saith he knowes the highway leading from East Grinton to West Grinton and the gate that hangs in the said highway (for that he repaired the said gate) and knowes alsoe the elme tree growing near thereto / and that the East Grinton inhabitants repairs the said highway to about six yards short of the east side of the said gate and that he hath knowne the same during all the time of his remembrance being borne att Swale hall mill and living there and att Swale hall during / all his time and that the place in the said highway to which the inhabitants of East Grinton repairs is a part of the boundary

261 In May 1708 James Naylor, aged 36, gave evidence for Lord Wharton at the trial in London (LI Misc 536, Ward MSS, notebooks vol. 6, f. 116v).

262 Parchment, 650mm x 730mm.

betwixt the said places and saith <that> the said boundary begins att the river of Swale crosses the end of the Holme and soe crosses the said highway / about six yards east from the said gate and a little west of Cuddy Spout and soe up the ridge to Benson or Ginny Wrenn cowhouse and soe along the east side of Ox Intack wall upon the moor to the east corner of the said intack and soe in a direct line to the White / stone on the How and soe to the Gray Stone on the Greets and then to where heaven water deales betwixt Harkaside Moors and Wensleydale and soe westwards as heavenwater deales to Gibbon hill and then <to> Browney gill head where begins the boundary betwixt Harkaside / and Whitaside and downe Browney gill to Scabby Yate and soe to the river of Swale. /

6: To the sixth interrogatory saith that he knowes the lead mines now wrought by the defendant Marriott called How Groves and alsoe knowes the lead mines lately opened and claimed by the Lord Wharton adjoyning near thereto and saith that the same mines opened by the / Lord Wharton are reputed to ly within the moores and commons of Harkaside or West Grinton and this <deponent> verily beleives they doe and that they are not within the commons of East Grinton or the boundary thereof saith that the said mines lye about one hundred yards distant / from each other and that the said boundary goes betwixt them saith that the Lord Wharton opend the said mines in September 1705 and that the said boundary was ridd by the Lord Whartons agents and severall other neighbours (whereof this deponent was one) before / the opening of the said mines and saith that publick <notice of> rideing {of} the said boundary was given in the church yard of the parish church of Grinton upon Sunday next before {the} riding thereof after divine service) and beleives that the same notice was alsoe given in / Reeth markett and this deponent further saith that the said Lord Wharton and his tennants and other inhabitants of Harkaside and West Grinton att the time of riding the said boundarys and the said Lord Wharton and his ancestors and their tennants and other the inhabitants / there during all this deponents knowledge have been in possession of the moores of Harkaside and West Grinton aforesaid comprized or included within the boundarys soe as aforesaid described by this deponent by useing the said commons or moores in eating the herbage with their / horses cattle and sheep getting of ling and turfes and peates and digging of stone as they had occasion and saith he beleives and all along hath beleived that the said Lord Wharton and his ancestors had right to digg for and gett lead or coales upon any part of the said moores and / any place within the said boundarys and might have done soe if they pleased saith that he this deponent and some other inhabitants of West Grinton and Harkaside once lead away some turfes from <of> Harkaside Moor that some of the East Grinton men had graved there. /

7: To the seaventh interrogatory saith that the greater part of the lands in Harkaside and West Grinton lyes within the mannor of Helaugh and constablery of Reeth saith that he hath heard his father whose name was Christopher Naylor and who farmed Swale / hall lands of Sir Soloman Swale say that he appeared att the Lord Whartons court att Helaugh for the said Sir Soloman Swales lands there and paid essoins for himself att the said court. /

10: To the tenth interrogatory saith that the coppyhold or customary lands att Harkaside att the death of the tennants descend to all the sons alike according to the custome within the mannor of Helaugh. /

15: To the fifteenth interrogatory saith that the inhabitants of Harkaside that are within the constablery of Reeth payes the land tax with the towne of Reeth and that they doe and performe watch and ward there when required /

Thomas Milner of Harkaside in Swaledale yeoman aged fifty six yeares sworne and examined. /

1: To the first interrogatory saith he knowes the complainant the Lord Wharton and the defendant Marriott and did knowe Philip Lord Wharton deceased saith he dyed about eleven yeares agoe. /

4: To the fourth interrogatory saith he knowes the towne and constablery of Reeth saith it lyes within the mannor of Helaugh and saith the lordship of Swaldale the mannor of Helaugh and the constablery of Reeth extend not onely to the north of the river Swale but to the south / side of the said river as farr as heaven water deales betwixt Wensleydale and Swaledale saith he hath knowne the said places for forty yeares and upwards and never heard that the river Swale was ever reputed a boundary to limitt or divide the said lordship of Swaledale / mannor of Helaugh and vill or constablery of Reeth or any of them. /

5: To the fifth interrogatory saith he knowes Harkaside and West Grinton and the moores and commons thereunto belonging saith the same are not divided by any boundarys the one from the other but are one and the same place saith he knowes Grinton or East Grinton and the moors / thereof saith he both hath heard of and for forty yeares past hath knowne a boundary dividing Grinton or East Grinton from Harkaside or West Grinton saith the said boundary divides not onely the moores or commons of Harkaside or West Grinton from those of East Grinton / but the same boundary extends throughout and divides likewise the inclosed lands of each respectively saith he hath knowne the highway or road leading from East Grinton to West Grinton and Harkaside and the gate in the said road and the elme tree growing near the said / gate for forty yeares past saith the inhabitants of East Grinton have during all this deponents remembrance repaired or amended the said road or highway allmost to the said gate but noe further saith the place to which the said inhabitants of East Grinton repairs the / said highway is one of the boundermarkes dividing East Grinton from West Grinton and Harkaside saith the said boundary begins att the river Swale and crosses a peice of land called the Holme and soe to the place crosse the highway to which East Grinton repairs the / same being east of the said gate and west of Cuddy Spout from thence up the bounder ridge to Benson cow house and from thence to the east corner of the Ox Intack from thence crosse Hughes Ditch in a straight line to the White Stone in the How and from thence in a / direct line to the Gray Stone in the Greets and soe up to where heavenwater deales between Harkaside and West Grinton moores and Wensleydale saith the boundary betwixt Harkaside or West Grinton and Wensleydale proceeds thence as heavenwater deales to Gibbon hill / and soe to Browney gill head saith that the boundary which divides Harkaside and West Grinton from Whitaside begins att the top of Browney gill and proceeds downe Browney gill to Scabby Yate and soe to the river Swale saith he hath the greater reason to / know the said boundarys for that this deponent when he was young used to look after his fathers horses cattle and sheep which went on Harkaside or West Grinton commons least they should stray <and> upon that occasion his father often used to show him and tell him of / the said boundarys and about twenty eight yeares agoe the inhabitants of Harkaside and West Grinton rode the said boundary betwixt Harkaside or West Grinton and East Grinton in the same manner as this deponent hath described the same beginning att the river / Swale and ending where heavenwater deales above the Gray Stone in the Greets att which riding this deponent was present and one was Mr Timothy Blackburne deceased then one of the principall inhabitants of East Grinton saith that riding was occasiond by one / Richard Harker then of East Grinton who had pulld some ling upon West Grinton and Harkaside Moores which they would not suffer him to carry away but ledd it away themselves and gave it away to James Hird a poor man of Harkaside to thatch his / house withall /

6: To the sixth interrogatory saith he knowes the lead mines now or lately wrought by the defendant Mr Marriott called How Groves and alsoe the lead mines lately opend by the complainant the Lord Wharton or his agents near adjoyning thereto saith / that the workes soe opened and claimed by the Lord Wharton lye within the moores and commons of Harkaside and West Grinton and the boundarys thereof and not within the moores and commons of East Grinton and the boundarys thereof saith the said workes / lye about eighty yards distant one from the other and the said bounder <line or> mark is betwixt them but rather nearer to the defendants workes then to the complainants to the best of this deponents judgement saith the complainant the Lord Wharton did not open the said workes till / after the riding <of> the said boundarys saith the said boundarys were rode about two yeares agoe by about fourscore men together whereof this deponent was one and the said boundarys were generally agreed unto by those present att the said rideing saith he was / present in Grinton church yard immediately after divine service on Sunday before the said riding when publick notice was given of this riding and heard that the like notice was given in Reeth market the markett day before and this deponent saith that the said Lord Wharton and / his tennants and other inhabitants of Harkaside and West Grinton att the time of the last menciond riding of the boundarys and the said Lords ancestors and their tennants and other inhabitants there during all this deponents remembrance have been in possession of the / moores of Harkaside and West Grinton aforesaid comprized or included within the said boundarys soe as aforesaid described by this deponent by enjoying the eatage of the herbage there with their horses cattle and sheep getting of ling and turfes and digging of stones as / they had occasion and saith he beleives that the said Lord Wharton and his ancestors might lawfully have digged and gott lead and coale upon any part of the said moores within the said boundarys if they had pleased and this deponent saith that the tennants of / East Grinton never in this deponents time either <that he knows or heard of> pulld ling graved peates or digged stones upon the said moores of Harkaside and West Grinton but they were disturbed and interrupted by the inhabitants of Harkaside and West Grinton and particularly the / ling soe as aforesaid pulled by Richard Harker of East Grinton upon Harkaside Moor was taken away by the Harkaside men and given to Hird as aforesaid saith that about three or four score yeares agoe one Roger Lee of East Grinton having graved turves / upon Harkaside or West Grinton moor the Harkaside and West Grinton men would not suffer him to lead them away telling him he had noe right there and thereupon the said Lee acknowledged he had done them wrong and paid them a six pence / as an acknowledgement and for leave to carry them away. /

7: To the seaventh interrogatory saith that all Harkaside and West Grinton and the moores thereof have been att the time of this deponents knowledge thereof reputed to lye within the mannor of Helaugh saith there are some freeholders <there> which doe not / appear att Helaugh mannor court but most part of the freeholders and all the coppyholders and the Lord Whartons tennants there appear att that court. /

8: To the eighth interrogatory saith there are diverse grounds called intacks in Harkaside aforesaid adjoyning on the common or moore there which have been reputed to have been taken of from the said moor but not within the memory of man saith the tennants / thereof pay a rent for the same to the lord of the mannor of Helaugh but how much this deponent knowes not. /

10: To the tenth interrogatory saith the coppyhold or customary lands in Harkaside and West Grinton are descendible to all the sons alike of the ancestors dying seized thereof which descent is according to the custome of the mannor of Helaugh but this deponent never / heard of any such custome in East Grinton. /

13: To the thirteenth interrogatory saith he knowes the pinfold att Reeth saith that when the Harkaside and West Grinton men have occasion to impound any cattle they impound them in Reeth pinfold. /

14: To the fourteenth interrogatory saith he knowes severall of the complainants tennants in the mannor of Helaugh and vill of Reeth which have noe lands in Harkaside or West Grinton or farme any lands there which have to this deponents knowledge for twenty yeares past / by times as they had occasion gott stone upon Harkaside Moor and ledd them away to Reeth without <either asking permission of the inhabitants or suffering> any interrupcion from them that this deponent knowes or heard of particularly Henry Machell Arthur Hutchinson John Loansdale and Roger Alderson saith one / Francis Whipp of Reeth used to put his sheep upon Harkaside common from Reeth and kept them there without disturbance saith the said Whipp had noe land of his owne att Harkaside or West Grinton onely farmed some lands there belonging Swale hall saith / there is a lane leading from Helaugh to Harkaside called Neatherd Lane or Noutherd Lonning saith the tradition goes and soe is still beleived in the neighbourhood that it gott that name by the Helaugh mens useing to drive their neat or cattle from / Helaugh to Harkaside common to depasture there. /

15: To the fifteenth interrogatory saith there are lands in Harkaside and West Grinton belonging the complainant <the Lord Wharton> which are distinguished by the names of old lands and new lands being the same distinguishing names betwixt other lands in other parts of / the said mannor of Helaugh saith Harkaside pays all assessments whatsoever (saving to the poor and church) with the towne of Reeth and keep watch and ward there in their turne with Reeth and this deponent hath kept watch himselfe att Reeth for his / lands att Harkaside. /

James Hird of Harkaside in Swaledale yeoman aged sixty four yeares and upwards sworne and examined. /

1: To the first interrogatory saith he knowes the defendant Mr Marriott but knowes not the present Lord Wharton nor knew the late Lord Wharton in his life time. /

4: To the fourth interrogatory saith he knowes the vill and constablery of Reeth and that the same lyes within the mannor of Helaugh and that the lordship of Swaledale mannor of Helaugh and the vill and constablery of Reeth lyes as well on the south side of the / river of Swale as on the northside thereof and that the river of Swale was never reputed a boundary during his time that he knowes or heard of to divide the same or any of them. /

5: To the fifth interrogatory saith that he knowes Harkaside and West Grinton and the moores and commons thereunto belonging that the same are not divided by any boundary but have been allways accounted during his time one and the same place knowes Grinton / or East Grinton and the moores thereof saith that there is a boundary that divides the same from Harkaside and West Grinton and that the same boundary divides as well the inclosed grounds as the moores and commons belonging to the said places saith he knowes the / highway leading from East Grinton to West Grinton or Harkaside and the gate in the said highway and the elmetree near adjoyning and that the inhabitants of East Grinton repair the said highway near to the elme tree and a little west of Cuddy Spout and saith / he hath knowne <of> the said places for above sixty yeares and that the place in the <said> highway to which the inhabitants of East Grinton repaires is a part of the said boundary and that the said boundary betwixt East Grinton and Harkaside or West Grinton begins at the / river of Swale and crosses the Holme into the road betwixt Cuddy Spout and the elme tree to where Grinton men repaires the said way and soe up the bounder ridge to Benson cow house and then to the east corner of Ox Intack from thence crosse Hughes Ditch to the White / stone in the How and from thence to the Gray Stone on the Greets and soe as farr as the heaven water deales which divides Harkaside Moores from Wensleydale and so along westwards as heavenwater deales to Gibbon hill and to Gibbon hill standard and from thence to / Browney gill head which divides Harkaside Moores from Whitaside and soe down Browney gill to Scabby Yate and soe to the river Swale. /

6: To the sixth interrogatory saith he knowes the lead mines now wrought by the defendant Marriott called How Groves knowes alsoe the lead mines lately opened and claimed by the Lord Wharton near thereto and saith that the lead mines or workes soe opened by the Lord / Wharton lye within the moores and commons of Harkaside and not within the moores of East Grinton and that the said workes are about fourscore yards distant from each other and that the boundary dividing the said moores lyes between the said workes saith that the said / lead mines were opened by the Lord Wharton about two yeares agoe and that the boundarys above sett forth was ridd by the said Lord Whartons agents and a great many neighbours (whereof this deponent was one) the day before the opening of the said workes / and beleives and hath heard that publick notice was given of the riding thereof and this deponent saith that the said Lord Wharton and his tennants and other the inhabitants of Harkaside and West Grinton att the time of riding <the said boundary> and the said lords ancestors and the / tennants and other the inhabitants there during all this deponents remembrance have been in possession of the moores of Harkaside and West Grinton aforesaid comprized and included within the said boundarys soe as aforesaid described by this deponent by enjoying / the eatage of the herbage there with their horses cattle and sheep getting of ling and turves and digging of stone as they had occasion and saith that the said Lord Wharton and his ancestors has for thirty yeares and upwards wrought for and gott [lead] considerably / att Old Gang within the mannor of Helaugh on the north side the river of Swale and beleives they might lawfully have digged and gott lead and coales upon any part of the said moores within the said boundary if they had pleased saith that about twenty / eight yeares agoe one Mathew Mires of East Grinton since dead was imployed by Richard Harker then alsoe of East Grinton and also dead to pull ling for the said Richard Harker upon Harkaside Moor and that the inhabitants of Harkaside hearing /

[membrane v *verso*]

thereof went and took the said ling away and gave it to this deponent to thatch his house with and saith that some of the East Grinton men has had turves taken from them by the Harkaside men about three or four yeares since which they had graved upon / Harkaside Moor. /

7: To the seaventh interrogatory saith that Harkaside and West Grinton and the moores thereof have been during all his time reputed part and parcell of the mannor of Helaugh and that the same (save some part of West Grinton) lyes within the constablery of Reeth / saith that there are lands lying in Harkaside which are called my lords owne lands farmed by the Hirds att six pounds thirteen shillings and four pence per annum saith that the owners of Swalehall lands and part of Ralph Wensley lands Mr Hartley Ralph / Addison Christopher Raper John Hird and Anthony Hird are freeholders and appear att Helaugh court saith that there are coppyhold lands in Harkaside both old land and new land the tennants whereof appears likewise att the Lord Whartons court att Helaugh /

10: To the tenth interrogatory saith that the coppyhold lands in Harkaside att the death of the tennants descends to all the sons alike according to the custome within the mannor of Helaugh and that there is no such custome within the pretended mannor of Grinton /

13: To the thirteenth interrogatory this deponent saith that he knowes the pinfold att Reeth and that the inhabitants of Harkaside usually impound their cattle there. /

Alice Wife of Thomas Milner[263] of Harkaside aged fifty yeares sworne and examined /

6: To the sixth interrogatory saith that above three yeares agoe John Hird of Harkaside deceased being then about seaventy yeares old told this deponent that when he was a lad and went to the freeschool he knew two sheep taken up as strays or waifes / upon a place called Carrgate head[264] on Harkaside common and that they were carryed to the then Lord Whartons bayliffe as strayes. /

14: To the fourteenth interrogatory saith she hath alsoe been told by the said John Hird that James Galloway of Helaugh deceased used to carry his ewes onto Harkaside Moor to wean the lambs from them and this deponent saith she hath seen the goods both / of Jeoffrey Loansdale and John Loansdale of Helaugh feeding on Harkaside common and alsoe Francis Whip of Reeth hath had his goods going there. /

James Close[265] of Whitaside in Swaledale yeoman aged fifty nine yeares or thereabouts sworne and examined. /

1: To the first interrogatory saith he knowes the present Lord Wharton and the defendant Marriott and knew the old Lord Wharton who dyed about eleaven yeares agoe. /

2: To the second interrogatory saith he knowes the dale called Swaledale saith that the same lyes within the honor or liberty of Richmond and Richmondshire and in the wapentacks of Gilling West and Hang West that the extent of the said dale is from Stollerston Stile in the east / to Hollow Mill Crosse on the west and from the height where heaven water deales adjoyning upon Wensleydale to where heaven water deales adjoyning upon Arkingarthdale and that the Lord Wharton and his ancestors have been during all the time of this deponents / remembrance reputed cheife lords of Swaledale and none other that ever he heard of that there are courts leet once a year att Helaugh and Mukarr within the said dale for the said Lord Wharton and noe other courts kept there that he knowes of and that the said / Lord Wharton and his ancestors have all along during his time had and enjoyed the royalties of hunting fishing fowling within the said dale and appoints game keepers there and knowes that there have been acknowledgements paid to the said lords bayliffe for / fishing there. /

4: To the fourteenth[266] interrogatory saith he knowes the vill and constablery of Reeth and that the same lys within the mannor of Helaugh saith that the lordship of Swaledale mannor of Helaugh and constablery of Reeth extend and lye as well on the south side as / northside of the river of Swale and that the <said> river of Swale was never reputed a boundary during all his time that ever he heard of to limitt and divide the same or any of them. /

5: To the fifth interrogatory saith that he knowes Harkaside and West Grinton and the moores and commons thereunto belonging that the same are not divided by any boundary but have all along been reputed to be one and the same place knowes Grinton or East / Grinton and the moores thereof saith that there is a boundary to divide the same from Harkaside and West Grinton and that the said boundary divides as well the moores as the inclosed grounds belonging to both the said places knowes

263 Alice wife of Thomas Milner of Harkaside swore an affidavit on 17 May 1706 (E207/112/1).

264 In her affidavit, Alice Milner refers to 'Carlgate head'.

265 In May 1708 James Close, aged 59, gave evidence for Lord Wharton at the trial in London (LI Misc 536, Ward MSS, notebooks vol. 6, f. 117r).

266 'Fourteenth' written in error for 'fourth'.

the highway leading from / East Grinton to Harkaside and West Grinton and the gate that usually hangs in the said highway and the elme tree near thereto saith that the East Grinton inhabitants repaires the said highway to near the elmtree that the boundary dividing East Grinton / from Harkaside and West Grinton begins att the river Swale crosses the Holme to the said highway near the elmtree and soe up the ridge to Benson cowhouse and from thence to the east corner of Ox Intack wall from thence crosses Hughes Ditch to the White Stone / in the How and from thence to the Gray Stone in the Greets and soe to where heaven water deales that divides Harkaside Moor from Wensleydale and soe along as heaven water deales Westwards to Gibbon hill and from thence to the top of Browney gill where begins / the boundary betwixt Harkaside and Whitaside and soe downe Browney gill to Scabby Yate and soe to the river Swale and this deponent saith that about forty seaven or forty eight yeares agoe this deponents grandfather being then about seaventy yeares of / age told him of and shewd him the said severall boundarys above by this deponent described. /

6: To the sixth interrogatory saith that he knowes the lead mines or workes now wrought by the defendant Marriott called How Groves saith alsoe that he knowes the lead mines or workes lately opened and claimed by the Lord Wharton near adjoyning thereto saith that / the same lye within the moores and commons of Harkaside or West Grinton and not within the moores and commons of East Grinton that the said mines or workes ly about one hundred yards from each other and about an equall distance from the said boundary / the same lying betwixt them that Mr Marriotts workes lys nearer the said boundary then the Lord Whartons that the Lord Wharton opened the said workes about two yeares agoe and saith that the said boundary was ridd by the said Lord Whartons agents and / severall neighbours (whereof this deponent was one) before the opening of the said workes and this deponent saith that the said Lord Wharton and his tennants and other inhabitants of Harkaside and West Grinton att the time of riding the said boundarys and the / said lords ancestors and their tennants and other the inhabitants there during <all> this deponents remembrance have been in possession of the moores of Harkaside and West Grinton aforesaid comprised or included within the said boundarys so as aforesaid described by / this deponent by enjoying the eatage of the herbage there with their horses cattle and sheep and getting of ling peates and turves and digging of stone there as they had occasion and saith that Thomas Summers and James Alderson about four yeares agoe were / imployed by Mr Hartley a steward or agent to the Lord Wharton to digg and sink two shafts or pitts for lead oar in the inclosed lands att Harkaside called my Lord Whartons owne lands and leased to one John Hird there but then in William Arundells possession / and they did sink the same accordingly and this deponent saith that he hath seen the said pitts but knowes not whether they gott any oare there or not and this deponent further saith that about five or six yeares agoe he was sent for by Mr West the present viccar of / Grinton to digg peates for him (having been imployed by him severall times <before> to digg peates upon Grinton moor) this deponent then told the said Mr West that the {said} place upon Grinton moor where <he> formerly digged for him was inconvenient and spoiled his worke tooles / and then told him he would digg peates for him on Harkaside Moor Mr West then replyed he could not gett peates there without leave from the Harkaside men because he had noe right there this deponent then answered he was not soe ill loved amongst his parishoners but / that he might gett leave of them for asking and soe Mr West bid this deponent go home and he would send for him again after he had gott leave soe to do and immediately after he sent again for this deponent and told him he had gott <leave from> Ralph Wensley and George Miller two / Harkaside men to gett peates upon Harkaside Moor and desired him to sett on and gett peats for him there which he did and saith that about twenty eight yeares agoe one Richard Harker an inhabitant in East Grinton and since dead having pulld ling upon Harkaside / moor this deponent see some Harkaside men lead the said ling away and gave it to James Hird to lay upon his house /

7: To the seaventh interrogatory saith that Harkaside and the most part of West Grinton and the moores thereof lys within the mannor of Helaugh and constablery of Reeth. /

10: To the tenth interrogatory saith that the coppyhold lands in Harkaside and West Grinton descends to all the sons alike at the death of the tennants according to the custome of the said mannor of Helaugh and that there is noe such custome within the pretended / mannor of Grinton. /

13. To the thirteenth interrogatory saith he knowes the pinfold belonging to the township of Reeth that the same stands att Reeth towne end and that the tresspassing cattle belonging to Harkaside are usually impounded therein. /

14: To the fourteenth interrogatory saith that he hath pulld ling upon Harkaside Moor for one James Loansdale of Helaugh and that the same was led to Helaugh for the said James Loansdales use without interrupcion and that John Loansdale of Helaugh / about thirty yeares agoe gott stones upon Harkaside Moor to build a cowhouse withall att Helaugh and that James Loansdale alsoe of Helaugh gott stones upon Harkaside Moor to build with att or near Helaugh towne and saith that he hath seen both the / said John Loansdale and James Loansdale drive cattle from Helaugh downe a laine to Harkaside Moor and saw the same cattle left there to depasture which same is called Nout herd Loning and hath heard his foreelders say that the said lane was soe / called because Helaugh men used to drive their cattle that way to Harkaside common to pasture there. /

Thomas Harrison[267] of Harkaside in Swaledale miner aged sixty eight yeares or thereabouts sworne and examined. /

1: To the first interrogatory saith he knowes the complainants and defendant and did know Philip late Lord Wharton deceased who dyed about eleven yeares agoe. /

2: To the second interrogatory saith the complainant the Lord Wharton and his ancestors have all this deponents time been reputed cheife lords of Swaledale saith there are but two courts have been kept in all Swaledale during his time <and> both for the Lord Wharton thone for / the mannor of Helaugh thother for the mannor of Mewkarr saith that the said Lord Wharton and his ancestors and noe other that this deponent {heard} <knowes> of have all his time claimed the royalties of hunting fishing fowling and the like within the said dale or / lordship of Swaledale and appointed and kept gamekeepers for that purpose who sufferd none to fish or fowle without the said lords license. /

3 & 4: To the third and fourth interrogatorys saith that he knowes and for all the time of his remembrance hath knowne the towne and constablery of Reeth and the mannor of Helaugh saith all the said towne and constablery lyes within the mannor of Helaugh and / that both the said mannor and constablery extend themselves south on the river Swale as farr as heaven water deales between Wensleydale and Swaledale and northwards of the said river as farr as heavenwater deales betwixt Arkingarthdale and Swaledale saith the / river Swale in all his time was never that he heard of reputed a boundary to the said mannor of Helaugh or constablery of Reeth or either of them. /

5: To the fifth interrogatory saith that he hath knowne Harkaside and West Grinton and the moores and commons thereto belonging even from his childhood saith

267 Thomas Harrison of Harkaside, aged 68, swore an affidavit on 18 October 1707 (E207/115/1).

they are not nor have been in his time divided but accounted one and the same place and in like manner hath knowne East Grinton and / itts moores saith there are antient boundaries which divide the moores commons and inclosures of West Grinton and Harkaside from the moores <commons> and inclosures of Grinton or East Grinton saith the said boundarys are as antient as this deponent can remember and hath heard antient / men whilst he was young talk of and describe the same saith the said boundarys beginn att the river Swale and goe up part of the Holme to the highway and crosses the highway (att the place to where East Grinton repaires that highway to) betwixt the elmtree and Cuddy / Spout soe go's along the great ridge up to Benson cowhouse from thence to the east nook of the Ox Intack from thence crosses Hughes Ditch on the moor in a direct line to the White Stone in the How and soe on in a direct line from the White Stone in the How to the Gray Stone / in the Greets and soe up to where heaven water deales saith that the boundary betwixt Wensleydale and the said West Grinton and Harkaside Moores gos on from thence as heaven water deales to the top of Gibbon hill and from thence to Browney gill head and saith the / boundary betwixt Whitaside and Harkaside goes downe from Browney gill head to Scabby Yate and soe to the river Swale. /

6: To the sixth interrogatory saith he knowes the lead mines now wrought by the defendant called Howgroves and alsoe those lately opened by the Lord Wharton adjoyning thereto saith the Lord Whartons workes lye within the boundarys of the mannor and constablery of / West Grinton and Harkaside and not within the boundarys of the moores or commons of East Grinton saith the said severall workes lye about fourscore or one hundred yards distant from each other the said boundary running betwixt them saith the Lord Whartons said / workes were not opened till after the riding <of> the boundary about two yeares agoe and this deponent saith that the said Lord Wharton and his ancestors and their tennants and other the inhabitants of Harkaside and West Grinton for all the time this deponent can / remember have been in possession of the moores of Harkaside and West Grinton aforesaid included within the boundarys aforesaid by enjoying the eatage of the herbage of those moores and commons pulling of ling graving of turves getting of stones and the lords / tennants and coppyholders and most part of the freeholders of West Grinton and Harkaside appearing att Helaugh mannor court and paying rents to the lord thereof and saith the Lord Wharton and his ancestors have for many yeares past gott great / quantitys of lead at the Old Gang and other workes in that part of Helaugh mannor lying on the northside of the river Swale and further saith that the present Lord Whartons steward about four or five yeares agoe caused a shaft to be sunk in the said lords owne / inclosed grounds in Harkaside which is leased to John Hird in order to find lead but whether the miners found any lead {oar} or not this deponent knowes not saith he beleives the said Lord Wharton and his ancestors might lawfully have digged for lead on any part of / Harkaside and West Grinton moores if they had pleased saith that he hath knowne East Grinton men have <sometimes> pulld ling and graved turfes on West Grinton and Harkaside Moore but the West Grinton and Harkaside men never that this deponent heard of sufferd them / to carry such ling or turves away but either burnt the same or ledd it away to Harkaside or West Grinton themselves. /

7: To the seventh interrogatory saith that the most or greatest part of West Grinton and Harkaside have been all this deponents time reputed parcell of the mannor of Helaugh and constablery of Reeth but never heard that any part of Harkaside or West Grinton were reputed / part or parcell of the <said> pretended mannor of Grinton or East Grinton. /

8: To the eigth interrogatory saith he knowes severall parcells of ground called the intacks adjoyning on Harkaside Moor and which this deponent verily beleives lye within the mannor of Helaugh in the possessions of Anthony Hird Ralph Addison and Ralph Wensley for which this / deponent hath heard the tennants pay certain rents to the Lord Wharton but this deponent cannot perticularise such rents. /

10: To the tenth interrogatory saith that the coppyhold or customary lands in Harkaside and West Grinton upon death of all ancestors (having two or more sons) descends to all the sons alike according to the custome of the mannor of Helaugh. /

13: To the thirteenth interrogatory saith the Harkaside and West Grinton men have all this deponents time impounded their goods in Reeth pinfold and noe other that this deponent ever heard of. /

14: To the fourteenth interrogatory saith that {about} fifty yeares agoe one Mathew Metcalfe of Reeth deceased who was this deponents uncle showed this deponent a place in Ralph Wensleys inclosed grounds in Harkaside leading to Harkaside Moor which he said / had formerly been an antient way for cart and carriage for Reeth men to load the turfes peats and ling which they used to gett on a place called Reeth peat common on Harkaside Moor saith that the said way hath been used as a way in this deponents time / and this deponent saith that about thirty two yeares agoe this deponent lead from Harkaside Moor to Reeth three of four score threaves of ling for James Hutchinson of Reeth which the said James Hutchinson had gott on Harkaside Moor and noe body questioned this deponent for the / same saith he knowes that the said James Hutchinson had lands att Reeth but knowes not of any he had in Harkaside or West Grinton. /

11 & 15: To the eleaventh and fifteenth interrogatorys saith that the inhabitants of Harkaside have all his time paid their proporcion of the modus for the tythes of the hay and corne and alsoe their land tax together with Reeth and alsoe kept watch and ward att / Reeth and this deponent saith he was constable of Reeth about fifteen or sixteen yeares agoe and warnd the inhabitants of Harkaside to watch and ward att Reeth which duty they performed accordingly. /

[membrane vi *recto*][268]

John Bland of Whitaside in Swaledale aged sixty five yeares and upwards sworne and examined /

4: To the fourth interrogatory saith he knowes the vill and constablery of Reeth and beleives that the same lye within the mannor of Helaugh and that the mannor of Helaugh and constablery of Reeth lyes on both the north and south sides / of the river of Swale and that the said river was never reputed a boundary to divide the lordship of Swaledale mannor of Helaugh and constablery of Reeth or any of them during all his time that ever he heard of. /

5: To the fifth interrogatory saith that he knowes Harkaside and West Grinton and the moores and commons thereunto belonging that the same were allways accounted one and the same place saith he knowes Grinton and East Grinton and the moores and commons belonging thereto and that there / is a boundary to divide the same from Harkaside and West Grinton which divides as well the inclosed grounds as the moores betwixt them saith he knowes the highway leading from East Grinton to Harkaside or West Grinton and the gate in the said road and the / elmtree standing near it and that the Grinton men repaires the said highway near to the said gate and saith he hath knowne all the above menciond places for fifty yeares and saith that

[268] Parchment, 650mm x 720mm.

the said boundary which divides East Grinton from Harkaside and West Grinton / begins att the river Swale and crosses the Holme into the road betwixt the elmtree and Cuddy Spout to where the inhabitants of East Grinton repaires and soe up the ridge to Benson cowhouse and then to the east corner of Ox Intack crosses Hughes Ditch to the White / stone in the How and from thence to the Gray Stone in the Greets and so to where heavenwater deales adjoyning upon Wensleydale and soe as heaven water deales to Gibbon hill and from thence to the top of Browney gill where begins the boundary betwixt / Harkaside Moor and Whitaside and downe Browney gill to Scabby Yate and soe to the river Swale. /

John Simpson of Whitaside in Swaledale husbandman aged seaventy yeares or thereabouts sworne and examined. /

1: To the first interrogatory saith he knowes the present Lord Wharton and did know Philip Lord Wharton. /

4: To the fourth interrogatory saith he knowes the towne and constablery of Reeth saith it lyes within the mannor of Helaugh and the said mannor and constablery extends to the south side of the river of Swale as well <as> on the northside as farr as heaven water deales on each / side of Swaledale never heard that the river Swale was a boundary or reputed boundary to limitt or divide the lordship of Swaledale mannor of Helaugh and towne and constablery of Reeth or any of them. /

5: To the fifth interrogatory saith he knowes Harkaside and West Grinton and the moores and commons belonging the same saith they have all his time been accounted as one place and undivided knowes Grinton or East Grinton and the moores thereof saith that he knowes / boundaryes which divide East Grinton from West Grinton and Harkaside and hath knowne the same for att least threescore yeares and saith the boundary divides not onely the moores and commons but the inclosed grounds of each saith he hath knowne the East Grinton / men repair the highway leading bewixt East Grinton to West Grinton almost to the gate and elmtree which grows near the gate saith the place to which the East Grinton men repair that way is one of the boundermarkes betwixt East Grinton and West Grinton and / Harkaside saith the said boundary begins att the river Swale crossing part of the Holme and crossing the <said> highway att the place whither East Grinton repaires that way then gos on betwixt the gate and Cuddy Spout by the great ridge to Benson cowhouse thence to the / east corner of the Ox Intack from thence in a streight line crosse Hughes Ditch to the White Stone in the How from thence <streight> to the Gray Stone in the Greets and beyond that to where heaven water deales and from thence to the boundary betwixt Harkaside and / West Grinton moores and Wensleydale run westwards as heaven water deales to Gibbon hill and soe to Browney gill head saith the said Browney gill from the head thereof downe to Scabby Yate and soe to the river Swale is the boundary {be}/ betwixt Harkaside and Whitaside saith he was att the riding of the said boundary with a great many others about two yeares agoe but was not att nor heard or knew of any riding thereof before saith that he hath heard antient men say that / the lands within the said boundary was all within the Lord Whartons liberty. /

George Wharton[269] of Bankheads near Spring End in Swaledale aged forty yeares and upwards sworne and examined. /

1: To the first interrogatory saith he knowes the complainant the Lord Wharton but knowes not Mr Marriott nor knew the old Lord Wharton. /

[269] In May 1708 George Wharton, aged 40, gave evidence for Lord Wharton at the trial in London (LI Misc 536, Ward MSS, notebooks vol. 6, f. 117r).

2: To the second interrogatory saith he knowes the dale called Swaledale that the same lys within the wapentacks of Gilling West and Hang West saith that the extent of the said dale is from Stollerton Stile in the east to Hollow Mill Crosse on the west and from Arkingarthdale on the north / to Wensleydale on the south that the Lord Wharton and his ancestors have been during all his time reputed cheife lords of Swaledale that he knowes two courts kept in Swaledale for the Lord Wharton (vizt.) att Helaugh and Mewkarr and that there is noe other court kept within the said dale / that he knowes or ever heard of saith that the said Lord Wharton and his ancestors have during all his time claimed and enjoyed the royalties of hawking hunting fishing and fowling within the said dale by appointing game keepers within the same and hindring people to hunt fish or fowle / <there> without lycense from him or them. /

4: To the fourth interrogatory saith he knowes the vill and constablery of Reeth and that the same lyes within the mannor of Helaugh and that the lordship of Swaledale mannor of Helaugh and constablery of Reeth lyes on both north side and south side of the river of Swale and that the / river of Swale was never reputed a boundary during all his time to divide the same or any of them that ever he heard of. /

7: To the seaventh interrogatory saith that Harkaside and West Grinton and the moores thereof have all along during his time been reputed part and parcell of the mannor of Helaugh and constablery of Reeth and was never reputed part and parcell of the pretended mannor of Grinton that he / hath heard of saith he knowes the Lord Whartons owne lands lying att Harkaside farmed by the Hirds but does not know the yearly rent thereof saith also that there are severall freehold tennants in Harkaside and West Grinton that appear att the Lord Whartons court att Helaugh / but knowes not what free rents they pay the said lord saith that Ralph Wensley Christopher Raper William Arundell and his three brothers and Mr Hartley are coppyholders in Harkaside and doe appear att Helaugh court and did suit and service there and that the said coppyholders hold / their land by coppy of court roll from the mannor court att Helaugh. /

8: To the eigth interrogatory saith that he knowes severall parcells of land called intacks lying att Harkaside and adjoyning upon the moor there that the same have been inclosed before the time of this deponents remembrance and beleives the ocupyers thereof payes some rent to the Lord / Wharton for the same but what the severall rents are this deponent cant sett forth. /

10: To the tenth interrogatory saith that when any coppyhold tennants in Harkaside and West Grinton dye one third part of their lands goes to the widdow during the time she continues widdow of the said tennant and noe longer and the other two parts descends to all the sons alike / according to the custome of the mannor of Helaugh and that there is noe such custome within the pretended mannor of Grinton /

14: To the fourteenth interrogatory saith that he hath heard his father Richard Wharton an antient man (since dead) above twenty yeares since and severall other old men say that the inhabitants of Reeth Helaugh and Parke hall had a right to put their cattle on Harkaside / moor and pull ling digg peates grave turfes and gett stones there saith that theres a laine leading from the towne of Helaugh to Harkaside Moor called Noutherd Loneing and that they told him the said laine gott that name from the Helaugh men driving their cattle to Harkaside / common and saith that his father alsoe told him that there was an old way which lead from Reeth to Harkaside and soe to a place upon Harkaside moor where Reeth men used to gett peates and turves called Reeth Peat Moore and that the said old way begun att the gate near / Reeth bridge and soe crosse the Swale to Harkside and soe to the moore and that the inhabitants drove their cattle that way to Harkaside Moor to depasture and alsoe lead their ling peats and turves that way but saith that the said way was never made use of during all / his time. /

22: To the twenty second and last interrogatory saith that he lives att Spring End on the south side of the river Swale and that there are coppyhold lands there for which the tennants pay to Lord Wharton six pounds one shilling per annum antient coppyhold rent and that there / are freehold lands there for which the tennants payes five pounds nineteen shillings as a free rent to the Queen and saith that all the freeholders there except one man appear att the Lord Whartons court att Helaugh for both the freehold and coppyhold lands there / and saith that there is one Mr Walbanke that has coppyhold lands att Whitaside and that he or his tennants appear att Helaugh court and doe suit and service there for the same. /

George Metcalfe of Reeth in Swaledale miner aged sixty eight yeares and upwards sworne and examined. /

2: To the second interrogatory saith he knowes Swaledale and that it lys within the honor of Richmond and Richmondshire saith there are two court leets kept in the said dale the one att Helaugh th'other att Mewkarr both for the Lord Wharton but never heard of any other saith he / hath knowne the said dale for fifty yeares and upwards and during all his knowledge thereof the Lord Wharton and his ancestors have claimed and enjoyed the royalties of hunting fishing and fowling throughout the whole dale and <no> other that ever this deponent heard of and saith / that none that this deponent ever heard of presumed to hunt fish or fowle within the said dale without license from the Lord Wharton or his ancestors and they appointed game keepers (to witt) Ralph Freer and others to look after the preservacion of the said royalties /

4: To the fourth interrogatory saith he knowes the towne and constablery of Reeth and that the same lys within the mannor of Helaugh and lordship of Swaledale saith the said constablery extends not onely on the north side of the river Swale but alsoe on the south side of / the said river as farr as heavenwater deales saith the river Swale in all this deponents time was never that this deponent heard of reputed a boundary {for} <to> the said mannor of Helaugh or constablery of Reeth or either of them. /

5: To the fifth interrogatory saith he knowes Harkaside and West Grinton and the moores and commons thereto belonging and alsoe Grinton or East Grinton and the moores thereof saith the said West Grinton and Harkaside hath all the time of this deponents knowledge thereof / (viz[t].) for fifty yeares and upwards been accounted one and the same place and divided from Grinton or East Grinton by a certain boundary hereafter menciond saith the said boundary divides both the moores and inclosures of West Grinton and Harkaside from the moores / and inclosures of East Grinton and begins att the river Swale and goes up to and crosses the way betwixt Cuddy Spout and the elmtree and soe up the ridge to Jinny Wrenn cowhouse and up by the Ox Intack wall to the east nook or corner of the said wall and from thence cross / Hughes Ditch to the White Stone or gray stone in the How and from thence to the Gray Stone in the Greets and soe to where heavenwater deales From thence the boundary betwixt West Grinton and Harkaside Moores and Wensleydale proceeds westwards to Gibbon hill / and soe on as heavenwater deales to the head of Browney gill and saith that West Grinton and Harkaside are divided from Whitaside by the said Browney gill from the head thereof downe to Scabby Yate and soe to the river Swale saith he the better knowes the said / boundarys for that when this deponent was fifteen or sixteen yeares old his father had both beastes and sheep which used to depasture on West Grinton and Harkaside Moores and commons and his said father shewed him severall times the said boundaryes. /

12: To the twelfth interrogatory saith he knowes the mill in West Grinton or Harkaside belonging to the owners of Swaledale saith his said father who dyed about twenty yeares agoe told this deponent about forty yeares agoe the said mill was built by the owners of / Swalehall but by leave of the Lord Wharton and upon condicions one whereof was that the Reeth men when they could not have their corne grinded att Reeth mill might have liberty to grind the same att Swalehall mill for halfe mulcture and this deponent saith he hath / knowne Reeth men grind corne att Swalehall mill for half mulcture and perticularly one William Hutchinson by name. /

13: To the thirteenth interrogatory saith he hath knowne Reeth pinfold most of his time and saith the West Grinton and Harkaside men have allways used the same for impounding their goods and none other pinfold. /

14: To the fourteenth interrogatory saith there is visible marks to this day of a way for cart and carriages which runs through Ralph Wensleys grounds and other mens grounds in Harkaside and is now inclosed in those grounds and hath not been open in his time but saith the generall repute and / tradition is that the same was a way which the Reeth men used to fetch ling and turves with their carts from a place on Harkaside Moor called Reeth Peat Moore to Reeth. /

16: To the sixteenth interrogatory saith that about forty yeares agoe one John Barningham deceased wrought some part of the lead workes att Harkaside under Sir Solomon Swale and the said John Barnigham severall times told this deponent that he was att an agreement with Sir Solomon / about working the same and was to have an article or lease thereof but that when he required Sir Solomon to execute such article or lease Sir Solomon told him he could not make him any lease thereof because he did not know how soon the Lord Wharton might come and claime / his right there. /

Christopher Barningham[270] of Arkingarthdale in the county of Yorke miner aged fifty six yeares or thereabouts sworne and examined /

2: To the second interrogatory saith he knowes Swaledale that the reputed extent or boundary thereof is from Stollerston Stile on the east to Hollow Mill Crosse on the west and from Arkingarthdale on the north to where heavenwater deales adjoyning upon Wensleydale on the south / and that the Lord Wharton and his ancestors have been reputed during all his time cheif lords thereof saith that he knowes the Lord Wharton usually holds courts att Helaugh and Mukarr within the said dale once a year and never heard of any other courts kept within the said dale / saith that the Lord Wharton and his ancestors during all his time have enjoyed the royalties of hunting fowling and fishing there and appoints game keepers there to preserve the game. /

4: To the fourth interrogatory saith that he knowes the vill and constablery of Reeth and that the same is reputed to lye within the mannor of Helaugh saith that the mannor of Helaugh and constablery of Reeth lys as well upon the south side as north side of the river of Swale and extends / southward as farr as to where heavenwater deales adjoyning upon Wensleydale and saith during all his time the said river of Swale was never reputed a boundary to divide the lordship of Swaledale mannor of Helaugh and vill and constablery of Reeth or any of them that ever / he heard of /

16: To the sixteenth interrogatory saith that he very well remembers that about thirty eight yeares agoe John Barningham this deponents father was imployed by Sir Soloman Swale in working lead mines att Harka upon Harksaide moor and att Grinton How

[270] In May 1708 Christopher Barningham, aged 56, gave evidence for Lord Wharton at the trial in London (LI Misc 536, Ward MSS, notebooks vol. 6, f. 117r).

and att / Birks on Whitaside moor and wrought the said mines for some time att some certain agreement betwixt them but gott little and then he took all the said places of the said Sir Solomon Swale to make tryalls for lead and was to give the said <Sir> Solomon every eight pigg of lead for his / share and this deponent saith that whilst his said father was working att Grinton How and Harka hill one James Fryar an agent for the old Lord Wharton came to this deponent and bid him tell his father to desist working there and one Robert Barker since dead and then a steward / again for the old Lord Wharton hearing as this deponent beleives of the said bargain or agreement that his father had made with the said <Sir> Solomon Swale came to this deponents fathers house in Arkingarthdale and in this deponents presence and hearing told this deponents father that / he had heard of this agreement with Sir Solomon Swale about working the lead mines on Harkaside Moor and bid him beware what he did for that his master the Lord Wharton claimed an interest upon those moores and commons and therefor he did then discharge him to proceed or / work any further upon those moores and thereupon this deponents father ordered this deponent to fetch away their work tooles from the said mines which he did and carryed the same to another liberty and this deponent saith that after the said Barker had discharged his said / father from working the said mines he applyed himselfe to the said Sir Solomon Swale and told him what the said Barker had said and that he had discharged him and in this deponents presence and hearing desired the said Sir Solomon Swale to give him bond or some other / security to indempnify him for soe doing but Sir Solomon then replyed he thought his word and promise was sufficient enough without writing and said he would give noe bond or security and soe his father wrought noe more there. /

Francis Hutchinson[271] of Reeth in Swaledale yeoman aged fifty four yeares sworne and examined. /

1: To the first interrogatory saith he knowes the complainant and hath seen the defendant Marriott did know Philip Lord Wharton deceased and saith he dyed about eleaven yeares agoe /

2: To the second interrogatory saith he knowes the dale called Swaledale saith it lys in the honor of Richmond and Richmond shire and in the wapentacks of Gilling West and Hang West saith the mannor of Helaugh and the towne of Reeth and its constablery lyes all in / the Wapentack of Gilling West saith the Lord Wharton and his ancestors are and have been reputed cheif lords of Swaledale saith the said lord and his ancestors usually have kept court leets for the mannors of Mewkarr and Helaugh once a year and noe other courts have / in this deponents time been kept in any part of the said dale saith the Lord Wharton and his ancestors as lords of Swaledale have claimed and enjoyed the royalties of hunting fishing and fowling throughout the said dale and hath heard and beleives appointed game keepers for / the same. /

4: To the fourth interrogatory saith that he knowes the vill and constablery of Reeth saith the same lys in the mannor of Helaugh and extends it selfe south on the river Swale as farr as heaven water deales between Wensleydale and West Grinton and Harkaside Moor /

[membrane vi *verso*]

and north of the said river as farr as heavenwater deales betwixt Reeth Moores and Arkingarthdale saith he never heard that the river Swale was ever accounted a

271 In May 1708 Francis Hutchinson, aged 50, gave evidence for Lord Wharton at the trial in London (LI Misc 536, Ward MSS, notebooks vol. 6, f. 117r).

boundary to limitt the said lordship of Swaledale or the mannor of Helaugh or the vill and constablery / of Reeth or any of them and this deponant hath knowne them for forty yeares past. /

5: To the fifth interrogatory saith that he knowes Harkaside and West Grinton and the moores and commons thereto belonging and never knew of any boundary that divides them but they have been all along accounted as one and the same towne or place saith / he knowes Grinton and East Grinton and its moores saith he hath of long time heard that there were and are boundaryes which divide the said East Grinton and its moores and inclosed grounds from West Grinton and Harkaside and the inclosed grounds and moores belonging / the same but did not know the perticular markes of the said boundaries before the same were rode about two yeares agoe. /

6: To the sixth interrogatory saith that he knowes the lead mines called the How Groves now or lately wrought by the defendant Mr Marriott and alsoe the lead workes lately opened by the complainant near adjoyning thereto saith the said lead workes soe opened by the / complainant are reputed to lye within the moores or commons of Harkaside and West Grinton saith the complainant and his ancestors and their tennants and other inhabitants in West Grinton and Harkaside att the time of and long before the said riding were in possession of the / moores and commons of Harkaside and West Grinton by eating the herbage of the said commons and moores and getting turves peats stone and ling there and saith he never heard that the tennants of East Grinton ever enjoyed the said moores or commons of Harkaside / and West Grinton without disturbance but hath heard that when ever they took upon them to get ling or turfes there the Harkaside and West Grinton men never suffered them to lead the same away. /

7: To the seaventh interrogatory saith that during all his time all West Grinton and Harkaside have been reputed to lye within the mannor of Helaugh and the greater part thereof in the constablery of Reeth the rest lying in Grinton constablery saith the complainant / Lord Wharton hath lands of his owne in Harkaside and there are alsoe coppyhold lands in Harkaside and West Grinton holden by coppy of court roll of the mannor of Helaugh and there are alsoe severall freehold lands there for which the tennants thereof appear and doe suit / and service att the court holden for the said mannor of Helaugh but the said severall lands this deponent is not enabled to particularise /

10: To the tenth interrogatory saith that when a coppyholder dyes seized of any coppyhold lands in Harkaside or West Grinton the same descend according to the custome of the mannor of Helaugh to all his sons alike but if he leaves a wife his widdow by the said custome hath a / third part thereof soe long as she continues his widdow and noe longer but never heard there was any such custome in the pretended mannor of Grinton. /

11: To the eleventh interrogatory saith that for near forty yeares past he hath <knowne> an annuall modus or prescript rent of seaven pounds two shillings and two pence for tyth corne and hay paid to the impropriator of the parish of Grinton for the township of Reeth and saith / the hamlett of Harkaside as part of the township of Reeth pays yearly twenty shillings and eight pence or near thereabout as their proporcion of the said modus. /

12: To the twelfth interrogatory saith the owners of Swalehall lands have all this deponents remembrance been called for their lands there att the courts held for the mannor of Helaugh and the tennants thereof have usually appeared for them or essoyns for them att those courts and this deponent is / the better enabled to depose herein for that he this deponent being an inhabitant of Reeth seldome omitts appearing att that court himselfe and hath observd the same and never heard that the owners of Swalehall lands ever appeard att any other courts or ever held any courts / themselves in Harkaside or West Grinton or ever had any freehold or coppyhold tennants there or any free or customary rents within the same. /

13: To the thirteenth interrogatory saith the inhabitants of Harkaside and West Grinton have all along made use of Reeth pinfold to impound goods in and paid their share for repairing the same and have noe pinfold att Harkaside or West Grinton. /

14: To the fourteenth interrogatory saith he hath knowne Arthur Hutchinson and Henry Machell both of Reeth about fourteen yeares agoe gett stones on Harkaside Moor and lead them to Reeth without any disturbance that this deponent heard of and saith he dos not beleive / they have any lands in Harkaside. /

15: To the fifteenth interrogatory saith the inhabitants of Harkaside in Reeth constablery are joyntly taxed and pay all assessments with the towne of Reeth except to the church and poor and formerly about twenty yeares agoe paid jointly with them to / the poor too and saith they keep watch and ward att Reeth in their turn or pay others for doing it for them. /

Henry Matchell[272] of Reeth in Swaledale aged fifty yeares and upwards sworne and examined. /

1: To the first interrogatory saith he knowes the Lord Wharton and Mr Marriott and knew the old Lord Wharton who dyed about eleven yeares agoe. /

2: To the second interrogatory saith he knowes Swaledale saith that all the north side of the dale lys within the wapentack of Gilling West and part of the south side alsoe (vizt.) Harkaside Crackpott Spring End Sattron and some coppyhold lands / belonging to one Mr Walbanke in Whitaside being about tenn acres or upwards and alsoe the rest of the lands in the said dale (except Grinton Whitaside and Summer Lodge) are alsoe in Gilling West and the last / menciond places in Hang West saith the reputed boundary of the said dale is from Stollerton Stile on the east to Hollow Mill Crosse on the west and from the height of the moor next Arkingarthdale to where heavenwater deales on the north to the / height of the moor where heaven water deales next to Wensleydale on the south. That the Lord Wharton and his ancestors during all his time have been reputed cheif lords of the said dale saith that the Lord Wharton keeps court leets att / Helaugh and Mukar within the said dale and never heard of any other courts kept there and that the Lord Wharton and his ancestors have during all his time claimed the royalties of hunting fishing and fowling within the same and knowes one / Edmond Harker who lives att Cracky pott hall is now a keeper under the Lord Wharton to look after the deer that comes within the dale and hath heard one Anthony Fryer is alsoe a game keeper there to hinder any people from hunting hawking / fowling and fishing there and hath heard and beleives that there are severall persons who pay acknowledgements to the lord or his bayliffe for liberty of fishing within the river of Swale. /

4: To the fourth interrogatory saith that he knowes the vill and constablery of Reeth and that the same lys within the mannor of Helaugh and that the lordship of Swaledale mannor of Helaugh and constablery of Reeth lyes as well on the south as on / the north side of the river of Swale and reaches southwards to where heaven water deales adjoyning to Wensleydale upon Harkaside Crackpott Sattron and Spring End moores and that there are some coppyhold lands in Whitaside on the south side of the / river

272 In May 1708 Harry Machell, aged 51, gave evidence for Lord Wharton at the trial in London (LI Misc 536, Ward MSS, notebooks vol. 6, f. 117r).

that belongs to the mannor of Helaugh and that the said river of Swale was never reputed a boundary that ever he heard of to divide the same or any of them. /

5: To the fifth interrogatory saith he knowes Harkaside and West Grinton and the moores and commons thereof and that there is noe boundary to divide the same but have been during all his time reputed one and the same place knowes alsoe / East Grinton and the moores thereof and that there is a boundary to divide the same from West Grinton and Harkaside and that the same boundary divides as well the inclosed grounds as the moores belonging to both the said places saith he knowes / the highway leading from East Grinton to West Grinton and the gate therein and the elmtree growing near thereto and that he hath knowne all the said places for twenty years and upwards and that the boundary dividing the said East Grinton from / West Grinton and Harkaside begins att the river Swale and crosses the Holme to the said highway betwixt Cuddy Spout and the elmtree and soe up the bounder ridge to Jenny Wrenn cowhouse and from thence to the east corner of Ox Intack and from thence / to the White Stone on the How and from thence to the Gray Stone on the Greets and from thence to the height of the moores where heavenwater deales adjoyning on Wensleydale and saith that his grandfather about twenty yeares agoe and who was then a / very old man and since dead told this deponent that the boundary aforesaid was the old boundary betwixt East Grinton and Harkaside and West Grinton and the moores and commons thereof /

6: To the sixth interrogatory saith that he knowes the lead mines now wrought by the defendant Marriott called the How Groves and alsoe the lead mines lately claimed and opened by the Lord Wharton near thereto and saith <he beleives> the said mines / lately claimed and opened by the Lord Wharton lye within the moores and commons of Harkaside and West Grinton and the boundary thereof and not within the moores and commons of East Grinton and the boundary thereof and that the said / mines lye about one hundred yards distance from each other and that the boundary goes up betwixt the said mines saith that the said lead mines were opened by the Lord Wharton about two yeares agoe and that the <said> boundary was rid / by the Lord Whartons agents and severall neighbours att Harkaside (whereof this deponent was one) before the opening the said workes and that publick notice of riding the said boundary was given att Reeth upon a markett day and saith that the Lord / Wharton and his freehold and customary tennants att Harkaside and West Grinton and other the inhabitants there were in the possession of the said moores and commons there by pulling ling getting peates turves and stone and graseing their / cattle there and saith he hath heard that about four yeares agoe Thomas Sommers and James Alderson took a bargain of one of Lord Whartons agents to sink for lead in the Lord Whartons owne lands att Harkaside farmed by one Hird for that / the place there seemed hopefull because some oar appeared above ground and the said Sommers and Alderson did work some small time there and this deponent hath seen the said workes but not proving well was left of as he beleives. /

7: To the seventh interrogatory saith that the lands att Harkaside and West Grinton hereafter menciond as is reputed and as he beleives lyes within the mannor of Helaugh and constablery of Reeth (viz[t].) Dikas lands belonging to Mr Hutchinson Christopher / Rapers lands part of Leonard Galloways lands John Hirds lands Anthony Hirds lands the lands belonging to Mr Hartley formerly Richard Closes part of Ralph Wensleys lands belonging to Robert Robinson and Luke Robinson William / Arundell and his three brothers lands lands now belonging to Mr Blackburne called Smithsons lands a close belonging to Mr Dickinson formerly one John Blaids and is the better enabled to depose herein and sett forth the same for that he has been constable / of Reeth and collected the constable assessments of and from the occupyers of the lands above sett forth and saith that the said lands att Harkaside and West Grinton lying within the said mannor of Helaugh and constablery of Reeth above sett forth is farr / the greater part of the said township or hamlett of Harkaside and West Grinton and saith that part of the aforesaid lands are coppyhold lands and att death or alienacion payes a fine to the Lord Wharton (viz[t].) for old land a twenty penny fine and for / new land a fifteen penny fine as coppyhold lands in other parts of the mannor of Helaugh doe and that the freehold tennants which are in Harkaside and which hold of the mannor of Helaugh appears att thelords court att Helaugh and payes / some free rent but knowes not what they pay. /

10: To the tenth interrogatory saith that all the coppyhold lands in Harkaside and West Grinton att the death of the tennants descends to all the sons alike according to the custome within the mannor of Helaugh and that there is noe such / custome within the reputed mannor of Grinton that ever he heard of. /

11: To the eleaveth interrogatory saith that he knowes that there is a modus of seaven pounds two shillings and two pence per annum paid by the township of Reeth for tyth of hay and corne to the impropriator of Grinton and that some of the inhabitants of / of[273] Harkaside as part of the township of Reeth payes one pound and eight pence part thereof (viz[t].) Richard Close now Mr Hartleys lands pays six shillings Christopher Rapers payes eight shillings and four pence Hirds lands three shillings eight pence Lord / Whartons lands in Hirds possession eight pence Ralph Addison three pence Anthony Hird eleaven pence Mr Dickenson tenn pence in all one pound and eight pence. /

13: To the thirteenth interrogatory saith that he knowes the pinfold att Reeth and that the inhabitants of Harkaside and West Grinton usually impound their cattle there or att Helaugh and that there is noe pinfold att Harkaside /

14: To the fourteenth interrogatory saith that his grandfather above twenty yeares agoe and who was then a very old man and since dead told this deponent that the Reeth men had liberty on Harkaside Moor to gett peats and what else they wanted as much as / upon their owne moor and saith that about tenn yeares agoe he this deponent cutt and wonne free stone upon Harkaside moor to build part and repair his dwellinghouse att Reeth with all and lead the same away and used it in building his said / house as aforesaid and had noe disturbance or was molested by any person for soe doing and remembers that about the same time he saw one Arthur Hutchinson of Reeth lead stones from Harkaside Moor to Reeth for his owne use and never heard of any / interrupcion that the said Hutchinson had for soe doing and alsoe hath heard that Roger Alderson of Reeth about seaven yeares agoe gott stones there and lead them to Reeth for his owne use without any interrupcion and saith that he hath heard severall / antient people say perticularly James Galloway now alive that theres an antient way leading from Helaugh to Harkaside Moor called Nout Herd Loneing and supposes it gott that name by the Helaugh men driving their cattle to Harkaside Moor to / depasture and saith he hath heard Francis Chalder an old man say (now alive) that the Harkaside men have a sheep rake upon the moore att Reeth. /

15: To the fifteenth interrogatory saith that Harkaside joynes and contributes in the land tax and constable sess to Reeth and saith that the inhabitants of Harkaside keeps watch and ward there when required /

273 Word repeated in the original.

James Hutchinson[274] of Reeth in Swaledale smith aged fifty nine yeares or thereabouts sworne and examined /

1: To the first interrogatory saith he knowes the complainants and the defendant Mr Marriott and did know Philip late Lord Wharton who dyed about eleven yeares agoe. /

2: To the second interrogatory saith he knowes Swaledale saith it lyes in the honour of Richmond and Richmondshire and part of it in the wapentack of Gilling West and part in the wapentake of Hang West saith the Lord Wharton and his ancestors have / all this deponents time been reputed cheif lords of Swaledale saith he never heard of any courts kept in the said dale saving for the Lord Wharton in his mannors of Helaugh and Mukarr and saith court leets are kept there usually once in the year saith / the said Lord Wharton and his ancestors have all along claimed the royalties of fishing hunting and fowling throughout the whole dale and appointed game keepers for the preservacion thereof /

4: To the fourth interrogatory saith he knowes the vill and constablery of Reeth saith the same lys within the mannor of Helaugh saith the lordship of Swaledale the mannor of Helaugh and the constablery of Reeth extend as farr as to where heaven-water / deales betwixt Wensleydale and Swaledale on the south side of the river Swale and to where heavenwater deales on the north side thereof betwixt Arkingarthdale and Swaledale saith the said river of Swale hath never that this deponent heard of / been accounted a boundary to limitt or divide the said dale mannor of Helaugh or constablery of Reeth or any of them. /

5: To the fifth interrogatory saith he knowes Harkaside and West Grinton and the moores and commons thereto belonging saith the same lye together and undivided by any boundarys and have been and are reputed as one and the same place saith he knowes / Grinton or East Grinton and the moores thereof saith he knowes the boundary which divides Grinton or East Grinton from Harkaside and West Grinton and hath knowne the said boundary for forty yeares past saith the said boundary divides {as well} the / inclosed lands as well as the moores and commons of East Grinton from the inclosed lands and moores of West Grinton and Harkaside saith he knowes the road or highway leading from East Grinton to West Grinton and Harkaside and the gate in the said way / and the elmtree growing near it saith the said boundary begins att the river Swale crosses the Holme to the said highway betwixt Cuddy Spout and the elmtree and soe up the ridge to Benson cowhouse or Jinny Wrenn cowhouse and from thence gos up by the / Ox Intack wall to the east corner thereof and soe over the moor in a direct line crosse Hughes Ditch to the White Stone in the How and from thence in a direct line to the Grey stone in the Greets and soe to the height of the moor where heaven water / deales saith the boundary of Harkaside and West Grinton proceeds from thence westwards as heaven water deales to Gibbon hill and from Gibbon hill as heaven water deales to the head of Browney gill saith <from thence> the said Browney gill divides Harkaside / and West Grinton from Whitaside downe Scabby Yate and soe to the river Swale saith the said boundarys soe as aforesaid described have been all this deponents time reputed by antient men the true boundarys betwixt the said places. /

6: To the sixth interrogatory saith he knowes the lead mines now wrought by the defendant Marriott called How groves and alsoe those workes lately opened by the complainant which lye near the said Mr Marriotts workes saith the said workes soe

274 In May 1708 James Hutchinson, aged 60, gave evidence for Lord Wharton at the trial in London (LI Misc 536, Ward MSS, notebooks vol. 6, f. 117r).

opend by / the Lord Wharton lye within the moores and commons of Harkaside and West Grinton and not within the moores and commons of East Grinton saith the said boundary runs betwixt them and the workes are about one hundred yards distant the / one from the other saith that the said Lord Whartons workes were not open'd till two or three days after the said riding of the said boundarys on the Lord Whartons behalfe saith the said riding was about two yeares agoe and publick /

[membrane vii *recto*][275]

notice given thereof before hand both in Reeth markitt on a markitt day and in the church yard of Grinton {church} on Sunday imediately after divine service saith the said boundaries were / then ridd in the manner this deponent hath described them that there were att least fifty men att the said riding who all agreed the same to be the right boundarys betwixt the aforesaid places and saith / that the complainant the Lord Wharton and his ancestors and their tennants of Harkaside and West Grinton and other the tennants freeholders and inhabitants there have been all this deponents remembrance / in the possession of the said moores of Harkaside and West Grinton and enjoyed the same by eating the herbage thereof getting ling peates turfes and stone thereon and saith that about three or four yeares agoe James Alderson / and Thomas Summers by order of or agreement with Mr Hartley agent for the complainant did dig and try for lead oar in a place of the complainants owne ground in Harkaside farmed by the Hirds saith that some East Grinton men / about three yeares agoe got turfes on Harkaside or West Grinton moors but the West Grinton men would not suffer them to lead 'em away. /

7 & 10: To the seaventh and tenth interrogatorys saith that Harkaside and West Grinton and the moores thereof have been all this deponents remembrance reputed part and parcell of the mannor of Helaugh and most part of it in the / constablery of Reeth saith that there are both lands of the Lord Whartons owne and alsoe diverse coppyhold and frechold lands in Harkaside and West Grinton which pay coppyhold rents and free rents to the Lord Wharton and / appear att his court att Healaugh but cannot particularise all the said lands and rents but saith that the coppyhold lands pays upon death of tennant or alienacion a twenty penny fine for such as is knowne by the name / of old lands and a fifteen penny fine for that called new as doe other the coppyhold lands in the rest of the mannor of Helaugh and this deponent saith that when a coppyholder in Harkaside or West Grinton dyes seised the / lands descends to all his sons alike but if he leaves a widdow a third part of the lands fall to her share as her third or dower whilst she continues his widdow and noe longer and upon her second marriage or death that / third part descends to all the sons alike saith there is a parcell of land of about tenn acres of ground lying near the river Swale att Whitaside belonging to one Walbanke holden of the mannor of Helaugh by coppy of court / roll which is subject to the said custome of discent as the other coppyhold lands in Helaugh mannor but saith he never heard of any such custome in the pretended mannor of Grinton /

11: To the eleventh interrogatory saith that there is an annuall modus or prescript rent for tyth hay and corne paid by the township of Reeth to which the hamlett of Harkaside contributes a part but how much this deponent / cannot sett forth. /

12: To the twelfth interrogatory saith that the owners of Swalehall have been all this deponents time called att the court held for the mannor of Helaugh as freehold

[275] Parchment, 570mm x 700mm.

tennants within the same and saith the occupyers or tennants thereof / have appeard att those courts for the same or sent their essoins but never <appeard at any other court that this deponent ever heard of saith he never {heard}> heard that the owners of Swalehall ever kept any court or courts in Harkaside or West grinton or have had any freehold or coppyhold tennants or / any free of customary rents within the same. /

13: To the thirteenth interrogatory saith that the inhabitants of Harkaside and West Grinton when they have occasion to impound cattle make use of Reeth pinfold and contributes towards repair of the same. /

14: To the fourteenth interrogatory saith that this deponent and his father before him and severall other coppyhold tennants of the Lord Wharton of the mannor of Helaugh inhabitants in Reeth have intercommoned with the / tennants of Harkaside and West Grinton by getting of ling and stone upon Reeth Peat Moor part of Harkaside common and leading the same to Reeth without interrupcion saith that this deponents father long since dead gott / turfes there and allways asserted his right to doe soe and this deponent and Arthur Hutchinson of Reeth not living have gott stones there without interrupcion altho none of them had or have any lands in Harkaside or West / Grinton saith there is a lane <or way> called Noutherd Loning leading from Helaugh to Harkaside for driving of cattle from Helaugh to Harkaside Moores and for fetching of fuell or such like from thence to Helaugh. /

15: To the fifteenth interrogatory saith that the tennants of Harkaside are joyned and contribute to and with the tennants of Reeth in all taxes and assessments saving the church and poor. /

Francis Chalder of Reeth in Swaledale aged sixty eight yeares and upwards sworne and examined. /

2: To the second interrogatory saith that he knowes the dale called Swaledale that the same lys within the honor of Richmond and Richmond shire and within the wapentacks of Gilling West and Hang West that the extent / or reputed boundary of the said dale is from Hollow Mill Crosse in the west to Stollerston Stile in the east and from Arkingarthdale on the north to Wensleydale on the south and that the present Lord Wharton and his /ancestors have during all his time been reputed cheif lords of the said dale and that the Lord Wharton keeps court leet and court baron every year att Helaugh for the mannor of Helaugh and keeps court att Mukarr for the / mannor of Mukarr and never heard that any other courts had been kept in the said dale saith that the present Lord Wharton and his ancestors have all along during his time claimed and enjoyed the royalties of hunting / fishing and fowling in and through the said dale saith that about fifty four yeares agoe he see John Kirton then bayliffe for the old Lord Wharton within the said dale and Leonard Hutchinson of Helaugh an / agent for the said Lord Wharton hunt a red deer upon Harkaside Moor and killd the said deer in the river of Swale for the lords use saith that about twenty two yeares agoe this deponent serving then as a / juror at Helaugh court one Richard Close then of Harkaside being alsoe a juror with him the said Richard Close was presented for fishing with a shakle net in the river of Swale adjoyning upon Harkaside / and saith he remembers very well that the jury then fined the said Richard Close tenn shillings for the said fishing. /

11: To the eleaventh interrogatory saith he knowes that the township of Reeth pays a modus or prescript rent for tyth of corn and hay to the impropriator of Grinton of seaven pounds two shillings two pence / per annum and saith that the inhabitants of Harkaside pays a part of the said modus with the inhabitants of Reeth but what part thereof this deponent cannot sett forth. /

12: To the twelfth interrogatory saith he knowes the mill at West Grinton and Harkaside belonging to the owners of Swale hall and hath heard his grandfather Mathew Chalder who was a very old man above ninety / yeares of age and since dead above fifty eight yeares agoe say and relate to this deponent that when the said mill was to be built the Lord Whartons ancestors alleadging the place where the said mill was to stand belongd / to the wast which he claimd there and therefor it was agreed by both the partyes that in consideracion of the mill being sett there the Lord Whartons miller att Reeth mill should have free liberty to grind severall quarters / of corne att Swalehall mill yearely mulcture free and this to be done when the said Reeth mill was frozen up or wanted water or was any way out of order and further saith that his said grandfather att that time alsoe / told this deponent that the Reeth men had a right upon Harkaside Moore to put their cattle horses and sheep to depasture there but had discontinued the same by reason of their sheep beasts and horses being stollen from them upon / the said moores in the time of the late troubles in King Charles the First time and that the said moores lay something farr of and shewd this deponent an old way which leads from Reeth to Harkaside Moor that the inhabitants / of Reeth used for driving their cattle and sheep to and from <thence> and alsoe told this deponent that the inhabitants of Reeth being debarred from digging peates or graving turfes or pulling ling upon the moor adjoyning upon the northside / of Reeth the same being stinted pasture the said men of Reeth were thereby forced to fetch all their ling peates and turves from a place on Harkaside Moor called Reeth Peat Moore. /

13: To the thirteenth interrogatory saith he knowes the pinfold belonging to Reeth and that the inhabitants of Harkaside usually impound their tresspassing cattle therein and payes towards repairing the said pinfold. /

Mathew Alderson[276] of Reeth in Swaledale yeoman aged fifty three yeares and upwards sworne and examined. /

1: To the first interrogatory saith he knowes the complainant and defendant and did know Philip late Lord Wharton deceased. /

2: To the second interrogatory saith he knowes Swaledale saith it lyes in Richmond shire and in the wapentacks of Gilling West and Hang West saith it is bounded by Stollerston stile on the east and Hollow Mill Crosse on the west / Arkingarthdale on the north and Wensleydale on the south saith the present Lord Wharton and his ancestors have been reputed cheife lords thereof and that they have usually and yearely kept two courts in the said dale to witt / one for Helaugh mannor and the other for Mewkarr and this deponent never heard of any other court kept in the said dale saith the said lord or his ancestors have all his time claimed the royalties of fishing fowling and hunting / within the said dale and this deponent was att one of the courts holden att Helaugh twenty yeares agoe or upwards when one Richard Close being one of the jury was presented upon his owne informacion or confession for fishing in / Swale river without lycense from the then Lord Wharton and fined for the same tenn shillings and William Arundell of Harkaside about twelve yeares agoe paid to the said lord two pounds for greenheugh and two pence for / license to fish in the said river. /

4: To the fourth interrogatory saith he knowes the vill and constablery of Reeth and saith it lyes in the mannor of Helaugh saith the mannor of Helaugh and the said vill and constablery of Reeth have all his time been reputed to extend / not onely to the northside of the river Swale but alsoe to the south side of the said river as farr

276 In May 1708 Mathew Alderson, aged 55, gave evidence for Lord Wharton at the trial in London (LI Misc 536, Ward MSS, notebooks vol. 6, f. 117r).

as heaven water deales betwixt Wensleydale and Harkaside and West Grinton moors and as farr as heaven water deales betwixt / Crackpott Sattron and Spring End and the moores thereof and the said dale called Wensleydale saith he never knew or heard of the river Swale being a boundary to the said mannor of Helaugh or constablery of Reeth or either of / them altho this deponent hath knowne the said mannor and constablery for forty yeares past. /

5: To the fifth interrogatory saith he knowes Harkaside and West Grinton and the moores and commons thereto belonging saith they are not divided but reckond one place or hamlett knowes Grinton or East Grinton and the moores thereto / belonging saith there is a boundary which divides the said East Grinton from West Grinton and Harkaside and it not onely divides the moores and commons of East Grinton from the moores and commons of West Grinton and / Harkaside but divides the inclosed grounds and lands of each respectively saith he hath knowne the said boundary for forty yeares past and that the same begins att Swale and crosses the Holme to the road betwixt East Grinton and / West Grinton short of the said gate in the said road and soe betwixt the said gate and Cuddy Spout up the ridge to Benson cowhouse thence to the east corner of Ox Intack wall and from thence crosse Hughes Ditch upon the moor up in a / direct line to the White Stone in the How and from thence <straight> to the Gray Stone in the Greets and up to where heaven water deales saith as heaven water deales betwixt Harkaside Moores and Wensleydale unto Gibbon hill is the reputed boundary / betwixt these moores and that dale unto the head of Browney gill and saith Browney gill downe Scabbey Yate and soe to the river Swale is the boundary betwixt Harkaside and Whitaside. /

6: To the sixth interrogatory saith he knowes the lead mines now wrought by the defendant Marriott called How groves and those lately opened by the complainant or his agents near thereto saith those opend by the complainant lye within the boundary / of West Grinton and Harkaside and those of the defendant within the boundary of East Grinton saith the said complainants workes were not opened till after the riding of the said boundaryes about two yeares agoe and publick notice of the said / riding was given in Reeth markett and alsoe in Grinton church yard on Sunday after divine service and saith the said riding was performed according to the said boundarys as before described by this deponent saith that the Lord Wharton / and his ancestors and their tennants att Harkaside and West Grinton and other inhabitants there att the time of the riding the boundarys aforesaid and as long before as this deponent can remember have been in possession of the said / moores and commons of Harkaside and West Grinton by the tennants and other the inhabitants getting of ling turfes and stones there depasturing their cattle and the like and this deponent saith <that> the present Lord Whartons agents or / some by their order two or three yeares agoe opend some inclosed grounds being his owne lands att Harkaside in Hirds possession to search for lead. /

7 & 10: To the seaventh and tenth interrogatorys saith that all Harkaside and West Grinton and the moores thereof have been all this deponents time reputed to lye within and to be part and parcell of the lordship of Swaledale and mannor / of Helaugh and saith not all but most part of West Grinton and Harkaside are reputed to be in Reeth constablery saith there are diverse coppyhold or customary lands in Harkaside and West Grinton and diverse freehold / lands alsoe but this deponent cannot rightly distinguish them but saith the coppyhold or customary land there if it be of that part called new land the same payes a fifteen penny fine to the lord of {the lord of} the mannor of Helaugh / upon death or alienacion of a tennant if old land a twenty penny fine saith the said lands upon death of the ancestor descend to all the sons alike but if he leave a widdow she has one third thereof during soe long as she continues

his / widdow and noe longer and then it descends in like manner to all the sons which is according to the custome of the mannor of Helaugh but noe such custome in the pretended mannor of Grinton /

11: To the eleaventh interrogatory saith that Harkaside tennants pay a part or share of the modus for tyth corne and hay with which Reeth is charged but how much the same is knowes not /

13: To the thirteenth interrogatory saith that Harkaside and West Grinton people <have> made use of Reeth pinfold as long as this deponent can remember to impound goods in and this deponent remembers that about twenty yeares agoe the township / of Reeth was presented and fined att Helaugh court for not keeping the said pinfold in repair and saith that the inhabitants of Harkaside paid a proporcion of that fine. /

14: To the fourteenth interrogatory saith he hath heard it commonly said by antient men of Reeth that are dead (viz[t].) James Robinson and Mathew Metcalfe that the Helaugh and Reeth tennants used formerly till the late Civill Wars / to intercommon with the tennants of Harkaside and West Grinton and hath been shewn by them an antient way in a parcell of ground called the Dikehouse in West Grinton which had been shutt up or inclosed before this deponents time / which way the said old men told this deponent the Reeth men used to drive their cattle up to West Grinton and Harkaside common to depasture there and saith they alsoe shewed this deponent the laine called Noutherd Loning / {Harkaside} leading from Helaugh to Harkaside Moor which the said antient men said the Helaugh tennants used to drive their cattle through to {the} Harkaside Moores to depasture and this deponent saith he this deponent about / seaventeen yeares agoe did lead stone which had been gott on Harkaside Moor for Henry Matchell of Reeth from the said moor to Reeth but saith neither the said Henry Matchell nor this deponent were disturbed therein nor had either / of them any land in Harkaside or West Grinton. /

15: To the fifteenth interrogatory saith that the tennants of Harkaside pay all manner of assessments with Reeth saving to the church and poor and did pay with them to the poor formerly till an Act of Parliament relating to the / maintainance of the poor separated them in that respect from Reeth /

Arthur Hutchinson of Reeth in Swaledale yeoman aged fifty yeares and upwards sworne and examined. /

1: To the first interrogatory saith he knowes the present Lord Wharton and hath seen Mr Marriott and knew the old Lord Wharton who dyed about eleaven yeares agoe. /

3: To the third interrogatory saith he knowes the mannor of Helaugh and that the lordship of Swaledale and Helaugh have been during all his time enjoyed by the Lord Wharton and his ancestors and that the same has been during all the time aforesaid / been called by the same name and beleives the reason of its being called soe is because the river Swale runs through the said dale /

7: To the seaventh interrogatory saith that all Harkaside and West Grinton and the moores thereunto belonging are reputed to lye within the mannor of Helaugh and the greatest part thereof within the vill and constablery of Reeth and that / during all his time the same were never reputed part and parcell of the mannor of Grinton saith he heard Anthony Hird say that he and one John Hird farmed lands of the Lord Wharton att Harkaside but did not tell him the yearly / vallue saith he knowes that there are freehold lands lying at Harkaside and West Grinton belonging to the severall persons herein after menciond <and> appears att the Lord Whartons courts att Helaugh (viz[t].) Anthony Hird John Hird Mr /

[membrane vii *verso*]

Hartley Ralph Wensley Ralph Addison John Blaides William Arundell and payes free rents to the said Lord Wharton but what perticular free rent each of them payes this deponent cannot sett forth and / that these persons following holds coppy-hold lands in West Grinton and Harkaside by coppy of court roll of the mannor of Helaugh (vizt.) Ralph Wensley Christopher Raper William Arundell John Arundell Mr Hartley / and that they pay the said Lord Wharton a yearely rent but what they pay knowes not and appeares att the said Helaugh court and saith that part of these coppyhold lands are old lands and part new lands the old land / payes a twenty penny fine to the lord of the mannor att death or alienacion and the new land payes a fifteen penny fine att the time aforesaid (vizt.) for every penny rent twenty pence fine and for every penny rent / fifteen pence fine.

13: To the thirteenth interrogatory saith that he knowes the pinfold att Reeth and that the inhabitants of Harkaside impound their trespassing cattle either att Reeth or Helaugh pinfold which lyes more convenient for them. / saith that this deponents goods once trespassed into the inclosed grounds att Harkaside and they were impounded att Helaugh pinfold. /

14: To the fourteenth interrogatory saith that he being an inhabitant in Reeth about twelve yeares agoe digged freestone upon Harkaside Moor and ledd the same to Reeth to build his house with and that about the / same time he ledd stones alsoe from Harkaside Moor for one Henry Matchell of Reeth that he had digged there and saith that he alsoe imployed one Luke Robinson now alive an inhabitant in Harkaside to help him to / digg for and gett the stones above menciond and that when they were getting the stones this deponent wanting some water to drink found some old peat holes and he askt the said Luke Robinson who digged those peat holes and he / replyed that part of the said moor was called Reeth Peat Moor and those pitts had been digged by some of the Reeth men /

George White[277] of Gunnerside in Swaledale aged forty eight yeares and upwards sworne and examined. /

4: To the fourth interrogatory saith he knowes the vill and constablery of Reeth and that the same lyes within the mannor of Helaugh and that the lordship of Swaledale mannor of Helaugh and constablery of Reeth extends and lyes as / well on the southside as on the northside of the river of Swale and extends southwards to Wensleydale and that the said river of Swale was never reputed a boundary to limitt and divide the said lordship of Swaledale mannor of {Helau}/ Helaugh and constablery of Reeth or any of them that ever he heard of /

6: To the sixth interrogatory saith he knowes the lead mines now wrought by the defendant Marriott called How groves and alsoe the lead mines lately opened by the Lord Wharton near adjoyning saith the said last mencioned mines lye within / the moores and commons of Harkaside and West Grinton and not within the moores and commons of East Grinton and that the said mines lye about fourscore yards from each other saith that the said lead mines were opened by the Lord / Wharton about two yeares agoe and notice was given of rideing the said boundary betwixt Harkaside and West Grinton and East Grinton and the same was ridd some short time before opening the said workes as this deponent heard and / beleives saith that the Lord Wharton and his tennants att Harkaside and West Grinton and other the

277 In May 1708 George White, aged 49, gave evidence for Lord Wharton at the trial in London (LI Misc 536, Ward MSS, notebooks vol. 6, f. 117r).

freeholders and inhabitants there were in possession of the said moores and commons att Harkaside and West Grinton before the / opening of the said mines by pulling of ling digging of peats graving of turfes and grazing their cattle there and saith that thirty yeares agoe this deponent being a miner was working in some lead workes in Arkingarthdale / with one James Crathorne alsoe a miner and discoursing of working mines on Harka Moor the said Crathorne then said he wondred the Lord Wharton did not make a further tryall there for that he had made a tryall for getting lead upon / the said Harka moor for the said old Lord Wharton above tenn yeares before that time. /

Richard Gibson[278] of Reeth in Swaledale husbandman aged seaventy yeares or thereabouts sworne and examined /

1: To the first interrogatory saith he knowes the complainant and did know Philip Lord Wharton deceased. /

2: To the second interrogatory saith he knowes Swaledale and hath knowne the same for near threescore yeares saith the Lord Wharton and his ancestors have been all this deponents time reputed cheif lord or lords of the said dale saith / the said lords have usually kept court leets and court barons att Helaugh and Mewkarr once in the year and this deponent never knew or heard of any other courts kept in the said dale and saith that the said lords of Swaledale / have all along claimed the royalties of fishing fowling and hunting and other royalties and priviledges in the said dale and appointed gamekeepers there saith the present Lord Wharton and the late Philip Wharton / have during this deponents time had lead mines att the Old Gang in Helaugh mannor and other places in the said dale as Lownythwaite and Fryerfold within the said mannor of Helaugh on the north side of the river / Swale which were rich mines and great quantitys of lead <oar> gott thereout and this deponent saith that about forty yeares agoe this deponent knowes that the late lord Wharton had two or three shafts sunk for / lead near to the place where the smithy house now stands upon Harka hill in Harkaside Moor on the south side of the said river Swale and hath seen lead oare gott out of those lead workes by one James Crathorne / deceased and others under him which said Crathorne told this deponent he had taken the said workes of Mr Philip Swale and Mr John Gunter agents or stewards for the said late Lord Philip Wharton but what quantity of / oar was gott or how long wrought this deponent cannot perticularly sett forth but saith the lead oar which was gott there was smelted att Marrick mill the Lord Wharton then having noe smelt mills of his owne. /

4: To the fourth interrogatory saith he knowes and hath knowne the vill and constablery of Reeth for threescore yeares past having lived all or most of his time therein saith the said vill and constablery of Reeth hath all / all[279] his time been reputed to lye in the mannor of Helaugh and lordship of Swaledale saith the said lordship of Swaledale and the said mannor of Helaugh and the said vill and constablery of Reeth extend not onely to the northside / of the river Swale but alsoe to the southside of the said river as farr as to where heaven water deales betwixt Wensleydale and Swaledale saith he never in all his time heard that the river of Swale was called or reputed a / boundary to the said lordship of Swaledale mannor of Helaugh and vill and constablery of Reeth or to any of them. /

5: To the fifth interrogatory saith he knowes Harkaside and West Grinton and the moores and commons thereto belonging saith they are not divided by any boundary

278 In May 1708 Richard Gibson, aged 70, gave evidence for Lord Wharton at the trial in London (LI Misc 536, Ward MSS, notebooks vol. 6, f. 116r).

279 Word repeated in the original.

th'one from the other but are and have been all along / accounted one and the same place saith he knowes Grinton or East Grinton and the moores thereof saith there is a boundary which divides the said East Grinton from West Grinton and Harkaside and saith the said / boundary divides not onely the moores and commons of Harkaside or West Grinton from those of East Grinton but the said boundary extends throughout and divides likewise the inclosed lands of each respectively saith / he knowes the road or way leading from East Grinton to West Grinton and Harkaside and alsoe the said gate in the way and the elm tree growing near it saith the inhabitants of East Grinton have all the time / repaired or amended the said road or way as farr as <to> betwixt Cuddy Spout and the said gate and noe further saith he hath knowne the said boundary for betwixt fifty and sixty yeares past saith the said boundary / begins att the river Swale and crosses the Holme to the place in the said way or road to which East Grinton repaires the way and the said boundary crosses the way there and go's betwixt Cuddy Spout and the elmtree / up the great ridge to Benson cowhouse thence to the east corner of Ox Intack wall thence in a direct line cross Hughes Ditch to the White Stone in the How from thence in a streight line to the Gray Stone in the Greets / and then to where heavenwater deales and saith that when he this deponent was but a young man he hath been told of the said boundary by diverse very old <men> as Richard Fothergill who was this deponents godfather and / James Robinson that those were the antient boundermarkes betwixt the said places and used to shew this deponent the same when he went with them to [gett][280] coales saith the boundary betwixt West Grinton and Harkaside / moores and Wensleydale runs westward as heavenwater deales from above the Gray Stone in the Greets to Gibbon Hill and soe to the head of Browney gill and the boundary betwixt Harkaside and Whitaside is from / the head of the said Browney gill downe the said gill to Scabbey Yate and soe to the river Swale. /

6: To the sixth interrogatory saith he knowes the lead mines now wrought by the defendant Mr Marriott called How Groves and the lead workes opened near them by the Lord Whartons agents saith those of the Lord Whartons / are within Harkaside and West Grinton moores and not within the moores of East Grinton saith he was not att the riding of the said boundary on the Lord Wharton's behalfe about two yeares agoe but heard publick notice / given thereof both in Grinton church <yard> and Reeth markett before the said riding saith that the present Lord Wharton and his ancestors and their tennants and other freeholders and inhabitants of Harkaside and West Grinton / have been in possession of the moores of Harkaside and West Grinton as long as this deponent can remember by depasturing their cattle horses and sheep thereon getting ling peates turfes and stones there and by the said late Lord / Wharton getting of lead on Harkahill as aforesaid and by the present Lord Wharton or his agents sinking for lead about three yeares agoe in his owne inclosed grounds in Harkaside in Hirds possession saith that he never knew or / heard that East Grinton men enjoyed any common right upon Harkaside or West Grinton moores without disturbance but hath known that when any of them have pulld ling or graved turves or peates thereon they were / not suffered to carry the same away but the West Grinton and Harkaside men have either burnt the same or lead it away themselves. /

7: To the seaventh interrogatory saith that all Harkaside and West Grinton and the moores thereof within the boundarys afore sett forth by this deponent have been all this deponents remembrance reputed to lye in the lordship of / Swaledale and within the mannor of Helaugh and not within the pretended mannor of Grinton or East

[280] Word omitted in the original.

Grinton and saith most of Harkaside and West Grinton hath in like manner been reputed part and parcell of the yownship / or constablery of Reeth. /

13: To the thirteenth interrogatory saith that Harkaside and West Grinton make use sometimes of Reeth pinfold and sometimes of Helaugh pinfold (according to their conveniences) for impounding their cattle and have done so / all this deponents time and knowes that they contribute to the repair of Reeth pinfold and hath heard they doe the like to Helaugh. /

14: To the fourteenth interrogatory saith that he hath knowne diverse of the lords tennants of the towns of Helaugh and Reeth intercommon with the tennants of Harkaside and West Grinton particularly he hath seen / James Robinson and James Alderson both of Helaugh (and who had noe land att Harkaside or West Grinton) severall times drive their goods about forty yeares agoe downe Neat or Noutherd Loning leading from Helaugh / to Harkaside common to depasture there and were not that ever this deponent heard of disturbed by Harkaside or West Grinton [men][281] for soe doing alsoe this deponent saith one Francis Whip of Reeth about four or five yeares agoe depastured / his sheep on Harkaside common without any disturbance that this deponent heard of and the said Whip had noe lands in Harkaside or West Grinton and saith that all this deponents time it hath been the common reputacion that / Helaugh and Reeth tennants had as much right on Harkaside and West Grinton moores as the tennants of Harkaside or West Grinton had. /

Nicholas Cowpland[282] of Reeth in Swaledale aged forty five yeares and upwards sworne and examined /

4: To the fourth interrogatory saith that he knowes the vill and constablery of Reeth and that the same lys within the mannor of Helaugh and that the lordship of Swaledale mannor of Helaugh and constablery of Reeth lyes / as well on the south side as the north side of the river of Swale and extends southwards to Wensleydale saith that the said river of Swale was never reputed a boundary to divide the said lordship of Swaledale mannor of Helaugh / and constablery of Reeth or any of them that ever he heard of /

5: To the fifth interrogatory saith he knowes Harkaside and West Grinton and the moores and commons thereunto belonging and that the same is allways accounted one and the same place having noe boundary to divide it knowes / Grinton or East Grinton and the moores thereof and that there is a boundary to divide the same from Harkaside and West Grinton which divides as well the inclosed grounds as the moores and commons thereof saith he knowes the / road leading from East Grinton to Harkaside and West Grinton and the gate in the said road and the elm tree near thereto and saith the said boundary begins att the river Swale and crosses the Holme into the said highway / betwixt the said gate and Cuddy Spout and soe up the ridge to Jenny Wrenn cowhouse and to the east nook or corner of Ox Intack and crosses Hughes Ditch upon the moor and soe to the White Stone in the How and from / thence to the Gray Stone on the Greetes and soe to where heaven water deales adjoyning upon Wensleydale and soe along as heaven water deales to Gibbon hill and from thence to the top of Browney gill where begins the boundary / betwixt Harkaside Moor and Whitaside and downe Browney gill to Scabbey Yate and soe to the river of Swale /

6: To the sixth interrogatory saith he knowes the lead mines now wrought by the

281 Word omitted in the original.

282 In May 1708 Nicholas Coupland, aged 45, gave evidence for Lord Wharton at the trial in London (LI Misc 536, Ward MSS, notebooks vol. 6, f. 117v).

defendant Marriott called the How Groves and alsoe the lead mines lately opened by the Lord Wharton adjoyning near thereto saith that the said last / menciond mines lye within the moores and commons of Harkaside and West Grinton and the boundary thereof and not within the moores and commons of East Grinton and the boundary thereof and that the said mines lye about / fourscore yards distant from each other and that the said boundary goes betwixt them saith that the Lord Wharton opened the said mines about two yeares agoe and that the boundary was ridd by the Lord Whartons agents / and a great many more (whereof this deponent was one) some little time before the said mines were opened by the said Lord Whartons agents and that publick notice was given of riding the said boundary in Reeth markett the / markett day before by the Lord Whartons bayliffe saith that the Lord Wharton and his ancestors and their tennants and other the freeholders and inhabitants of Harkaside and West Grinton have been in possession of the said moores / att Harkaside and West Grinton during all the time of this deponents remembrance thereof by their grasing cattle and pulling ling graving turves getting peates and stones upon the said moores and beleives they were never out of / possession of the same. /

7: To the seaventh interrogatory saith that Harkaside and West Grinton and the moores thereof are reputed part and parcell of the mannor of Helaugh & that the greatest part thereof lyes in the vill and constablery of Reeth & was / never reputed part and parcell of the pretended mannor of Grinton that this deponent ever heard of saith that the Lord Wharton has lands att Harkaside farmed by the Hirds and that there are severall coppyholders and freeholders / that lyes there which lyes within the mannor of Helaugh and doe suit and service att the courts held there but what particular rents fines and services they pay and performe to the lord of the said mannor for their said lands / this deponent knowes not /

13: To the thirteenth interrogatory saith he knowes the pinfold belonging to the township of Reeth that the same stands on the north side of Reeth and hath seen Harkaside men bring cattle from thence & impound their cattle / therein. /

14: To the fourteenth interrogatory saith he knowes that Mathew Harrison and Francis Whipp two inhabitants in Reeth hath put sheep to graze upon Harkaside Moors as belonging to Reeth and has used the same for / near twenty yeares and this deponent saith the inhabitants of Harkaside have the like liberty of turning their sheep horses and cattle upon the moores and commons belonging to Reeth on the north side of Swale & saith / that he hath seen Harkaside men put sheep upon Reeth Moor and he himselfe has fetched Reeth mens sheep from Harkaside Moor and saith that he lead six score foot of freestone about twelve yeares agoe / from Harkaside Moor for one Roger Alderson of Reeth which stone the said Roger Alderson had gott or caused to be gott in the quarries upon Harkaside Moor and saith that about thirty five yeares agoe / this deponents father John Cowpland since dead pulled ling upon Harkaside Moor and this deponent led the same to Reeth to thatch his fathers house withall att Reeth. /

[membrane viij *recto*][283]

Hugh Stones of Arkingarthdale in the county of Yorke miner aged about fifty yeares sworne and examined /

6: To the sixth interrogatory saith that one Christopher Naylor late of Swalehall deceased who marryed one James Crathornes daughter took this deponent along with him to

[283] Parchment, 650mm x 730mm.

Harka upon Harkaside Moor about seaven / yeares agoe and shewed this deponent two shafts or mines which lye near the nick of Harka about eighty yards distant from the place where there has been a house since built and the said Christopher Naylor then told this deponent that the said / Crathorne had formerly sunk those shafts by agreement with the late Lord Whartons stewards or agents to try for lead and that Crathorne had found lead ore there and the said Christopher Naylor desired this deponent to go to the Lord Whartons steward / and see if he could agree about working the said shafts again and told this deponent if he could agree about working the said shafts again {and told this deponent if he could agree} with the steward he the said Naylor would goe share therein with this deponent. /

Anthony Hird[284] the elder of Harkaside in Swaledale labourer aged forty eight yeares sworne and examined. /

4: To the fourth interrogatory saith he knowes the towne and constablery of Reeth saith the same lys within the mannor of Helaugh saith the said mannor and constablery extends to the southside of the river Swale as farr as to where heaven water deales as well as to / the northside of the said river saith the river of Swale was never in this deponents time that he heard of reputed a boundary to limitt or divide the said mannor or constablery or either of them. /

5: To the fifth interrogatory saith he knowes Harkaside and West Grinton and the moores and commons thereunto belonging saith they are undivided but are accounted one and the same place knowes alsoe Grinton or East Grinton and the moores thereof saith the / said East Grinton is divided from West Grinton and Harkaside by certain boundarys saith the said boundarys divide both the moores and inclosures of West Grinton and Harkaside from those of East Grinton saith the said boundary begins att the river Swale and / crosses a parcell of ground called the Holme to the laine or highway leading from East Grinton to West Grinton and crosses the said laine or way about eight yards eastwards of the elmtree near the gate and soe goes up the great ridge to Jenny Wrenn cowhouse and soe / to the east corner of Ox Intack wall and from thence over the moor cross to Hughes Ditch in a streight line to the White Stone in the How and from thence to the Grey stone in the Greetes and soe up to the height of the moor where heavenwater deales saith there / is a boundary from thence betwixt Harkaside and West Grinton moores and Wensleydale which runs as heavenwater deales to Gibbon hill currock and on to above Browney gill head and saith there is a boundary betwixt Harkaside and Whitaside which / is from where heavenwater deales over against Browney gill head down Browney gill to Scabby Yate and soe to the river Swale saith he hath knowne the said boundary for thirty five yeares past having been shewn to him <soe long time> by his father now dead who / was then an antient man and he told this deponent they were the antient boundarys betwixt those places. /

6: To the sixth interrogatory saith he knowes the lead mines called the How groves now wrought by Mr Marriott and alsoe the lead workes near them opened by the Earle of Whartons agents saith the said Earle of Whartons said workes lye on Harkaside and / West Grinton moores and not upon East Grinton moores saith the same lye much att the same distance from the one side of the said boundary as Mr Marriotts workes doe from the other side thereof and that the said severall workes are about fourscore yards distant / thone from the other saith the Lord Whartons workes were not opened till after the boundary afore sett forth by this deponent was ridd on behalfe of the Lord Wharton saith he this deponent was att the riding thereof about

[284] Anthony Hird of Harkaside, yeoman, aged 48, swore an affidavit on 18 October 1707 (E207/115/1).

two yeares agoe and the said boundarys / were <then> ridd as this deponent hath described them saith that the present Earle of Wharton and his tennants and other the freeholders and other inhabitants of Harkaside and West Grinton att the time of opening the said workes and riding the said boundary and / the said Lord Whartons ancestors lords of the mannor of Helaugh and their tennants and other the freeholders and inhabitants in Harkaside and West Grinton have been all this deponents time before the said riding in the possession of the said moores and commons / of Harkaside and West Grinton without disturbance or interrupcion saith he knowes that about twenty eight yeares agoe one Mathew Myres of East Grinton pulld some ling upon Harkaside and West Grinton moor for one Richard Harker of East Grinton but the / Harkaside and West Grinton tennants would not suffer Harker to lead it away but gave this deponent leave to lead it away to a house this deponents brother was building in Harkaside and it was used for thatching the said house. /

7: To the seaventh interrogatory saith that all Harkaside and West Grinton and the moores thereof have during this deponents remembrance been reputed to lye within the mannor of Helaugh and the greater part of Harkaside and West Grinton in the township and constablery of Reeth /

13: To the thirteenth interrogatory saith the inhabitants of Harkaside and West Grinton usually impound goods sometimes in Reeth pinfold and sometimes att Harkaside pinfold as those pinfolds stand with their convenience but never att any other pinfold that this deponent hath heard of /

Christopher Hall[285] of East Grinton in Swaledale aged thirty five yeares sworne and examined /

4: To the fourth interrogatory saith he knowes the towne and constablery of Reeth saith the same is reputed to lye within the lordship of Swaledale and mannor of Helaugh saith the said lordship of Swaledale as alsoe the said mannor of Helaugh and township and / constablery of Reeth extends not onely on the northside of the river Swale as farr as <to where> heaven water deales betwixt Arkingarthdale and Swaledale but they alsoe extend on the south side of the said river as farr as <to where> heavenwater deales betwixt Wensleydale & Swaledale / saith the river Swale was never in this deponents time that he hath heard reputed a boundary to limitt or divide the said lordship of Swaledale mannor of Helaugh and towne and constablery of Reeth or any of them. /

5: To the fifth interrogatory saith he knowes Harkaside and West Grinton and the moores and commons thereto belonging saith there is noe boundary <that> divides them but they are accounted one place knowes alsoe Grinton or East Grinton and the moores thereof saith / there is a boundary which divides Grinton or East Grinton from West Grinton and Harkaside saith the said boundary divides not onely the moores and commons of Harkaside or West Grinton from those of East Grinton but the said boundary extends throughout / and divides likewise the inclosed grounds of each respectively saith the said boundary begins att the river Swale and crosses the low end of the Holme to the road leading from East Grinton to West Grinton and crosse the road about eight yards eastwards of the elmtree growing / near the gate in the road up the ridge to Jenny Wrenn cowhouse and from thence to the east corner of the Ox Intack wall and from thence to {the} Hughes Ditch on the moor crosse that ditch in a straight line to the White Stone in the How and from thence directly to the / Gray Stone in the Greets and from thence to the head of the Greets where

285 Christopher Hall, a freeholder of East Grinton, yeoman, swore an affidavit on 17 May 1706 (E207/112/1).

heaven water deales saith he hath knowne the said boundary for fourteen yeares and upwards and was att two severall ridings thereof th'one by the present Sir Solomon Swale whilest he was in / possession and before Mr Marriott evicted him and the other riding there was by the present Earle of Whartons agents about two yeares agoe saith att the first menciond riding Sir Solomon Swale himselfe was present and the said boundaryes were ride att both those / ridings according as this deponent hath described them. /

6: To the sixth interrogatory saith he knowes the lead mines called How Groves now wrought by the defendant Mr Marriott and alsoe the workes opened near them by the Lord Whartons agents saith the said Lord Whartons said workes lye within the moores of West Grinton & / Harkaside and the boundaryes thereof and not within the moores of East Grinton and the boundaryes thereof and saith they lye about four-score yards distant from the said Mr Marriotts workes and the said boundary betwixt them about an equall distance from each of / the said workes saith that during all this deponents remembrance the Lord Wharton and his ancestors and their tennants in Harkaside and West Grinton and other the freeholders and inhabitants there have been in possession of the said moores and commons of Harkaside and / West Grinton by depastureing their cattle and getting turfes peates ling and stone thereon saith he never knew that the East Grinton tennants enjoyed the said commons of Harkaside and West Grinton without interrupcion and this deponent saith that he this deponent being / an East Grinton man and having once occasion to gett peates on Harkaside and West Grinton moor he asked leave of John Milner Francis Milner and Anthony Hird tennants of Harkaside who gave their consent to this deponent soe to doe and told this deponent they / thought he need not apply himselfe to any more for leave but this deponent when he had gott the peates and was leading them away old John Hird of Harkaside mett this deponent and asked him by what authority he gott peats there and this deponent told him he had leave / from the aforesaid three persons to which the said Hird replyed he the said Hird had as good right to the common as they and therefore this deponent should not carry them away without his leave but this deponent prevailed with him to suffer him to carry them away. /

John Close[286] of East Grinton in Swaledale miner aged fifty five yeares sworne and examined. /

4: To the fourth interrogatory saith he knowes the vill and constablery of Reeth and that it is reputed to lye within the mannor of Helaugh saith the lordship of Swaledale mannor of Helaugh and towne and constablery of Reeth are reputed to extend not onely to the northside of the / river of Swale but alsoe to the south side of the said river as farr as to where heaven water deales betwixt Wensleydale and Swaledale saith the said river of Swale was never in this deponents time before this cause (between the complainants and defendants) reputed a boundary to limitt or divide / the lordship of Swaledale mannor of Helaugh and towne and constablery of Reeth or any of them that this deponent heard of. /

5: To the fifth interrogatory saith he knowes Harkaside and West Grinton and the moores and commons thereto belonging and never heard of any boundary that divides the same but are one and the same place saith he knowes East Grinton and the moores thereof saith he knowes there / is a boundary dividing East Grinton from West Grinton and Harkaside and hath <knowne> the same for forty yeares and

[286] John Close of East Grinton, yeoman, aged 55, swore an affidvait on 18 October 1707 (E207/115/1). In May 1708 John Close (no age given) gave evidence for Lord Wharton at the trial in London (LI Misc 536, Ward MSS, notebooks vol. 6, f. 117v).

upwards saith the said boundary divides not onely the commons and mooores of West Grinton and Harkaside from those of East Grinton but the same divides alsoe the / inclosed grounds or lands of each respectively saith he knowes the road or highway leading from East Grinton to West Grinton and Harkaside and the gate in the said road and the elmtree growing near the said gate saith the tennants or inhabitants of East Grinton have all his / time repaired the said highway as far as the said gate & noe further and this he knowes the better for that he this deponent hath been surveyor of the highways in East Grinton and hath orderd the repair thereof soe farr and hath been att the repairing thereof severall times and / saith the said boundary begins att the side of the river Swale and crosses the Holme and the said highway about seven or eight yards east of the elmtree and from thence up a ridge of ground to Benson cowhouse thence to Ox Intack nook from thence up <on> the moor cross / Hughes Ditch in a line to the White Stone in the How thence directly up to the Gray Stone in the Greets and thence to the height of the Greets to where heavenwater deales and saith there is a boundary betwixt Harkaside and West Grinton and Wensleydale where begins / the <said> height of the Greets and runs westwards as heavenwater deales to Gibbon hill and soe to the height of Browney gill saith there is alsoe a boundary <which> divides Harkaside and West Grinton from Whitaside beginning att the height of Browney gill and runs downe the said Browney gill to / the river Swale saith this deponent hath heard from antient men both of East Grinton <West Grinton> Harkaside and Whitaside that the said boundarys have all along before this deponents time <been> reputed antient boundarys betwixt the aforesaid places /

6: To the sixth interrogatory saith he knowes both the lead mines called How Groves now wrought by the the defendant Mr Marriott and alsoe the lead mines or workes lately opend near them by the Lord Whartons agents saith the said workes opend by the Lord Wharton lye within the moores & / commons of Harkaside and West Grinton and the boundary thereof and not within the moores and commons of East Grinton or the boundary thereof saith the said Lord Whartons workes are distant from Mr Marriotts mines about fourscore yards saith both the said workes and mines lye / much att an equal distance from the said boundary but thinks the Lord Whartons workes are rather nearer to it then Mr Marriotts saith the said workes of the Lord Whartons were opend about two yeares agoe but not till after the said boundary was ridd by the Lord Whartons agents and others / saith that the Lord Wharton and his ancestors and their tennants of Harkaside and West Grinton and other the Freeholders and inhabitants there have been in possession of the said moores and commons of Harkaside and West Grinton ever since this deponent can remember by depasturing / their cattle thereon and getting ling peats turfe and stone there and saith that the late Lord Whartons agents severall times in severall yeares above twenty yeares agoe have <& within twenty yeares past> cutt feld and gott wood in Arundells ground being coppyhold and upon the intacks and severall other / places in Harkaside for useing in his lead workes att the Old Gang in Helaugh lordship and saith that about six moneths agoe one Luke Robinson who is now a miner for the defendant Mr Marriott in those mines called How Groves told this deponent that about thirty yeares agoe / he wrought and gott lead oar on Harkaside Moor for the Lord Wharton saith he hath knowne Ralph Appleby Christopher Hall Thomas Hawkins Richard Harker Christopher Chambers and William Lee all of East Grinton gott some of them ling and some of them peats / within this deponents memory on Harkaside and West Grinton moores but the West Grinton and Harkaside tennants allways disturbed and opposed them and sometimes burnt the same and sometimes ledd the same away to Harkaside or West Grinton. /

7: To the seaventh interrogatory saith that all Harkaside and West Grinton and the moores thereof have all this deponents time been reputed part and parcell of the mannor of Helaugh and saith that all Harkaside and West Grinton is <alsoe> reputed to be in the township and constablery of / Reeth save some part of West Grinton which is reputed to be in East Grinton constablery saith that the Lord Wharton hath lands of his owne in Harkaside and West Grinton leased to the Hirds and there are alsoe diverse coppyhold and customary lands there and alsoe / freehold lands but under what rents the tennants hold them or what certain fines the coppyhold or customary tennants pay to the lord this deponent cannot sett forth. /

10: To the tenth interrogatory saith it is common repute that the coppyhold or customary lands in West Grinton and Harkaside descend from the father to all his sons alike and if the tennants dye seized leaving a widdow such widdow enjoyes a third part of such lands / whilst she continues his widdow and noe longer according to the custome of the mannor of Helaugh saith there is noe such custome in the mannor of Grinton or East Grinton that this deponent ever heard of. /

14: To the fourteenth interrogatory saith that all the time of this deponents remembrance it has been commonly reputed that the complainants tennants of the mannor of Helaugh and towne of Reeth have allways had right to intercommon with the tennants of Harkaside / and West Grinton upon Harkaside and West Ginton moores and alsoe that the tennants of Harkaside and West Grinton had the like right to intercommon with the tennants of Helaugh and Reeth upon Helaugh and Reeth Moores. /

James Galloway[287] of Reeth in Swaledale aged sixty five yeares & upwards sworne & examined /

2: To the second interrogatory saith that he knowes the dale called Swaledale and that the same lyes within the Wapentacks of Gilling West and Hang West and that the extent <& boundary> of the said dale is from Stollerston Stile on the east to Hollow Mill Crosse on the west and from Arkingarth / dale on the north to Wensleydale on the south and that the Lord Wharton and his ancestors during all this deponents remembrance have been reputed cheife lords of the said dale and that there are courts kept by the Lord Wharton att Helaugh and Mukarr for / his two mannors there within the said dale and never any other courts kept in the said dale that ever this deponent heard of saith that the Lord Wharton and his ancestors during all the time aforesaid have claimed and enjoyed the royalties of hunting foweling and / fishing within the said dale and that when any deer comes into the dale they belong to the Lord Wharton and if any hunts fowles or fishes there without license they are fined att the lords courts and saith that the Lord Wharton & his ancestors have / during all this deponents time appointed and had gamekeepers within the said dale to look after the preservacion of the game within the same. /

4 & 14: To the fourth and fourteenth interrogatorys saith he knowes the vill and constablery of Reeth and that the same lyes within the mannor of Helaugh and that the lordship of Swaledale mannor of Helaugh & vill and constablery of Reeth lye as well on the southside as <on> / the northside of the river Swale and extends southwards as farr as <to> the height of the moor where heavenwater deales adjoyneing upon Wensleydale and saith that the river of Swale was never during his time reputed a boundary to limitt or divide the same or any / of them for he saith he hath heard it said by his father and severall antient men inhabitants in Reeth that there was a common feild belonging to Reeth Helaugh and Harkaside which lay part upon the north side of

[287] In May 1708 James Galloway, aged 69, gave evidence for Lord Wharton at the trial in London (LI Misc 536, Ward MSS, notebooks vol. 6, f. 116v).

the river of Swale and part on the south / side there adjoyneing upon Harkaside and fenced or inclosed from Harkaside Moor in which common feild the inhabitants of Reeth Helaugh and Harkaside had severall shares or parts and mowed and gott the hay there and when they came to eat the fogg / or aftergrowth of the said common feild then the said inhabitants put their cattle to eat in common and the said cattle had liberty to goe in the said feild as well on the south side as on the north side of the said river and saith he hath heard the said old men alsoe say and relate / that the inhabitants of Reeth Helaugh and Harkaside finding it inconvenient in removing their cattle sheep and horses from the common on the north side of Swale to the common on the south side of Swale on Harkaside (because the way ledd through the said common feild / and the cattle sheep and horses trod downe the meadow and made spoyle therein) therefor they agreed to make a laine to drive their said cattle to and fro as they had occasion which said laine was & is yet called Noutherd Loning and further saith that he remembers / part of the said common feild divided into closes in his time and that the last inclosed part thereof lyes on the south side of the said river att Harkaside and saith that one Richard Close has a part of the said inclosure for his part and one Ralph Addison had another part / thereof and Anthony Hird another part all inhabitants then of Harkaside and further saith that Reeth Low Moor lying upon the north side of the river of Swale is a stinted pasture and saith that the inhabitants of Harkaside that are the Lord Whartons freehold & coppyhold / and leasehold tennants have their shares and stints there according to their rents of their lands att Harkaside upon the said Reeth Low Moor as well as the inhabitants of Reeth & Helaugh and has during all his time used and enjoyed the same as they had occasion / as well as the Helaugh and Reeth men useing Harkaside Moor and hath heard that formerly they were reckoned all one and saith that Roger Alderson of Reeth gott both ling and stones upon Harkaside Moor and led the same to Reeth for his own use there / without any disturbance that ever he heard of and that the Reeth and Helaugh men have liberty to gett stones ling peates and turves upon Harkaside Moor as they have occasion and saith that theres a part upon Harkaside Moor called Reeth Peat Moor which the / inhabitants of Reeth used for that purpose. /

5: To the fifth interrogatory saith he knowes Harkaside and West Grinton and the moores and commons thereunto belonging saith that there was never any boundary to divide the same that ever he heard of but during all his time were accounted one and the same place / saith he knowes Grinton or East Grinton and the moores thereof and that theres a boundary to divide the same from West Grinton or Harkaside which boundary divides as well the inclosed grounds as the moores thereof saith he knowes the road leading from East / Grinton to Harkaside or West Grinton and the gate in the said road & the elmtree there and that the inhabitants of East Grinton repairs the said highway to a place bewtixt Cuddy Spout and the elmtree and that the boundary runs att the same place where Grinton / repaires to and that he hath knowne the same for forty yeares and upwards saith that the said boundary dividing East Grinton from Harkaside and West Grinton begins att the river of Swale crosses the Holme into the road near the elmtree to where Grinton / repaires and soe up the ridge to Jenny Wrenn or Benson cowhouse and soe to the east corner of Ox Intack and from thence crosses Hughes Ditch to the White Stone in the How and to the Gray Stone on the Greets <& to the height of the Greets> where heaven water deales dividing Swaledale from / Wensleydale and soe along as heaven water deales to Gibbon hill currock and from thence to Browney gill head which divides Harkaside Moor from Whitaside and soe downe Browney gill to Scabby Yate and soe to the river of Swale /

[membrane viij *verso*]

6: To the sixth interrogatory saith he knowes the lead mines called How Groves now wrought by the defendant Marriott & alsoe knowes the lead mines lately opened <& claimed> by the Lord Wharton near adjoyning saith that the said mines opened by the Lord / Wharton lye within the moores and commons of Harkaside and West Grinton and the boundary thereof and not within the the moores and commons of East Grinton and the boundary thereof and that the said lead workes lye about four-score yards distance / from each other that the said boundary goes up betwixt them saith that the said workes were opened by the Lord Wharton near two yeares agoe and that the said boundary was ridd the day before the opening of the said workes by the Lord / Whartons agents and a great many neighbours and others (whereof this deponent was one) and that publick notice was given of riding the said boundary att Reeth markett cross upon the markett day next before the riding thereof and in / Grinton church yard the Sunday before after divine service by the Lord Whartons bayliffe and saith that the Lord Wharton and his ancestors and their tennants and other the freeholders and inhabitants in Harkaside and West Grinton have / been in possession of the moores and commons there during all this deponents remembrance by useing thereof and the freeholders and tennants grazing their cattle thereon and pulling of ling getting of peates turves and stones there and saith he as / well knowes that the Lord Whartons agents about four yeares agoe caused two shafts or workes to be opened att Harkaside in the grounds belonging to the said Lord Wharton there and farmed by one Hird & made tryall to gett oare but gott little and that the / reason for trying there was that some oar was found near the surface of the ground and saith that the Lord Wharton might as this deponent beleives have tryed & wrought for lead oar upon any part of Harkaside Moor and saith he remembers that the Lord / Whartons bayliffe about thirty yeares agoe cutt downe severall parcells of wood in William Arundells ground being coppyhold lands att Harkaside for his mines and mills att Helaugh. /

7: To the seaventh interrogatory saith that Harkaside and West Grinton and the moores thereof have <been> during all his time reputed part and parcell of the lordship of Swaledale and mannor of Helaugh and that the greatest part thereof is part and parcell of the constablery of / Reeth & that the same was never reputed part and parcell of the pretended mannor of Grinton saith that there are lands lying at Harkaside called the lords owne lands farmed by the Hirds and payes six pounds thirteen shillings four pence rent per annum and that / there are severall coppyhold or customary lands lying there which holds of the Lord Whartons mannor of Helaugh by coppy of court roll & payes rent but knowes not what and saith that there are both old lands & new lands coppyhold or customary there the / old lands payes a twenty penny fine & the new lands payes a fifteen penny fine to the lord att death or alienacion & there are alsoe severall freeholders there which alsoe appears att Helaugh court & does suit & service & payes some free rent to the lord / but what rent they pay this deponent cannot sett forth /

11: To the eleaveth interrogatory saith there is a modus or prescript rent of seaven pounds two shillings and two pence paid for tythe of corne & hay by the township of Reeth to the impropriator of Grinton yearly & that Harkaside payes one pound & eight pence to the / towne of Reeth part thereof & saith he hath knowne this paid during all the time of his remembrance. /

12: To the twelfth interrogatory saith that for thirty yeares last past he hath heard the owners of Swalehall called as freehold tennants att the court at Helaugh & hath seen one John Blades & after him Christopher Naylor who were tennants at Swalehall appear for the owners / thereof & payes essoynes for them. /

13: To the thirteenth interrogatory saith that he knowes the pinfold att Reeth & that the inhabitants of Harkaside usually impound their tresspassing cattle there saith he remembers that the township of Reeth about six yeares agoe was fined att Helaugh / court for the said pinfold being out of repair & that the inhabitants of Harkaside within the said township of Reeth contributed with Reeth in paying that fine. /

15: To the fifteenth interrogatory saith he remembers there was a suit depending betwixt the old Lord Wharton and his customary or coppyhold tennants within the mannor of Helaugh touching & concerning the inhansed rents within the said mannor / saith that Marke Close or Richard Close his son & William Arundell coppyhold tennants in Harkaside contributed with the other coppyholders within the said mannor of Helaugh towards manageing the said suit and saith that Harkaside joyned / with Reeth in payment of all taxes (except church & poor) & that the inhabitants of Harkaside keeps watch & ward att Reeth when required. /

Ralph Addison[288] of Harkaside in Swaledale aged sixty five yeares & upwards sworne & examined. /

4: To the fourth interrogatory saith he knowes the vill & constablery of Reeth and the same lyes within the mannor of Helaugh and that the lordship of Swaledale mannor of Helaugh & the said constablery of Reeth lyes as well on the south side as on the north side of the river of / Swale & that the same extends southwards as farr as where heavenwater deales adjoyneing upon Wensleydale and saith that the said river of Swale was never reputed a boundary to limitt or divide any of them till of late that this suit began that ever he heard of /

5: To the fifth interrogatory saith he knowes Harkaside & West Grinton & the moores thereof & that the same is not divided by any boundary but always (during his time) accounted one & the same place knowes Grinton or East Grinton & the moores thereof & the same are / divided by a boundary from Harkaside or West Grinton which divides the inclosed grounds as well as the moores belonging to both the said places saith he knowes the road leading from East Grinton to West Grinton or Harkaside & the gate therein & the elmtree near there / and that East Grinton men repaires the said highway to near the elmtree & hath knowne all the above menciond places for fifty yeares & upwards & that the boundary begins att the river of Swale crosses the Holme into the highway a little east of the said elmtree & soe up / the Bounder ridge to Jenny Wrenn cowhouse & from thence to the east corner of Ox Intack & from thence crosses Hughes Ditch to the White Stone on the How & from thence to the Gray Stone on the Greets <& soe to the topp of the Greets> to where heavenwater deales which divides Swaledale from / Wensleydale & soe along as heaven water deales to Gibbon hill & soe to the top of Browney gill where begins the boundary betwixt Harkaside Moor & Whitaside & downe Browney gill to Scabby Yate & soe to the river Swale. /

6: To the sixth interrogatory saith he knowes the lead mines lately wrought by the defendant Marriott called How Groves & alsoe knowes the lead mines lately opened & claimed by the Lord Wharton near thereto & that these said last menciond workes lye within the moores / and commons of Harkaside and West Grinton & not within the moores & commons of East Grinton & that the same lye within eighty yards from each other or thereabouts & that the said boundary runs up betwixt the said

[288] In May 1708 Ralph Addison, aged 66, gave evidence for Lord Wharton at the trial in London (LI Misc 536, Ward MSS, notebooks vol. 6, f. 117r).

mines & nearer to Mr Marriotts workes then to / Lord Whartons saith that the said workes were opened <by the Lord Wharton> about two yeares agoe & that the said boundary was ridd some short time before opening the said workes & was ridd by the Lord Whartons agents & severall others (whereof this deponent was one) according to the said boundary / above described & that publick notice was given of rideing thereof by the said Lords bayliffe before the said riding and saith that the Lord Wharton & his ancestors & their tennants & other the freeholders and inhabitants within Harkaside & West Grinton during <all> the time of / this deponents remembrance had been in possession of the said moores & commons there by depasturing <their> cattle graving turves pulling ling & digging peates & stone thereon & was allways in possession thereof and saith that the Lord Whartons agents about three or four / yeares agoe caused two shafts to be sunk & tryall made for lead oare in the Lord Whartons owne lands in Harkaside farmed by John Hird but beleives they gott little or noe oar there saith that about twenty eight yeares agoe one Richard Harker of East Grinton / having pulld some ling upon Harkaside Moor the inhabitants of Harkaside fetched the same away & gave it to James Hird a poor man of Harkaside to thatch his house with & saith that when any of the inhabitants of East Grinton came to pull ling or gett turves / or peates upon Harkaside Moor they were allways disturbed <& hinderd > by the Harkaside men for soe doing and saith that the Harkaside men sometimes gave leave to the Grinton men & perticularly to Mr West viccar of Grinton for soe doing /

7: To the seaventh interrogatory saith that all Harkaside & West Grinton (except four or five houses as this deponent beleives) lyes within the mannor of Helaugh & constablery of Reeth & that it was never reputed to lye within the pretended mannor of East Grinton & beleives the / said four or five inhabitants appears att the sheriffe turn & att noe other court that this deponent ever heard of saith that there are lands lying att Harkaside called lords owne lands of six pounds thirteen shillings four pence per annum & farmed by Anthony & / John Hird & they appear & do's service att Helaugh court & that there are severall freehold tennants there which payes free rents to the lord but what they severally pay this deponent knows not saith that he has a small freehold att Harkaside for which he / payes to the lord two pence per annum as a free rent & that he & the other freehold tennants appears alsoe att Helaugh court & does suit & service there and alsoe saith there are severall coppyhold or customary tennants att Harkaside which payes yearly / rents to the Lord Wharton but what rents they pay yearly this deponent knowes not but saith att death or alienacion they pay fines to the lord (viz[t].) for old land a twenty penny fine & for new land a fifteen penny fine as this deponent hath heard & beleives /

8: To the eigth interrogatory saith he knowes there are severall peices of ground lying att Harkaside called Intacks which have formerly been taken of & inclosed from the moores there as he beleives & now occupyed by James Hird Anthony Hird & William Arundell & beleives / they pay yearly rents to the Lord Wharton for the same but knowes not what they pay but saith the same have been inclosed long before this deponents time /

10: To the tenth interrogatory saith that when any of the coppyhold tennants att Harkaside & West Grinton dye the widdow of such tennants has a third part of the lands soe long as she continues his widdow & when she dyes or marrys again that / third part together with the other two parts descends to all the sons alike as the rest of the coppyhold lands within the other parts of the said mannor of Helaugh doe & that there is noe such custome within the pretended mannor of Grinton that / ever this deponent heard of /

11: To the eleaveth interrogatory saith that there is a modus or prescript rent paid yearely for tythes of corne & hay by the township of Reeth to the impropriator of Grinton & that Harkaside contributes & payes a part thereof with the towne of / Reeth but how much the said modus is yearely or what part thereof Harkaside payes yearely this deponent knowes not certainly. /

13: To the thirteenth interrogatory saith he knowes Reeth pinfold and that Harkaside men usually impound their cattle therein /

14: To the fourteenth interrogatory saith that he knowes that Arthur Hutchinson and Henry Machell both of Reeth gott freestone upon Harkaside Moor and led the same to Reeth and never heard that they were disturbed for soe doing and saith that / Francis Whipp of Reeth had goods several times depasturing upon Harkaside Moor without any interrupcion. /

15: To the fifteenth interrogatory saith that Harkaside men payes land tax constable sess & window tax to the towne of Reeth & that the inhabitants of Harkaside keeps watch and ward att Reeth when required /

Richard Etherington of Fremington in the county of Yorke miner aged sixty yeares & upwards sworne & examined. /

14: To the fourteenth interrogatory saith that about fifty three yeares agoe he lived with his uncle Michaell Appleby who then lived in Reeth & was a coppyhold tennant under the Lord Wharton within the mannor of Helaugh and hath heard his said uncle say / that he had severall times gott peates & ling upon Harkaside Moor & brought the same to Reeth for his owne use & hath heard him alsoe say that severall other coppyholders in Reeth usually gott ling and peates upon Harkaside Moor & brought the same to Reeth / for their owne use hath alsoe heard his said uncle say that the Harkaside men put their cattle upon Reeth Moor which was a stinted common and this deponent beleives they used that liberty because the Reeth men had liberty to gett peates & ling upon Harkaside / Moor but saith he dos not remember that any Reeth men gott any ling or peats there in his remembrance. /

Nicholas Buckle[289] of Bowz in Arkingarthdale aged fifty yeares or thereabouts sworne & examined. /

2: To the second interrogatory saith that he knowes Swaledale that the same lys in Richmond shire & extends from Hollow Mill Crosse on the west to Stollerston Stile on the east and from the height of the mountains next Arkingarthdale on / the north to the height of the mountains next Wensleydale on the south and that the Lord Wharton & his ancestors during all the time of this deponents remembrance have been reputed cheife lords of the said dale and saith that the Lord / Wharton keeps courts yearely att Helaugh & Mukar within the said dale and never heard that any other courts were kept there besides and that the Lord Wharton & his ancestors during his time have usually claimed the royalties of hunting / fishing & fowling in & throughout the whole dale and has heard that when any deer came into the said dale they were reputed to belong to the Lord Wharton & when they were in Arkingarthdale then they belonged to the king & has alsoe / heard that they usually appointed & kept gamekeepers within Swaledale for preserving the game there & hath alsoe heard that his grandfather had liberty to fowle <through the whole dale> & beleives that he had that liberty from the Lord Whartons ancestors never / hearing of any other lords in Swaledale that ever pretended to grant <any> such liberty but the Whartons. /

4: To the fourth interrogatory saith he knowes the vill & constablery of Reeth that the same lyes within the mannor of Helaugh saith that the lordship of Swaledale mannor

289 Nicholas Buckle of Harkaside, miner, swore an affidavit on 1 June 1706 (E207/112/2).

of Helaugh & the said vill & constablery of Reeth extends on the south / side of Swale as well as on the north side thereof & extends southwards to where heaven water deales adjoyning upon Wensleydale saith that the river of Swale was never reputed a boundary during his time that ever he heard of till of late that this suit / begunn to limitt & divide the lordship of Swaledale mannor of Helaugh & the said vill & constablery of Reeth or any of them /

6: To the sixth interrogatory saith that he knowes the lead mines lately wrought by the defendant Marriott called How Groves & alsoe knowes the lead mines lately opened & claimed by the Lord Wharton near adjoyning saith that the said last menciond / workes are reputed to lye within the moores & commons of Harkaside & West Grinton & the boundary thereof & not within the moores & commons of East Grinton & the boundary thereof and that the <said> lead workes lye about fourscore yards distant from / each other saith that the Lord Wharton opened the said workes about two yeares agoe & that there was a boundary rid by the said Lord Whartons agents & severall others between Harkaside & West Grinton & East Grinton some short time before opening the / said workes & that publick notice was given of riding the said boundaryes by the Lord Whartons bayliffe and saith that the Lord Wharton & his ancestors & his & their coppyhold tennants of Harkaside & West Grinton & other the freeholders & inhabitants / there have been in possession of the said moores <& wastes> of Harkaside & West Grinton during all the time of this deponents remembrance by their grazing their cattle pulling ling graving peates & turves & getting stones thereupon & saith that the Lord Wharton / & his ancestors being cheife lords of the said dale called Swaledale & mannor of Helaugh might have digged & wrought for lead (as this deponent) beleives upon any the moores and commons within the said dale or mannor aforesaid if they had thought fitt /

7: To the seaventh interrogatory saith that Harkaside & West Grinton & the moores thereof have been during all his time reputed part & parcell of <the mannor of> Helaugh & that part of Harkaside & West Grinton lyes within the constablery of Reeth. /

12: To the twelfth interrogatory saith that it has been the constant report & discourse in the neighbourhood att Harkaside & this deponent has often heard it said that the owners of Swalehall are the first called att Helaugh court /

13: To the thirteenth interrogatory saith he knowes the pinfold att Reeth & that Harkaside men usually impound their cattle therein /

14: To the fourteenth interrogatory saith that he hath heard one James Thomson a very old man & since dead above twenty yeares agoe say in this deponents fathers house att Helaugh that there had been a herd kept formerly to look after / cattle belonging to Helaugh & that the said herdsman usually drove cattle from Helaugh downe a little laine crosse the river Swale to Harkaside Moor to depasture there & that the said laine was for that reason called Noutheard Loning saith / that he hath heard one Elizabeth Galloway who was the deponents mothers sister & who lived att Helaugh & was a very old woman of above eighty yeares of age & dyed above twenty yeares agoe often say that she helped to milk her fathers / ewes who lived att Helaugh after the lambs were weaned of them and hath heard his mother often say that she has gone severall times with her said sister Elizabeth Galloway to milke her fathers ewes upon Harkaside Moor & that they went / over the river Swale att Helaugh Hippinstones by the Stubbins near the west end of Harkaside to the Moor. /

Roger Alderson of Reeth in Swaledale aged fifty one yeares and upwards sworne and examined /

4: To the fourth interrogatory saith that he knowes the vill & constablery of Reeth that the same lyes within the mannor of Helaugh saith that the lordship of Swaledale mannor of Helaugh & vill & constablery of Reeth extends & lyes as well on the / south side of the river of Swale as on the north side of the said river & that the same extends southwards as farr as where heavenwater deales adjoyning upon Wensleydale and that the said river of Swale was never reputed a boundary that / ever he heard of till within twelve moneths last past to divide the said lordship of Swaledale mannor of Helaugh & vill & constablery of Reeth or any of them. /

11: To the eleaventh interrogatory saith that there is a modus paid by the township of Reeth for the tythes of hay & corne to the impropriator of Grinton of about seaven pounds per annum & saith that Harkaside contributes and / payes about twenty shillings per annum part thereof to the towne of Reeth. /

[membrane ix *recto*][290]

13: To the thirteenth interrogatory saith he knowes the pinfold belonging to the township of Reeth and that the same stands near the said towne and that the inhabitants of Harkaside usually impound their / tresspassing cattle therein /

14: To the fourteenth interrogatory saith that about twelve yeares agoe this deponent gott or caused to be gott one hundred foot of freestone upon Harkaside Moor and ledd the same to Reeth to build his house with / and saith that about twenty yeares agoe one James Robinson then of Reeth and this deponents uncle who was a very old man about fourscore and tenn yeares of age and since dead having frequent discourse with / this deponent told this deponent that the Reeth men had common right upon Harkaside Moor and that if the Reeth men did not make use thereof the same would <in time> be lost and thereupon this deponent gott the said freestone / above menciond for his owne use as above without any contradiction or disturbance and alsoe told this deponent that he and one Mathew Metcalfe another old man of Reeth and alsoe dead had gott peates formerly upon Harkaside Moor and / that Reeth inhabitants has as much Right there as upon Reeth Moor /

15: To the fifteenth interrogatory saith that he hath been constable of Reeth for four yeares together and during that time he allways collected of such of the inhabitants of Harkaside as are within the constablery of Reeth their proporcion / towards the constables assessments att Reeth and has alsoe gathered hearth money att Harkaside when constable of Reeth and hath seen the inhabitants of Harkaside keep watch and ward att Reeth. /

George Butterfield of Melback in Swaledale yeoman aged sixty nine yeares or upwards sworne and examined. /

4: To the fourth interrogatory saith he knowes the towne & constablery of Reeth saith the same lyes within the mannor of Helaugh and hath knowne the same ever since he was fourteen yeares / of age saith the said dale mannor of Helaugh & towne & constablery of Reeth extend not onely on the north side of the river Swale but alsoe on the south side of the said river as farr as to where heaven water deales betwixt Wensleydale and / Swaledale saith the river Swale was never that this deponent heard of reputed a boundary to limitt or divide the lordship of Swaledale mannor of Healaugh and vill or constablery of Reeth or any of them till of late that this suit begunn. /

5: To the fifth interrogatory saith he knowes Harkaside and West Grinton and the moores and commons belonging the same and alsoe Grinton or East Grinton and

290 Parchment, 630mm x 750mm. The upper part of the *verso* is badly rubbed making the first six lines of text difficult to read.

the moores thereof saith there is noe boundary betwixt West Grinton and Harkaside / but they are & have all this deponents time been accounted one place but saith there is a boundary which divides East Grinton from West Grinton and Harkaside which boundary divides not onely the moores and commons of East Grinton from the moores / and commons of West Grinton and Harkaside but the said boundary extends throughout and divides the lands & inclosed grounds of each respectively saith the said boundary begins att the river Swale & crosses the Holme and the highway betwixt / Cuddy Spout & the elmtree & soe up the ridge to Jenny Wrenn cowhouse & soe on <to> the moor by the Ox Intack wall to the east corner of the said wall soe over the moor crosse Hug<h>es Ditch to the White Stone in the Howe & from thence to the Gray Stone in the / Greetes & soe up to the height of the Greets where heaven water deales betwixt Wensleydale & Swaledale then the boundary divideing Harkaside & West Grinton together with the {other} <said> dale called Swaledale from the other adjoyning dale called / Wensleydale proceeds westwards from the said height of the Greets as heaven water deales to Gibbon hill & soe to the height of Browney gill and saith the boundary which divides Harkaside & West Grinton from Whitaside proceeds from the head of Browney gill / downe the said gill to Scabbey Yate & soe to the river Swale saith he hath knowne the said boundarys ever since he was a youth having been shewn to him to be such boundarys by diverse old men whilst he was young & hath many times observed / them since & therefore beleives them to be true boundaryes. /

6: To the sixth interrogatory saith he knowes the lead mines now wrought by the defendant Marriott called How Groves and the lead mines or workes lately opened by the Lord Wharton or his agents within about an hundred yards of the said defendants workes saith / the said Lord Whartons workes are within the moores & commons of Harkaside & West Grinton & the boundary thereof {& the boundary} & the said boundary passes about midway betwixt the said workes saith the Lord Whartons said workes were not opend / till the day after the riding the said boundarys saith this deponent was att the riding of the said boundary about two yeares agoe on the Lord Whartons behalfe & saith the said riding was according to the boundarys & markes above described / and sett forth by this deponent but this deponent was not <att> nor knowes of any former or other riding of the said boundary. /

Christopher Hird of Whitaside in Swaledale aged fifty yeares or upwards sworne & examined. /

4: To the fourth interrogatory saith he knowes the vill & constablery of Reeth & that the same lyes within the mannor of Helaugh that the lordship of Swaledale mannor of Helaugh & constablery of Reeth lyes as well on the south side as on the / north side of the river Swale & extends southwards to the height of the moores adjoyneing upon Wensleydale and that the said river of Swale was never reputed a boundary that ever he heard of till this suit begunn to limitt & divide the said / lordship of Swaledale mannor of Helaugh & constablery of Reeth or any of them. /

5: To the fifth interrogatory saith he knowes Harkaside & West Grinton & the moores & commons thereunto belonging & that the same is not divided by any boundary but have been during his time allways accounted one & the same place saith / he knowes Grinton and East Grinton & the moores thereof & that there is a boundary dividing the same from West Grinton & Harkaside which boundary divides as well the inclosed grounds as the moores betwixt the same saith he hath heard the / same reputed a boundary as aforesaid for thirty yeares last past & hath knowne all the said places from his youth having lived all his time near thereto saith he knowes the highway leading from East Grinton to Harkaside & West Grinton & the gate there / and the elm tree growing nigh thereto and that the said boundary begins att the river

Swale & soe into the road or way betwixt Cuddy Spout & the elm tree & soe up the bounder rigg to Benson cowhouse or Jenny Wrenn cowhouse & soe to the corner / of Ox Intack & crosses Hughes Ditch on the moor to the White Stone in the How & soe to the Gray Stone on the Greets & up <to> the height where heaven water deales dividing Swaledale from Wensleydale & soe along westwards as heavenwater deales to / Gibbon hill & from thence to the topp of Browney gill which divides Harkaside moores from Whitaside & soe downe Browney gill to Scabby Yate & soe to the river of Swale /

6: To the sixth interrogatory saith he knowes the lead mines called How Groves wrought by defendant Marriott & alsoe the lead mines lately opened & claimed by the Lord Wharton near adjoyning saith the said lead mines lately opened by the / Lord Wharton lye within the moores & commons <of Harkaside & West Grinton & the boundary thereof & not within the Moors & commons> of East Grinton & the boundary thereof & that the said boundary comes up betwixt the said mines but nearer to Mr Marriotts then the Lord Whartons saith that the said lead workes were opened by the / Lord Wharton about two yeares agoe & that the said boundary was ridd some little time before opening the same by the Lord Whartons agents & severall other neighbours (whereof this deponent was one) & agreed unto as this deponent beleives hearing / nothing to the contrary & saith that publick notice was given of riding thereof by the Lord Whartons bayliffe in Grinton church yard on the Sunday next before after divine service saith that the Lord Wharton & his ancestors & their coppyhold / tennants att Harkaside & West Grinton & other the Freeholders & inhabitants there have been in possession of the moores & commons of Harkaside & West Grinton during all his remembrance by grasing their cattle pulling ling graving peats & / turves & getting stones thereupon and saith that about eleaven or twelve yeares agoe he was imployed by Anthony Fryer a steward or agent for the then Lord Wharton to carry wood that the said Fryer had caused to be felled in some / coppyhold lands att Harkaside then belonging to Richard Close & saith he carryed the said wood from Harkaside to Helaugh side of the water to build a smelting mill belonging to the Lord Wharton & standing upon the moores above Helaugh /

John Dowson of the Levell House in Swaledale miner aged sixty six yeares sworne & examined. /

4: To the fourth interrogatory saith he hath knowne the vill & constablery of Reeth for thirty nine yeares past saith it lyes within the mannor of Helaugh & that both the said mannor & constablery extend not onely on the north side of the river Swale / but alsoe on the south side of the said river as farr as <to> heaven water deales betwixt Wensleydale and Swaledale saith that in all this deponents time the river Swale was never reputed a boundary to <limit or> divide the said mannor towne & constablery or any of them / that this deponent heard of. /

5: To the fifth interrogatory saith Harkaside & West Grinton are reputed one & the same place & are not divided by any boundary th'one from the other saith he knowes & hath knowne the said Harkaside & West Grinton & the moores & commons thereto / belonging & alsoe Grinton or East Grinton & the moores thereof for thirty nine yeares past saith there is a knowne boundary which divides East Grinton from Harkaside & West Grinton saith the said boundary divides not onely the moores & commons of / Harkaside & West Grinton from those of East Grinton but the said boundary extends throughout & divides the inclosed grounds <alsoe> of each respectively saith he knowes the highway or road leading from East Grinton to West Grinton & Harkaside & the / gate in the said way west of Cuddy Spout & the elm tree growing near the gate saith the East Grinton men have during his time repaired the said way to near the said gate & noe further saith the said boundary begins att the river Swale goes to the / Holme & crosses the said highway att the place to which East Grinton repaires & soe in att a gate betwixt Cuddy Spout & the elmtree & soe up the dividing ridge to Jenny Wrenn cowhouse & from thence to the corner or nook of Ox Intack & then / crosse Hughes Ditch upon the moor & soe directly to the White Stone in the How and from the White Stone in the How directly to the Gray Stone in the Greets & soe up to the height of the Greets to where heaven water deales betwixt Wensleydale & / Swaledale saith there is alsoe a boundary which divides West Grinton & Harkaside moores from Wensleydale beginning att the said height of the Greets proceeding Westward as heaven water deales to Gibbon hill & soe to Browney gill standard & the / boundary betwixt Harkaside & West Grinton & Whitaside goes from Browney gill standard downe Browney gill to Scabby Yate & thence to the river Swale saith he hath knowne all the said boundarys for thirty nine yeares having been showne the / same soe long since by severall antient men long since dead & having been a great part of his time employed in working mines on the said moores & commons & hath frequently observed the said bounder markes /

6: To the sixth interrogatory saith he knowes the lead mines or workes now wrought by Mr Marriott called How Groves & alsoe the workes lately opened near them by the Lord Whartons agents saith those workes of the Lord Whartons lye within / the moores & commons of Harkaside & West Grinton & the boundary thereof & not within the moores & commons of East Grinton that the said severall workes lye att about fourscore or one hundred yards each from the other & att near an equall / distance from the said boundary saith he was att the riding of the said boundarys about two yeares agoe & the same was ridd according to the descripcion this deponent hath before given of them and saith the Lord Whartons said workes were not opend / till the day after the said riding saith that the Lord Wharton & his tennants in Harkaside & West Grinton & other the tennants & freeholders there were att the time of the said riding & his ancestors & their tennants & other the freeholders & tennants / there ever since this deponent can remember have been in possession of the said moores & commons of Harkaside & West Grinton by grazing their cattle horses & sheep thereon getting ling stone peates & turfes there and the late Lord Philip Wharton as Lord / of the mannor of Helaugh hath severall times to this deponents knowledge & in diverse yeares above thirty five yeares agoe & since cutt wood in diverse coppyhold grounds in Harkaside belonging to the Arundells & Richard Close & carryed the same / to & use it att the smelting mill & lead workes att the Old Gang near Helaugh and the present Lord Wharton or the late lords trustees did the like in the said Richard Closes grounds in Harkaside about tenn or eleaven yeares agoe saith that he never knew or / heard that the tennants of East Grinton ever had any enjoyment upon the moores of Harkaside & West Grinton without disturbance or interrupcion & saith he knowes that one Christopher Hall of East Grinton about three or four yeares agoe had graved some / peates upon Harkaside Moor but the tennants of West Grinton & Harkaside would not suffer them to carry them away. /

James Metcalfe of Thorpe under Stone in the county of Yorke miner aged fifty four yeares & upwards sworne & examined. /

4 & last: To the fourth & last interrogatorys saith he knowes the vill & constablery of Reeth that the same lyes within the mannor of Helaugh saith that the lordship of Swaledale mannor of Helaugh & constablery of Reeth extends & lyes on the south side of the river of Swale / as well as on the north side of the said river saith that the said river of Swale was never reputed a boundary that ever he heard of till lately that this suit was begunn to limitt & divide the said lordship of Swaledale Mannor of Helaugh & the said vill & constablery of / Reeth or any of them and further saith that Thorpe under Stone the towne where this deponent now lives lys on the south side of the

said river of Swale & is within the mannor of Ravensworth which lyes on the north side of the said river opposite to Thorpe & / saith that during all his time the said river of Swale was never reputed a boundary that ever he heard of to divide the said township of Thorpe from the said mannor of Ravensworth. /

James Blaides of Richmond in the county of Yorke yeoman aged eighty seaven yeares sworne & examined /

4: To the fourth interrogatory saith he knowes the vill & constablery of Reeth & hath knowne the same most of his time being borne near Helaugh & having lived there till within these two or three yeares last saith the said vill & constablery of Reeth lyes within the / mannor of Helaugh saith the lordship of Swaledale mannor of Helaugh & towne & constablery of Reeth extends not onely on the north side of the river Swale but on the south side alsoe of the said river as farr as where heavenwater deales between Wensleydale / & Swaledale saith that untill this suit began this deponent never heard that the said river of Swale was reputed a boundary to the said lordship of Swaledale mannor of Helaugh & town & constablery of Reeth or any of them. /

William Kirton of Helaugh in Swaledale aged seaventy eight yeares & upwards sworne & examined /

2: To the second interrogatory saith that he knowes the dale called Swaledale & that the same lyes in the Wapentacks of Gilling West & Hang West & that the extent & reputed boundarys thereof is from Hollow Mill Crosse on the west to Stollerstone Stile on / the east & from where heavenwater deales on the height of the mountaines adjoining upon Arkingarthdale on the north to the height of the mountaines where heavenwater deales adjoyning upon Wensleydale on the south side that the present / Lord Wharton & his ancestors <during> all the time of this deponents remembrance have been reputed cheif lords of the said dale & that there have been during all the time aforesaid courts frequently kept att Helaugh & Mukarr within the said dale by the Lord Wharton / and his ancestors & never heard of any other courts kept within the said dale saith alsoe that during all the time aforesaid the present Lord Wharton & his ancestors have allways claimed & enjoyed the royalties of hawking hunting fishing & fowling / in & through the whole dale lying within the mannors of Helaugh & Mukarr & usually had kept a rainger or raingers to preserve & look after the deer that was in the dale & kept gamekeepers there to preserve other game within / the same & hath seen some men pay an acknowledgement att Helaugh court for liberty of fishing in the said river of Swale /

4: To the fourth interrogatory saith that he knowes the vill & constablery of Reeth & that the same lys within the mannor of Helaugh & saith that the lordship of Swaledale mannor of Helaugh & constablery of Reeth extends & lyes as well on the / south side of the said river of Swale as on the northside thereof and saith that the said river was never reputed a boundary (that ever he heard of till this suit was begunn) to limitt & divide the said lordship of Swaledale mannor of Helaugh & vill & / constablery of Reeth or any of them. /

John Galloway[291] of Helaugh in Swaledale aged fifty three yeares & upwards sworne & examined. /

6: To the sixth interrogatory saith that the Lord Wharton & his tennants & other the tennants & inhabitants of Harkaside & West Grinton & his ancestors <& their tennants> & other the tennants & inhabitants there have been in the possession of the moores & / commons att Harkaside & West Grinton during all the time of his

[291] In May 1708 John Galloway, aged 53, gave evidence for Lord Wharton at the trial in London (LI Misc 536, Ward MSS, notebooks vol. 6, f. 117r).

remembrance by pulling ling graving peates & turves & digging & getting of stones & grazeing their cattle horses & sheep upon the said moores & commons & saith that he hath often heard / his father whose name was Daniell Galloway & who was above eighty five yeares of age when he dyed often say that <all> Helaugh men used to drive their cattle & sheep from Helaugh down a laine to Harkaside Moor to depasture there & that / the said laine thereby gott the name of Noutherd Loneing and further saith he very well remembers that betwixt thirty & forty yeares agoe he being then a carryer of oar for the old Lord Wharton att his mines then att the old Gang on the / north side of Swale was filling up oar att a place called the Levell & one James Fryer (who is since dead) who was then an agent or steward for the said Lord Wharton att the said workes came to this deponent & told him he might come noe more to fill / that day because he & one Robert Barker since dead another steward for the said Lord Wharton were going to Harkaside to discharge one John Barningham who was then as he said working <in> the lead mines att Harka upon Harkaside / Moor from working any longer in those workes. /

James Galloway of Helaugh in Swaledale yeoman aged thirty four yeares sworne & examined /

2 & 4: To the second & fourth interrogatorys saith he knowes the vill & constablery of Reeth saith the same lyes within the mannor of Helaugh saith he hath knowne the said mannor & constablery ever since he can well remember saith the said mannor & constablery extends not onely on the / north side of the river Swale but on the south side of the said river alsoe saith he never heard that the river Swale was reputed a boundary to limitt or divide the said mannor & constablery or either of them nor doth this deponent beleive the same to be a boundary thereto for that this / deponent hath been severall times a jury man att Helaugh court when severall tennants in Harkaside who appear & doe suit & service att Helaugh court have been on the same jury & hath knowne the Lord Whartons bayliffe cutt wood on the Arundells coppyhold land in Harkaside & / carry it to his lead workes att the Old Gang near Helaugh & saith there is one John Loansdale of Helaugh hath a pasture which adjoynes part of it on the north side of the river Swale towards Helaugh & part of it on the south side of the said river towards Harkaside the river / running through it saith it is used as one pasture & his cattle going therein passe from the one side to th'other side of the said pasture as they please saith he hath often heard his aunt Elizabeth Forster who dyed about sixteen yeares agoe & was about ninety five yeares / old when she dyed say that her father James Galloway who lived att Helaugh & was this deponents grandfather used whilst she was a young woman <to> drive his goods down the Noutherd lane & crosse Swale on to Harkaside Moor to depasture there & that she used year by / year to milk her fathers ewes which depastured upon Harkaside Moor saith he hath heard that the Lord Whartons Stewards did grant licenses to severall Harkaside tennants to fish in the river Swale for which this deponent hath seen them pay two pence / a peice by way of acknowledgement for such license and saith he hath heard James Fryer deceased who was Lord Whartons bayliffe say that he had taken fishing rods from severall men that fished in the river Swale without license from the Lord Wharton /

[membrane ix *verso*]

Robert Buckle of Helaugh in Swaledale aged fifty five yeares & upwards sworne & examined /

4: To the fourth interrogatory saith he knowes the vill & constablery of Reeth & that the same lyes within the mannor of Helaugh saith that the lordship of Swaledale mannor of Helaugh & constablery of Reeth lyes as well on the south / side of Swale as on the north side of the said river & that the same extends Southward to the height of the moores adjoyneing upon Wensleydale saith that the said river of Swale was never reputed a boundary during his time that / ever he heard of till this suit was begunn to limitt or divide the said lordship of Swaledale mannor of Helaugh & constablery of Reeth or any of them. /

William Blaides[292] of East Grinton in Swaledale aged forty eight yeares & upwards sworne & examined. /

4: To the fourth interrogatory saith he knowes the vill & constablery of Reeth that the same lyes within the mannor of Helaugh & that the lordship of Swaledale mannor of Helaugh & vill & constablery of Reeth lyes as well on the south / side of Swale as on the north side thereof & that the same extends southwards as farr as to where heaven water deales adjoyning upon Wensleydale saith that the said river of Swale was never during his time reputed a boundary that ever he / heard of to limitt & divide the said lordship of Swaledale mannor of Helaugh & vill & constablery of Reeth or any of them till of late that this suit was begunn /

5: To the fifth interrogatory saith he knowes Harkaside & West Grinton & the moores & commons thereof that they are allways accounted one & the same place & noe boundary to divide them & alsoe knowes East Grinton & the moores thereto / belonging saith theres a boundary to divide the same from Harkaside & West Grinton which divides as well in the inclosed grounds as the moores saith he knowes the highway leading from East Grinton to Harkaside & West Grinton & the / gate standing in the said way & the elm tree growing near the same & has knowne all the said places from his infancy being born near Swalehall & the inhabitants of East Grinton repaires the said boundary near to the gate & that the / said boundary crosses the road att the said place and saith the said boundary beginns att the river of Swale crosses the Holme into the road a little west of Cuddy Spout & east of the gate & up the bounder ridge to Benson cowhouse or Jenny / Wrenn cowhouse & soe along the wall of Ox Intack upon the moor to the east corner of the said wall & crosses Hughes Ditch in a line to the White Stone on the How & from thence in a line to the Gray Stone on the Greets & soe to the top of / the Greets to where heavenwater deales which divides Swaledale from Wensleydale & soe on Westwards as heavenwater deales to Gibbon hill & soe to the top of Browney gill where begins the boundary betwixt Whitaside & Harkaside / moores & soe down Browney gill by the east side of the Horse pasture to Scabby Yate & soe to the river of Swale saith that when he was a schoolboy one Stephen Wheelhouse then an antient man & since dead who was then schoolmaster / att Fremington told this deponent of the boundary above sett forth & described by this deponent & he hath alsoe heard his father who lived att Swalehall & who dyed above twenty yeares agoe & was an agent for old Sir Solomon Swale / often relate the said boundary & tell him of it severall yeares agoe. /

6: To the sixth interrogatory saith he knowes the mines called How Groves now wrought by the defendant Marriott & alsoe the lead mines lately opened & claimed by the Lord Wharton near thereto saith that the said last menciond mines / lyes within the moores & commons of Harkaside & West Grinton & the boundary thereof and not within the moores & commons of East Grinton & the boundary thereof saith that the said workes were opened by the Lord Whartons agents about two / yeares agoe & that the said boundary was ridd some short time before the said workes were

292 In May 1708 William Blaydes, aged 48, gave evidence for Lord Wharton at the trial in London (LI Misc 536, Ward MSS, notebooks vol. 6, f. 117r).

opened & saith that publick notice of riding the same was given by the Lord Whartons bayliffe in Reeth markett & that severall men on both / north & south side of the river of Swale were present & ridd the said boundary & this deponent rid alsoe with them which said boundary was then rid as above described by this deponent saith that the Lord Wharton & his tennants att / Harkaside & West Grinton & other the freeholders tennants & inhabitants there <and the said Lord Whartons ancestors & their tennants & other the freeholders tennants & inhabitants there> have been in possession of the moores & commons att Harkaside & West Grinton during all the time of his remembrance by depasturing their cattle / horses & sheep & pulling ling graving peates & turves & digging & getting stones thereupon and saith that the present Lord Wharton about three or four yeares agoe caused a shaft or two to be digged or sunk in part of his <owne> lands att / Harkaside in John Hirds possession & made tryall there for getting lead oar but beleives they gott some oar but very little there saith that about twenty yeares agoe one Richard Harker of East Grinton pulled or caused to be pulld / some ling upon Harkaside Moor & the men of Harkaside hearing thereof led the same away & gave it to a poor man to thatch his house with hath alsoe heard that several other inhabitants of Grinton were disturbed & hindred by / the Harkaside men from getting ling peates & turves upon Harkaside Moor & that this deponent himselfe never durst goe to gett ling peates or turves there because they allways said they had noe right to soe to doe /

7: To the seaventh interrogatory saith that Harkaside & West Grinton & the moores thereof have been during all his time reputed part & parcell of the mannor of Helaugh & that the greatest part thereof lyes within the townshipp of Reeth /

8: To the eigth interrogatory saith that he knowes two parcells of ground lying att Harkaside adjoyning upon the moores in possession of Anthony Hird & John Hird called the intacks that the same have been inclosed & taken of the said moores / long before this deponents remembrance saith he hath heard & beleives the said Hirds payes a yearely rent to the Lord Wharton for the same but what perticular rent this deponent knowes not /

10: To the tenth interrogatory saith that when any coppyhold tennants dyes seized in Harkaside the third part of the lands goes to the widdow whilest she keepes widdow & the other two parts descends to all the sons alike & after the / death of the widdow or her marrying again her third part descends alsoe to all the sons alike according to the custome within the mannor of Helaugh & this deponent saith that he & his three brothers after the death of their father had a / close in Harkaside which was coppyhold lands & held of the mannor of Helaugh which descended alike to him & his three brothers. /

12: To the twelfth interrogatory saith that Sir Solomon Swale was all along called att Helaugh court the first man during this deponents remembrance <and> hath heard John Blaides (this deponents father being then his tennant) att Swalehall say that he / appeared att the said court for him. /

21: To the twenty first interrogatory saith he remembers his father near thirty yeares agoe living then in Harkaside brought an action against one Christopher Rapier alsoe of Harkaside & both coppyholders there in Helaugh court for not / making his fence sufficient in his said coppyhold lands att Harkaside & had a verdict against him in that suit. /

James Lonsdale the elder of Harkaside in Swaledale stockinger aged forty five yeares or thereabouts being one of those people called Quakers having repeated the solemn affirmation according to the statute in such case / made & provided being examined affirmeth as followeth. /

4: To the fourth interrogatory this affirmant saith he knowes the towne & constablery of

Reeth & saith it is reputed to lye within the mannor of Helaugh saith the said lordship of Swaledale mannor of Helaugh & constablery of Reeth extend not onely on / the northside of the river of Swale but alsoe on the southside of the same river as farr as to the boundaryes which divide Wensleydale from Swaledale saith he never heard that the river Swale was reputed a boundary to limitt or divide the / said mannor or constablery or either of them from any other mannor or place till [xxx] lately since this suit began nor doth this affirmant beleive the same to be a boundary to the said mannor or township & constablery or either or any / of them. /

14: To the fourteenth interrogatory saith that about thirty yeares agoe this affirmants father since dead who lived then att Helaugh gott ling upon Harkaside Moor soe much as thatched halfe his barne att Helaugh which ling this / deponent helped to lead from Harkaside Moor to Helaugh & neither he nor his father were obstructed or disturbed therein by the tenannts of Harkaside and saith that about five yeares agoe this deponent gott both ling and stones upon Harkaside / Moor which he led away to Helaugh & used in building an house there saith he was not disturbed or opposed by the tennants of Harkaside therein onely this deponent not being then sure whether he might not be opposed in it by one Luke / Robinson of Harkaside whom this deponent lookd upon as a troublesome neighbour sent to acquaint him with this deponents intencions to gett ling & stone upon those moores & sent alsoe to acquaint two of the Arundells of Harkaside therewith /

James Lonsdale the younger of Helaugh in Swaledale yeoman aged twenty three yeares & upwards being one of those persons comonly called Quakers & having repeated solemnly the affirmacion appointed by Act of / Parliament in such case & being thereupon examined affirmeth as followeth /

4: To the fourth interrogatory saith he knowes the towne & constablery of Reeth & saith the said towne & constablery is reputed to lye within the mannor of Helaugh saith that the lordship of Swaledale mannor of Helaugh & town & / constablery of Reeth & every of them extend nor onely on the north side of the river of Swale but alsoe on the south side of the said river as farr as to where heavenwater deales betwixt Wensleydale & Swaledale saith the river / of Swale was never that this affirmant heard of reputed to be a boundary to limitt or divide the said lordship of Swaledale mannor of Helaugh and the said towne & constablery of Reeth or any of them save that the same have been talked of since / the comencement of this suit by some persons to be a boundary but this affirmant saith he doth not beleive the same to be any boundary or limitt to the said lordship mannor or constablery for that this deponent knowes that the Lord / Wharton lord of the mannor of Helaugh hath not onely lands of his owne in Harkaside (which said Harkaside lyes on the south side of the river of Swale) the tennants of which appear & doe suit & service att the court holden for / the mannor of Helaugh but there are alsoe diverse coppyholders & freeholders in Harkaside which not onely pay certain coppyhold rents & free rents for their lands in Harkaside to the said Lord Wharton (as this deponent beleives[293]as / lord of the <said> mannor which alsoe appear & doe suit & service att the said court as doe alsoe the tennants & inhabitants of Crackpott Spring End & Satteron which three last places lye all on the south side of the river Swale & are / within the constablery of Reeth & further for that John Lonsdale of Helaugh this affirmants father hath a parcell of pasture ground which lyes part of it on the south side of the said river & part of it on the north side thereof / through which the river Swale doth run & the whole lyes within the mannor of Helaugh & constablery of Reeth & the cattle that

293 This bracket is not closed.

depasture therein passe att times from the one side to the other thereof through the river Swale / (except when there is some high flooding of the river) there being a track over the same for their passage & noe fence between the one side & the other thereof & for that that this deponent hath heard from antient people att Helaugh that / the tennants of Harkaside used formerly to drive their cattle downe a lain called Noutherd Laine over the said river up to Harkaside Moor to depasture there /

James Alderson of Helaugh in Swaledale yeoman aged fifty yeares & upwards sworne and examined. /

4: To the fourth interrogatory saith he knowes the vill & constablery of Reeth that the same lyes within the mannor of Helaugh saith that the lordship of Swaledale mannor of Helaugh & towne & constablery of Reeth & every of them / extends not onely on the north side of the river of Swale but alsoe on the south side of the said river as farr as heaven water deales betwixt Wensleydale & Swaledale saith that the river of Swale was never reputed a boundary / that this deponent heard of to limitt or divide the said lordship of Swaledale mannor of Helaugh & town & constablery of Reeth or any of them save that the same hath been talked on by some persons since the comencement of this suit / saith he do's not beleive the same to be a boundary or limitt to the said lordship mannor or constablery for that this deponent knowes that the lord of the mannor of Helaugh hath not onely lands of his owne in Harkaside / (which Harkaside lyes on the south side of the said river of Swale) the tennants of which doe suit and service att Helaugh court but there are alsoe diverse coppyholders & freeholders in Harkaside which not onely pay certain coppyhold & / free rents for their lands in Harkaside to the said Lord Wharton as this deponent beleives as lord of the mannor which alsoe appear & doe suit & service att the said court as doe alsoe the <tennants &> inhabitants in Crakepott Spring End & Sattron which / three last menciond places lye all on the south side of the said river of Swale & are within the constablery of Reeth & extends on the south side of the said river /

Anthony Fryer[294] of Low Row in Swaledale yeoman aged fifty seaven yeares sworne & examined. /

2: To the second interrogatory saith he knowes the dale called Swaledale & hath knowne the same for near fifty yeares saith the same lyes in the honour of Richmond & Richmondshire & in the Wapentakes of Gilling West & Hang West that / East Grinton Whitaside & Spring End part of the said dale are reputed to lye in Hang West & the rest of the dale in Gilling West saith there have not been any courts kept within the said dale that ever this deponent heard of save for the mannors [of][295] / Helaugh & Mukarr which both belong to the Lord Wharton & courts are kept att both those places for the most part once a year att least saith that the Lord Wharton & his ancestors have all this deponents time claimed the royalties of / hunting fishing & fowling within the same & appointed game keepers to preserve the same and this deponent is one of the present Lord Whartons said game keepers & when any deer <happen to> come into the said dale the Lord Wharton claimes them / as his & when any persons fish in the said river of Swale betwixt Harkaside & Helaugh on either side of the river & from the west side of Whitaside to the head of the dale the said game keepers may seize their rods unlesse they have / licenses from the Lord Wharton and saith the tennants of Harkaside Crackpott and Sattron when they appear att Helaugh court usually pay to the lord two pence apeice for license to fish in the said river. /

294 Anthony Fryer of Low Row, a copyhold tenant of Helaugh, aged 56, swore an affidavit on 17 May 1706 (E/207/112/1).

295 Word omitted in the original.

4: To the fourth interrogatory saith he knowes the vill & constablery of Reeth saith the same lyes within the mannor of Helaugh saith the lordship of Swaledale extends not onely on the north side of the river Swale but alsoe on the / southside of the said river as farr as to where heavenwater deales betwixt Wensleydale and Swaledale & soe doth the mannor of Helaugh & constablery of Reeth att Harkaside Crack pott Satteran & Spring End all which places lye in the / said constablery of Reeth & mannor of Helaugh & on the south side of the river of Swale saith he never heard that the river of Swale was reputed a boundary to limitt or divide the said lordship of Swaledale mannor of Helaugh / & constablery of Reeth or any of them nor ever heard the same talked of as a boundary <to> {of} any of them till within those two yeares since this suit begun & saith he verily beleives the same is noe boundary to the said mannor / & constablery of Reeth or either of them for that the tennants of Harkaside Cracke pott Spring End & Sattron appear att Helaugh court & doe suit & service there & the lord of the mannor of Helaugh by the custome of the said mannor hath / power to cutt & take wood growing upon any of the coppyholders lands att Harkaside Crack pott & Satteron & this deponent hath knowne the lords of the said mannor cutt {down} wood in those places & use the same att their smelt mills & / lead works. /

5 & 6: To the fifth & sixth interrogatorys saith that during all the time of this deponents knowledge of Harkaside & West Grinton which is above forty yeares they have been reputed one and the same place saith the Lord Wharton & his ancestors / lords of the mannor of Helaugh & their tennants & other the freeholders & inhabitants there have during all this deponents remembrance been in the possession of the moores & commons belonging to Harkaside & West Grinton & had the enjoyment / thereof by grazing their goods getting of ling peates turves & stones thereon & saith that about fourteen or sixteen yeares agoe one Mr Hutchinson of Gilling then owner of an house & lands att Harkaside called Dikehouse being about to / inclose a parcell of Harkaside Moor or Common by agreement with the Harkaside tennants (in lieu of which he was to keep a common bull or stone horse for them) & the late Lord Wharton being acquainted therewith sent this / deponent to the said Mr Hutchinson to forbid him making the said inclosure the which this deponent did & thereupon the said Mr Hutchinson desisted & soe a stop was put to the same /

Mathew Harrison of Reeth in Swaledale yeoman aged fifty nine yeares & upwards sworne & examined. /

4: To the fourth interrogatory saith he knowes the vill & constablery of Reeth & that the same lyes within the mannor of Helaugh and saith that the lordship of Swaledale mannor of Helaugh & vill & constablery of Reeth extends & lyes on both north & / south side of the river of Swale & extends on the south side of the said river as farr as to where heavenwater deales adjoyning upon Wensleydale saith that he never heard the river of Swale ever reputed a boundary till of late that this suit begun / to limitt & divide the said lordship of Swaledale mannor of Helaugh & vill & constablery of Reeth or any of them and saith his reasons that induces him to beleive that the river of Swale is not a boundary there are because there are severall / freehold & coppyhold tennants att Harkaside on the south side of Swale who appears att the Lord Whartons courts att Helaugh & payes their free rents coppyhold rents fines gressomes & other dutys for fishing there as the rest of the freeholders / & coppyholders within other parts of the said mannor doe & alsoe the freeholders & coppyholders att Crackpott Sattron & Spring End which likewise lyes on the south side of the said river <doe> {did} alsoe appear att Helaugh courts & pay their free rents coppyhold / rents fines gressomes & other dutys there as the freeholders coppyholders & other inhabitants of Harkaside & other places within the mannor of Helaugh doe. /

Roger Colvill[296] late of Fremington in Swaledale esq. aged sixty five yeares or thereabouts sworne & examined /

4: To the fourth interrogatory saith he knowes the towne & constablery of Reeth saith the same lyes within the mannor of Helaugh saith the said mannor & constablery extends on the south side of the river of Swale as well as on the north side / thereof & that he never heard till since this suit began that the river Swale was soe much as talked of to be a boundary to the said mannor or constablery & this deponent lived att Fremington very near the said towne of Reeth for about forty / yeares of his time & saith that all Harkaside (that is within the constablery of Reeth) have all his time paid their assessments with Reeth & doe the duty of watch & ward att Reeth. /

[membrane x *recto*][297]

Leonard Raw of Reeth in Swaledale yeoman aged fifty six yeares & upwards sworne & examined /

To the fourth & fif<teen>th interrogatorys saith he knowes the vill & constablery
4 & 15: of Reeth that the same lyes within the mannor of Helaugh & that the lordship of Swaledale mannor of Helaugh & vill & / constablery of Reeth extends & lyes on both north & south Sides of the river of Swale & extends southwards to where Swaledale adjoynes upon Wensleydale saith that the river of Swale was never / reputed a boundary that ever he heard of till of late that this suit was begunn to limitt & divide the lordship mannor & constablery or any of them and further saith that he has been constable of Reeth for five yeares last past / & that during all that time he has collected & gathered the constable assessments of the inhabitants of Harkaside that are within the constablery of Reeth & alsoe of all the inhabitants of Crackpott & of all the / inhabitants att Spring End except one house & an estate which formerly belonged to Edward Bradrick which said house & estate lyes in Mukarr constablery & those severall places lying on the south side of Swale & the inhabitants / appearing att Helaugh court & belonging to the constablery of Reeth as aforesaid which lyes on the north side of the said river are his reasons which induces him to beleive that the said river is noe boundary to divide the mannor & / constablery as aforesaid /

To the seaventh interrogatory saith that Harkaside & West Grinton & the moores
7: thereof has been during his knowledge thereof reputed part & parcell of the mannor of Helaugh & that the greatest part thereof lys within / the vill & constablery of Reeth & never heard that they were reputed part & parcell of the pretended mannor of East Grinton saith that there are lands lying att Harkaside which belongs to the Lord Wharton & called his own lands & now / farmed by Anthony & John Hird att six pounds thirteen shillings four pence per annum as this deponent hath heard & beleives & that all the inhabitants att Harkaside & West Grinton both freeholders & coppyholders except / John Blaides George Miller & Robert Waller appeares att Helaugh courts for their lands att Harkaside & West Grinton & this deponent hath collected the constable assessments of all the inhabitants of Harkaside except / the persons above excepted. /

Ralph Ellott of Parke hall in Swaledale yeoman aged sixty four yeares or near it sworne & examined /

296 In May 1708 Roger Colvile Esq, aged 66, gave evidence for Lord Wharton at the trial in London (LI Misc 536, Ward MSS, notebooks vol. 6, f. 116v).

297 Parchment, 520mm x 610mm.

4 & 14: To the fourth & fourteenth interrogatorys saith that he hath knowne the towne & constablery of Reeth for about fifty yeares past having lived all his time in the said constablery saith the said towne & constablery lye within the mannor of / Helaugh & both the mannor of Helaugh & constablery of Reeth extend not onely on the north side of the river Swale but alsoe on the south side of the said river as farr southwards as to where heaven water deales betwixt Wensleydale & Swaledale / and this deponent saith the river Swale was not in all this deponents memory ever reputed a boundary to the said mannor or constablery that this deponent heard of (nor hath this deponent ever heard the said river soe much as talked of for such boundary / till this suit begun & saith the reason why he beleives the same to be noe such boundary is because he knowes that the present Earle of Wharton hath & the late Lord Wharton had diverse lands & tennements on the south side of the river / Swale of their own the tennants whereof appear & doe suit & service att the mannor court of Helaugh and alsoe there are diverse coppyholders & freeholders in Harkaside Crackpott Satteron & Spring End which places lye all on the south / side of the said river & doe suit & service att the Helaugh court and saith he hath heard about twenty yeares agoe James Thompson a man then of sixty yeares of age (since deceased) Say that the tennants of Helaugh & Reeth had / a right to intercommon with the tennants of Harkaside & West Grinton upon West Grinton & Harkaside moores & that the tennants of Harkaside & West Grinton had a right to intercommon with the tennants of Helaugh & Reeth upon / Helaugh & Reeth Moores & that the tennants of Helaugh herdsman used to drive their cattle down Noutherd Loning over Swale & soe to Harkaside moores to depasture /

Christopher Robinson of Fremington in the county of Yorke yeoman aged sixty yeares or thereabouts sworne & examined /

4 & 15: To the fourth & fifteenth interrogatorys saith he {hath} knowes the town & constablery of Reeth & hath knowne the same during all the time of his remembrance being borne very near the same & living there all his time saith the same / lys within the mannor of Helaugh & that the mannor & constablery aforesaid lys on both north & south side of the river of Swale saith that the said river of Swale was never reputed a boundary during his time that ever he heard of till of / late that this suit begun to limitt & divide the said mannor & constablery or either of them saith that during all the time aforesaid Harkaside hath paid their land tax & constable assessments to the towne of Reeth & likewise keeps / watch & ward att Reeth when required <or payes> for the same (and because Harkaside which lyes on the south side of the river Swale as dos alsoe Crackpott Satteron & Spring End & that the freeholders & coppyholders of all the said places / appearing & doing suit & service att Helaugh court) are the reasons which make this deponent beleive the said river is noe boundary to limitt & divide the said mannor & constablery as aforesaid. /

John Lonsdale of Helaugh in Swaledale aged seaventy two yeares & upwards being one of the people called Quakers took the solemn affirmation pursuant to the Act of Parliament in that case made & provided & being / examined affirmeth as followeth. /

4: To the fourth interrogatory affirmeth he knowes the vill & constablery of Reeth that the same lys within the mannor of Helaugh saith that the lordship of Swaledale mannor of Helaugh & vill & constablery of Reeth lys on both south & north / side of the river of Swale & extends southwards to where heaven water deales on the height of the moores which devides Swaledale from Wensleydale and that the <said> river of Swale was never reputed a boundary during his time that ever he heard of / till this suit was begun to limitt & divide the said lordship mannor & constablery or any of them saith that he enjoyes a pasture close being coppyhold lands which lys in the mannor of Helaugh & descended to this affirmant from his father which / pasture lys on both north & south side of the said river & that part which lys on the south side adjoynes upon Harkaside & that the cattle which depastures therein gos through the river of Swale att their pleasure & saith that the inhabitants of / Harkaside Crackpott Satteron & Spring End (which places lye on the south side of the said river) appears & does suit & service att the Lord Whartons courts att Helaugh. /

14: To the fourteenth interrogatory saith that about forty yeares agoe in this affirmants fathers time his said father John Lonsdale caused to be digged & gott upon Harkaside Moor severall quantitys of stone att severall times & that the said stones / were brought over the river Swale for this affirmants fathers use att Helaugh and this affirmant saith that <since that> time he has gott both ling & stone upon Harkaside Moor for his owne use & led the same to Helaugh where he lives saith he remembers / that his father used to put sheep upon Harkaside Moor <the sheep this affirmant himselfe put on Harkaside & hath put both sheep & horses upon Harkaside Moor> to depasture there & sometimes drove the same down Noutherd Loning & sometimes over att Scabby Wath all which were done without any interruption from the Harkaside tennants and / hath heard some people say that the said loneing gott the name of Noutherd Loning by the inhabitants driving their cattle to & fro downe the same to the commons aforesaid. /

Christopher Lonsdale of Helaugh in Swaledale yeoman aged eighty yeares sworne & examined /

4: To the fourth interrogatory saith he knowes the vill & constablery of Reeth that the same lys within the mannor of Helaugh saith the said mannor of Helaugh & towne & constablery of Reeth extends both on the south side & north side of the said river Swale / & hath known the same for seaventy yeares past & never heard (till within these two moneths past) that the river Swale was reputed a boundary to limitt or divide the said mannor towne & constablery or any of them /

Richard Hodgson of Kersey Green in the parish of Kirby Ravensworth in the county of Yorke yeoman aged seaventy two yeares & upwards sworne & examined. /

4: To the fourth interrogatory saith he knowes the town of Reeth & that the same lys within the mannor of Helaugh saith that the lordship of Swaledale & mannor of Helaugh lyes on both north & south sides of the river of Swale saith he / hath known the same for upwards of forty yeares & that the said river of Swale was never reputed a boundary that ever he heard of to limitt & devide the said lordship & mannor as aforesaid and further saith that about forty yeares agoe one / William Hodgshon this deponents brother bought a house in the towne of Grinton & this deponent was along with his said brother both when he bought the said house & when he paid the purchase money for the same att Grinton & being then in / company with one William Copperthwaite & Christopher Chambers two inhabitants of Grinton & discoursing of the priviledges that belonged to the said house which was to gett ling peates & turves for fireing upon the moores belonging to / Grinton & how farr the boundarys extended the said Copperthwaite & Chambers then declared that the river of Swale was noe boundary upon the west of the said towne of Grinton. /

6: To the latter part of the sixth interrogatory saith that about forty yeares agoe he very well remembers his said brother William Hodgson & Christopher Chambers both above menciond & both since dead were imployed by the old Lord / Wharton the present Lord Whartons ancestor as they informed him to worke lead mines upon Harkaside Moor on the west of Grinton How & this deponent was imployed by his said brother & Chambers to work there in the said mines & he / wrought some short time with them there & as he remembers wrought there about a fortnight. /

Isabell wife of **Francis Whipp** of Reeth in Swaledale husbandman aged eighty yeares & upwards sworne & examined. /

4 & 14: To the fourth & fourteenth interrogatorys saith she knowes the town & constablery of Reeth & saith the said town & constablery is reputed to lye within the mannor of Helaugh saith the said mannor of Helaugh & constablery of Reeth lys / on both sides of the river of Swale and this deponent tho' she hath knowne the same ever since she can remember never heard that the river of Swale was ever reputed a boundary to the said mannor or constablery or either of them saith that / this deponents mother who dyed above forty yeares agoe & was sixty yeares old when she dyed often told this deponent that her father Mathew Alderson of Reeth being since dead used to depasture his steers & other cattle upon / Harkaside Moor & that she this deponents said mother used to goe once or twice a week to look after them. /

Francis Whipp[298] of Reeth in Swaledale aged seaventy five yeares or thereabouts sworne & examined /

2: To the second interrogatory saith he knowes the dale called Swaledale that the same lyes in the wapentacks of Gilling West & Hang West & that the extent or boundary thereof is from Stollerstone Stile on the east to Hollow Mill Cross on / the west & from Arkingarthdale on the north to Wensleydale on the south saith that the whole dale lys within the parish of Grinton & that theres one forest or free chace in the said dale two mannors (vizt.) Helaugh mannor & Mukar / mannor three constablerys (vizt.) Reeth Grinton & Mukar & that the towne & hamletts following are within the said dale (vizt.) Reeth [x x x][299] & alsoe Fremington Castle Rucroft Sturfitt hall Helaugh Park hall Kirton Mellbecks Low Raw / Smallbarr Blaides Winteringes Pott Ing Gunnerside which places lyes on the north side of the river of Swale & Spring End Crakepott & Harkaside which lyes on the south side of the said river & are all in the constablery of Reeth & mannor / of Helaugh and that East Grinton Cogden Whitaside & Sumer Lodge which lyes on the south side alsoe of the river of Swale & are within the constablery of Grinton & that Mukarr Thwaite Angram Keld Rash high Oxnop low Oxnop / Satteron these lyes on the south side & Ravenseat East stonesdale & West stonesdale Fryth Cracky pott hall Calvett house Ivelett Dikeheads & part of Gunnerside which said last places are on the north side of Swale & are within the constablery of / Mukar and saith that the Lord Wharton & his ancestors have been reputed during all his time cheife lords of the said dale & that there are courts kept yearely att Helaugh & Mukar for the Lord Wharton & noe other courts kept during {all} his time/ within the said dale saith that he never knew or heard of any other lord or lords within the said dale but the Lord Wharton & his ancestors & that they have used the royalties of hunting fishing & fowling within the same & hath been a jury man / att Helaugh court where people have been fined for fishing without license in the river Swale above Arkle beck foot & that when any deer comes into the dale they are reputed to belong to the Lord Wharton & remembers the killing of / red deer in the dale & brought to Reeth to the old Lord Wharton whilst he was there. /

4 & 14: To the fourth & fourteenth interrogatorys saith he knowes the vill & constablery of Reeth having lived there above fifty yeares & that the same lyes within the mannor of Helaugh & that the said vill & constablery extends & lyes on both / north & south sides of the river of Swale & extends southwards as farr as to the topp of the moores

[298] Francis Whipp of Reeth, yeoman, swore an affidavit on 17 May 1706 (E207/112/1). In May 1708 Francis Whipp, aged 75, gave evidence for Lord Wharton at the trial in London (LI Misc 536, Ward MSS, notebooks vol. 6, f. 117r).

[299] One word has been erased.

adjoyning upon Wensleydale and that the <said> river of Swale was never during his time reputed a boundary that ever he heard of till of / late since this suit begun to divide the said lordship of Swaledale mannor of Helaugh & the said town of Reeth from Harkaside or any of them for that Reeth & Harkaside have been accounted all one township during his time & saith that he / this deponent liveing all his time att Reeth about forty yeares <agoe> pulled & gott eighty threave of ling upon Harkaside Moor & lead the same to Reeth for his owne use without asking leave or being hindred or disturbed by the Harkaside / inhabitants or any other & hath had severall goods in severall yeares depastured & summered on Harkaside Moor without any disturbance saith he gott ling & depastured his cattle there having <there> a right soe to doe being both / a freehold & coppyhold tennant att Reeth. /

7: To the seaventh interrogatory saith that Harkaside & West Grinton & the moores thereof have been during all his time reputed part & parcell of the mannor of Helaugh & that the greatest part thereof lys within the constablery of Reeth saith he / knowes the lands called the lords owne lands lying att Harkaside & now farmed by the two Hirds but not the yearely rent that they pay for the same & that there are alsoe severall coppyhold lands lying att Harkaside the / tennants whereof pays yearly coppyhold rents to the Lord Wharton & payes alsoe the said lord fines att death or alienacion (vizt.) for that which is called the old land a twenty penny fine which is for every penny rent twenty penny fine & for the new land / for every penny rent fifteen penny fine & the said coppyhold {lands} <tennants> doe alsoe appear & doe their suit & service att the said Lord Whartons coppyhold courts held for the said mannor of Helaugh saith that there are severall freeholders att Harkaside which alsoe appear / & dos suit & service att Helaugh court. /

11: To the eleventh interrogatory saith that the vill & township of Reeth payes a yearely modus of seaven pounds two shillings & two pence for tyth of hay & corn to the Impropriator of the parish of Grinton & that Harkaside contributes & payes a part / of the same but the particular summe this deponent dos not now remember /

Richard Wensley of Featham in Swaledale yeoman aged ninety two yeares sworne & examined /

4: To the fourth interrogatory saith that he knowes the towne & constablery of Reeth saith the same lys within the mannor of Helaugh saith the said mannor & constablery extends on the south side of the river of Swale as farr as to Wensleydale & / on the north side of the said river as farr as to Arkingarthdale saith he hath lived all his time nigh the river Swale & never heard that it was reputed a boundary to the said mannor towne & constablery or any of them nor can this deponent / imagine how it should be reckoned a boundary to the said mannor towne & constablery for that the same extends soe farr on both sides of the said river & that Harkaside Crackpott & Spring End lye all on the south side of the same / river & are all in the said constablery of Reeth & mannor of Helaugh. /

[membrane x *verso*]

Edward Chirry of Fetholme in Swaledale aged sixty three yeares & upwards sworne & examined. /

4: To the fourth interrogatory saith he knowes the vill & constablery of Reeth & that the same lyes within the mannor of Helaugh and that the lordship of Swaledale mannor of Helaugh & vill & / constablery of Reeth extends & lyes as well on the south side of the river of Swale as on the north side thereof & that the same lys & extends southwards to the top of the moores adjoyning upon Wensley / dale & that the said river of Swale was never reputed a boundary till the suit began that ever he heard of to limitt or divide the said mannor & constablery or either of them & saith he hath knowne the same / about fifty yeares & hath lived in the said dale all his time and saith that the reasons which induces him to beleive that the said river is noe boundary are that there are severall places on the south side of / the river Swale where are both coppyhold lands & freehold lands which belongs to the Lord Wharton & have been all his time reputed within the mannor of Helaugh & constablery of Reeth (vizt.) att Harkaside / Satteron Crackpott {Satteron} & Spring End & saith that Harkaside is within the constablery of Reeth mannor of Helaugh & remembers that about sixteen yeares agoe one Mr Hutchinson of Gilling having / purchased an estate att Harkaside of one John Miller this deponent was along with the said Mr Hutchinson att Harkaside where he proposed to the inhabitants there that if they would consent to let him inclose / a small peice of Harkaside Common & lay it to the said estate he had purchased of Miller he would in consideracion thereof be obliged to keep a common bull for the use of the said Harkaside men which was then agreed / unto & amongst them and one Anthony Fryer who was then steward to the old Lord Wharton as this deponent beleives hearing thereof came there & in this deponents presence & hearing told the said Mr Hutchinson & the / Harkaside men that in case they inclosed the said peice of common without leave or lycense from the said Lord Wharton (who he said was lord there) he would throw the said inclosure down again or words to the like effect / upon which the said Mr Hutchinson declared he would not meddle therein & soe the said agreement took noe effect. /

William Simpson of Helaugh in Swaledale miner aged sixty three yeares sworne & examined. /

4: To the fourth interrogatory saith he knowes the vill & constablery of Reeth & hath knowne the same for fifty yeares saith the said towne & constablery have all this deponents time been reputed to lye within / the mannor of Helaugh & that the mannor of Helaugh & town & constablery of Reeth extend on the south side of the river of Swale as farr as to the boundarys of Wensleydale as well as on the north side of the said / river and this deponent saith <in> all his time (untill of late that this suite was begun) he never heard the least mention of the said river Swale being a boundary to the said mannor of Helaugh or towne & constablery of Reeth / or any of them nor doth this deponent beleive that the said river can with any reason be reputed a boundary to the said mannor or constablery for that not onely Harkaside but alsoe Crackpott Satteron & Spring End / which all lye in Swaledale & on the south side of the river Swale are part of the constablery of Reeth & had all this deponents time been reputed to lye within the said mannor & all the tennants or inhabitants / thereof except two or thee houses in Harkaside or West Grinton appear & doe suit and service att Helaugh court. /

13: To the thirteenth interrogatory saith that the cattle & goods belonging to West Grinton & Harkaside have all this deponents time been usually impounded in Reeth pinfold the tennants of Harkaside & West Grinton / having noe other pinfold. /

14: To the fourteenth interrogatory saith that the complainants tennants of the mannor of Helaugh have frequently above twenty yeares agoe used to drive their cattle down Noutherd Loning to Harkaside Moore / to depasture them there & the tennants of West Grinton & Harkaside have alsoe made use of the commons of Reeth & Helaugh for depasturing their cattle thereon but not soe frequently within twenty yeares past / as before. /

Robert Stoddard of Reeth in Swaledale aged sixty two yeares & upwards sworne & examined /

4: To the fourth interrogatory saith he knowes the vill & constablery of Reeth that the same lyes within the mannor of Helaugh & that the lordship of Swaledale mannor of Helaugh & constablery of Reeth lyes as well on the / southside of Swale as on the northside of the said river & that the same extends southwards to the height of the moores adjoyning upon Wensleydale & that the said river was never reputed a boundary that ever he heard / of till of late that this suit begun to limitt & divide the said mannor & constablery or either of them & his reasons are that Swale cannot be a boundary from Arkle beck foot to the top of the dale because the Lord Wharton / hath landes on both sides of the said river which belongs to his mannors of Helaugh & Mukar & that the said river runs downe in the midst of the said mannors. /

12: To the twelfth interrogatory saith that about the time the suit was depending betwixt Sir Solomon Swale & Mr Marriott this deponent having discourse with one John Watson who lived in East Grinton & since dead / about the said suite & alsoe of severall other matters relating to the said Sir Solomon Swale & his ancestors the said John Watson then told this deponent that he remembered the building of Swalehall & that he had heard that old / Sir Soloman Swale asked leave to build the mill att Swalehall of the old Lord Wharton & that Reeth miller has liberty to grind corne att the said Swalehall mill in the drought of summer when Reeth mill wanted / water or when it was frozen up in winter and the said John Watson att the same time told this deponent that he remembered that coales were very scarse in Grinton Moor and that the Grinton men asked leave of / the Harkaside men to grave & gett peates & turves for firing upon Harkaside Moor. /

13: To the thirteenth interrogatory saith that he knowes Reeth pinfold which stands on the north side of the towne of Reeth & that the inhabitants of Harkaside usually impound their tresspassing cattle therein /

Margarett Buckle of East Grinton widdow aged ninety yeares att least sworne & examined. /

14: To the fourteenth interrogatory saith that James Galloway long since dead was this deponents father & lived att Helaugh & had noe lands att Harkaside & when this deponent was a young girl she remembers her / father had ewes pastureing on Harkaside Moor & this deponent & her sister Elizabeth who is long since dead milkd their father's ewes whilst they went on the said Harkaside Moor as aforesaid /

Mary Garth of Whitaside in Swaledale widdow aged eighty yeares & upwards sworne & examined. /

4: To the latter part of the fourth interrogatory saith that she hath lived att Whitaside during all her time but about three yeares & during all that time she never heard the river of Swale reputed / a boundary till about three or four moneths since to limitt & divide the lordship of Swaledale mannor of Helaugh & vill & constablery of Reeth or any of them. /

Anthony Kirton of Oxnop in Swaledale aged seaventy two yeares & upwards sworne & examined /

2: To the second interrogatory saith he knowes the dale called Swaledale saith the same lyes within the wapentakes of Gilling West & Hang West that the extent or reputed boundary of the said dale / is from Stollerston Stile to Hollow Mill Crosse & from Arkingarthdale to Wensleydale & that the Lord Wharton & his ancestors have been during all his time reputed lords of Swaledale & keeps courts yearly att / Helaugh & Mukar for the mannors of Helaugh & Mukar & never knew any court kept besides in the said dale and that the Lord Wharton & his ancestors have all along claimed & enjoyed the royaltys / of hunting fowling & fishing through the whole dale & saith he remembers that severall persons in his time have been put to trouble by the Lords of Swaledale for killing deer & hares & moor game / within the dale & that the said lords appoints & keeps game keepers to preserve the said severall games within the said dale & that severall men has been fined att Helaugh courts for fishing without licenses / in the river of Swale & this deponent saith that he hath served in the jury att Helaugh court when such persons were fined. /

4: To the fourth interrogatory saith he knowes the vill & constablery of Reeth that the same lys within the mannor of Helaugh & that the lordship of Swaledale mannor of Helaugh & vill & constablery of Reeth / lyes on both north & south sides of the river of Swale & extends as farr southwards as to Wensleydale & saith that the said river has never been reputed a boundary that he heard of till this suit was / begun to divide the said mannor & constablery or either of them & saith he cannot beleive Swale to be a boundary because the Lord Whartons lands lye on both sides the said river which belongs to his two / mannors of Helaugh & Mukarr. /

James Harker of Satteron in Swaledale yeoman aged seaventy eight yeares sworne & examined /

4: To the fourth interrogatory saith he knowes the vill & constablery of Reeth saith the same is within the mannor of Helaugh & both the said mannor & constablery lys as well on the south side of the river of Swale as on / the north thereof & extends on the south as farr as to where heaven water deales betwixt Wensleydale & Swaledale saith that till this suit began he never heard that the said river was ever reputed a boundary to / limitt or divide either the said mannor of Helaugh towne & constablery of Reeth or either or any of them neither doth this deponent beleive the said river can with any colour or reason be reputed a boundary to any of / them because Harkaside Crackpott Satteron & Spring End lye all on the south side of the said river & have all this deponents time been reputed to be in the constablery of Reeth & mannor of Helaugh (he this / deponent having lived all his time att Satteron[300] & Helaugh & Reeth lye both on the north side of the said river. /

Margarett Carter of Whitaside in Swaledale spinster aged sixty yeares sworne & examined. /

4: To the fourth interrogatory saith she was born & bred & lived all her time att Whitaside in Swaledale & untill this suit begun she never heard that the river Swale was accounted a boundary to the / lordship of Swaledale mannor of Helaugh & towne & constablery of Reeth or any of them /

14: To the fourteenth interrogatory saith that she hath within these twenty or thirty yeares seen James Lonsdale of Helaugh lead stones from Harkaside Moor (which he had gott there) to Helaugh to build / his house withall & hath seen John Lonsdale of the same towne of Helaugh bear ling which he had gott on Harkaside Moor to Helaugh & thatch his cowhouse att Helaugh therewith /

William Dunn of Grinton in Swaledale aged fifty five yeares or thereabouts sworne & examined. /

4: To the fourth interrogatory saith he knowes the vill & constablery of Reeth that the same lyes within the mannor of Helaugh & saith that the lordship of Swaledale mannor of Helaugh & vill & constablery / of Reeth lyes as well on the south side of the river of Swale as on the north side thereof & that it extends southward to Wensleydale hath knowne the same above forty yeares & never heard the river / of Swale reputed a boundary till this suit was begun to limit & divide the said lordship mannor or constablery or any of them saith he has been a fisherman during all his time & never heard that the Swale / was a boundary from the top of the dale to Grinton /

5: To the fifth interrogatory saith he knowes Harkaside & the moores & commons thereunto belonging & alsoe Grinton & the moores thereunto belonging saith he

300 This bracket is not closed.

knowes a boundary which divides the said places / & which boundary divides as well the moores as the inclosed grounds thereto belonging saith he knowes the highway leading from Grinton to Harkaside & the gate in the said highway & the elmtree / standing near thereto & that the inhabitants of Grinton repaires the said highway near to the said elmtree & saith that the said boundary which divides the said Grinton from Harkaside begins att the river of Swale / crosses Naylors holme into the highway betwixt that elmtree & Cuddy Spout & soe up the bounder ridge to Jenny Wrenn cowhouse & along the side of the Ox Intack wall upon the moor to the east corner of the said / intack wall & crosses Hughes Ditch to the White Stone on the How & from thence to the Gray Stone in the Greets & soe up to where heaven water deales which divides Swaledale from Wensleydale & soe / along as heaven water deales to Gibbon hill & from thence to the topp of Browney gill which divides Harkaside from Whitaside & soe downe Browney gill to Scabby Yate & soe to the river of Swale saith / he was acquainted & told of this boundary forty yeares agoe by one William Robinson of Harkaside then an old man since dead & hath been often told of the said boundary since. /

Joseph Clarkson of Satteron in Swaledale yeoman aged fifty one yeares & upwards sworne & examined /

2: To the second interrogatory saith he hath knowne Swaledale for forty yeares past saith the Lord Wharton & his ancestors have been all his this deponents remembrance reputed cheif lords of Swaledale / & claimed the royaltyes of hunting fishing & fowling therein & a right to all the deer that came into the said dale & have appointed game keepers for those purposes who have seized doggs guns / fishing rods & fishing nets & taken them from such as have used them without license from the lords of Swaledale. /

[membrane xi *recto*][301]

4: To the fourth interrogatory saith that he knowes the vill & constablery of Reeth saith it lys in the mannor of Helaugh <and both the mannor of Helaugh> & vill & constablery of Reeth extend on the south side of the river of Swale where / heaven water deales betwixt Wensleydale & Swaledale & on the north side of the said river to Arkingarthdale saith that he never heard that the river Swale was reputed a boundary to limitt or divide the said / mannor of Helaugh & towne & constablery of Reeth or any of them nor ever knew it so much as talked of to be a boundary till within a month past nor doth this deponent beleive theres any / reason for accounting it a boundary there because the said mannor lys on both sides the said river & soe doth the constablery of Reeth & the mannor of Mewkarr alsoe which reaches up to the head of Swaledale. /

Edmond Harker of Cracky pott hall in Swaledale aged seaventy two yeares or thereabouts sworne & examined. /

2: To the second interrogatory saith he knowes the dale called Swaledale & that the same lys within the Wapentacks of Gilling West & Hang West & that the Lord Wharton & his ancestors have been for all his time cheife lords of the / said dale and that the said lords have during all his time kept courts att Helaugh & Mukar within the said dale & never heard of any other courts kept there and that the <said> Lord Wharton & his father have during all his time had & / enjoyed the royalties of hunting fishing & fowling within the said dale & has kept game keepers for preserving of game in and

301 Parchment, 520mm x 620mm.

through the whole dale and this deponent saith that he was a game keeper under the old Lord Wharton / & is now a keeper to this present lord to look after the red deer that comes into the dale which has allways been reputed a free chace & has killed & taken red deer within the said dale for the Lord Whartons use. /

4: To the fourth interrogatory saith he knowes the vill & constablery of Reeth & that the same lyes within the mannor of Helaugh saith that the mannor of Helaugh & vill & constablery of Reeth extends & lyes on both north & / south sides of the river of Swale & extends southwards as farr as where heaven water deales adjoyning upon Wensleydale & saith that the river of Swale was never talked of as a boundary to limitt or divide the said mannor or / constablery during his time that ever he heard of till this present day that he has heard it talked on & thinks it a sufficient reason that Swale cannot be a boundary because the Lord Wharton hath lands on both sides of the said / river which appears att his lordships courts as above. /

Ralph Waggett of Reeth in Swaledale miner aged fifty yeares sworne & examined /

4: To the fourth interrogatory saith he knowes the vill & constablery of Reeth saith the same lyes within the mannor of Helaugh & that the said mannor & constablery extend both on the north & south sides of the river Swale saith that before this suit / begun this deponent never heard the river Swale soe much as talked on to be a boundary to the said mannor or constablery. /

11: To the eleventh interrogatory saith there is a modus of about seaven pounds per annum for tyth hay & corne paid by the township of Reeth & saith the hamlett of Harkaside as being part of the said township contributes a part thereto / but how much this deponent cannot certainly tell having never collected the same himselfe. /

15: To the fifteenth interrogatory saith he hath been severall yeares collector of the land tax & poll money within twenty yeares past & saith Harkaside was allways during his being collector thereof joyned with Reeth / in one & the same assessments bills to those aids & saith that the tennants of Harkaside doe the duty of watch & ward att Reeth in their course. /

Ralph Simpson of Feetholme in Swaledale yeoman aged seaventy nine yeares & upwards sworne & examined /

4: To the fourth interrogatory saith he knowes the vill & constablery of Reeth that the same lyes within the mannor of Helaugh & that the lordship of Swaledale mannor of Helaugh & vill & constablery of Reeth lyes on both north / and south sides of the said river & extends southwards to where heaven water deales adjoyning upon Wensleydale & that the river of Swale was never during his time talked of as a boundary to limitt & divide the said lordship mannor / constablery or any of them till about a fortnight since that he heard of and saith his reasons which make him beleive Swale not to be a boundary are because that Harkaside Crackpott Satteron & Spring End which / places lye on the south side the river & all lye within the mannor of Helaugh & the inhabitants of the said severall places which have lands there appear att Helaugh court & dos suit & service there & that the said places / are also within the constablery of Reeth which town of Helaugh & Reeth lyes on the north side of the said river. /

Anthony Close of Bank head in Swaledale yeoman aged seaventy seaven <yeares> sworne & examined. /

4: To the fourth interrogatory saith he hath knowne the vill & constablery of Reeth for threescore yeares saith the said towne & constablery lys within the mannor of Helaugh & that the said mannor of Helaugh & constablery of Reeth / extend not onely on the north side of the river Swale to Arkingarthdale but alsoe on the south side thereof to Wensleydale saith that in all his time he never heard the river Swale reputed a boundary to limitt the said mannor / of Helaugh or towne & constablery of Reeth or either of them and this deponent cannot beleive it to be any boundary to the said mannor of Helaugh or to the town & constablery of Reeth because both the said mannor & constablery / lye on both sides the said river as aforesaid & because the said river is noe boundary to the mannor of Mukarr in the said dale & the mannor of Mukar lys alsoe on both sides the said river. /

James Tiplady of Smarber in Swaledale aged fifty five yeares & upwards. /

4: To the fourth interrogatory saith he knowes the towne & constablery of Reeth & that the same lys within the mannor of Helaugh saith that the lordship of Swaledale mannor of Helaugh & constablery of Reeth lyes on both south / & north sides of the river of Swale & extends southwards to where heaven water deales adjoyning upon Wensleydale & saith that he never heard the river of Swale talked of as a boundary till this suit begun to divide the said mannor / & township & beleives the said river is noe boundary because there are severall lands both freehold & coppyhold lying att Harkaside Crackpott Satteron & Spring End which belongs to the Lord Wharton & the tennants thereof appear att the said / lords courts att Helaugh & are within the constablery of Reeth which said severall places lyes on the south side of the river & Reeth lys on the north side thereof. /

Thomas Spenceley of Helaugh in Swaledale husbandman aged fifty yeares & upwards sworne & examined. /

4: To the fourth interrogatory saith he knowes the vill & constablery of Reeth saith it lyes within the mannor of Helaugh & both the said mannor & constablery extend as well on the south side of the river Swale as farr as to Wensleydale & on the / north side thereof to Arkingarthdale saith he hath knowne the river Swale & the mannor of Helaugh & towne & constablery of Reeth for about forty yeares past & never in all that time knew or heard that the said river hath / been ever reputed a boundary to limitt or divide the said mannor or towne or constablery or either or any of them & saith he doth not beleive the said river can with any reason be supposed any such boundary because Harkaside / & Crackpott & Satteron & Spring End which lye all on the south side of the said river are within the constablery of Reeth & mannor of Helaugh. /

Anthony Fothergill of Feetholme in Swaledale husbandman aged forty eight yeares sworne & examined /

4: To the fourth interrogatory saith he knowes the vill & constablery of Reeth & that the same lyes within the mannor of Helaugh & that both the said mannor & constablery extends as well on the south side as on the north side of the / river of Swale & extends southwards as farr as Wensleydale saith he hath knowne the said mannor of Helaugh & the said river & the towne of Reeth for about forty yeares & that the said river during all that time was never reputed a / boundary that ever he heard of till this suit began to limitt & divide the said lordship of Swaledale mannor of Helaugh & constablery of Reeth or any of them & saith he dos not beleive the said river can be a boundary because that this / deponent hath served upon the jury att Helaugh court severall times in severall yeares with severall inhabitants of Harkaside Crackpott Satteron & Spring End which places lye on the south side of the river & this deponent lives on the / north side & the said river runs betwixt Helaugh towne & the said places menciond above and this deponent alsoe saith that he had a close lying att Fetholme up the river above Helaugh & adjoyning upon the same river & part of the / said close was taken away with the said river in a great flood & laid on the contrary side being on the south side & this deponents brother now occupyes both the parts of the said close & the river runs betwixt the said parts soe divided by the said river /

George Storey of Ivelett in Swaledale husbandman aged seaventy four yeares sworne & examined /

2: To the second interrogatory saith he knowes the dale called Swaledale being born & having lived most part of his time therein saith that during all his time the late Lord Philip Wharton & the present Earl of Wharton have / been reputed cheif lords of all Swaledale saith there have been court leets kept for the mannors of Helaugh & Mukarr during all his time but noe other courts in any part of the said dale that this deponent heard of saith the said / cheif lords of Swaledale have all his time claimed to have & enjoy the chases for deer & royalties of hunting fowling & fishing throughout the whole dale & appointed game keepers there for those purposes. /

4: To the fourth interrogatory saith he knowes the vill & constablery of Reeth saith the same towne & constablery lys within the mannor of Helaugh saith the lordship of Swaledale mannor of Helaugh & constablery of Reeth & every / of them not onely extend to the north side of the river Swale but the same extends to the south side of the said river as farr as to where heaven water deales betwixt Wensleydale & Swaledale saith he never heard in all his time / that the said river was ever reputed a boundary to limitt or divide the said lorship of Swaledale mannor of Helaugh & the said vill & constablery of Reeth or any of them nor doth this deponent beleive that the same was ever / discoursed of <as> a boundary to any of them unlesse some people since this suit began have talked of it to be soe & saith he doth not beleive there is any ground to repute it a boundary to the said dale or mannor or constablery / for this deponent saith that Harkaside Crackpott Satteron & Spring End lye all on the south side of the said river & have all along been reputed to be within the constablery of Reeth & mannor of Helaugh (Reeth & Helaugh / lying on the north side of Swale) & the tennants of all those places appear att Helaugh court & doe suit & service there. /

5: To the fifth interrogatory saith that about sixty yeares agoe this deponent being sent with a letter from one Michaell Wharton of Wharton Hall who was then steward to the late Philip Lord Wharton with orders to / deliver it to Roger Co[l]vill of Fremington esq. dead who was then steward of the <said> late Lord Whartons courts of Helaugh & Mukarr & father to Roger Colvill now living this deponent found the said <late> Mr Colvill in company with Sir Solomon Swale / deceased grandfather to the present Sir Solomon att one Douglasse house in Low Fremington after this deponent had delivered the letter he heard the said Mr Colvill relate to old Sir Solomon Swale that West Grinton & Harkaside was within / the Lord Whartons liberty & that the boundary betwixt West Grinton & Harkaside & East Grinton liberty began att the river Swale & soe went up to the highway near the gate & soe up to the White Stone / in the How & soe to the Grey stone in the Greets & that Harkaside & West Grinton boundary went on as heaven water deales betwixt Wensleydale & Swaledale to the head of Browney gill & down Browney gill to Scabby Wath & this / deponent saith that the said Sir Solomon Swale said nothing to the contrary nor seemd to deny the same and this deponent saith he hath seen severall of the said bounder markes but doth not know them all /

14: To the fourteenth interrogatory saith that the same day that this deponent carryed the said letter to Mr Colvill he mett one John Carter of Reeth with an horse load of turfes bringing them to Reeth & this deponent saith that he inquiring / of Carter where he had gott those turfes Carter told this deponent he had gott them on Harkaside Moor for he said it was in my Lord Whartons Liberty & that Reeth tennants might gett turfes thereon if they would or to that effect /

William Pratt of Fetham in Swaledale husbandman aged fifty <seaven> yeares & upwards sworne & examined. /

4: To the fourth interrogatory saith he knowes the towne & constablery of Reeth & that the same lyes within the mannor of Helaugh & that the lordship of Swaledale mannor of Helaugh & constablery of Reeth extends & lyes on both / north & south sides of the river of Swale and saith that he hath knowne the said places & the said river of Swale for upwards of forty yeares having lived all his time in the dale & never heard the said river called a boundary till / this suit <was> begun to limitt & divide the lordship of Swaledale mannor of Helaugh & constablery of Reeth or any of them & saith he beleives Swale is noe boundary att the low side of the mannor of Helaugh because where / he lives is above the towne of Helaugh & towards the upper end of the said mannor & there the Lord Wharton hath coppyhold lands on both north & south side of the said river & hath heard & beleives that the said Lord Wharton / hath lands on both sides the said river lower downe as farr as the said mannor of Helaugh gos. /

James Raw of Low Raw in Swaledale yeoman aged fifty eight yeares & upwards sworne & examined /

4: To the fourth interrogatory saith he knowes the towne & constablery of Reeth & that the same lys within the mannor of Helaugh saith the mannor of Helaugh & constablery of Reeth lys on both north & south sides the river of / Swale & extends southwards to Wensleydale hath knowne the said places & the said river for above forty yeares being borne in the said dale & having lived all his time there & never heard the said river talked of as a / boundary till this suit begun to divide the said mannor & constablery or either of them saith the reasons that makes him beleive that Swale is noe boundary because the Lord Wharton hath lands in severall places on the south side / the river of Swale the tennants whereof appears att Helaugh court & dos suit & service there /

Ralph Peacock of Reeth in Swaledale husbandman aged fifty six yeares & upwards sworne & examined /

4: To the fourth interrogatory saith he knowes the vill & constablery of Reeth saith the same lys in the mannor of Helaugh <and both the mannor of Helaugh> & constablery of Reeth extends as well on the south side as on the north side of the river of Swale & never / (till this contest) heard that the said river was reputed a boundary to the said mannor or constablery & beleives the same to be noe boundary because that the tennants of Harkaside Crackpott & Spring End which places lye on / the south side of the said river lye within the said constablery of Reeth <and mannor of Helaugh> & are assessd with the tennants of Reeth in the constable assessments. /

James Parkin of Pott Ing in Swaledale husbandman aged fifty <four> yeares & upwards sworne & examined /

4: To the fourth interrogatory saith he knowes the towne & constablery of Reeth & that the same lys within the mannor of Helaugh saith that the said mannor & constablery lys on both north & south side of the river of Swale saith / that he was born in Swaledale & lived there all his time & never heard the river Swale talked on as a boundary till within two moneths last past to limitt & divide the said mannor & constablery or either of them & saith Swale cannot / as this deponent beleives be a boundary to divide the mannor of Helaugh because there are severall coppyhold lands lying att Harkaside belonging to the Lord Wharton on the south side of Swale the tennants whereof appears att / Helaugh court and this deponent saith that he hath cutt wood att severall times in severall yeares for twenty yeares last past in the said coppyhold lands att Harkaside & in other coppyhold lands higher up in the dale alsoe in / the south side of the said river for the Lord Whartons mills & mines on the north side of the said river. /

James Spenceley of Whitaside in Swaledale rough mason aged sixty seaven yeares sworne & examined on the complainants behalfe /

14: To the fourteenth interrogatory saith that above forty yeares agoe one James Robinson of Reeth deceased shewd this deponent a way which lay in Harkaside on the south side of the river Swale which the said Robinson told this deponent / was a cartway formerly used by Reeth men for leading turves & ling from Harkaside Moor to Reeth & he alsoe told this deponent that the tennants of Reeth had a Right to gett ling & turfes on Harkaside Moor. /

[membrane xi *verso*]

Thomas Wharton of Brockabanke in Swaledale husbandman aged sixty eight yeares sworne & examined. /

4: To the fourth interrogatory saith he knowes the vill & constablery of Reeth & hath knowne the same all his time saith the said towne & constablery lys within the mannor of Helaugh & saith the mannor / of Helaugh & constablery of Reeth extends not onely on the north side of the river Swale to the boundarys of Arkingarthdale but alsoe on the south side of the said river for severall miles in the said dale & up south / wards to where heaven water deales betwixt Wensleydale & Swaledale saith he never heard (till since this suit begun) that the river Swale was ever reputed a boundary to limitt or divide the said mannor or / constablery or either of them & this deponent beleives there can be noe ground or reason to repute the said river a boundary to the said mannor or constablery for that Harkaside Crackpott Sattron / & Spring End which all lye on the south side of the said river have all this deponents time been reputed to lye in the constablery of Reeth & mannor of Helaugh & hath knowne John Lonsdale of / Helaugh servant severall yeares since lead stones from Harkaside which he beleives he had gott upon Harkaside Moor because this deponent saw the leader cart in Harkaside coming towards / Helaugh with stones. /

Henry Clarkson of Crackpott in Swaledale husbandman aged fifty eight yeares & upwards sworne & examined /

4 & 15: To the fourth & fifteenth interrogatorys saith he knowes the towne & constablery of Reeth that the same lys within the mannor of Helaugh & saith that the said mannor & constablery lys on / both south & north sides of the river of Swale & extends southwards to Wensleydale saith he hath knowne the said mannor & constablery & the said river of Swale for forty yeares & that during all / that time the said river was never reputed a boundary that ever he heard of till this suit begun to devide the said mannor & constablery or either of them saith he was two yeares collector of the / land tax for the constablery of Reeth & that the inhabitants of Harkaside <were taxed> with Reeth assessments & he collected their proporcion of the said land tax of the said inhabitants of Harkaside and / because Harkaside which lyes on the south side of Swale is taxed with Reeth which lyes on the north side thereof is his reason which makes him beleive & think that Swale is noe / boundary to devide the same. /

James Wharton of Low Raw in Swaledale husbandman aged sixty five yeares sworne & examined /

4: To the fourth interrogatory saith he knowes the towne & constablery of Reeth that the same lyes within the mannor of Helaugh & that the said mannor of Helaugh & constablery of Reeth / lyes on both south & north sides of the river of Swale & extends southwards as farr as Wensleydale saith he was born in Swaledale & lived all his time there & never heard the river of Swale reputed / a boundary till this suite begun to divide the said mannor & constablery or either of them saith that the Lord Wharton hath coppyhold lands att <Harkaside> Crackpott Satteron & Spring End which lyes on the south / side of Swale & are in the constablery of Reeth & mannor of Helaugh & hath alsoe some coppyhold lands att Whitaside & saith he hath knowne George Raw & John Clemison both of Low Raw pay / each of them two pence per annum to the Lord Wharton for liberty of fishing in Swale & saith alsoe that James Lonsdale of Helaugh about eight or tenn yeares since gott both stones & ling upon / Harkaside Moore & brought the same to Helaugh for his own use there without any interruption that ever he heard of & alsoe saith that he remembers the Lord Wharton caused wood to be cutt in the / coppyhold lands att Whitaside & Crackpott & the same was carryed to his mills & mines on the north side the said river & that the Lord Wharton hath lands on both sides the river to the high end of the / dale & saith for these things above menciond he cannot think Swale is a boundary to divide the said mannor & constablery as above /

Thomas Chirry of Fetholme in Swaledale smelter aged fifty seaven yeares sworne & examined /

4: To the fourth interrogatory saith he knowes the vill & constablery of Reeth saith the same lys within the mannor of Helaugh & that the said mannor & constablery extend on the south side of the river Swale as / farr as to Wensleydale & on the northside of the said river to Arkingarthdale saith he never heard till this suit began that the river Swale was ever reputed a boundary to limitt or divide the lordship of / Swaledale mannor of Helaugh & towne & constablery of Reeth or any of them alsoe this deponent hath lived all his time in Swaledale & doth not beleive the said river can be a boundary to any of the said / places for that Harkaside Crackpott Satteron & Spring End all lyeing on the south side of <the said> river are and have been all along reputed to lye within the constablery of Reeth & mannor of Helaugh & the tennants of / those places under the Lord Wharton appear att Helaugh court & doe suit & service there. /

Ralph Wensley of Riddings in Swaledale miner aged fifty nine yeares sworne & examined. /

4: To the fourth interrogatory saith he knowes the towne & constablery of Reeth & that the same lys within the mannor of Helaugh saith that the lordship of Swaledale mannor of Helaugh & towne & constablery / of Reeth extends & lyes on both south & north sides of the river of Swale & saith he was borne in the said dale & lived all his time there & never heard the said river talked of as a boundary till this suit was begun / to divide the said lordship mannor & constablery or any of them & further saith that the Lord Wharton having lands & common right on both sides the river are the reasons that makes him think Swale is noe boundary / to divide the said lordship mannor & constablery as above. /

John Metcalfe[302] of Gunnerside in Swaledale aged fifty yeares sworne & examined /

4: To the fourth interrogatory saith he knowes the vill & constablery of Reeth that the same lys within the mannor of Helaugh saith the said mannor & constablery extend not onely on the northside of the / river Swale but alsoe on the southside thereof as farr as to Wensleydale boundarys saith he hath lived all his time in Swaledale & (untill this suit was begun) never heard that the river Swale was / reputed a boundary to the lordship of Swaledale mannor of Helaugh & towne & constablery of Reeth or any of them saith he cannot imagine why the said river should be reputed a boundary as aforesaid for / that Harkaside Crackpott Satteron & Spring End lye all on the southside of the river Swale & are within the mannor of Helaugh & constablery of

302 John Metcalfe of Swaledale, miner, swore an affidavit on 14 November 1707 (E207/115/1).

Reeth & my Lord Whartons tennants there appear att / Helaugh court & doe suit & service <therein> & this deponent hath lands lying on both sides the said river lyeing in the constablery of Mukarr but in the mannor of Helaugh for which he doth suit & service att Helaugh / court. /

Anthony Wharton of Feetham in Swaledale miner aged thirty eight yeares & upwards sworne & examined /

4: To the fourth interrogatory saith he knowes the towne & constablery of Reeth & that the same lys within the mannor of Helaugh & saith that the lordship of Swaledale mannor of Helaugh & constablery of / Reeth lyes on both south & north sides of the river of Swale & that during all his remembrance he never heard Swale reputed a boundary to limitt or divide the said lordship mannor & constablery or any of them / till this suit began saith that the Lord Wharton having lands lying on both sides of the said river for tenn miles together cannot beleive Swale to be a boundary to devide the said lordship & he hath / alsoe led wood from Harkaside out of William Arundells pasturing which was cutt & felled there being coppyhold lands & brought the same to the Lord Whartons mines on the northside the Swale / and this deponent being fishing severall times in the river Swale betwixt Helaugh & Harkaside & in severall other places in the said river on the south side thereof was threatened by James Fryer & his / father who were bayliffes to the Lord Wharton within his mannor of Helaugh and the said bayliffes have catched him both on the north & south sides of the said river & when they soe catched him / discharged him from fishing on either side and this deponent further saith that several fishers in summer time came & lodged att his fathers house att Brockabanke within the mannor of Helaugh / & when the said bayliffes heard of them the said bayliffes came & discharged them from fishing in the river of Swale & for these reasons last mencioned be beleives Swale cannot be a boundary as above /

William Wilkin of Helaugh in Swaledale blacksmith aged forty eight yeares sworne & examined /

4 & 14: To the fourth & fourteenth interrogatorys saith he hath knowne the vill & constablery of Reeth for eighteen yeares past saith the same lys within the mannor of Helaugh & that both the said / mannor of Helaugh & constablery of Reeth extend both on the northside & southside of the river Swale saith he never heard that the river of Swale was ever accounted a boundary to the / lordship of Swaledale mannor of Helaugh & towne & constablery of Reeth or any of them before this suit begun saith he beleives the same cannot be a boundary to any of the said places for / that there are lands lying on both sides of the said river which have all this deponents time of knowing the said towne & constablery been reputed to lye in the said mannor of Helaugh & constablery of Reeth / and perticularly on the southside of the said river att Harkaside & saith that James Lonsdale the elder & James Lonsdale the younger both of Helaugh have ledd stones from Harkaside Moor to / Helaugh & used them there & William Arundells sons & Thomas Miller an exciseman have been fined att Helaugh court for fishing in the river Swale without license from the lord. /

Daniel Addison of Helaugh yeoman aged fifty two yeares sworne & examined. /

4 & 14: To the fourth & fourteenth interrogatorys saith he knowes the towne & constablery of Reeth that the same lys within the mannor of Helaugh & that the lordship of Swaledale mannor / of Helaugh & towne & constablery of Reeth lys on both south & north sides the river of Swale & extends southwards as farr as Wensleydale saith he was borne in the dale & lived all his time there / & never heard Swale talked of as a boundary till this suit began to limitt & divide the said lordship mannor & constablery or any of them saith that the Lord Wharton having lands att Harkaside / Crackpott Satteron & Spring End which lyes on the south side the Swale the

tennants whereof appears att Helaugh court & dos suit & service there & hath heard one James Thomson a very old man now dead / who lived in Helaugh severall times say that the Lord Wharton had good right upon Harka & that Sir Solomon Swale had a sheeprake upon Calva being Helaugh Common on the northside of Swale & for / these reasons he cannot think Swale can be a boundary and saith that theres a loning called Noutherd Loning lying from Helaugh to the river Swale & hath heard old men say that Helaugh men used to / drive their cattle that way to Harkaside Common to depasture there. /

Thomas Hutchinson[303] of Reeth in Swaledale yeoman aged thirty five yeares sworne & examined. /

2: To the second interrogatory saith that he knowes the dale called Swaledale & that the present Lord Wharton & the Late Lord Philip Wharton have during his time been reputed cheif lords of the / said dale saith the said lords have kept court leets & court barons yearly for the mannors of Helaugh & Mukarr but noe courts else are or have been kept in the said dale that he ever heard of saith / the said lords have all his time claimed the royalties of hunting fishing & fowling within the said dale & this deponent & his brother & others not having license used to fish in the night in the river / Swale from the foot of Arklebeck over ag[ainst] Cuddy Spout & soe upwards least the Lord Whartons game keepers should meet with them fishing there. /

4: To the fourth interrogatory saith he knowes the vill & constablery of Reeth saith it lys within the mannor of Helaugh & saith the lordship of Swaledale mannor of Helaugh & constablery of Reeth & / every of them extend not onely on the north side of the river Swale but alsoe on the south side of the said river as farr as from the said river southwards to the confines of Wensleydale saith he never heard / that the river Swale was ever reputed to be a boundary to limitt or divide the said lordship of Swaledale mannor of Helaugh & town & constablery of Reeth or any of them saith he doth not beleive / there can be any grounds to repute the said river as a boundary to the said lordship mannor towne or constablery because the said lordship of Swaledale lys on both sides the said river & Harkaside Crackpott / Satteron & Spring End which lye on the south side of the said river are reputed to be in the constablery of Reeth & mannor of Helaugh. /

[membrane xii *recto*][304]

5: To the fifth interrogatory saith he knowes Harkaside & West Grinton & the moores & commons thereof & hath knowne the same above twenty yeares saith the same are accounted as one place & not divided the one from the other / by any boundary saith he knowes Grinton or East Grinton & the moores thereof saith there is a boundary dividing Grinton or East Grinton from Harkaside & West Grinton saith the said boundary dos not onely / divide the moores & commons of Harkaside or West Grinton from those of East Grinton but the said boundary doth extend throughout & divides the inclosed lands of each respectively saith he knowes the highway or road / leading from East Grinton to West Grinton & Harkaside knowes alsoe the gate in the said road & the elmtree growing near it & hath knowne the same ever since he was a boy saith the said boundary begins att the river Swale / over against Arkle beck foot crosses the end of the Holme to the said road & soe gos betwixt Cuddy Spout and

303 In May 1708 Thomas Hutchinson (no age given) gave evidence for Lord Wharton at the trial in London (LI Misc 536, Ward MSS, notebooks vol. 6, f. 117r).

304 Parchment, 580mm x 680mm.

the said elmtree up the great ridge to Benson cowhouse & soe to Ox Intack nook & from thence cross the moor over Hughes Ditch in a streight / line to the White Stone in the How & from thence in a direct line to the Gray Stone in the Greets & soe up to where heaven water deales saith the reason he hath to beleive the same to be true & remarkable boundermarkes is that he hath been / shewne the same whilst he was young by severall antient neighbours for true boundermarkes as aforesaid. /

6: To the sixth interrogatory saith he knowes the lead mines called Howgroves now wrought by the defendant Marriott & alsoe the lead mines or workes lately opened by the complainants agents lying near thereto (vizt.) about seaventy or eighty / yards distance saith the complainants said workes lye within the moores & commons of West Grinton & Harkaside & the boundarys thereof & not within those of Grinton or East Grinton & the boundary line runs betwixt the complainants & defendants said workes / saith the complainants said workes were not opened till after the riding of the aforesaid boundary about two yeares agoe & publick notice was given of the riding att Reeth markett crosse <on> the markett day before saith the said boundary was ridd from Swale / to the height of the Greets by the perticular boundermarkes as afore described by this deponent saith he this deponent was att the said riding <& saith at the said riding> they proceeded from the height of Grinton Greets westward to Browney gill head & downe Browney gill to / Scabby Yate & soe to the river Swale saith he beleives the Lord Wharton & his ancestors & their tennants & other the freeholders & inhabitants of Harkaside & West Grinton have been all this deponents time in possession & had enjoyment / of the moores & commons of Harkaside & West Grinton by eating the herbage thereof with their horses cattle & sheep getting of ling stones turfes & peates there & by disturbing & hindering the tennants of East Grinton from getting of ling / & turfes on the said moor & saith that about twenty five yeares agoe he very well remembers his this deponents mother having given this deponents brother James Hutchinson who dwelt at Leyburn seaven or eight sheep which his / said brother had driven to Layborn the said sheep strayed from Leyborn & were taken up by one Miles Freer bayliffe or agent for the late Lord Wharton who told this deponent he had taken them up on Harkaside Moor as waifes for the said / Lord Whartons use & this deponents mother paid the said Freer groats a peice for restoring the said sheep again. /

7: To the seaventh interrogatory saith that Harkaside & West Grinton & the moores thereof have been all this deponents knowledge & remembrance thereof reputed to lye within the lordship of Swaledale & mannor of Helaugh & not within the / mannor of Grinton or East Grinton saith the complainant the Lord Wharton hath Leasehold lands in Harkaside & West Grinton farmed by Anthony Hird & John Hird under the yearely rent of twenty nobles or thereabouts saith that / Sir Solomon Swale Christopher Raper Ralph Addison Marmaduke Hartley & Ralph Wensley are freeholders within Harkaside & West Grinton & are called att Helaugh court saith that Ralph Wensley Marmaduke Hartley Christopher / Rapier William Arundell John Arundell James Arundell & George Arundell his brethren are coppyholders there & are all called att the said court saith all the said persons appear att the said court & doe suit & service att the said court (except / Sir Solomon Swale who hath not appeared there in this deponents time that he can remember) but whether the said freeholders & coppyholders appear their for their freehold or coppyhold lands or for both this deponent cannot tell nor / doth he know what perticular free rents or coppyhold rents they pay to the lord of the said mannor of Helaugh /

8: To the eighth interrogatory saith he knowes severall lands or inclosures adjoyning to the said moores of Harkaside & West Grinton called intacks which belong to the complainant the Lord Wharton & to which Anthony Hird & John Hird are tennants / but the same have not been taken in within this deponents memory nor doth this deponent know what rents the said Hirds pay for the same to the complainant saith that one Ralph Addison a poor man about twenty yeares agoe by / consent of the neighbourhood took in about halfe an acre from the said moores but whether he had the consent of the lord or not this deponent cannot tell but saith the same is of little or noe vallue. /

10: To the tenth interrogatory saith that if a coppyholder in Harkaside or West Grinton dye seized of any coppyhold or customary lands there leaving severall sons all the sons inherit the same alike according to the custome of the mannor of Helaugh & / if he leave a widdowe such widdow enjoys by the said custome a third part <of> such lands during the time she continues his widdow but noe longer. /

11: To the eleventh interrogatory saith there is & during all his time hath been a yearly modus or prescript rent for tyth corne & hay of seaven pounds two shillings two pence payable by the township of Reeth & saith that the tennants both of / freehold lands & coppyhold lands in Harkaside contribute their parts thereto amounting to about twenty shillings. /

13: To the thirteenth interrogatory saith he knowes the pinfold belonging to the township of Reeth & saith the cattle or goods belonging to Harkaside & West Grinton are usually impounded therein. /

14: To the fourteenth interrogatory saith that for twenty yeares past he hath knowne a laine called Noutherd Lane leading downe from Helaugh towards Harkaside which he hath heard was formerly used for driving Helaugh tennants cattle / through from Helaugh to Harkaside Moor to pasture & for Harkaside tennants cattle to be driven through to pasture on Reeth Moor & hath heard that Reeth Moor was formerly kept as a stinted pasture & whilst it was soe the inhabitants of / Harkaside had cattle gates therein according to their stints. /

15: To the fifth<eenth> interrogatory saith that the inhabitants of Harkaside in Reeth constablery are assessed & pay with Reeth both the land taxe & constable sesse as being part of the township of Reeth & saith that Harkaside tennants keep watch & ward / att Reeth. /

William Hutchinson of Reeth in Swaledale aged forty five yeares & upwards sworne & examined /

4: To the fourth interrogatory saith he knowes the towne & constablery of Reeth & that the same lys within the mannor of Helaugh saith that the lordship of Swaledale mannor of Helaugh & town & constablery of Reeth lyes on both north & south sides / of the river of Swale & extends southwards to where Swaledale adjoynes upon Wensleydale saith he was born att Reeth & hath lived there all his time & never heard Swale menciond as a boundary till of late that this suit began to divide the said / lordship mannor & constablery or any of them. /

5: To the fifth interrogatory saith he knowes Harkaside & West Grinton & the moores & commons thereof & that they are & have been all his time accounted one & the same place & theres noe boundary to divide the same saith he knowes Grinton or East / Grinton & the moores thereof & that theres a boundary which divides the same as well on the moores as in the inclosed grounds from Harkaside or West Grinton saith he knowes the road leading from East Grinton to Harkaside & West Grinton & the gate standing in / the road & the elmtree growing near it & that the inhabitants of East Grinton repaires the said highway near to the said elm tree & hath knowne the same during all the time of his remembrance & saith that the place in the said highway that Grinton men / repaires to is one of the bounder markes & that the said boundary begins att the river of Swale over against Arkle beck foot & soe into the <said> highway a little east of the said elmtree & where Grinton repaires to & soe up the ridge to Jenny Wrenn cowhouse & soe / to the corner of Ox Intack & soe to the White Stone on the How & from thence to the Gray Stone on the Greets & soe to where heaven water deales which divides Swaledale from Wensleydale & soe along the topp of the Greets as heaven water deales to Gibbon / hill & from thence to the topp of Browney gill which divides Harkaside from Whitaside & soe downe Browney gill to Scabby Yate & soe to the river of Swale. /

6: To the sixth interrogatory saith he knowes the lead mines called How Groves now wrought by the defendant Marriott & alsoe the lead mines or workes lately opened & claimed by the Lord Wharton near adjoyning saith that the lead workes soe opened / by the Lord Wharton lyes within the moores & commons of Harkaside & West Grinton & the boundary thereof & not within the moores & commons of East Grinton & the boundary thereof & that the said workes lye about fourscore yards from each other & that / the said boundary runs up betwixt the said severall workes saith that the Lord Wharton caused the said workes to be opened about two yeares agoe & that the said boundary was ridd some short time before the said workes were opened & publick notice was given of / riding the said boundary by the Lord Whartons bayliffe in Reeth markett upon a markett day & in Grinton church yard after divine service was ended & the said boundary was ridd as above described by this deponent by the Lord Whartons agents & / near fourscore more neighbours & the same was agreed unto to be a true boundary by most of the persons soe riding saith he hath heard of the said boundary severall yeares agoe by severall old men that were neighbours saith that the said Lord / Wharton & his ancestors & their tennants att Harkaside & West Grinton & other the tennants & inhabitants there were & have been in possession of the moores & commons att Harkaside & West Grinton during all the time of this deponents remembrance / by their pulling of ling getting peats turves & stones & grazing & depasturing of their cattle thereupon & saith about twenty five yeares agoe this deponents mother gave to this deponents brother James Hutchinson who then lived att Layborn betwixt six & / tenn ew sheep but the certain number this deponent cannot now remember & the said ew sheep happened to stray from Layborn & this deponents mother hearing that Miles Fryer who was then bayliffe to the old Lord Wharton within his mannor of Helaugh / had taken up such like sheep senth[305] this deponent & his brother Thomas Hutchinson to the said Miles Fryer to looke if they were the said ewes that she had given her son James & finding they were the same & that the said Fryer had tar'd the same about / the necks as waifes asked the said Fryer where he took the same up & he saith upon Harkaside Moor & he made this deponent pay four pence a head for loosing them. /

7: To the seaventh interrogatory saith that the greatest part of the Harkaside & West Grinton lyes within the mannor of Helaugh & constablery of Reeth & that the Lord Wharton hath lands there which are lett to Anthony & John Hird att six pounds / thirteen shillings & four pence per annum & called lords owne lands and saith Marmaduke Hartley gentleman Christopher Raper Mr Thomas Dickenson Ralph Wensley Ralph Addison & John Hird have freehold lands in Harkaside & are called att Helaugh / court & beleives they pay free rents to the Lord Wharton but knowes not what free rent they severally pay & saith Sir Solomon Swale in alsoe called att Helaugh court amongst the freehold tennants saith that William Arundell & his three brothers Mr / Hartley Christopher Raper have all of them coppyhold lands lying att Harkaside & holds the same by coppy of court roll of the Lord Whartons mannor of Helaugh & each of them payes coppyhold rents to the Lord Wharton yearely but

305 For 'sent'.

what perticular / rents knowes not but saith that the said coppyhold tennants att death of tennant or alienacion pays for old land a twenty penny fine & for new land fifteen penny fine to the lord of the mannor of Helaugh. /

11: To the eleventh interrogatory saith that the township of Reeth pays yearly a modus of seaven pounds two shillings & two pence for tyth corne & hay to the impropriator of Grinton parish & that Harkaside payes above twenty shillings of that summe to Reeth /

13: To the thirteenth interrogatory saith he knowes Reeth pinfold & that Harkaside men usually impound their tresspassing cattle therein & that the said pinfold stands on the north side of the said towne of Reeth. /

James Boneson of Sturtforth hall in Swaledale fisherman aged seaventy yeares sworne & examined /

4: To the fourth interrogatory saith he knowes the vill & constablery of Reeth & hath knowne the same for above fifty yeares saith the said towne & constablery lye within the mannor of Helaugh & that both the said mannor & constablery extend not onely on the north side of the / river Swale but on the south side of the said river alsoe as farr as to Wensleydale saith that he never in all his time heard that the river Swale was accounted a boundary to the said mannor or constablery or either of them nor doth this deponent beleive the same can be / reputed a boundary thereto by reason the said mannor & constablery lye on both sides the said river & he hath used to fish in the said river for fifty yeares past & never heard any mencion of its being a boundary nor ever was disturbed of fishing in the / said river of either side of it save by the Lord Whartons bayliffe or steward & saith he hath had his fishing rodd taken from him severall times by the Lord Whartons officers both on the south side & north side of the said river for fishing therein / without license from the Lord Wharton twenty yeares ago & in severall yeares since. /

Dorothy Metcalfe of Reeth in Swaledale spinster aged sixty five yeares & upwards sworne & examined. /

4 & 14: To the fourth & fourteenth interrogatorys saith she knowes the vill & constablery of Reeth being born there & having lived all her time there & that the same lys within the mannor of Helaugh saith that the lordship of Swaledale mannor of Helaugh & towne & / constablery of Reeth lys on both north & south sides of the river of Swale & saith that during all her time she never heard the said river talked of as a boundary to limitt & divide the said mannor township & constablery or any of them & that her father Mathew / Metcalfe had lands both in Reeth & Harkaside & hath heard her father say that Reeth & Harkaside were allways reckoned as one township because Harkaside paid their sesses with Reeth & had common right upon each others moores saith her father / sold his lands att Harkaside to one John Hirds ancestors for thirteenscore pounds & that the said John Hird now enjoyes the same and this deponent saith that she heard Christopher Hird (who had the said lands att Harkaside which her father sold) say that he had / gates on Reeth Moor in right of these lands & had thirteen nobles bid for his right of those gates but would not take it saith her father was about one hundred yeares old when he dyed & she never heard him speak of Swale as a boundary & beleives if it / had been a boundary he would have att sometime spoke of it saith she remembers that near fifty five yeares since she saw John Kirton a bayliffe to the old Lord Wharton Bryan Metcalfe of Gunnerside in Swaledale & Leonard Hutchinson of / Helaugh in Swaledale hunt a red deer in Harkaside & killed the same in the river Swale beneath Harkaside & pulled him out on the north side & they then said they hunted & killed the said deer for the old Lord Whartons use. /

Thomas Fothergill of Kirton in Swaledale miner aged forty yeares & upwards sworne & examined /

4: To the fourth interrogatory saith he knowes the vill & constablery of Reeth saith it lys within the mannor of Helaugh saith that both the said mannor & constablery extend not onely on the north side but alsoe on the south side of the river Swale as farr as to the height of the moors / dividing Wensleydale from Swaledale saith that untill this suit began he never heard that the river Swale was reputed a boundary to limitt or divide the said mannor or constablery or either of them altho he hath knowne the said mannor & constablery from his youth & doth / not beleive theres any reasonable grounds to repute the said river a boundary to the said mannor & constablery for that most part of the hamletts of Harkaside Crackpott Satteron & Spring End lying all on the south side of the said river are within the said constablery & mannor and / this deponents brother Richard Fothergill hath a parcell of meadow ground which lys on both sides of the said river & within the said constablery & mannor & the river Swale runs through the same. /

Richard Hutchinson of Rucroft in Swaledale gentleman aged thirty nine yeares sworne & examined. /

2: To the second interrogatory saith he knowes the dale called Swaledale & that the same lys within the Wapentacks of Gilling West & Hang West saith that the extent or boundary of the said dale is from Hollow Mill Cross on the west to Stollerston Stile on the east & from / Arkingarthdale on the north to Wensleydale on the south & that the Lord Wharton & his ancestors have during all his time been reputed cheif lords of the said dale & that there are courts kept yearly att Helaugh & Mukar for the said mannors in Swaledale for the Lord / Wharton & noe other courts kept in the dale that ever he heard of & that the Lord Wharton & his ancestors as Lords of Swaledale have during his time had & enjoyed a free chase for red deer in Swaledale <& appointed> & keeps keepers there & other game keepers for preserving {of }/ of other games therein & hath heard that the said game keepers hinders persons from fowling fishing & hunting within the said dale without license from the said Lord Wharton. /

4: To the fourth interrogatory saith he knowes the vill & constablery of Reeth that the same lys within the mannor of Helaugh & that the lordship of Swaledale mannor of Helaugh & vill & constablery of Reeth lyes on both south & north sides of the river of Swale & extends / southward to Wensleydale & saith he hath lived all his time in the dale & never heard the river of Swale accounted a boundary to divide the said lordship mannor & constablery or any of them neither dos he beleive Swale can be a boundary for that the Lord Wharton / hath both freehold & coppyhold lands att Harkaside & alsoe att Crackpott & Spring End which places lye on the south side of Swale & are within the mannor of Helaugh & constablery of Reeth & the tennants <there> appears att Harkaside court & doe suit & service there & this / deponent hath collected the land tax att Harkaside Crackpott & Spring End as belonging to the constablery of Reeth he being collector of the sess severall times for the towne & constablery of Reeth & hath seen Arthur Hutchinson of Reeth about twelve yeares agoe lead / freestone in a cart to Reeth for the said Arthur Hutchinsons use which said Arthur told this deponent he gott the said stones upon Harkaside Moor & further saith he hath seen William Arundell who lives att the west end of Harkaside pay two pence att Helaugh court for / lycense to fish in the river of Swale. /

John Peacock of Reeth in Swaledale yeoman aged fifty four yeares sworne & examined /

2: To the second interrogatory saith he knowes the dale called Swaledale & hath known the same ever since his youth being borne & having lived all his time therein saith the Lord Wharton & his ancestors have been during his time all along reputed cheife lords / of the said dale saith the said lord & his ancestors have usually kept courts yearely for Helaugh & Mukar <mannors> & this deponent never heard of any other courts kept in the said dale saith the said cheif lords of Swaledale have all his time claimed & enjoyed free chace for deer & the / royalties of hunting fishing & fowling within the said dale. /

4: To the fourth interrogatory saith he hath known the vill & constablery of Reeth all his time saith the said vill & constablery are within the mannor of Helaugh saith the lordship of Swaledale mannor of Helaugh & constablery of Reeth extend not onely on the north side of the / river Swale but alsoe on the south side of the said river as farr as to Wensleydale side saith in all his time till since this suit begunn he never heard that the river Swale was soe much as talked on or reputed a boundary to limitt or divide the said lordship mannor town / & constablery or any of them neither doth this deponent beleive the same can be in reason reputed a boundary to the said lordship mannor town or constablery for that the said dale lys on both sides the said river & soe doth the said mannor & constablery too perticularly Harkaside / Crackpott Sattron & Spring End lye all on the south side of Swale & are in the constablery of Reeth & mannor of Helaugh. /

James Pratt of Gunnerside in Swaledale miner aged fifty four yeares sworne & examined /

2: To the second interrogatory saith he knowes the dale called Swaledale & that the extent or boundary thereof is from Hollow Mill Cross on the west to Stollerston Stile on the east & from Arkingarthdale on the north to Wensleydale on the south & that the Lord Wharton / & his ancestors have during all his time been reputed lords of Swaledale and that the Lord Wharton keeps courts yearly att Helaugh & Mukarr for the said mannors which are in the said dale & never heard of any other courts kept there & has heard & beleives that / Lord Wharton {that the said Lord Wharton} & his ancestors has all along during his time claimed & enjoyed the royalties of hunting fishing & fowling within the said dale & has game keepers for preserving the severall games therein /

[membrane xii *verso*]

4: To the fourth interrogatory saith he knowes the vill & constablery of Reeth & that the same lys within the mannor of Helaugh & that the lordship of Swaledale mannor of Helaugh & towne & constablery of Reeth lys on both north & south sides of / the river of Swale & extend southwards to Wensleydale saith that he hath lived about forty yeares in the said dale & never heard Swale talked of as a boundary till this suit was began to divide the said lordship mannor & constablery or any of them / saith he knowes the Lord Wharton has coppyhold lands att Helaugh which this deponent has farmed formerly & has paid the coppyhold rents for the said lands to the Lord Whartons bayliffe of the mannor of Helaugh & has alsoe paid seses / for the said lands att Harkaside with the towne of Reeth & that there are lands lying att Gunnerside where this deponent now lives in possession of John Metcalfe & that part of the said lands lyes on the north side of Swale & part on the / south side & the said Metcalfes cattle gos over the Swale from one part to another att their pleasure & for these reasons this deponent beleives the river Swale is noe boundary as above. /

William Harrison of Barugh in Swaledale miner aged fifty six yeares sworne & examined /

2: To the second interrogatory saith that he knowes Swaledale & that the Lord Wharton & his ancestors have all the time of this deponents remembrance been reputed cheif lords thereof & claimed & enjoyed the royalties of hunting fishing & / fowling therein & saith the said Lords have yearly kept courts in the mannors of Helaugh & Mukarr & this deponent never heard or knew of any other courts kept in the said dale. /

4: & 14: To the fourth & fourteenth interrogatorys saith he knowes the vill & constablery of Reeth saith the same lyes in the mannor of Helaugh & that both the said mannor & constablery lye on both the north & south sides of the river Swale & / this deponent saith he never heard till since this suit begann that the river Swale was ever reputed a boundary to the said mannor town or constablery or any of them nor doth this deponent beleive the same can with any reason be / reputed a boundary to the said mannor or constablery or either of them for that his father was an old man when he dyed & beleives if the said river had been reputed a boundary in his time he this deponent should have heard him / talk of it but never did hear him mention it as such & saith there are lands lying att Harkaside Crackpott Spring End Oxnop on the south side of the <said> river & are within the constablery of Reeth & mannor of Helaugh & John Metcalfe / of Gunnerside hath a parcell of ground att Gunnerside which lyes part of it on the north side of Swale & other part on the south side of the said river & both sides of the said land lye within the said mannor of Helaugh and / constablery of Reeth & John Lonsdale of Helaugh hath likewise a close att Helaugh part whereof lyes on the north side & part on the south side of Swale & both parts lys in the said mannor & constablery & the cattle which depasture / in both the said closes passe from one side to the other as they please & saith the Lord Whartons tennants of Helaugh & Reeth (as this deponent hath been told) by his uncle Mathew Metcalfe (who was ninety nine yeares old when he / dyed) had right to gett turfe ling peates or stones on Harkaside Moor /

James Gorton of Low Row in Swaledale aged thirty one yeares being one of the people called Quakers took the solemne affirmacion according to the Act of Parliament & being examined affirmeth as followes /

4: To the fourth interrogatory affirmeth that he knowes the vill & constablery of Reeth & that the same lys within the mannor of Helaugh saith that the mannor of Helaugh & constablery of Reeth lyes on both the north & south sides of the river of Swale / saith he hath knowne the said places above for about fifteen yeares & during all that time never heard Swale reputed a boundary to divide the mannor of Helaugh & constablery of Reeth or either of them till this suit began that he / remembers. /

William Plewes[306] of Richmond in the county of Yorke gentleman aged sixty eight yeares & upwards sworne & examined /

4: To the fourth interrogatory saith he hath knowne the vill & constablery of Reeth for forty five yeares past & saith the same lys in the mannor of Helaugh saith the lordship of Swaledale mannor of Helaugh & township & constablery of Reeth / extend both on the north side & south side of the river Swale saith that in all his time (till since this suit begun) this deponent never heard that the river Swale was ever reputed a boundary to limitt or divide the said lordship of Swaledale / mannor of Helaugh or towne & constablery of Reeth or any of them & this deponent doth not beleive there can be any just grounds to call the said river a boundary to the said lordship mannor or constablery for that the lordship or mannor of / Mukarr lying in the said dale lys on both sides of the said river & there are lands in Harkaside Crackpott Satteron which lye on the south side of the river Swale & have been all

306 In May 1708 William Plewes, aged 68, gave evidence for Lord Wharton at the trial in London (LI Misc 536, Ward MSS, notebooks vol. 6, f. 116v).

this deponents knowledge thereof reputed to be within the said mannor / of Helaugh & constablery of Reeth. /

15: To the fifteenth interrogatory saith that about thirty yeares agoe this deponent being then an attorney att law was imployed by the customary or coppyhold tennants of the mannor of Helaugh & township of Reeth in a suit brought / against them by the late Philip Lord Wharton relating to fines which the said late Philip Lord Wharton demanded of them for a rent called the enhansed rents & saith that William Arundell & Marke Close who were then tennants of / coppyhold or customary lands in Harkaside joyned & contributed to the maintaining of that suit for their said lands in Harkaside with the other coppyhold tennants of the mannor of Helaugh & township of Reeth & that there is a / decree between the lord & tennants in that suit. /

John Scott of Reeth in Swaledale husbandman aged fifty one yeares & upwards sworne & examined /

2: To the second interrogatory saith that he knowes the dale called Swaledale & that the same lys within the Wapentacks of Gilling West & Hang West & that the extent or boundary of the said dale is from Stollerston Stile on the east to / Hollow Mill Crosse on the west & from where heaven water deales adjoyning upon Arkingarthdale on the north to the height of the moores where heaven water deales adjoining upon Wensleydale in the south & that the Lord Wharton & / his ancestors have during all his time been called cheif lords of Swaledale & that there are courts kept yearly att Helaugh & Mukarr within the said dale by the Lord Wharton & never <knew> or heard of any other courts kept in the said dale / and that the Lord Wharton & his ancestors have during all this deponents time claimed & enjoyed the royalties of fishing fowling & hunting through the dale & knowes Edmond Harker hath been a keeper for the Lord Wharton to look / after the red deer that are in the said dale & that there are other keepers to look after the preservacion of other games there. /

4: To the fourth interrogatory saith he knowes the vill & constablery of Reeth & that the same lys within the mannor of Helaugh & that the lordship of Swaledale mannor of Helaugh & vill & constablery of Reeth lys on both north & south / sides of the river of Swale & extends southwards to where heaven water deales adjoyning upon Wensleydale saith he hath knowne the severall places & the river of Swale for upwards of forty yeares & during all that time never heard Swale / talked of to be a boundary to limitt & divide the lordship of Swaledale mannor of Helaugh & towne & constablery of Reeth or any of them till about twelve moneths since that this suit was begun neither does he beleive that Swale can be / a boundary to divide the said mannor & town & constablery because the Lord Wharton hath lands att Harkaside Crackpott Satteron & Spring End which are on the south side of Swale & are within the mannor of Helaugh & constablery of Reeth / and this deponent is the better enabled to depose herein because he having been collector of the land tax for the towne & constablery of Reeth hath gathered the said tax att Harkaside Crackpott Satteron & Spring End as being part & parcell / of the said towne & constablery of Reeth. /

5: To the fifth interrogatory saith he knowes Harkaside & West Grinton & the moores thereof that they are not divided but are accounted one & the same place saith he knowes Grinton & East Grinton & the moores thereof & that there's a boundary to divide / the same from Harkaside & West Grinton which divides the inclosed grounds as well as the moores saith he knowes the highway leading from East Grinton to Harkaside & West Grinton & the gate therein & the elmtree growing near thereto / and that the inhabitants of East Grinton repaires the said highway to near the <said> tree saith that the boundary begins at the river of Swale ne[a]r against Arklebeck foot & crosses Naylors holme into the said highway a little east of the elmtree / & soe up the ridge to Jenny Wrenn cowhouse & soe to the east corner of Ox Intack

& crosses Hughes Ditch & soe to the White Stone in the How & from thence to the Gray Stone on the Greets & soe to where heaven water deales which divides / Swaledale from Wensleydale. /

6: To the sixth interrogatory saith he knowes the Groves called How Groves now wrought by the defendant Marriott & the lead mines or workes lately opened by the Lord Wharton near adjoyning saith that the Lord Whartons workes lye within the moores & / commons of Harkaside & West Grinton & the boundary thereof & not within the moores & commons of East Grinton & the boundary thereof & saith that the said workes lye about fourscore yards distant from each other & the <said> boundary runs up / betwixt them & that the said Lord Wharton begun to open the said workes about two yeares agoe & the said boundary was ridd by the Lord Whartons agents & severall others the day before the said workes were opened & that publick notice of the said / riding was given in Reeth markett upon a markett day and saith that the Lord Wharton & his ancestors & their tennants & other the tennants & inhabitiants att Harkaside & West Grinton have been in possession of the said moores & commons att / Harkaside & West Grinton during all the time of this deponents remembrance by grazing & depasturing their cattle sheep & horses pulling ling graving peates & turves & getting stones thereupon & saith that Arthur Hutchinson Henry Machell / & Roger Alderson being coppyhold tennants at Reeth about twelve yeares agoe gott stone upon Harkaside Moor & ledd the same to Reeth for their severall uses in building & repairing their severall houses there & saith he hath heard that / the inhabitants of East Grinton have been interrupted & hindered from getting ling & peates upon Harkaside Moor & that the Harkaside men have burnt severall times ling & peats that the Grinton men had pulled & digged there & not / suffered them to carry the same away. /

7: To the seaventh interrogatory saith that Harkaside & West Grinton & the moores thereof have been during all the time of his remembrance thereof reputed part & parcell of the lordship of Swaledale mannor of Helaugh & vill & constablery of Reeth / & that there are lands att Harkaside belonging to the Lord Wharton & called the lords owne lands now farmed by Anthony Hird & John Hird att Six pounds thirteen shillings four pence yearely rent & that there are coppyhold lands att Harkaside / now in possession of Ralph Wensley Christopher Raper William Arundell George Arundell James Arundell & John Arundell & Mr Hartley & that the said coppyhold lands lyes within the mannor of Helaugh & are held by coppy of court roll from / the said mannor & that the severall coppyhold tennants there att death or alienacion payes fines to the Lord Wharton (vizt.) for old lands a Twenty penny fine & for new lands a fifteen penny fine & that they pay yearely coppyhold / rents to the lord but what rents in perticular this deponent knowes not and further saith that Mr Hutchinson Ralph Wensley Christopher Raper Mr Hartley & Ralph Addison are freehold tennants there & are called & appears att Helaugh / court saith he beleives they pay some free rents to the said lord but knowes not what they are. /

10: To the tenth interrogatory saith that the coppyhold or customary lands att Harkaside & West Grinton are descendible to all the sons alike att the death of the tennants & that the widdowes have a third part during the time they continues widdowes / but if they marry again they loose their part & saith that the same is according to the custome within the mannor of Helaugh & that theres noe such custome within the pretended mannor of East Grinton /

11: To the eleventh interrogatory saith he knowes that theres a modus or prescript rent of seaven pounds two shillings & two pence paid yearely for tyth of corn & hay by the township of Reeth & that Harkaside payes above twenty shillings / part of the said modus. /

12: To the twelfth interrogatory saith that he hath heard Sir Solomon Swale called severall times att Helaugh court as a freehold tennant within the said mannor /

13: To the thirteenth interrogatory saith he knowes the pinfold belonging to the town of Reeth & that the inhabitants of Harkaside usually impound their tresspassing cattle therein /

15 latter pt: To the latter part of the fifteenth interrogatory saith that the inhabitants of Harkaside which are within the township of Reeth have during all this deponents remembrance paid land tax & constable assessments with Reeth & that they / keep {their} watch & ward att Reeth when required or paid for the same. /

Marmaduke Hartley of Middleton Tyas in the county of Yorke gentleman aged forty one yeares sworne & examined. /

4 & 7: To the fourth <& seventh> interrogatorys saith he knowes the town & constablery of Reeth & that it lys within the mannor of Helaugh saith he is owner of both freehold & coppyhold lands in Harkaside which are within the constablery of Reeth & mannor / of Helaugh & appears & dos suit & service att Helaugh court for both the said freehold & coppyhold lands saith the freehold lands is about the yearly vallue of seaven pounds per annum for which freehold land this deponent pays a yearely / free rent to the lord of the mannor of Helaugh & to the best of this deponents remembrance the said free rent is nine pence per annum saith his coppyhold lands are part of them Helaugh mannor new lands & part of it Helaugh mannor / old land saith he pays coppyhold rents for the said lands to the lord of the mannor of Helaugh & to the best of his remembrance the same amounts to about twenty shillings per annum but how much above he cannot be positive and / part of the said coppyhold rent is enhansed rent for which this deponent pays noe fine upon death or alienacion but for the rest of the said coppyhold lands a twenty penny fine for the old Land & a fifteen penny fine for the new / lands are paid to the lord of the said mannor upon death or alienacion & to the best of this deponents remembrance the same amounts to thirteen pounds odd moneys but the certain summe he cannot tell because he hath not of late / yeares looked into his writings saith he letts the said coppyhold lands for twenty pounds per annum or near thereabouts & saith he never heard that the river Swale was reputed a boundary to limitt or divide the lordship / mannor of Helaugh or town & constablery of Reeth or any of them till of late since this suit began /

James Metcalfe of Gunnerside in Swaledale miner aged fifty five yeares & upwards sworne & examined. /

2: To the second interrogatory saith he knowes the dale called Swaledale that the same lys in the wapentacks of Gilling West & Hang West & that the reputed boundary thereof is from Stollerston Stile on the east to Hollow Mill Cross on the west & / from the height of the moores adjoyneing upon Arkingarthdale on the north to the height of the moores adjoyning upon Wensleydale on the south & that the Lord Wharton & his ancestors have been during all his time reputed cheife lords of Swaledale / and that the Lord Wharton keeps courts yearly at Helaugh & Mukarr for the said mannors within the said dale & never heard <of> or knew {of} any other courts kept there and that the Lord Wharton & his ancestors have had & claimed the royalties of fishing / fowling & hunting through the whole dale & that James Miller Thomas Miller & Mathew Alderson all dead were keepers in this deponents remembrance under the old Lord Wharton & that Edmond Harker is now a keeper there for looking after & / taking care of the red deer within the said dale & taking care of preserving other games therein. /

4: To the fourth interrogatory saith he knowes the vill & constablery of Reeth & that the same lys within the mannor of Helaugh & that the lordship of Swaledale mannor of Helaugh & <vill &> constablery of Reeth lye on both north & south sides of the / river of Swale & extends southwards to the top of the moores which divides Swaledale & Wensleydale saith he hath knowne the said places during all the time of his remembrance & never heard the said river of Swale accounted as a boundary to / divide the said lordship mannor or constablery or any of them till of late that this suit begun neither does he think that Swale can be a boundary to divide the said mannor & constablery because this deponent hath felled & cutt down severall parcells / of wood in eight or nine severall places being coppyhold lands belonging to the Lord Wharton on the south side of the river of Swale in Swaledale & hath severall times been fowling with one John Parke who said he had lycense from the Lord / Wharton upon Crackpott Satteron & Whitaside moores on the southside the said river & was not hindered or interrupted by any person for soe doing. /

George Wharton of Low Spenceley in Whitaside in Swaledale yeoman aged sixty three yeares sworne & examined /

4: To the fourth interrogatory saith he knowes the town & constablery of Reeth & saith the said towne & constablery lye in the mannor of Helaugh saith the lordship of Swaledale mannor of Helaugh & towne & constablery of Reeth extend not onely on / the north side of the river Swale but on the south side alsoe of the said river as farr as to where heaven water deales betwixt Wensleydale & Swaledale saith he hath lived forty yeares att Low Spenceley in Whitaside aforesaid & hath knowne the said / lordship mannor & constablery for all that time & never heard (till since this suit began) that the river Swale was reckoned or reputed a boundary to limitt or divide the said lordship of Swaledale mannor of Helaugh or town & constablery / of Reeth or any of them & this deponent saith he doth not beleive the said river can be a boundary to any of them for that the said lordship of Swaledale lys on both sides the said river as doth alsoe the said mannor & constablery & that / lands lying att Harkaside Crackpott Satteron & Spring End on the southside of the said river Swale are in the constablery of Reeth & mannor of Helaugh. /

[membrane xiij *recto*][307]

James Liddell of Heggs in Swaledale yeoman aged forty four yeares sworne & examined. /

4: To the fourth interrogatory saith he knowes the vill & constablery of Reeth that the same lys within the mannor of Helaugh & that the lordship of Swaledale mannor of Helaugh & vill & constablery of Reeth lys on both the north & south / sides of the river of Swale & extends southwards to Wensleydale & saith he was born in the dale & lived all his time there & never heard the said river talked of as a boundary till about fourteen days since to limitt & divide the said / lordship mannor & constablery or any of them saith he knowes that there are lands lying att Harkaside Satteron & Spring End on the south side of Swale which are part of the mannor of Helaugh & constablery of Reeth which lys on the north side of the said / river for which reason he cannot think Swale a boundary to limitt & divide the said mannor & constablery as above. /

Giles Simpson of Winterings in Swaledale husbandman aged forty five yeares & upwards sworne & examined. /

4: To the fourth interrogatory saith he knowes the vill & constablery of Reeth saith the same lye in the mannor of Helaugh & that the mannor of Helaugh {& vill} & constablery of Reeth extends not onely on the northside of the {said} river Swale

307 Parchment, 650mm x 730mm.

but alsoe on the south side of / the said river as farr up the moores <as> to Wensleydale saith he was borne att & lived all his time in the mannor of Helaugh & never heard till late since this suit began that the river Swale was ever reputed a boundary to limitt or divide the said mannor of Helaugh town / & constablery of Reeth or either or any of them and this deponent saith he doth not beleive the said river can be thought a boundary for that there are lands <lying> in Harkaside Crackpott & Spring End all on the south side of the river Swale & are within the constablery of / Reeth & mannor of Helaugh & this deponent hath been both assessor & collector of the land tax in the year one thousand six hundred & ninety one & saith the said lands in Harkaside Crackpott & Spring End were assessed together with Reeth as being a part of the / said township & constablery & in the mannor of Helaugh & the assessments paid accordingly to this deponent & the tennants thereof appear & do suit & service att Helaugh court saith his father Thomas Simpson of Winterings is eighty two yeares old & very / infirme & lame not able to travill & this deponent hath heard him often say since this suit began <that> the river Swale was never in his time reputed a boundary to the said mannor or constablery. /

Edward Wray of Fremington in the county of Yorke white smith aged sixty one yeares & upwards sworne & examined /

4: To the fourth interrogatory saith he knowes the vill & constablery of Reeth & that the same lys within the mannor of Helaugh & that the lordship of Swaledale mannor of Helaugh & vill & constablery of Reeth lys on both north & south sides of the river of Swale & / beleives it is soe because a great many Harkaside men appeares att Helaugh court & pays their land tax & constable assessments with Reeth. /

Anthony Metcalfe of Gunnerside in Swaledale yeoman aged sixty yeares being one of that perswasion called Quakers having first repeated the solemne affirmacion appointed in such cases by Act of Parliament /

2: To the second interrogatory saith he hath knowne Swaledale & hath lived most of his time in it & that the late Lord Philip Wharton & the present Earle of Wharton have during his time been reputed cheif lords thereof saith the said lords have kept courts for the mannors of / Helaugh & Mukarr & knowes not nor hath not heard of any other court kept in the said dale saith the said lords have during his time claimed the royalties of hunting fishing & fowling within the said dale & appointed game keepers to look after the deer that came into the said / dale & for preserving other sorts of game there to witt James Milner of Thwait Thomas Milner Mathew Alderson & Edmond Harker /

4: To the fourth interrogatory saith he knowes the vill & constablery of Reeth saith the same lys within the mannor of Helaugh & that both the said mannor & constablery extend on the southside & northside of the river Swale on the northside to Arkingarthdale & on the south / side to Wensleydale saith that during his time he never heard that the river Swale was reputed a boundary to the said mannor or constablery (till since this suit began) saith he beleives the said river is noe such boundary for that there are lands <in> Harkaside Crackpott & Spring / End on the southside of Swale which are or have been all his remembrance reputed to lye in the constablery of Reeth & mannor of Helaugh and this affirmant hath been often att Helaugh court & hath seen the tennants of such lands there appear & doe suit & service / at {Helaugh} that court saith that he hath heard both his father & grandfather (both dead[308] discourse of the late Lord Whartons boundarys but never heard them mencion Swale to be a boundary. /

308 This bracket is not closed.

Mathew Clemitson of Fetham in Swaledale miner aged forty five yeares sworne & examined. /

4: To the fourth interrogatory saith he knowes the vill & constablery of Reeth saith the same lys in the mannor of Helaugh saith the lordship of Swaledale mannor of Helaugh & constablery of Reeth extend not onely on the northside of the river Swale but the same extend to the southside / alsoe of the said river as farr as to where heaven water deales betwixt Wensleydale & Swaledale saith he was borne att Feetham & hath lived all his time there or att Low Raw in Swaledale saith that in all his time he never heard that the river Swale <was> ever reputed a boundary to / limitt or divide the said lordship of Swaledale mannor of Helaugh & the said vill & constablery of Reeth or any of them and this deponent saith he doth not beleive the said river can be a boundary to the said lordship mannor & towne & constablery of any of them for that the lordship of / Swaledale mannor of Helaugh & constablery of Reeth lye on both the north & south sides of the said river & the Lord Wharton lord of the mannor of Helaugh hath lands lying att Harkaside Crackpott Satteron & Spring End on the south side of the said river which are in the constablery / of Reeth & the tennants thereof appear & doe suit & service att Helaugh mannor court & saith he knowes that for above thirty yeares together this deponents father (whom this deponent doth not know whether he be alive or dead) having for four yeares past been gone out of the / dale & doth not know where he is) used to appear att Helaugh mannor court & paid two pence att every court to the Lord Whartons stewards for license for fishing in the river Swale by vertue of which license his father used to fish on both sides the said river Swale throughout / the whole dale & was never disturbed therein that this deponent ever heard of. /

James Close of Whitaside in Swaledale miner aged forty nine yeares & upwards sworne & examined /

2: To the second interrogatory saith he knowes the dale called Swaldedale that the same lys within the wapentaks of Gilling West & Hang West & that the antient reputed boundary of the said dale is from Stollerston Stile on the east to Hollow Mill Cross of the west & from Arkingarthdale on / the north to Wensleydale on the south & that the Lord Wharton & his ancestors have been during all this deponents remembrance reputed cheif lords of the said dale & that there are courts kept yearely by the Lord Wharton att Helaugh & Mukarr for the said mannors which are in the / dale & never heard or knew of any other courts kept there & that they have during all his time claimed & enioyed free chace in the said dale & has keepers there for looking after & preserving the red deer therein & enjoyed the royalties of fishing fowling & hunting therein & / has game keepers alsoe there for preserving the other games therein & beleives none dare hunt fish or fowl therein without license from the said Lord Wharton. /

4: To the fourth interrogatory saith he knowes the vill & constablery of Reeth & that the same lys within the mannor of Helaugh & that the lordship of Swaledale mannor of Helaugh & vill & constablery of Reeth lys on both north & south sides of the river of Swale & extends Southwards / to where heaven water deales adjoyning upon Wensleydale saith he was borne in the dale & lived all his time therein & never heard the said river reputed or accounted a boundary till within this twelve monthes last past to limitt & divide the said lordship mannor vill & constablery / or any of them neither dos he beleive the said river is a boundary to divide the same as above for that this deponent has a close or parcell of ground which is freehold lands lying att Whitaside but not in the mannor of Helaugh & lys part on the north side of Swale & part on / the southside of the same river & that his cattle gos through the same river at their pleasure to depasture on each side & that this deponent lived for twenty yeares together with one John Close this deponents uncle)[309] who dyed about twenty yeares agoe & in his lifetime the said John / Close farmed grounds att Crackpott which is on the southside of Swale & in the constablery of Reeth & mannor of Helaugh part whereof was freehold lands & belonged then to one Richard Wharton & the other part was customary lands as he beleives which belonged then to / Margarett Waller & both the said farmes laid part on the northside of Swale & part on the southside thereof & that the said John Closes cattle went through the river & depastured in the said severall farmes on both sides the river as they liked & this deponent has since his said uncles / death farmed the said grounds belonging to the said Margarett Waller & during the time he soe farmed them {the} <his> cattle went through the Swale & depastured on both sides as they liked /

5: To the fifth interrogatory saith he knowes Harkaside & West Grinton & the moores & commons thereunto belonging and that the same are not divided by any boundary but are accounted one & the same place knowes alsoe Grinton or East Grinton & the moores thereof & that the same are / divided by a boundary from Harkaside & West Grinton which boundary divides as well the inclosed Grounds as the moores & commons saith he knowes the highway leading from East Grinton to Harkaside & West Grinton & the gate therein & the elmtree near thereto & hath knowne the same / during all his remembrance and saith that the said boundary begins att the river Swale over against Arklebeckfoot & crosses Naylors holme into the said highway betwixt the gate & Cuddy Spout & soe up the bounder ridge to Jenny Wrenn or Benson cowhouse & soe along the side of Ox Intack wall / to the east corner of the said intack & crosses Hughes Ditch on the moor & soe to the White Stone on the How & from thence to the Gray Stone on the Greets & so to the topp where heaven water deales which divides Swaledale from Wensleydale & soe along as heaven water deales to Gibbon hill / & from thence to the topp of Browney gill which divides Harkaside moores from Whitaside & soe down Browney gill to Scabby Yate & soe to the river of Swale. /

6: To the sixth interrogatory saith he knowes the lead mines called How Groves now wrought by the defendant Marriott & alsoe knowes the lead workes lately opened & claimed by the Lord Wharton & saith the said last menciond workes lye within the moores & commons of Harkaside & West Grinton / & the boundary thereof & not within the moores & commons of East Grinton & the boundary thereof & saith the said workes lys about eighty yards distance from each other to the best of this deponents judgement & that the said boundary gos up betwixt the said workes & saith that the said workes / were opened by the Lord Wharton about two yeares agoe & that the said boundary was ridd some short time before opening the same by the Lord Whartons agents & severall others (whereof this deponent was one) & that publick notice was given of riding the said boundary in Reeth markett / by the Lord Whartons bayliffe on a markett day saith that the Lord Wharton & his ancestors & their tennants & other the freeholders & inhabitants att Harkaside & West Grinton have been during all this deponents remembrance in possession of the moores & commons att Harkaside / & West Grinton by grazing & depasturing their cattle getting & digging stones peats & turves & pulling ling thereupon & saith he hath heard severall yeares agoe that waifes & estrays had been taken up upon Harkaside Moors & carryed to the Lord Whartons bayliffe att / Helaugh & hath likewise seen a grove or pitt that was sunk by Thomas Sumers & James Alderson in the inclosed grounds att Harkaside belonging to the Lord Wharton & then in William Arundells

309 This bracket is not opened.

possession which grove or pitt was soe sunk by the direcion of the Lord / Whartons agent or steward & a tryall made for lead oar there but beleives they gott little or noe oar there /

14: To the fourteenth interrogatory saith he hath seene James Loansdale the elder of Helaugh some years agoe lead both stones & ling from Harkaside Moor to Helaugh to build a cowhouse withal & hath seen young James Lonesdale of Helaugh several / times lead stones from Harkaside Moor to Helaugh & never heard that they were at any time interrupted for soe doing and hath seen a red cow belonging to widdow Lonesdale of Helaugh about 4 years agoe several times in the summer season goe / upon Harkaside Moore & depasture there and knowes the lane leading from Helaugh down to the river Swale & so to Harkaside Moore & that the same is now called Neat Hird laneing & has heard old men say it got that name by the Helaugh men / driveing their cattle to & fro to Harkaside moores therin /

George Chirry of Fetholm in Swaledale smelter aged sixty nine years & upwards sworn & examined /

4 To the fourth interrogatory saith he knows the vill & constablery of Reeth that the same lyeth within the mannor of Helaugh saith that the lordship of Swaledale mannor of Helaugh & vill & constablery of Reeth lyes on both north & south sides / of the river Swale & extends southwards to the heigth of the moors adjoyning upon Wensleydale saith he was born at Fetholm & lived all his time there & never heard Swale reputed a boundary to divide the said lordship mannor & constablery or any / of them till this suit began neither doth he beleive Swale is a boundary to divide the said mannor & constablery as above because he knowes there are lands belonging to the Lord Wharton lying at Harkaside Crakepott Satteran & Spring End which are / all on the south side of the river of Swale & are within the mannor of Helaugh & constablery of Reeth /

Richard Fothergill of Feetam in Swaledale husbandman aged forty years sworn & examined /

4 To the fourth interrogatory saith he knoweth the vill & constablery of Reeth saith the same lyes in the mannor of Helaugh saith the lordship of Swaledale mannor of Helaugh & township & constablery of Reeth extend not onely on the north side of the / river Swale but alsoe on the south side of the said river & as far on the south side of the said river as to Wensleydale saith he hath lived all his time in Swaledale near to Helaugh and (untill since this suite began) never heard that the river Swale was reputed / or so much as talked of to be a boundary either to the lordship of Swaledale mannor of Helaugh or township & constablery of Reeth nor doth this deponent beleive the same can be a boundary thereto because there are lands lying both at Harkaside / & Satteran on the south side of the said river & are within the township & constablery of Reeth & mannor of Helaugh the tennants whereof appear & do suite & service at Helaugh court and this deponent saith he hath a parcel of ground at <Feetham> {Reeth} and / which lyes part on the north side & other part of it on the south side of the said river and is all within the said constablery & mannor & one Wallbanke hath some coppyhold land in Whitaside & he or his tennant appears at the said court /

George Lonesdale of Helaugh in Swaledale husbandman aged forty seven years sworn & examined /

2 To the second interrogatory saith he knows the dale called Swaledale that the same lyeth within the wapentaks of Gilling West & Hang West and that the antient reputed boundarys of the said dale are from Stollertsone Stile on the east to / Hollow Mill Cross on the west & from Arkingarth dale on the north to Wensleydale on the south & that the Lord Wharton & his ancestours have beene reputed cheife lords

of Swaledale dureing all this deponents time & that there are courts leet / yearly at Helaugh & Mukar in the said dale by the Lord Wharton for the said mannors & never heard of any other courts kept in the said dale and that the Lord Wharton & his ancestours have dureing all this deponents time claimed & enjoyed the royaltys of / fishing fowling & hunting through the dale & appoints & keeps game keepers therein for preservation of the several games there & hath seene several persons agree to pay {an} acknowledgement at Helaugh court for licence to fish in the river Swale /

4 To the fourth interrogatory saith he knowes the vill & constablery of Reeth that the same lyes within the mannor of Helaugh & that the lordship of Swaledale mannor of Helaugh & vill & constablery of Reeth extends & lyes on both north & south sides of / the river of Swale & extends southwards to Wensley dale saith he was born at Helaugh & lived all his time there & never heard the said river spoke of as a boundary till of late {that} that this suite began to limit & divide the said lordshipp mannor or constablery / or any of them neither doth this deponent beleive the Swale can be a boundary there because there are coppyhold lands lying at Harkaside Crakepott Satteran & Spring End belonging to the Lord Wharton all on the south side of Swale which are / within the mannor of Helaugh & constablery of Reeth and that he hath known John Lonsdale & James Lonsdale both of Helaugh gett ling & stones upon Harkaside Moor & lead the same to Helaugh side for their uses there & never heard of any interruption / but saith John Lonsdale hath lands on the south side of Swale /

Thomas Carter of Kerton in Swaledale taylor aged forty three years sworne & examined /

2 To the second interrogatory saith he knows Swaledale & hath known the same all his time being born & bredd there saith the Lord Wharton & his ancestours have beene reputed lords of the said dale all his time & claimed & enjoyed the royaltys of hunting / fishing & fowling there & appointed game keepers to preserve the said royalties & game /

4 To the fourth interrogatory saith he knowes the vill & constablery of Reeth & that the same lyes within the mannor of Helaugh saith the lordshipp of Swaledale mannor of Helaugh & vill & constablery of Reeth extend not onely to the north side of the river Swale / but alsoe on the south side of the said river up to Wensleydale saith that untill this contest began he never heard that the river Swale was ever reputed a boundary to limitt or divide the said lordshipp mannor & constablery or any of them and this deponent / doth beleive the said river cannot be a boundary to any of them for that the said lordshipp mannor town & constablery lye on both sides the said river and there are lands in Harkaside Whitaside Crakepott Satteran & Spring End all on the south side of / of [310] the said river the tennants whereof appear & doe suit & service at Helaugh mannor court and saith that the tennants of those lands are assesed for the land tax with Reeth & pay there assesments accordingly. /

Ambrose Clemetson of Law Raw in Swaledale smelter aged thirty seven years sworn & examined /

2 To the second interrogatory saith he knows Swaledale & that the Lord Wharton & his ancestors have been during all his time reputed cheif lords of the said dale and that there are courts kept yearly in the said dale by the Lord Wharton at Helaugh & / Mukar for the said mannors & never heard of any other courts kept within the said dale saith the Lord Wharton & his ancestors have dureing all his time claimed & enjoyed the royaltys of fishing fowling & hunting in & through the whole dale and has / game keepers for looking after the several games within the said dale &

310 Word repeated in the original.

for preservation of the same & saith that he hath several times fished on both sides the river Swale in most parts of Swaledale with his father & that his father paid an / acknowledgement at Helaugh court for fishing in the said river /

4 To the fourth interrogatory saith he knows the vill & constablery of Reeth that the same lyes within the mannor of Helaugh saith that the Lordshipp of Swaledale mannor of Helaugh & constablery of Reeth lyes on both north & south sides of the river / of Swale & extends southwards to where heavenwater deals adjoyning upon Wensleydale and further saith he was born & lived all his time in the dale & never heard Swale talked on as a boundary till this suit began to devide the said lordshipp mannor and / constablery or any of them neither does he beleive that Swale is a boundary there because there are coppyhold lands belonging the Lord Wharton lying at Harkaside Crakepott Satteran & Spring End all on the south side of Swale which are in the mannor / of Helaugh & constablery of Reeth the tennants whereof apppears at Helaugh court & does suit & service there /

[membrane xiij *verso*]

James Milner of Calvert House in Swaledale yeoman aged seventy one years sworn & examined /

4 To the fourth interrogatory saith he knowes the vill & constablery of Reeth & that the same lyes in the mannor of Helaugh & saith that the said mannor & constablery extend both on the north & south sides of Swale & as farr on the / south side thereof as <is> Wensleydale saith he hath lived all his time save seven years in the said dale & never heard till this suite began)[311] that the river Swale was ever reputed a boundary to the said mannor or townshipp & constablery / of Reeth or any of them and this deponent saith he does not beleive the said river can be a boundary to the said mannor for that there are lands in Harkaside Whitaside Crakepott Satteran & Spring End on the the south side of the / said river which lye in Helaugh mannor & the tennants thereof appear & do suit & service at Helaugh court and so doth this deponent for lands of his own which lye both on the north & south sides of the said river /

Edward Bland of Whitaside in Swaledale miner aged fifty two years or thereabouts sworn & examined /

6 To the sixth interrogatory saith that about twelve years agoe he was imployed by the agents of the old Lord Wharton to carry several quantatys of wood which was cutt in the grounds at Harkaside then in William Arundells possession & he carried the said wood to / Helaugh side of Swale to the said Lord Whartons lead mills there & was paid for carrying the said wood thither by Joseph Etherton a steward for the said Lord Wharton & saith that William Arundell then told this deponent that the said wood was cutt in the said William / Arundells freehold Land at Harkaside & that he had sold the said wood to the said Lord Whartons agents and that about three years agoe he by the direction of the present Lord Whartons agents was imployed to cutt & did cutt several quantities of wood in coppyhold Land / lying at Spring End Satteron & Whitaside all on the south side of Swale which belonged to the Lord Wharton & beleives the same was carried to & made use of at the Lord Whartons mills & mines on the north side of Swale /

Cornelius Frier of Low Raw in Swaledale yeoman aged thirty nine years sworn & examined /

311 This bracket is not opened.

2 To the second interrogatory saith he knowes the dale called Swaledale saith it lyes in the honor of Richmond & Richmond shire & part of it in the wapentak of Gilling West & part of it in Hang West saith that Grinton Whitaside & Summer Lodge lying on the south side of / the river Swale are reputed to lye in Hang West & the rest of the said dale on the south side of the said river & {alsoe} <all> on the north side of the said river within the said dale are reputed to lye in Gilling West saith that court leets are kept usually once a year for the mannors of Mukar / & Helaugh within the said dale whereof the Earle of Wharton is lord & no other courts have during this deponents remembrance been kept for any other places in the said dale saith the court for the mannor of Helaugh generally takes up two days one for the tennants of the old / lands the other for the new lands in the said mannor saith the Lord Wharton & his ancestours have all this deponents time been reputed cheif lords of Swaledale & have claimed & enjoyed the royalties of the forests & chaces & of hunting fishing & fowling throughout the said dale / & appointed gamekeepers for the deer that hapen to come into the dale & for preserving the game there /

4 To the fourth interrogatory saith he knowes the vill & constablery of Reeth saith the said vill & constablery lys within the mannor of Helaugh saith the lordshipp of Swale dale mannor of Helaugh & vill & constablery of Reeth extend not onely on the north side of the river / Swale but the same extends to the south side of the said river alsoe as far on the south side of the said river as to where heaven water deales betwixt Wensleydale & Swaledale saith he was born at Low Raw about a mile & an half from Helaugh town & hath lived all his / time there & never heard that the river Swale was ever reputed a boundary to limitt or divide the said lordshipp of Swale dale mannor of Helaugh & the vill & constablery of Reeth or any of them till since this suit was commensed saith he doth not beleive the said river / can be a boundary to any of the said places by reason that the Lord Wharton hath & his ancestours have all this deponents time had lands lying for severall miles together both on the north & south side of the said river which lye in the two mannors of Meuker & Helaugh / reaching to the head of Swaledale & the tennants thereof appear & do suite & service at these courts /

5 To the fifth interrogatory saith he knowes Harkaside & West Grinton & the moors & commons thereunto belonging saith they are not divided by any boundary the one from the other but are one & the same place or Hamlett saith he knows Grinton or East Grinton & the moors / thereof saith he knows that there is a boundary dividing the said Grinton or East Grinton from the said Harkaside & West Grinton which boundary doth not onely divide the moors & commons of Harkaside & West Grinton from those of East Grinton but the said boundary / extends throughout & divides likewise the inclosed lands of each respectively saith he knows the highway or road leading from East Grinton to West Grinton & Harkaside & knows the gate in the said road there & the elm tree growing near the said gate & hath known the / same from his youth saith the tennants or inhabitants of East Grinton repair or maintain the said highway to near the said gate as this deponent hath heard & beleives & saith the place to which they repair the said way is reputed one of the marks of the said boundary / saith the said boundary so dividing the said Grinton or East Grinton from Harkaside or West Grinton begins over against Arkle beck foot and so from the river Swale cross the east end of the Holme & so crosse the way on the east side of the gate & elm tree / betwixt them & Cuddy Spout up the ridge to Benson cowhouse or Jenny Wren cowhouse & so on the east side of the Ox Intak wall to the east corner of the said intak from thence over the moor in a streight line crosse Hughs Ditch to the White Stone in the How / and from thence to the Grey stone in the Greets & so to where heaven water dealing parts Wensleydale &

Swaledale saith the boundary of Harkaside & West Grinton which divides the same together with the said dale called Swaledale from the other / adjoyning dale called Wensleydale proceeds westwards from the head of the Greets as heaven water deals to Gibbon hill standard & from thence to Browney gill head and the boundary which divides Harkaside & West Grinton from the hamlett or place / called Whitaside proceeds from the head of Browney Gill down the said gill to Scabby Yate & so to the river Swale /

6 To the sixth interrogatory saith he knowes the lead mines now wrought by the defendant Marriott called the Howgroves & the lead mines or works lately opened by the Lord Whartons agents near adjoyning thereto saith the lead works so belonging to the said / Lord Wharton & opened by his agents lye within the moors & commons of Harkaside & West Grinton & the boundary thereof & not within the moors & commons of East Grinton & the boundary thereof saith the said severall works lye at about four score or an hundred / yards distance the one from the other & the boundary line or mark near the midst betwixt them saith that the aforesaid boundary so set forth by this deponent was about two years agoe ridd by the said Lord Whartons order before the said Lord Whartons said / works were opened and the same were ridd by a great number of people & agreed to be a true boundary by the antient people that were at the riding and publick notice was given of the said riding by this deponent at the markett crosse in Reeth on Friday / the markett day there next before the rideing and the riding was on the Wednesday next after saith that at the time of the said rideing & before the said wideing of the said boundary & opening the said lead works & the said Lord Wharton & his ancestours & their tennants / of Harkaside & West Grinton & other the freeholders & inhabitants there were in the actuall possession {&} <or> enjoyment of the said moors of Harkaside & West Grinton by depasturing their cattle thereon & getting lyng peats turfs & stone there & the said Lord Wharton / about three years agoe sunk a shaft in his owne grounds in Harkaside in John Hirds possession for lead oar & saith this deponents brother James Fryer deceased who was baylif both to the late Lord Wharton & the present Lord Wharton told this / deponent that the said lords had cutt & gott wood in Harkaside for his lead works at Helaugh as they did in other places in the said mannor /

7 12 To the seventh & twelfth interrogatorys saith that for his time all Harkaside & West Grinton have been reputed part & parcell of the mannor of Helaugh & most part thereof part & parcell of the constablery of Reeth saith there are lands lying & being in Harka / side commonly called demesne or leasehold lands belonging to the complainant the Lord Wharton in the occupation of John Hird & Anthony Hird of the yearly rent of six pound thirteen shillings & four pence payable to the said Lord Wharton saith there / are several freeholders in Harkaside which pay yearly free rents to the Lord Wharton for there freehold lands there to witt Ralph Addison two pence Anthony Hird one penny Marmaduke Hartley nine pence Christopher Rapier three shillings / one penny saith that there are alsoe divers coppyholders in Harkaside & West Grinton which hold their lands of the Lord Wharton by coppy of court roll of the complainants mannor of Helaugh & pay yearly coppyhold rents for the same to the Lord Wharton / & alsoe fines at death of tennant or alenation to witt Ralph Wensley five shillings per annum rent Christo[pher] Rapier six shillings and one penny Marmaduke Hartley one pound two shillings seven pence the four Arundells several respective coppyhold / rents yearly ammounting in all to eighteen shillings & four pence and saith that upon death of the tennant or alenation of the coppyhold land there is paid to the said lord of the mannor of Helaugh a twenty penny fine if it be old land & a fifteen penny fine it it be new / land saith that Ralph Wensleys land being new land pays on death or alenation three pound fifteen shillings fine

Christopher Rapier hath both new & old land the fine for both four pound thirteene shillings & nine pence Marmaduke Hartleys coppyhold and old / land & part new land the fine whereof is about sixteen pound and the fine for Arundels land betwixt nine & tenn pound & saith that all the said leasehold freehold & coppyhold tennants appear & do suit & service at Helaugh mannor court & saith that Sir / Soloman Swale as owner of Swalehall hath to this deponents knowledge for twenty years past been usually called at Helaugh court as a freehold tennant within the said mannor & called the first of the freehold tennants of the said mannor saith he never heard / that the owners of Swalehall never appeared at any other court nor ever heard that they held any court at Harkaside or West Grinton or had or hath any freehold or coppyhold tennants or any <free> rents or customary rents within that mannor. /

8 To the eighth interrogatory saith there are several lands or inclosiers adjoyning to the {mannor} <moors> of Harkaside & West Grinton commonly called by the name of the intacks to which Christopher Rapier Marmaduke Hartley & William Arrundel & Ralph Addison / are tennants & pay some small rents to the Lord Wharton & his ancestors lords of the mannor of Helaugh for the same but saith the said inclosiers have been before this deponents time /

9 To the ninth interrogatory saith he is bayliff to the present Lord Wharton for the mannor of Helaugh & employed in receiveing rents for his lordshipp in the said mannor and saith that the yearly rents {are} now payable to the said Lord Wharton for his demesne lands / in Harkaside & West Grinton & the yearly rents payable to him for the coppyhold & customary lands there & the free rents payable yearly to him for freehold lands there & the inhanced rents payable likewise yearly there are of greater value then other his / demesne lands & freehold coppyhold & customary rents in some other hamlets within the said mannor of Helaugh or vill of Reeth & of less value then his demesne lands & freehold coppyhold & customary rents in some other hamlets within the said mannor & vill of / Reeth and this he is the better enabled to depose for that he knowes the respective rents which are yearly payable to the said Lord Wharton out of the several & respective vills & hamlets within the said mannor & for what the same are paid which are as / followes (viz)

	For demesne lands yearly	Customary & coppy hold lands yearly	Free rents yearly	Enhansed rents yearly
	L - s - d	L - s - d	L - s - d	L - s - d
Helaugh	70-05-00	07-15-10 ½	01-11-07	00-00-00
Reeth	31-18-06	08-00-10 ½	04-09-03 ½	00-07-10
Harkaside	06-13-04	02-00-00	00-04-01	00-12-00
Park Hall	53-00-00	00-13-04	00-00-00	00-00-00
Kirton	00-00-00	06-03-09	00-00-00	00-00-00
Crakepott	00-00-00	01-01-09	00-13-10	01-00-08
Spring End	00-15-00	00-00-00	02-06-08	00-00-00
Blaides	00-00-00	04-00-04	00-00-00	00-00-00
Smarbarr	26-00-00	02-16-00 ½	00-00-00	00-00-00
Pott Ingg	00-00-00	03-17-06	00-00-00	00-00-00
Wintering gaites	00-00-00	01-18-06	00-00-00	00-02-00
Satteran	00-00-00	06-16-00	00-00-00	00-00-00
Ivelett	00-00-00	03-05-00	00-00-00	00-00-00

Calvert House	00-00-00	03-01-01½	00-00-00	03-01-02
Ravenseat	00-00-00	03-06-08	00-00-00	00-00-00
East Stonsdale	29-10-00	00-00-00	00-00-00	00-00-00
West Stonsdale	14-00-00	01-15-04	03-00-00	00-00-00

and this deponent saith that Harkaside Crackpott Spring End & Satteron lye on the south side of the river Swale & the rest of the said townes & hamletts lye on the north side of the said river & in the mannor of Helaugh /

10: To the tenth interrogatory saith that the coppyhold or customary lands in Harkaside & West Grinton are descendible to all the sons of the father alike onley if the father leave a widdow she hath a third part thereof soe long as she continues his / widdow & noe longer which is the custome of the mannor of Helaugh but this deponent never heard of any such custome in the pretended mannor of Grinton or East Grinton /

16: To the sixteenth interrogatory saith the present Lord Wharton & the late Lord Wharton his father have for twenty years past been in the actuall possession & enjoyment of the lead mines att the Old Gang Lownythwait / & Merryfeild within the mannor of Helaugh & northside of Swale & have had good gettings there. /

William Alderson of Kirkby Ravensworth in the county of Yorke miner aged thirty three yeares & upwards sworne & examined. /

4: To the fourth interrogatory saith he knowes the vill & constablery of Reeth that the same lys within the mannor of Helaugh & that the lordship of Swaledale mannor of Helaugh & vill & constablery of Reeth lyes on both north & south sides of / the river of Swale & extends southwards to the height of the moores which divides Swaledale from Wensleydale saith he was born att Reeth & lived all his time there or very near it & never heard Swale reputed a boundary to divide the said / lordship mannor & constablery or any of them neither does he beleive that Swale is a boundary there because he remembers very well that his father James Alderson since dead was about sixteen or seaventeen yeares agoe constable of / the towne of Reeth & that his father collected the constable assessments att Harkaside Crackpott & Spring End all on the south side of Swale as being part of the township & constablery of Reeth & saith that one Arthur Hutchinson of / Reeth severall yeares agoe gott stones upon Harkaside Moor & led the same to Reeth for his use there & that the severall inhabitants att Harkaside Crackpott & Spring End paid to Reeth for being excused from keeping watch there. /

Luke Wilson of Kirkby Ravensworth in the county of Yorke carpenter aged fifty five yeares sworne & examined. /

4: To the fourth interrogatory saith he knowes the vill & constablery of Reeth saith the same lyes in the mannor of Helaugh saith the mannor of Helaugh & constablery of Reeth extend not onely on the north side of the river Swale but / alsoe on the south side of the said river as farr as to Wensleydale saith he hath knowne the said mannor towne & constablery most of his time & never heard that the river Swale was reputed a boundary to limitt or divide the said mannor / towne & constablery or any of them till since this suit began & doth not beleive the said river can be a boundary to any of the said places for that there are lands lying at Harkaside Crackpott & diverse other places on the southside of / the said river the tennants whereof appear & doe suite & service att Helaugh mannor court & saith that above thirteen yeares agoe there were stones gott on Harkaside Moor & carryed to Reeth & used there in building one Anne / Hutchinsons house & saith that he knowes the said river Swale is noe boundary betwixt Thorp & Applegarth which lye a little below

Swaledale & one of the said townes lys on the north side & the other on the south side of the river / Swale. /

John Peacock of Spring End in Swaledale husbandman aged sixty eight yeares & upwards sworne & examined /

4: To the fourth interrogatory saith he hath knowne Swaledale the most part of his time & that the river of Swale was never reputed a boundary till within three weekes last past to divide the mannor of Helaugh & constablery of Reeth or <either>/ or any of them and the reasons that induces him to beleive it cannot be a boundary there are because he saith he hath coppyhold lands lying att Spring End which holds of the Lord Whartons mannor of Helaugh part whereof lys on / the south side of Swale & another part thereof lys on the north side of the said river & that the said lands is called old lands & payes a twenty penny fine to the said Lord Wharton & that he payes a coppyhold rent yearely of nine shillings three / pence halfe penny to the said Lord Wharton for his said lands & paid a fine nine pounds five shillings & tenn pence to the said lord when he purchased the same & he appears att Helaugh court & dos suit & service there. /

[membrane xiiij *recto*][312]

Margarett wife of Francis Charder[313] of Reeth aged sixty four yeares sworne & examined. /

4: To the fourth interrogatory saith she knowes the vill & constablery of Reeth saith the same lys within the mannor of Helaugh saith the mannor of Helaugh & constablery of Reeth extend to Arkingarthdale on the north side of the river / Swale & to Wensleydale on the south side of the said river saith she hath lived all or most part of her time at Reeth & Fremington & never heard (till of late since this suit began) that the river Swale was ever reputed a / boundary to limitt or divide the said mannor or towne & constablery of Reeth or either or any of them & saith she doth not beleive the said river can be a boundary thereto for that Harkaside Crackpott Spring End & Satteron which lye on the south side of the / river Swale have all this deponents remembrance been reputed to lye in the constablery of Reeth & mannor of Helaugh & this deponents husband was constable of Reeth for three yeares together about fifteen yeares agoe & collected his constable sesse & att all / those places on the southside of the said river as well as in other places in the said constablery on the north side thereof. /

6: To the sixth interrogatory saith she beleives the Lord Wharton & his ancestors have had the enjoyment of the moores & commons of Harkaside & West Grinton as long as this deponent can remember for she saith that her father James Kirton late of Fremington deceased / about three & forty yeares agoe had a steer which strayed from him & after the same had been for some time missing old James Miller of Dykehouse in Harkaside deceased sent word to James Freer who was bayliffe to the late Lord Wharton / that he had taken up a straye steer on Harkaside Moor for the Lord Whartons use & thereupon old James Freer fetched the same from him as the said James Freer told this deponent and her father and this deponents father hearing of the said steer [x x x][314] / being in old James Freers hands as a waife or stray as

312 Parchment, 660mm x 710mm. A small tear in the bottom right-hand corner of the sheet has resulted in minor damage to the text.

313 In May 1708 Margaret Charder, aged 63, gave evidence for Lord Wharton at the trial in London (LI Misc 536, Ward MSS, notebooks vol. 6, f. 117r).

314 One word has been erased.

aforesaid took this deponent along with him & paid old James Freer halfe a crowne for restoring the said steer to him again saith that about thirty yeares agoe a tree was brought downe / the river Swale & throwne by the river on Harkaside grounds & the said old James Freer seized the same for the said late Lord Whartons use. /

George Alderson of Helaugh in Swaledale miner aged thirty four yeares & upwards sworne & examined. /

6: To the sixth interrogatory saith that about tenn yeares agoe George Alderson then of Helaugh & since dead being this deponents father told this deponent that when he was a young man he lived as a servant with Christopher Harker of East Grinton & the said Christopher / Harker being then building a house att East Grinton sent this deponents father & one John Hodgson to digg stones upon Harkaside Moor for building the said house & accordingly they did goe & digg stones there but the Lord Whartons bayliffe & the inhabitants of Harkaside hearing / thereof came & hindred the said Christopher Harker for carrying the said stones away & the Harkaside men & the said lords bayliffe carryed the same to Harkaside & disposed of them there for their own uses and saith about twenty yeares agoe James Miller of / Dikehouse in Harkaside since dead took up a waife horse upon Harkaside Moor & in this deponents presence & carryed the said horse to old James Freer who was then bayliffe to the old Lord Wharton for his mannor of Helaugh as a waife. /

14: To the fourteenth interrogatory saith that about fourteen or fifteen yeares agoe he this deponent lived as a servant with James Galloway of the Banks near Helaugh on the north side of Swale & within the mannor of Helaugh & his said master James Galloway / during the time that he lived with him frequently put his sheep upon Harkaside Moor to depasture them there without any interrupcion or disturbance & that severall of the inhabitants of Harkaside frequently put beasts sheep & horses to depasture upon Helaugh / Moor or Reeth Moor without any interrupcion from the Helaugh or Reeth men & that the severall inhabitants of Harkaside & Helaugh drove their <said> cattle sheep & horses along Noutherd Loneing to each others commons and this deponent saith that Anthony Pratt / of Park Hall on the north side of Swale within the mannor of Helaugh usually put his horse to depasture upon Harkaside Moor having common right there & not upon Reeth Moor & this deponent living near the said Pratt & having occasion severall times to goe towards / Harkaside Moor was desired by the said Pratt about twelve yeares agoe to lead his horse from Park Hall down to Scabby Wath & soe through Scabby Yate & to put him upon Harkaside Moor to depasture there & saith he hath turned the said Pratts horse severall / times upon the said moores to depasture there & never heard of any interrupcion. /

Robert Waller of Dykehouse in Harkaside in Swaledale yeoman aged sixty one yeares sworne & examined /

5: To the fifth interrogatory saith he knowes Harkaside & West Grinton & the moores & commons thereto belonging saith they are not divided by any boundary from each other but are one & the same place knowes alsoe Grinton or East Grinton & the moores & commons / thereof saith there is a boundary which all this deponents time has been reputed to divide Grinton or East Grinton from Harkaside & West Grinton saith the said boundary divides not onely the moores or commons of Harkaside or West Grinton from those of East Grinton / but the same boundary extends throughout & divides likewise the inclosed lands of each respectively saith the said boundary begins att the river Swale & gos betwixt Cuddy Spout & the elm tree soe up the ridge to Benson cowhouse & soe takes the moor by Ox / Intack wall & to the East Nook of the Ox Intack & soe to the White Stone in the How & from the White Stone in the How to the height of the Greets & from the height of the Greets the boundary betwixt Harkaside & West Grinton moores & Wensleydale / proceeds Westwards to Gibbon hill & soe to Browney gill head & Browney gill downe to Scabby Wath is the boundary betwixt Harkaside & Whitaside saith he was borne in Harkaside & lived all his time there save part of one yeare & during all this deponents / remembrance the said boundermarkes have been reputed to be the true boundarys betwixt the said places aforesaid. /

6: To the sixth interrogatory saith that the tennants of East Grinton never that this deponent knowes or heard of had or claimed any enjoyment upon the moores of Harkaside or West Grinton without interrupcion or disturbance from the tennants of Harkaside or West Grinton / saving a rake for their cattle saith that the East Grinton tennants have severall times asked leave to gett turfes & ling on Harkaside Moor of the Harkaside tennants & have some of them graved turfes & pulled ling on Harkaside Moor without asking leave but when / they did soe without leave the Harkaside tennants never suffered them to carry the same away but disposed of the same themselves. /

7: To the seaventh interrogatory saith that the most part of Harkaside & the moores thereof are within the constablery of Reeth & that there are severall coppyholders in Harkaside which appear & doe suit & service att Helaugh court /

James Clarkeson[315] of Ivelett in Swaledale yeoman aged forty eight yeares sworne & examined. /

1: To the first interrogatory saith he knowes the present Lord Wharton & Mr Marriott. /

2: To the second interrogatory saith he knowes the dale called Swaledale & that the greatest part thereof lyes in the Wapentake of Gilling West & part of it lyes in the wapentack of Hang West & that the reputed boundarys of the said dale are from Stollerston Stile on the east / to Hollow Mill Crosse on the west & from Arkingarthdale on the north to Wensleydale on the south & that the Lord Wharton & his ancestors have been during all this deponents remembrance reputed cheif lords of the said dale & that there are courts kept yearely att Mukarr & / Helaugh in the dale for the Lords Wharton for the said mannors to witt att Helaugh one court kept there for the old lands as they are called & another court kept there for the new lands within that mannor and that the Lord Wharton & his ancestors having during all this / deponents remembrance held & enjoyed free chace for red deer in the said dale & alsoe enjoyes all other the royalties of hunting fishing & fowling through the whole dale & remembers that James Miller Thomas Miller George Alderson Mathew Alderson John Bonsfeild / Edmond Harker James Freer & Anthony Freer are or have been keepers under the late Lord Philip Wharton & the present Lord Wharton within Swaledale for looking after & preserving the red deer & other the severall games therein & hath knowne severall persons att / severall times hindred from fowling hunting & fishing within the said dale by the Lord Whartons bayliffe without lycense from the Lord Wharton soe to doe. /

4: To the fourth interrogatory saith he knowes the vill & constablery of Reeth that the same lys within the mannor of Helaugh & that the lordship of Swaledale mannor & vill & constablery of Reeth extends & lys on both south & north sides of the river of Swale / & that Harkaside Crackpott Satteron & Spring End which are in the constablery of Reeth & mannor of Helaugh & lye on the south side of the river of Swale & the moores & commons thereunto belonging extend & adjoyne upon Wensleydale on the south & saith he was / borne in Swaledale & lived all his time therein & till

315 James Clarkeson of Ivelett, yeoman, swore affidavits on 1 June 1706 (E207/112/2) and 18 October 1707 (E207/115/1). He describes himself as Lord Wharton's 'agent or overman' to inspect the workings (E207/112/1).

of late that this suit begann never heard the river Swale reputed a boundary to divide the said mannor of Helaugh & vill & constablery of Reeth or any of them and this deponent saith he cannot beleive the / river Swale is a boundary thereto because there is a parcell of pasture ground lying on the south side of Swale which belongs to Ivelett & Calvett house & Ivelett & Calvett house are on the north side of Swale which said parcells of land & the said places called Ivelett / & Calvett house are all in the mannor of Helaugh & further saith that William Smith & James Miller both of Calvett house & this deponent himselfe & Anthony Metcalfe att Ivelett John Metcalfe att Gunnerside Tamar Freer att Low Raw & Anthony Fothergill / & his brothers att Fetholme & John Lonesdale att Helaugh have all of them lands lying att the severall places abovemencioned which {places} <lands> lyes part on the north side & part on the south side of the river Swale & all within the mannor of Helaugh & the / cattle of the said severall tennants of the said severall parcells of land depasture in the said {severall} parcells of ground & crosse the river of Swale as they list from one part of the said grounds to the other and alsoe that there are severall coppyhold lands / lying att Harkaside Crakepott Satteron & Spring End all on the south side of Swale which are within the constablery of Reeth & mannor of Helaugh & Reeth & Helaugh are on the north side of the said river the tennants of all which said lands appear att / Helaugh court & doe suit & service there for those lands. /

5: To the fifth interrogatory saith he knowes Harkaside & West Grinton & the moores & commons thereunto belonging saith they are not divided by any boundary but are accounted all one saith he knowes Grinton or East Grinton & the moores thereof & that / theres a boundary which divides the same from Harkaside & West Grinton which boundary not onely divides the moores but the inclosed grounds of each respectively saith he knowes the highway leading from East Grinton to Harkaside & West Grinton / & the gate standing in the said highway & the elm tree growing near it & hath knowne the same during all his remembrance saith the said boundary beginns att the river Swale & crosses the Holme into the said highway near the said elmtree & soe up the bounderridge / to Benson cowhouse & from thence along the wall of the Ox Intack to the east corner of the said wall & soe crosses Hughes Ditch upon the Moor to the White Stone on the How & from thence to the Gray Stone on the Greets & soe to the topp of the Greets to where heaven / water deales which divides Swaledale from Wensleydale & soe along as heaven water deales to Gibbon hill & to Gibbon hill currock & soe to Browney gill head downe Browney gill near to Scabby Yate which Browney gill divides Harkaside from Whitaside / & soe to the river Swale saith he has knowne the said boundary as above is described for thirty yeares last past & beleives the same is a true boundary to divide both the said inclosures & the said moores & commons of the said severall places townships & / hamletts above menciond & hath heard the same described & named to be a boundary by severall antient men severall yeares before he saw or knew the same /

6: To the sixth interrogatory saith he knowes the lead mines called How Groves now wrought by the defendant Marriott & alsoe the lead workes lately opened & belonging to the Lord Wharton near adjoyning saith that the last menciond workes lye within the moores / & commons of Harkaside & West Grinton & the boundary thereof & not within the moores & commons of East Grinton & the boundary thereof and that the said severall workes lye about eighty yards distance from each other & that the said boundary runs / up betwixt the said workes saith that the Lord Wharton caused the said workes to be opened about two yeares since & that the said boundary was rid a day or two before the opening thereof by the Lord Whartons agents & severall others whereof this / deponent was one & generally agreed unto by the persons there present to be a true boundary & that publick notice of the riding the said boundary was given in Reeth markett upon the markett day next before and saith he was present with the Lord Wharton / about a yeare before the said riding when the said Lord Wharton & severall others ridd the said boundary from Swale to the topp of the Greets all along by the severall boundermarkes abovemenciond and saith that the Lord Wharton & his ancestors & / their tennants & other the freeholders & inhabitants of Harkaside & West Grinton have been during all the time of this deponents remembrance in the possession of the moores & commons att Harkaside & West Grinton by depastureing & grazing their / cattle sheep & horses & pulling ling digging of peates & stones & graving of turfes thereupon and saith that about three or four yeares agoe the Lord Whartons agents caused a grove or shaft to be sunk or digged for lead oar in the said Lord / Whartons inclosed grounds att Harkaside in Hirds possession but beleives they gott little oar therein but saith he was told by Thomas Somers & James Alderson who were imployed to sink the said shaft that there was lead oar found there near the / surface of the ground which caused the said Lords agents to make that tryall there. /

7 & 12: To the seaventh & twelfth interrogatorys saith that all Harkaside & West Grinton & the moores & commons thereunto belonging are reputed to be within the mannor of Helaugh & the greatest part thereof within the vill & constablery of Reeth saith there are / lands lying att Harkaside called demesne or leasehold lands belonging to the Lord Wharton & now farmed by Anthony Hird & John Hird att six pounds thirteen shillings & four pence yearely & that there are severall freehold lands alsoe lying att / Harkaside & West Grinton in the possession of Christopher Raper Marmaduke Hartley Ralph Addison & some others which he dos not remember the tennants whereof he beleives payes free rents yearely to the Lord Wharton & appeares att his lordships / courts att Helaugh & dos suit & service there for the said lands & saith alsoe that Sir Solomon Swale hath been during all this deponents remembrance called att Helaugh court as a freehold tennant within the same mannor and that there are severall coppyhold / tennants att Harkaside & West Grinton (viz[t].) Ralph Wensley Christopher Raper Mr Marmaduke Hartley William Arundell John Arundell James Arundell & George Arundell which hold coppyhold lands there by coppy of court roll of the mannor of Helaugh & / pays coppyhold rents yearely for the same to the said Lord Wharton but what perticular yearely coppyhold rents they severally pay this deponent cannot sett forth and saith the said severall tennants appeares att Helaugh court & dos suit & service there & att death or / alienacion the said severall coppyhold tennants payes fines to the said Lord Wharton (viz[t].) for old lands a twenty
8: penny fine & for new lands a fifteen penny fine. /

To the eighth interrogatory saith he knowes there are parcells of lands adjoyning upon the moores att Harkaside called intacks belonging to the Lord Wharton & are now in John Hird & Anthony Hirds possession & that he has heard & beleives they pay / yearly rent for the same to the said Lord Wharton but what particular rent this deponent knowes not but saith the same were inclosed long before his time. /

10: To the tenth interrogatory saith that the coppyhold or customary lands in West Grinton & Harkaside are descendible to all the sons alike att the death of the tennants dying seized & that the widdow of the tennants soe dying seized has a third part / of the said lands soe long as she continues widdow of such tennant according to the custome within the mannor of Helaugh & that theres noe such custome within the
15: pretended mannor of Grinton or East Grinton. /

To the fifteenth interrogatory saith that there was a suit begunn about thirty yeares agoe betwixt the old Lord Wharton & his coppyhold or customary tennants within the mannor of Helaugh which suit was about making the rents called inhansed rents / finable & that Christopher Raper William Arundell & Richard Close three coppyhold tennants att Harkaside joyned & contributed with the other coppyhold tennants within the mannor of Helaugh in maintaining the said suit and saith that the coppyhold lands / lying att Harkaside & West Grinton are distinguishable by the names of old lands & new lands as all other the coppyhold lands within the said mannor of Helaugh are and saith that Harkaside inhabitants which are in the constablery of Reeth payes their land tax / & constable assessments with Reeth. /

16: To the sixteenth interrogatory saith that the Lord Wharton & his ancestors have been in the possession & enjoyment of severall lead mines within the lordship of Swaledale mannor of Helaugh all the time of this deponents remembrance (to witt) att / Lownythwaite Old Gang & Freer Fold on the north side of Swale & att Spout Gill & Keld pasture on the southside of Swale & severall other places both on the north & southside of the said river & has made great profitt thereof & continues working att severall / of those mines. /

Luke Hutchinson of Newsham in the county of Yorke yeoman aged thirty eight yeares sworne & examined /

4: To the fourth interrogatory saith he knowes the vill & constablery of Reeth saith the same lys in the mannor of Helaugh saith the said mannor & constablery extend on both the south side & north side of the river Swale saith he was borne att Reeth & hath knowne the said mannor / & township & constablery of Reeth & the river Swale all his time & never heard till since this suit begann that the said river of Swale was ever reputed a boundary to limitt or divide the said mannor or township or constablery or any of them and this deponent saith he doth / not beleive the said river can be a boundary to the said mannor township or constablery for that about four or five & twenty yeares agoe this deponents brother James Hutchinson who then lived att Leyburne on the same side of the river Swale with Harkaside had tenn / sheep strayed from him then & were taken up & carryed to Miles Freer then bayliffe to the then Lord Wharton & this deponent & his brother William Hutchinson hearing of their being in the Lord Whartons bayliffes hands went to the said Miles Freer & challenged / the same sheep & paid him to the best of his remembrance five shillings for taking of them up & restoring the same again & the said Miles Freer told this deponent & his brother / that the said sheep were taken up as waifes on Harkaside Moor for the Lord Whartons use /

John Spenceley of Spring End in Swaledale yeoman aged thirty seaven yeares sworne & examined /

4: To the fourth interrogatory saith he knowes the vill & constablery of Reeth that the same lys within the mannor of Helaugh & saith that the lordship of Swaledale mannor of Helaugh & vill & constablery of Reeth extends & lyes on both north & south side of Swale & extends / southwards to where heaven water deales adjoyning upon Wensleydale saith he was borne att Whitaside in Swaledale & hath lived all his time in the said dale & never heard of the said river of Swale talked on as a boundary to limitt or divide the said lordship [vill] / & constablery or any of them till of late that this suit was begun saith he beleives the said river cannot be a boundary there because he knowes that there are coppyhold lands lying att Crackpott Satteron & Spring End all on the southside of Swale which belongs to / Lord Wharton & are within the mannor of Helaugh & constablery of Reeth both which last mencioned places are on the northside the said river. /

[membrane xiiij *verso*][316]

Joseph Ethrington of Hartford in the county of Yorke gentleman aged thirty yeares sworne & examined /

20: To the twentyth interrogatory saith that the parchment deed now to him produced att the time of his examinacion beginning with these words The Particulars of the Inhansed rents & dated the fifth day of March in the First Year of the / Raigne of our Soveraigne Lord King James the Second[317] was found by this deponent about two moneths since amongst severall writings & papers in the house lately belonging to Mr Philip Swale deceased who was in his life time an agent / or steward for Philip late Lord Wharton alsoe deceased which writings where the said deed was soe found were relating to the present Lord Whartons estate within the mannor of Helaugh in Swaledale and saith that the two witnesses / to the said deed being James Freer junior & Peter Hammond are both dead & beleives the names James Freer junior & Peter Hammond are the proper hand writing of the said James Freer junior & the said Peter Hammond & he this deponent being very well acquainted / with the hand writings of both the said witnesses in their life times. /

Leonard Brackenbury of Richmond in the countie of York surveyor of land aged fifty six yeares sworne & examined /

19: To the nineteenth interrogatory saith that he is by profession a surveyor of land and hath made that profession his study and practise for twenty years past saith the platt or map now produced & shewn to him att the time of his / examinacion containes a true account & descripcion to the best of his judgement beleife skill and informacion of all and every the places divisions & boundarys therein described & represented according to the table of explanacion & scale therein / incerted saith that he this deponent drew the said map or platt & took great care in settleing the same according to the best informacion he could gett of the boundarys moores grounds and places therein contained described & perticularized saith / he surveyed & measured the outmost boundarys of West Grinton & Harkaside & the utmost extent of the inclosed grounds therein as the same were related to him by severall judicious neighbouring inhabitants and fashioned and formed the / perticular feilds or closes within the utmost extent of the inclosures by inspecion as exactly as possibly he could forme them & according to the contents of acres of each severall feild or close by estimacion of those neighbouring inhabitants. /

/s/ Tho: Thomson
/s/ Tho: Gyll
/s/ Tho: Waite
/s/ Rob: Hilton

[membrane xv *recto*][318]

Counter interrogatoryes <exhibited> to bee administred to witnesses produced sworne & examined / in a cause depending in her majesties Court of Exchequer chamber <at Westminster> betweene the right honorable Thomas / Lord Wharton

316 Parchment, 660mm x 710mm.
317 i.e. 1685/6.
318 Parchment, 430mm x 340mm.

now Earle of Wharton complainant & Reginald Marriott esquire & other defendants on the / parte & behalfe of the defendant Marriott; /

I Are you a freeholder or inhabitant in Helaugh or within the township or mannor of Helaugh or within any parte thereof, or in what other place doe you now, or did you lately dwell, or / are you owner farmer or occupyer of any messuage house or lands lying within the said mannor of Helaugh, have you <now> or doe you claime <any> right of common upon any the moore & commons or wasts / within the said mannor, or within any parte or parcell thereof, or upon any the moores wasts of commons of West Grinton & Harkaside, doe you put your sheep beast or horse upon the said <moores or commons> / or any of them to graze or depasture, or have you or [x x x][319] doe you claime any right soe to doe, when you have occasion to use your right of common there, or if you have / any cattle may you or have <you> liberty or right of common within the said mannor to put them there, or doe you or have you right to dig for or get turves <ling> or peates within the said mannor / & what right of common have you or doe you claime within the said places or any or which of them Declare. /

II Did you ever know any of the freeholders or inhabitants within the mannor of Helaugh on the north side of the river Swale enjoy any right of common upon the moores or commons said / to belong to Harkaside or West Grinton who, & when, did any of the said freeholders or inhabitants on the north side of the said river claime the herbage or soyle of the moores & / commons to belong to them when, & by what right did they claime the same Declare /

III Doe you know the places called Harka alias Harkaside & West Grinton & the moores waists & commons thereunto adjoyning called Harka hill Harka Moor Grinton Moore Grinton / Greetes Grinton How or How Groves how long have you knowne the said places, or any of them, did you ever know that the late Lord Wharton or any of the present Lord Whartons ancestors or / himselfe ever wrought any mynes for lead or coale within or upon any of those places moors wasts or commons or in any other place within the mannor of Grinton, in what perticular / place, & when was the first tyme they wrought or began to work the same Declare /

4th Doe you know the present Sir Solomon Swale & did you know his father & grandfather & one Roger Hillary were they or any of them reputed or said to be lords of Grinton & which of them did any & which of / them worke for lead & coale at any tyme & for how long upon the moores wasts & commons of Harkaside or West Grinton or within any parte of the mannor of Grinton when & for how long did / they or any of them work for lead or coale there Declare /

5th Did you ever know any pretended bounders ridd betwixt Grinton Harkaside or West Grinton & Whitaside by whome was the same ridd & when was the first tyme of such rideing did you / ever know any publick notice given of rydeing the bounders between the townships of Grinton & Helaugh to each other & by which of them or when Declare /

6th Doe all the freeholders tennants & inhabitants of Harkaside & West Grinton doe suite & service at the complainants court [x x x][320] <at Helaugh>, or any, & which of them by what tenure doe they hold their lands that / doe suite & service ther[e] are all the said freeholders & inhabitants of Harkaside & West Grinton within the consta-

319 One word has been erased.

320 One word has been erased.

blery of Reeth or is there any parte of Harkaside & West Grinton that is within / any other & what constablery Declare /

7th Were you a witness in any <& what> cause now or late depending in the Court of Exchequer or in any other & what court where the lead mynes now in question or the mannor of Grinton were in contest or / dispute & when, did you then or at any other tyme depose make oath or affidavit that the places called Harkaside or Whitaside or either of them were parcell of or within the mannor / or lordshipp of Grinton or what other evidence did you then or at any other tyme give depose or make affidavit of concerning the same Declare etc /

/s/ Sam: Dodd[321]

[membrane xvi *recto*][322]

Ex parte defendant Marriott

Richard Close a witness before sworne on the parte of the Lord Wharton & others & <now> sworne & examined on the parte of Mr Marriott the defendant saith as followeth /

2 To the second interrogatory this deponent saith that he never knew any the free-holders or inhabitants within the mannor of Helaugh on the north side of the river Swale ever enjoyed any right of common upon the moores or commons said to belong to Harkaside or West Grinton save onely John Lonsdaile a / coppyholder who lived at Helaugh to have some lambs goeing upon the said commons of Harkaside at some tyme or tymes & as this deponent has heard & beleives got stones there for mending the water bank upon the north side of the river the said John Lonsdaile then haveing a peice of ground / on the south side of the said river which this deponent supposeth had been antiently left by the said river /

3 To the third interrogatory this deponent saith that he knowes the places called Harka Moore Grinton Moore Grinton Greetes Grinton How or How Groves & hath soe knowne them for Sixty yeares he this deponent haveing lived at Harkaside for fifty seaven yeares and saith that the / Lord Wharton nor any of his ancestors did ever worke any mynes of lead or coale within or upon any the said places moores or commons till about two yeares agoe & that was the first tyme of his workeing /

4 To the fourth interrogatory this deponent saith that he knew old Sir Solomon Swale & Mr Roger Hillarye & that they were reputed lords of the mannor of Grinton some said Sir Solomon Swale others said Mr Hillarye were the lord or lords thereof and saith that Sir Solomon Swale for some yeares wrought the / lead mynes on Harkaside as alsoe on Grinton How where Mr Marriott <now> workes & this deponent has heard that Mr Hillarye did alsoe worke the lead mynes on Harka hill /

5 To the fifth interrogatory this deponent saith that he never knew any boundary ridd betweene Grinton Harkaside or West Grinton till by the now Lord Wharton or his agents about two yeares since & never by him or any others before that tyme save that he this deponent hath heard that about forty yeares / since Sir Solomon Swale ridd bounders {bounders} thereabouts & remembers that he or his agents desired this deponent to goe along with him but this deponent did not goe nor doth he know what bounders he ridd at that tyme /

[321] For Sir Samuel Dodd, see p. 7, footnote 22 and profile in Appendix 3.

[322] Parchment, 690mm x 560mm.

6 To the sixth interrogatory this deponent saith that the [said] inhabitants of Harkaside which are tennants to Lord Whartons demesne lands who are at a rack rent & his coppyhold {land} tennants appeare at the Lord Whartons court at Helaugh & the freehold tennants in Harkaside which pay free rents to the Lord Wharton / viz the lands belonging to Swale hall about ninety pounds per annum & which pay a yearly free rent of four pence to the Lord Wharton other lands in the possession of Mr Marmaduke Hartley of the yearly [vallue] of about three pounds & pays nine pence per annum to the Lord Wharton & alsoe other / lands belonging to Ralph Addison of the yearly vallue of about three pounds & pays 2^{d} per annum to the Lord Wharton these said three freehold tennants used to appeare likewise at Helaugh court and this deponent further saith that there are alsoe crowne lands in the possession of the said Marmaduke / Hartley & of John Heard Anthony Heard Sir Henri Johnson Mr Dickinson Christopher Raper & Luke Robinson of the yearly value of about forty eight pounds tenn shillings all which last menciond lands the former occupyers thereof used to appeare forty yeares agoe at Arkingarthdale court & knowes <not> that they / have appeared at any other court {since} <since> that tyme further saith that there are other lands formerly belonging to the Priory of Bridlington viz Myles Close George Sympson Leonard Galloway as this deponent beleives & George Milner of about the yearly vallue of 29 li which last <4> persons usually appeared at / the sherriffs turne & not any other place that he know's off all which said severall places lye within the constablery of Reeth except the said lands belonging to Swale Hall Wensleys land & all those lands in West Grinton & alsoe all those lands formerly belonging to the Priory of / Bridlington lye within the constablery of Grinton saveing that Wensleys land in West Grinton which pays five shillings per annum coppyhold rent to the Lord Wharton lyes within the constablerye of Reeth within the mannor of Helaugh as this deponent beleives /

John Milner a witness formerly sworne on the complainants behalfe & now produced sworne & examined on the defendant Mr Marriotts behalfe /

3 To the third interrogatory this deponent saith that he very well knowes the moores or commons called Harka hill Harka Moore Grinton Moore Grinton Greetes Grinton How or Howgroves & has knowne the same for sixty yeares last past liveing all that tyme as a neare neighbour thereto and saith that / the present Lord Wharton nor any of his ancestors did ever worke any mynes of lead or coale upon the said moores or commons that ever he heard off till about two yeares agoe /

4 To the fourth interrogatory this deponent saith that he know's Sir Solomon Swale & knew his father Sir Henry & his grand father Sir Solomon & alsoe knew Roger Hillary & that they pretended to be lords of the mannor of Grinton & saith that old Sir Solomon Swale wrought mynes of lead upon Harkaside / & Grinton How for severall yeares together & alsoe the present Sir Solomon wrought mines of lead <there> & had considerable gettings till he was ousted by the {present} defendant Mr Marriott /

5 To the fifth interrogatory this deponent saith that he know's not of any boundary that was ridd betwixt Grinton West Grinton Harkaside or Whitaside except the boundary that was ridd about two yeares since by the said Lord Wharton saveing that he this deponent ridd some boundarys with old Sir Solomon Swale about fifty / yeares agoe which he claimed as lord of the mannor of Grinton beginning at Stallerston Style on the east & ending at Haverdale beck on the west & ridd parte of the said boundarye about eleaven or twelve yeares agoe with the present Sir Solomon /

6 To the sixth interrogatory this deponent saith that there are some inhabitants in Harkaside who hold crowne <land> & others in West Grinton who hold lands formerly belonging to the priory the certaine vallue whereof this deponent does not

know / nor does he know att what court they appear saith that he know's all the lands in West Grinton are within the constablery of Grinton except two closes of about seaven dayes moweing in the possession of Ralph Wensley /

Anthony Raper a witness formerly {sworne} & produced & examined for the complainant & now produced sworne & examined for the defendant Mr Marriott /

3 To the third interrogatory saith that he know's the places called Harkaside & West Grinton & know's Harka hill Grinton Moore Grinton Greetes Grinton How or How Groves & has knowne the same for nigh fifty yeares & all that tyme he this deponent did never know or heare that / the late Lord Wharton the present Lord Wharton nor any of his ancestors did ever worke any mynes for lead or coale within any of the said places moores or commons within Grinton but saith about twenty yeares agoe that the present Lord Wharton opened a myne or mynes on Harkaside Moore /

4 To the fourth interrogatory this deponent saith that he know's the present Sir Solomon Swale & did knowe his father & grand father & one Roger Hillarye & says that he has heard them all his tyme called lords of Grinton Says that they wrought for both lead & coales in the waists there Hillarye wrought for coales in East / Grinton libertyes & that old Sir Solomon Swale wrought for lead in the moores in question as alsoe did Mr Hillarye & the present Sir Solomon Swale & the Swales has ridd boundaryes as this deponent has heard & beleives & that the present Sir Solomon continued workeing mynes in the waists & commons of Harkaside & West Grinton / untill the said Mr Marriott recovered the said mynes & ousted the said Sir Solomon of his possession /

5 To the fifth interrogatory this deponent saith that about two yeares since this deponent heard that the present Lord Wharton ridd a boundarye betwixt Grinton Harkaside & West Grinton but never till then did ever heare that the present Lord Wharton or any of his ancestors did ever ride the said boundarye /

6 To the sixth interrogatory this deponent saith that besides the tennants of my Lord Whartons in Harkaside there are other tenures called {called} crowne lands the vallue this deponent know's not but saith that the tennants thereof doe not appear at Helaugh court nor does he know at what court they doe appear says there are alsoe / lands reputed to belong to the Priory of Bridlington the tennants whereof doe not appear at the court of Helaugh saith further that those priory lands [x x x][323] & the lands called Swale lands belonging to Swale Hall & all the lands in West Grinton except two closes in the possession of Ralph Wenslye lye within / the constablery of Grinton /

Christopher Raper a witness formerly examined for the complainant & now sworne & examined for the defendant Mr Marriott /

1 To the first interrogatory this deponent saith that he is an inhabitant freeholder & coppyholder in Harkaside under the Lord Wharton & had right of common on Harkaside Moore in right of his freehold & coppyhold lands but saith that within a month last past he released his common right to the complainant the Lord Wharton / but saith that his goods & sheep goe's on the said common by his Lordshipps leave /

3 To the third interrogatory this deponent saith that he know's Harka Hill Harka moor Grinton Moore Grinton Greetes Grinton How or Howgroves & has soe knowne the same for above fifty yeares & dureing all the tyme aforesaid he this deponent never knew or heard that the present Lord Wharton nor any of his ancestors wrought any / mynes for lead or coales within any of the said moores waists or commons in Harkaside & West Grinton or in the place in question till of late that the present Lord

[323] One or two words have been erased.

Wharton sunke & wrought for lead on Harkaside or West Grinton Common & saith
that the reason of such his knowledge is that this deponent has lived all / his tyme
4 nigh the place in question To the fourth interrogatory this deponent saith that he
very well know's the present Sir Solomon Swale & did know his grand father old
Sir Solomon Swale in his life tyme wrought & gott lead oare on Harkaside & West
Grinton / commons as alsoe did the present Sir Solomon for severall yeares together
& soe continued to worke till he was turned out of possession by the defendant Mr
Marriott who recovered the said mynes from the said Sir Solomon saith that the said
Mr Hillarye wrought & gott both lead & coale on Harka How Moore / and further
this deponent saith that old Sir Solomon Swale the said Hillarye & the present Sir
Solomon Swale all claimed to be Lords of Grinton /
5 To the fifth interrogatory this deponent saith that he never knew any boundarye ridd
betwixt Grinton Harkaside & West Grinton till about two yeares since that he heard
that the defendant the Lord Wharton & severall others ridd the said boundarye /
6 To the sixth interrogatory this deponent saith that all the tennants & inhabitants of
Harkaside & West Grinton doe Suite & service at the complainants court at Helaugh
except the crowne lands and saith that the lands & two or three houses in Grinton
& the commons belonging to Swale hall are in the constablerye / of Grinton except
7 Wensleys two closes To the seaventh interrogatory this deponent saith that he this
deponent did not at any tyme nor in any court make any affidavit that Harkaside &
Whitaside or either of them were parcell of or within the mannor or Lordshipp of
Grinton /

Ralph Wensley of West Grinton a witness formerly examined for the complainant
& now sworne & examined for the defendant Mr Marriott /
1 To the first interrogatory this deponent saith that he lives at West Grinton & is a
coppyholder there haveing two coppyhold closes & claimes right of common on
Harkaside Moore as belonging & in right of his said coppyhold closes saith he has
& does enjoy right of common on the said Harkaside or West Grinton Moore &
never / was disturbed since he had a tytle to the said coppyhold land /
2 To the second interrogatory this deponent saith that he {knows} do's not know that
ever any of the freeholders or inhabitants within the mannor of Helaugh on the north
side of the river Swale did ever enjoy <any> right of common upon the moores &
commons belonging to Harkaside or West Grinton /
3 To the third interrogatory this deponent saith that he knowes the places called Harka-
side or West Grinton & the moores thereounto adjoyning called Harka{side} <hill>
Harka Moore Grinton Moore Grinton Greetes Grinton How or Howgroves & has
soe knowne the same for above thirty yeares last past & in all that tyme / did never
know that the late Lord Wharton the present Lord Wharton nor any of their ances-
tors did ever worke any mynes for lead or coale within or upon any of the said
places waists moores or commons or in any other place within the reputed mannor
of Grinton till <that> about two yeares since the present Lord Wharton / opened
a myne or mines & wrought for [l]ead upon Harkaside Moore or West Grinton
Moore /
4 To the fourth interrogatory this deponent saith that he well know's the present Sir
Solomon Swale & did know his father Sir Henry & his grandfather Sir Solomon
& Roger Hillary saith that old Sir Solomon wrought for lead on the moores called
Grinton How as alsoe upon Harkaside or West Grinton Moore & gott in both /
places lead saith he does not remember Sir Henrys workeing either for lead or coale
but saith that the said Mr Hillary wrought for lead < on Grinton How> or for coales
on the east side of Grinton Moore {on Grinton How} and this deponent further
saith that the present Sir Solomon Swale wrought for lead on Grinton How / & on

Harkaside or Grinton Moore the place in question & gott severall quantitys of lead & continued workeing for lead in Harkaside or Grinton moores untill the defendant Marriott recovered the mynes from him & gott possession thereof saith that dureing the tyme that old Sir Solomon Swale Roger Hillarye / & the present Sir Solomon & dureing this deponents knowledge of their workeing they were called lords of the said mynes /

5 To the fifth interrogatory this deponent saith that he never knew of any boundarye ridd between Grinton Harkaside or West Grinton & Whitaside till about two yeares since that the present Lord Wharton as this deponent beleives caused a boundarye to be ridd of which rideing this deponent saith notice was given / in the church yard & this deponent with severall others were at the rydeing thereof /

6 To the sixth interrogatory this deponent saith that there are lands in Harkaside or West Grinton of severall tenures to witt freehold coppyhold the Lord Whartons owne lands crowne lands & priory lands some of the freeholders & all the coppyholders & the tennants of the Lord Whartons owne lands / doe suite & service at my lords court at Helaugh & as for the crowne lands priory lands & the other freehold lands the tennants thereof doe not appeare at Helaugh court nor do's he know at what court the crowne lands appeare & that the priory lands & the Swale lands are in the constablery of Grinton /

Miles Close of Harkaside sworne & examined on the parte & behalfe of the complainant & now sworne & examined on the parte & behalfe of the defendant Mr Marriott saith /

2 To the second interrogatory this deponent saith that he never knew any of the freeholders or inhabitants within the mannor of Helaugh on the north side of the river Swale use or enjoy any right of common within the moores & commons belonging to Harkaside & West Grinton nor claimes the herbage or soyle thereof except / old Francis Whipp of Reeth who farmed a small parcell of land at West Grinton put sheep for one yeare upon the said commons /

3 To the third interrogatory this deponent saith that he know's Harka hill Harka Moore Grinton Moore & the Greets called by some Grinton Greets Grinton How or Howgroves & has known the same for {about} <above> forty yeares saith that for all his tyme the Lord Wharton nor any of his ancestors did ever worke mynes for lead / or coale in any of the said places moores or commons or any other place within the mannor of Grinton as he beleives till about two yeares agoe /

4 To the fourth interrogatory this deponent saith that he knew old Sir Solomon Swale & Roger Hillarye & that they claimed to be lords of Grinton but there being a difference or dispute betwixt them this deponent cannott say which of them or whether either of them was taken to be lords thereof saith that Hillarye / wrought for coale upon East Grinton Moore & that old Sir Solomon Swale & his grandson the present Sir Solomon wrought mynes of lead upon Harkahill the place in question & the present Sir Solomon gott considerably there & he continued to worke there till the relator Mr Marriott recovered the same against him /

5 To the fifth interrogatory this deponent saith that he never knew or heard of any boundarye ridd betweene Grinton Harkaside West Grinton & Whitaside till about two yeares agoe by the present Lord Wharton /

6 To the sixth interrogatory this deponent saith that there are severall lands or tenures within Harkaside & West Grinton called freehold lands coppyhold lands crowne lands & priory lands & alsoe Swale lands which lye onely in West Grinton & which this deponent takes to be freehold lands saith / that all the coppyholders & some of the freehold tennants (besides the two Hirds who are tennants to my Lord Whartons owne lands there) appeare at his lordshipps court at Helaugh saith the rest doe not

appeare at Helaugh but know's not certainly where the rest doe or might to appeare save / that the owners or tennants of Swale lands are called at Helaugh saith that <all> the said coppyhold lands the Lord Whartons lands & some of the said freehold lands lye in the constablerye of Reeth & the rest in the constablery of Grinton /

Christopher Harker a witness formerly examined for the complainant & now sworne & examined on the defendants behalfe /

2 To the second interrogatory this deponent saith that he never knew any of the freeholders or inhabitants within the mannor of Helaugh on the north side of the river Swale enjoy any right of common <upon> the moores or commons belonging to Harkaside or West Grinton or claime herbage or soyle upon any / of the said moores or commons /

[membrane xvi *verso*]

3 To the third interrogatory this deponent saith that he know's the places called Harka alias Harkaside & West Grinton & the moores & waists called Harka hill Harka Moore Grinton Moore Grinton Greetes Grinton How or How Groves & has soe knowne the same for twenty four yeares & upwards saith that / dureing this deponents knowledge thereof he never heard that ever the late Lord Wharton the present Lord Wharton nor any of th<e>ir ancestors did ever worke any mynes for lead or coale within or upon any of the said places waists moores or commons or
within any other place within the mannor of Grinton [till] / two yeares since To
4 the fourth interrogatory this deponent saith that he know's the present Sir Solomon Swale did know old Sir Solomon Swale & Mr Roger Hillarye saith they were at severall tymes all reputed to be lords of Grinton & that they did all severally worke for lead upon Harkaside or West / =Grinton moores or commons and saith that [x x x][324] the present Sir Solomon Swale continued workeing there untill the said defendant Mr Marriott recovered the said mynes from him /

6 To the sixth interrogatory this deponent saith that the coppyholders & such of the freeholders as are within the constablery of Reeth doe appear at the Lord Whartons court at Helaugh the rest doe not appeare there this deponent saith that he ha's lands at West Grinton & does not appear at the / court but at the sherriffs turn

Robert Robinson formerly sworne & examined on the parte of the complainant sworne & examined on the parte & behalfe of the defendant Mr Marriott sayth as followeth /

3 To the third interrogatory this defendant saith that he know's the places moores & commons called Harkaside Harka Moore Grinton Moore Grinton Greetes Grinton How or Howgroves & ha's been a neighbour thereto for above fifty yeares last past & saith in all that tyme the late Lord Wharton the present / Lord Wharton or any of their ancestors did never worke any mynes for lead or coale upon any of the moores that ever this deponent knew or heard of till about two yeares agoe /

6 To the sixth interrogatory this deponent saith that parte of Harkaside & West Grinton lye in the constablery of Reeth & parte thereof lyes in the constablery of Grinton but what partes he know's not saith there are lands of different holdings <there> but know's <them> not all perticularly and saith that there are some / inhabitants in Harkaside & West Grinton who doe not appeare at Helaugh /

324 One or two words have been erased.

William Arundale a witness before sworne & examined on the parte of the complainant & now sworne & examined on the parte of the defendant Mr Marriott saith as followeth /

2 To the second interrogatory this deponent saith that for all his tyme vizt twenty yeares or thereabouts the inhabitants of the mannor of Helaugh on the north side of the river Swale did never enjoy or pretend to claime any right of common upon the moores & commons of Harkaside & West Grinton that ever he knew or heard off save two / persons who lived in Helaugh that gott stones there one whereof asked leave of some of the inhabitants of Harkaside & the other being neighbour was the reason as he beleives that he was not hindred or {beleives} disturbed /

3 To the third interrogatory this deponent saith that he has knowne the moore or common called Harkaside or Harka Moore haveing been a neighbour thereto for above twenty yeares last past saith that the Lord Wharton or any of his ancestors did not dureing the said tyme worke any mynes of lead or cole upon the said moores or commons till about two yeares last past /

4 To the fourth interrogatory this deponent saith that the present Sir Solomon Swale wrought mynes of lead upon the said moors of Harkaside or Harka Moore for some tyme & gott considerably there till the defendant Mr Marriott recovered the same & turned him out of possession /

6 To the sixth interrogatory this deponent < {saith} saith> that all the inhabitants & freeholders in Harkaside & West Grinton doe not suite & service at the Lord Whartons court at Helaugh as for such of them that he know's doe suite & service he refferrs to what he has so deposed upon his examination on the parte of {parte of the} complainants and for those / that doe <not> suite & service there he cannot name them all particularly or of what termes they hold saith that parte of Harkaside & West Grinton is within the constablery of Reeth & parte within the constablery of {West} Grinton but cannott particularize the same /

John Blaides a witness examined for the complainants & now produced sworne & examined for the defendant Mr Marriott /

2 To the second interrogatory this deponent saith that he never knew any of the freeholders or inhabitants within the mannor of Helaugh on the north side of the river Swale enjoy any right of common upon the moores or commons belonging to Harkasyde or West Grinton save that one Matthew Harrison of Reeth as this deponent / ha's heard gott peates upon a place called Reeth Peat Moore upon parte of Harkaside Moore but what right he had soe to doe this deponent know's not but beleives he was not disturbed nor does he know that any of the freeholders or inhabitants within the said mannor of Helaugh on the north side of the river Swale / either claimed the herbage or soyle of the said moores or commons upon Harkaside other then the said Matthew Harrison /

3 To the third interrogatory this deponent saith that he know's the places called Harkaside & West Grinton & the moores & waists thereunto belonging called Harka Moore Grinton Moore Grinton Greetes Grinton How or How groves & ha's soe knowne the same for forty yeares & upwards & in all that tyme did / never know that the late Lord Wharton the present Lord Wharton nor any of their ancestors did ever worke any mynes for lead or coales within or upon any of the places moores waists or commons or in any other place or places within the mannor of Grinton /

4 To the fourth interrogatory this deponent saith that he did know old Sir Solomon Swale & Mr Roger Hillarye & know's the present Sir Solomon saith that old Sir Solomon & the present Sir Solomon were reputed lords of Grinton & both wrought by their workemen & servants upon Harkaside Moore the place in question for lead / & gott considerable quantityes of lead there that the present Sir Solomon continued

workeing there till the defendant Mr Marriott recovered the mynes from him saith that the said Mr Hillarye wrought & gott coale upon Grinton Moore /

6 To the sixth interrogatory this deponent saith that all the coppyholders of Harkaside that belong & are the Lord Whartons tennants as al<so> some freeholders there doe Suite & service to his lordshipps court at Helaugh but the other freeholders doe not /

John Harker a witness before sworne & examined on the parte & behalfe of the Lord Wharton & other complainants & now sworne & examined on the parte & behalfe of the defendant Mr Marriott saith /

1 To the first interrogatory this deponent saith that he hath now a house & some lands in the mannor of Helaugh which he letts to John Roe at the yearly {vallue} rent of six pounds & which he has mortgaged to one Mr Smales but he this deponent receives the rent thereof /

James Robinson a witness formerly examined for the complainants & now sworne & examined for the defendant Mr Marriott /

1 To the first interrogatory this deponent saith that he is both a copyholder & freeholder in Reeth within the mannor of Helaugh but claimes noe common right within the said mannor /

3 To the third interrogatory this deponent saith that he know's Harka hill Harka Moore Grinton Moore & Grinton How & has soe knowne the same for about forty yeares last past and saith that till within this two yeares he this deponent dureing the tyme aforesaid did never either know or heare that the present Lord Wharton / nor any of his ancestors did ever worke for either lead or coales in any of the moores waists & places aforesaid /

4 To the fourth interrogatory this deponent saith that he know's the present Sir Solomon Swale & did know Roger Hillarye saith that the present Sir Solomon Swale & his father were reputed lords of Grinton & that it was comonly said & reported that the said Sir Solomon & his father wrought for lead on Harka hill /

James Naylor before sworne & examined on the parte & behalfe of the Lord Wharton & other complainants & now sworne & examined on the parte & behalfe of Mr Marriott & other defendants saith /

1 To the first interrogatory this defendant saith that he farm's the lands belonging Swale Hall which he do's not beleive to be within the mannor of Helaugh and saith that he has noe common right in any place within the said mannor of Helaugh /

3 To the third interrogatory this deponent saith he know's Harka hill Harka Moor Grinton Moore Grinton Greetes Grinton How or the Howgroves & ha's soe knowne the same for about twenty yeares & for all the tyme of his knowledge (he this deponent liveing for all that tyme at Swale hall) saith that the present Lord Wharton nor any of his <ancestors> did / ever worke any mynes of lead or coale upon any of the said moores & commons till about two yeares agoe /

4 To the fourth interrogatory this deponent saith that he know's the present Sir Solomon Swale & did know his father & alsoe Roger Hillarye saith that they were all reputed & called cheife lords and further saith that Sir Solomon wrought mynes of lead upon Harkaside Moore & gott considerably there till he was turned out by / Mr Marriott /

Thomas Milner a witness formerly examined on the complainants parte & now sworne & examined on the defendant Mr Marriotts parte /

3 To the third interrogatory this deponent saith that he know's the places called Harka hill Harkaside & West Grinton & the Waists & commons thereunto adjoyneing called Harka hill Harka Moore Grinton Moore Grinton Greetes Grinton How or How groves & ha's knowne the same for forty / yeares & upwards & saith that dureing all the tyme aforesaid this deponent did never know or heare that the late Lord Wharton

the present Lord Wharton nor any of their ancestors did ever worke any myne or mynes for lead or coale within or upon the said places moores & waists or commons
4 or in any other place within the mannor of Grinton till / of late To the fourth interrogatory this deponent saith that he know's the present Sir Solomon Swale & did know old Sir Solomon alsoe his grandfather & Sir Henry his father & Roger Hillarye saith that all the Swale's dureing this deponents remembrance were severally one after another reputed lord or lords of Grinton but cannot / remember that the said Hillary was reputed lord thereof saith that all the Swale's wrought from tyme to tyme for severall yeares for lead upon the moores waists & commons of Harkaside & West Grinton & won & gott severall quantities of lead upon the said commons & that the present Sir Solomon continued workeing there / untill the defendant Mr Marriott {continued} recovered the possession from him /

6 To the sixth interrogatory this deponent saith that all the coppyholders & most of the freeholders & inhabitants of Harkaside & West <Grinton> doe suite & service at the complainants court at Helaugh but some of the freeholders doe not saith that Harkaside inhabitants are within the constablery of Reeth & that West Grinton people are within the constablery of Grinton /

James Hird a witness before examined on the parte & behalfe of the Lord Wharton & his son complainants & now sworne & examined on the parte & behalfe of Mr Marriott & others defendants saith /

3 To the third interrogatory this deponent saith that he very well know's the moores or commons called Harka hill Harka Moore Grinton Moore Grinton Greetes Grinton How or Howgroves & hath knowne the same for about fifty yeares last past haveing for all that tyme been a near neighbour thereto & saith that the present / Lord Wharton nor any of his ancestors did ever in all that tyme worke any mynes of lead or cole upon any of the said moores that this deponent knew or heard till {about} within two years last past /

4 To the fourth interrogatory this deponent saith that he know's the present Sir Solomon Swale but did not know his father or grandfather saith he knew Roger Hillarye & that the said Sir Solomon & his ancestors for severall yeares wrought mynes of lead upon Harkaside or Harka Moore the place in question but what gettings they had this deponent know's not / saith that Sir Solomon continued to worke the said mynes till the defendant Mr Marriott recovered the same against him /

5 To the fifth interrogatory this deponent saith that he never knew or heard of any boundarye ridd betwixt Grinton Harkaside or West Grinton & Whitaside till about two yeares agoe by the present Lord Wharton /

Thomas Harrison a witness formerly sworne & examined for the complainant & now sworne & examined for the defendant Mr Marriott /

3 To the third interrogatory this deponent saith that he know's the places called Harka or Harkaside & West Grinton & the moores & waists thereunto adjoyneing called Harkahill Harka Moore Grinton Moore Grinton Greetes Grinton How or Howgroves & ha's knowne severall of the said places for fifty yeares & upwards & has / knowne them all for forty yeares & for those severall & respective tymes did <never> know or hear that the late Lord Wharton the present Lord Wharton nor any of the ancestors did ever worke any myne or mynes {o}fo<r> lead or coale within or upon any of the said places moores or commons or on any other place within the said mannor of Grinton /

4 To the fourth interrogatory this deponent saith that he know's the present Sir Solomon Swale & did knowe old Sir Solomon Swale his grandfather & Roger Hillary saith that old Sir Solomon wrought for lead on Harka hill within Harkaside Moore & won & gott severall quantity's of lead there saith that young / Sir Solomon wrought

there for many yeares & had very good gettings of lead & continued workeing till that the defendant Mr Mariott recovered the possession {of} <from> him does not remember that Hillary wrought for either lead or coale on Harkaside Moore /

5 To the fifth interrogatory this deponent saith that he did not know the boundarys ridd between Grinton Harkaside & West Grinton & Whitaside till the the present Lord Wharton caused the same to be ridd about two yeares agoe beleiv's publick notice was given of such rideing but whether such notice was given to the lords of Grinton this deponent know's not /

6 To the sixth interrogatory this deponent saith that there are lands of severall holdings in Harkaside & West Grinton to witt coppyhold lands in the possession of Christopher Raper Ralph Wensley John Blaides William John & George Arundells & another brother who were sons of William Arundell & have onely their faithers estates / saith there are severall freeholders within Harkaside & West Grinton to witt one Wensley & Hartley & some others & alsoe John Hird & another Hird which Hirds are the tennants of the Lord Whartons owne freehold lands & all which appeares at the Lord Whartons court at Helaugh as alsoe / some other freeholders but beleives Luke Robinson of Harkaside holds some priory lands in Harkaside do's not know whether he appeares at the Lords court at Helaugh or not & there are severall other inhabitants in Harkaside who this deponent do's not know at what courts they appeare /

John Bland before examined on the parte & behalfe of the Lord Wharton & other complainants & now sworne & examined on the parte & behalfe of Mr Marriott &
2 other defendants saith /
To the second interrogatory this deponent saith that he hath heard that some of the
4 inhabitants of Helaugh & Reeth have gott ling & peates upon Harkaside Moore /
To the fourth interrogatory this deponent saith that he know's old Sir Solomon Swale & alsoe the present Sir Solomon saith that they have been reputed & called lords of Grinton & that the present Sir Solomon did for some tyme worke mynes of lead
5 upon Harkaside Moore the place in question till he was ousted <by Mr Marriott> /
To the fifth interrogatory this deponent saith that he never knew or heard of any boundarye ridd betwixt Grinton Harkaside West Grinton & Whitaside till about two yeares agoe by the present Lord Wharton /

George Metcalfe a witness formerly examined for the complainants & now
2: sworne & examined for the defendant Mr Marriott /
To the second interrogatory this deponent saith that he never knew of any other freeholders or inhabitants (except his father & his selfe) {neither} <within> the mannor of Helaugh on the north side of the river Swale enjoy any right of common upon the moores & commons / of Harkaside & West Grinton or claime the herbage or soile of the said moores & commons to belong to them; But this deponent & his father haveing lands in Harkaside claimed & enjoyed both right of common & herbage in Harkaside moores & commons / but now this deponent has noe lands in Harkaside
3: nor claims any right of common there /
To the third interrogatory this deponent saith that he knowes the places called Harka alias Harkaside & West Grinton & the moores & Wastes called Harkahill Harka Moore Grinton Moore Grinton Greetes Grinton How or How Grooves & has soe known / the same for nigh fifty yeares & in all that time did never know that the <late> Lord Wharton nor the present Lord Wharton nor any of their ancestors did ever worke any mine or mines for lead or cole within or upon any of the said places & waists /
4: moors or commons or in any other place within the reputed mannor of Grinton; /
To the fourth interrogatory this deponent saith that hee knows the present Sir Solomon Swale & did know his father & Roger Hillary saith the said Sir Solomon

pretended or was reputed to be lord of Grinton & did worke for lead upon Harkaside Moor / & continued working for lead there untill the defendant Mr Marriott recovered possession from him /

5: To the fifth interrogatory this deponent saith that he never knew or heard of the rideing of any boundary betwixt Grinton Harkaside or West Grinton & Whitaside till about two yeares since that the present Lord Wharton rid a boundary / betweene the said places; /

[membrane xvij *recto*][325]

Francis Hutchinson a witness formerly examined for the complainants & now sworne & examined for the defendant Mr Marriott /

1 To the first interrogatory this deponent saith that he is an inhabitant in Reeth & within the mannor of Helaugh & has both freehold & coppyhold lands there but do's not claime any right of common or to digg for turves peates or ling in any parte of the said mannor of Helaugh /

3 To the third interrogatory this deponent saith that he know's the places called Harkaside & West Grinton & the moores & waists called Harka hill Harka Moore Grinton Moore Grinton Greetes Grinton How or Howgroves & has soe knowne the same for forty yeares last past & dureing / this deponents knowledge thereof this deponent never knew or heard that the late Lord Wharton the present Lord Wharton nor any of their ancestors ever wrought any myne or mynes for lead or coale within or upon any of these places moores or waists or in any other place within the / reputed mannor of Grinton till of late /

4 To the fourth interrogatory this deponent saith that he know's the present Sir Solomon Swale & did know his father & one Roger Hillaryc saith that he has heard at severall tymes they were all reputed lords of the reputed mannor of Grinton & has heard them called soe saith that the present / Sir Solomon wrought mynes for lead on Harkaside Moore & continued soe to doe till the defendant Mr Marriott recovered them from him beleives & it was reported that the present Sir Solomon had very good gettings of lead there /

5 To the fifth interrogatory this deponent saith that about two yeares <agoe> the present Lord Wharton caused a boundarye to be ridd between Grinton Harkaside or West Grinton & Whitaside but never heard of any other boundarye ridd between the said places saith that notice was given of the said rydeing /

Henry Maitchell before examined on the parte & behalfe of the Lord Wharton & others complainants & now sworne & examined on the parte of Mr Marriott & others defendants /

1 To the first interrogatory this deponent saith that he hath a house & a peice of ground lyeing at Reeth in the mannor of Helaugh now in his owne occupation & does claime some right upon some of the commons belonging to the said mannor of Helaugh /

James Hutchinson a witness formerly examined for the complainant & now sworne & examined for the defendant Mr Marriott /

1 To the first interrogatory this deponent saith that he lives at Reeth within the mannor of Helaugh & has coppyhold lands within the said mannor but claimes noe right of common therein haveing released the same /

325 Parchment, 620mm x 510mm. The *verso* is heavily rubbed and the last three depositions are almost wholly illegible except with the aid of ultraviolet light.

3 To the third interrogatory this deponent saith that he knowes the places called Harkaside & West Grinton & the moores & waists thereto adjoyning called Harkahill Harka Moore Grinton Moore Grinton Greetes Grinton How or How groves & has knowne the same for / forty yeares last past being a nigh neighbour thereto & liveing most of his tyme at Reeth in the mannor of Helaugh saith that dureing the tyme aforesaid the present Lord Wharton the late Lord Wharton nor any of their ancestors did ever that / this deponent heard of worke any myne or mynes for lead or coale within or upon any of the said places moores waists or commons or in any other place within the mannor of Grinton /

4 To the fourth interrogatory this deponent saith that he knowes the present Sir Solomon Swale & did knowe his father as alsoe Roger Hillary saith that Sir Solomon his father & Roger Hillarye were reputed & called at severall tymes lord or / lords of Grinton & that Sir Soloman wrought for & gott severall considerable quantitys of lead upon {Harka} Harkaside Moore the place in question & continued workeing there till the defendant Mr Marriott recovered the posession from him /

5 To the fifth interrogatory this deponent saith that till about two yeares agoe that the present Lord Wharton caused a boundarye to be ridd between Grinton Harkaside West Grinton & Whitaside he did never hear or know of any boundarye ridd betweene / the said places saith that publique notice was given or the said rydeing in the church yard of Grinton about two day's before such rydeing but whether any perticular notice thereof was given to the lord of Grinton this deponent know's not /

Francis Charder a witness formerly examined on the parte of the complainants & now sworne & examined on the parte & behalfe of Mr Marriott & other defendants /

1 To the first interrogatory saith that he hath some coppyhold or customarye lands in Reeth within the mannor of Helaugh but do's not claime any common right upon any the moores waists or commons within the said mannor /

3 To the third interrogatory this deponent saith that he know's the moores & commons called Harka hill Harka Moore & Grinton Moore & ha's known the same for about fifty yeares & that he never heard the Lord Wharton or any of his ancestors wrought any mines of lead or coale upon any of the said moores or <commons till of late> /

4 To the fourth interrogatory this deponent saith that old Sir Solomon Swale & the present Sir Solomon Swale wrought some mynes of lead upon Harkahill & mynes of coale upon Grinton Moore <for some time> & further know's not /

5 To the fifth interrogatory this deponent saith that he never knew any boundaryes ridd betwixt Grinton Harkaside or West Grinton & Whitaside till about two yeares agoe by the present Lord Wharton /

Matthew Alderson a witness formerly examined on the parte of the complainants & now sworne & examined on the behalfe of Mr Marriott & others defendants /

1 To the first interrogatory this deponent saith that he is a coppyhold or customary tennant at Reeth within the mannor of Helaugh & ha's customarye or coppyhold lands there but claimes noe common right or ha's any manner of cattle / that goes or depastures upon any the moores or commons in Harkaside or Reeth or in any other place within the mannor of Helaugh or reputed mannor of Grinton /

3 To the third interrogatory this deponent saith that he know's the places called Harkaside & West Grinton & the moores waists & commons thereto adjoyneing called Harkahill Harka Moore Grinton Moore Grinton Greetes Grinton How or / Howgroves & ha's knowne the same for forty yeares last past saith that the late Lord Wharton the present Lord Wharton nor any of their ancestors did ever dureing this deponents remembrance worke any myne or mynes for lead or coale within or upon any / of the said places moores waists or commons or in any other place within the

manner of Grinton till within these two yeares that the Lord Wharton gott into possession of the mines in question upon Harkaside Moore or some part thereof /

4 To the fourth interrogatory this deponent saith that he knowes the present Sir Solomon Swale & knew Roger Hillarye saith at severall tymes they were both reputed & called lords of Grinton & sometymes Sir Solomon would say he was lord & at other tymes the said Roger Hillarye / would say he was lord but as to the workeing of mynes & getting of lead on Harkaside & West Grinton this deponent know's nothing neither of their workeing or otherwise but saith he has heard the said Sir Solomon wrought & gott lead there /

5 To the fifth interrogatory this deponent saith that he did not know that ever any boundaries were ridd betwixt Grinton Harkaside or West Grinton or Whitaside till about two yeares since that the present Lord Wharton caused a boundarye to be ridd betweene the / said places saith that he heard notice given of such rydeing in Reeth markett & beleives like notice was given in Grinton church yard after divine service but whether any notice of such rideing was given to the lord of Grinton this deponent know's not /

Richard Gibson a witness formerly examined on the parte of the complainants & now sworne & examined on the parte of the defendants saith /

4: To the fourth interrogatory this deponent saith that he knowes the present Sir Solomon Swale and alsoe knew his father and Roger Hillary and that they were in their times called the lords of Grinton saith that he never knew or heard that they or any of them / wrought any mines of lead or cole upon any of the moores or commons belonging to Harkaside or West Grinton; /

Anthony Hird a witness formerly examined on the parte of the complainants & now sworne & examined on the parte of the defendants saith /

1 To the first interrogatory this deponent saith that he hath a little house in Harkaside but claimes noe right of common upon any of the said moors waists or commons of Harkaside or West Grinton or in any of the moors or commons / in or belonging to the mannor of Helaugh; /

3: To the third interrogatory this deponent saith that he knowes the places called Harkaside & West Grinton & the moores & commons thereunto adjoyning called Harka Hill Harka Moore Grinton Moore Grinton Greets / Grinton How or How Groves; & has soe known the same for thirty five yeares or thereabouts; & dureing that tyme the present Lord Wharton the late Lord Wharton nor any of their ancestors did ever worke any myne or mynes for either / lead or coale in any of the said places moores or waists that ever this deponent knew or heard of saith he has lived all his tyme in Harkaside or Reeth /

4 To the fourth interrogatory this deponent saith he do's not know Sir Solomon Swale but beleives his servants agents & workemen wrought <for> {& gott} lead on Harkahill or Harkaside Moore for sometyme but how long this deponent know's not /

Ralph Addison a witness formerly examined for the complainants & now sworne & examined for the defendants /

1 To the first interrogatory this deponent saith that he has an estate of inheritance in Harkaside paying thereout yearly to the lord two pence but claimes noe right of common upon any of the moores or commons in West Grinton & Harkaside or in any other place within the mannors / of Grinton or Helaugh haveing before his examination released his right of common to the present Lord Wharton /

3 To the third interrogatory this deponent saith that he know's the places called Harkaside & West Grinton & the moores & commons called Harkahill Harka Moore Grinton Moore Grinton Greetes Grinton How or Howgroves & ha's soe knowne the same for fifty yeares & upwards & dureing this / deponents knowledge thereof

the late Lord Wharton the present Lord Wharton nor any of their ancestors did ever worke any myne or mynes for lead or coale within or upon any of the said places waists moores or commons or in any other place within the mannor of Grinton that / ever this deponent knew or heard of till within these two yeares that the present Lord Wharton opened the mynes in question /

4 To the fourth interrogatory this deponent saith that he know's the present Sir Solomon Swale & alsoe Roger Hillarye saith Sir Solomon was reputed lord of the mannor of Grinton & did worke for lead on Harkahill the place in question & had good gettings of lead there & continued workeing till the relator Mr Marriott / recovered the mynes from him /

Nicholas Buckle a witness formerly examined on the complainant Lord Wharton's behalfe & now sworne & examined on the behalfe of Mr Marriott & others defendants saith /

3 To the third interrogatory this deponent saith that he know's Harkaside & West Grinton & the moores & commons called Harka hill Harka Moor Grinton Moore Grinton Greetes Grinton How or Howgroves & ha's knowne the same for about {fifty} forty yeares & that he do's not know that the present Lord / Wharton or any of his ancestors did ever worke any mynes of lead or coale upon any of the said moores or commons till about two yeares agoe /

4 To the fourth interrogatory this deponent saith that Sir Solomon Swale wrought mynes of lead upon Harkahill or Harkaside Moore for some tyme <but know's not for what tyme> & beleives he gott considerable quantity's of ore there /

5 To the fifth interrogatory this deponent saith that he never knew any boundarye rid betwixt Grinton West Grinton or Harkaside or the moores or commons thereof till about two yeares agoe by the present Lord Wharton /

James Blaides a witness formerly examined on the behalfe of the complainant the Lord Wharton & now sworne & examined on the behalfe of Mr Marriott & others defendants saith /

1 To the first interrogatory this deponent saith that he hath a house & lands lyeing at Blaides in the mannor of Helaugh & claimes right of common upon the moores & commons belonging to the mannor of Helaugh /

William Kirton a witness formerly examined for the complainants & now sworne & examined for the defendants /

3 To the third interrogatory this deponent saith that he know's the places called Harkaside & West Grinton & the moores waists & commons thereunto belonging & ha's soe knowne the same for sixty yeares last past & saith for the tyme aforesaid the present Lord Wharton the late Lord Wharton / nor any of their ancestors did ever worke any myne or mynes for lead or coale within or upon of the said places moores waists or commons in Harkaside or West Grinton or in any other place within the said mannor of Grinton till about two yeares last past that ever this deponent heard of /

4 To the fourth interrogatory this deponent saith that he know's the present Sir Solomon Swale & did know Roger Hillarye saith the common report was that Sir Solomon had a right to Harkaside Moore to worke there for lead /

5 To the fifth interrogatory this deponent saith that about two yeares since he with the Lord Whartons agents & others ridd a boundery betweene West Grinton Harkaside & Grinton but in all this deponents tyme the Lord Wharton nor any of his ancestors did ever ride the said boundery before / that ever this deponent knew or heard of /

Anthony Fryar a witness formerly examined on the complainants the Lord Whartons behalfe & now sworne & examined on the defendants behalfe saith /

3 To the third interrogatory this deponent saith that he know's Harkaside & West

Grinton & Harkahill <or> Harkaside Moore & ha's knowne the same betwixt forty or fifty yeares saith that the present Lord Wharton nor any of his ancestors did ever worke any mynes of lead or coale upon the said Harkaside / or Harkaside Moore or any of the moores & commons belonging to Grinton till about two yeares agoe that ever he knew or heard of /

4 To the fourth interrogatory this deponent saith that the present Sir Solomon Swale wrought mynes of lead upon the said Harkahill or Harkaside Moore & had good gettings there (as this deponent ha's heard) & enjoyed the same till Mr Marriott recovered the same against him & ha's heard / that Mr Marriott wrought alsoe mynes of lead there till the Lord Wharton entred upon him about two yeares agoe /

[membrane xvij *verso*]

Matthew Harrison a witnes formerly examined for the complainants & now presented sworne & examined for the defendants /

1 To the first interrogatory this deponent saith that he is a freeholder in Reeth within the mannor of Helaugh but claimes noe right of common within the said mannor of Helaugh haveing released his common right there as alsoe his <claime of> common right in West Grinton / & Harkaside unto the Lord Wharton /

3 To the third interrogatory this deponent saith that for forty yeares last the present Lord Wharton nor any of his ancestors did ever worke any myne or mynes {of} <for> lead or coale upon West Grinton or Harkaside Moore or in any parte thereof that ever this deponent <knowes or> heard of till about 2 yeares since /

4 To the fourth interrogatory this deponent saith that some tymes it was said that the present Sir Solomon Swale was lord of Grinton & at other tymes that Roger Hillarye was lord but whether they or either of them were lords this deponent knows not saith that / he ha's heard that Sir Solomon wrought mynes for lead on Harkaside Moore saith that he saw lead gott there in Sir Solomon's tyme & the workemen told him it was for Sir Solomon /

5 To the fifth interrogatory this deponent saith that he has heard that about two yeares agoe the present Lord Wharton caused a boundery to be ridd betweene {the said} West Grinton & Harkaside & Grinton & the commons thereto belonging never heard of any other boundery ridd betweene the said <places> /

Roger Colvill esquire a witness formerly sworne & examined for the complainants & now for the defendants saith /

3 To the third interrogatory this deponent saith that he know's the places called West Grinton & Harkaside & the moores thereto belonging & hath alsoe knowne the same for forty years last past saith that dureing that tyme this deponent never knew or heard that the present Lord Wharton / nor any of his ancestors did ever worke any myne or mynes for lead or coale upon the said places moores or waists till about two yeares since /

4 To the fourth interrogatory this deponent saith that he knew old Sir Solomon Swale's & ha's heard that the Swale's wrought mynes for lead on Harkaside Moore /

5 To the fifth interrogatory this deponent saith that he hath heard that about two yeares since the present Lord Wharton caused a boundary to be ridd between the moores of West Grinton Harkaside & Grinton but never heard of any other boundary ridd betweene the said <places> /

John Lonsdaile a witness formerly examined for the complainants & now by affirmacion being a Quaker & haveing taken the affirmacion on the parte of the defendant saith as followeth /

1 To the first interrogatory this affirmant saith that he is an inhabitant within the

mannor of of Helaugh & is a coppyholder by haveing coppyhold lands both on the north & south side of the river Swale claimes noe common right within the said mannor / of Helaugh or in Harkaside Moore haveing released the same /

3 To the third interrogatory this affirmant saith that he know's the places called Harkaside & West Grinton & the moores & commons thereto belonging & ha's soe knowne the same for sixty yeares last past saith that dureing that tyme the present Lord / Wharton nor any of his ancestors did ever worke any myne or mynes for lead or coale upon the said places moores waists & commons that ever <he> this affirmant knows or heard of till about two yeares since /

4 To the fourth interrogatory this affirmant saith that he know's the present Sir Solomon Swale & did know Roger Hillarye but can't say that ever <he> this affirmant heard them or either of them called lords of Grinton saith that he has / heard & beleives that the present Sir Solomon Swale wrought mynes for lead on Harka or Harkaside Moore /

Francis Whipp of Reeth a witness formerly examined on behalfe of the complainant the Lord Wharton & now sworne & examined on the defendants behalfe saith /

1 To the first interrogatory this deponent saith that hee hath both house & lands lyeing at Reeth in the said mannor of Helaugh saith he claimes noe right of common upon any the moores or commons belonging to the said places haveing assigned the same / over to the Lord Wharton but he has now some sheep going upon the said commons [. . .] /

3 To the third interrogatory this deponent saith that he knowes Harkaside & West Grinton & the moores & commons thereunto belonging & hath knowne the same for fifty yeares & never knew or heard that the present Lord Wharton / or any of his ancestors wrought any mines of lead or cole upon the said moores & commons until two yeares agoe. /

4 To the fourth interrogatory this deponent saith that the present Sir Solomon Swale & his ancestors were reputed lords of Grinton and Roger Hillary alsoe claimed to be lord thereof saith the said Sir Solomon wrought mines of lead upon Harkaside Moore & / old Sir Solomon wrought upon Grinton Moore for lead & cole & the said Hillary wrought for cole upon Grinton Moore /

5 To the fifth interrogatory this deponent saith that hee never knew or heard of any boundary rid betwixt Grinton Harkaside & West Grinton till about two yeares agoe that he heard of such boundary rid by the Lord Wharton /

6 To the sixth interrogatory this deponent saith that all the freeholders tennants & inhabitants of Harkaside & West Grinton doe suite & service at the Lord Whartons courts at Helaugh /

Ralph Waggitt a witness formerly examined upon the behalfe of the complainant the Lord Wharton & now sworne & examined on the defendants behalfe saith /

3 To the third interrogatory this deponent saith hee know's the places called Harkaside & alsoe the moores where the mines in question lye and hath knowne the same for about twenty yeares & never knew or heard that the present Lord Wharton or any of / his ancestors dureing the time aforesaid wrought any mines of lead or cole upon the said places /

4 To the fourth interrogatory this deponent saith that he has heard & beleives the said Sir Solomon was in possession of the said moores & wrought for lead there till about eleaven yeares agoe that the said Marriott recovered the same against him and after / that the said deponent wrought in the said mines for the said Marriott about a fortnight when he fell sick & went off /

Henry Clarkson a witness formerly examined on behalfe of the Lord Wharton & now sworne & examined on the behalfe of the defendants saith

1 To the first interrogatory this deponent saith that hee hath coppyhold lands in the mannor of Helaugh for which hee claimes common right within the moores & commons belonging to the said mannor. /

/s/ Tho: Thomson
/s/ Tho: Gyll
/s/ Tho: Waite
/s/ Rob: Hilton

Orders[326]

Michaelmas term, 4 Anne, 1705

E127/24 f.178r – order no. 8[327]

24 October 1705 – Between the Queen's Attorney General at the relation of Reginald Marriott esquire plaintiff and Edward Hartley Anthony Wharton and others defendants. By English bill

Upon the mocion of Mr Dodd of councell with her majesty & the said relator informeing the court that the bill was exhibited against the said plaintiffs for quieting the relator in the possession of severall lead mines in question lyeing in Grinton Whitaside and Harkaside in the county of York and that the said defendants had disturbed the said relator in his said possession of the said mines as appeared by the affidavit of the said relator now read in court the said Mr Dodd prayed the injuncion of this court for quieting the said relator in such possession of the said lead mines as he had at the time of exhibitting of the said bill or for three yeares before It is this day ordered by this court that an injuncion under the seale of this court shall be forthwith awarded against the said defendants their servants and agents for restraineing them from disturbeing the said relator in such quiet possession of the said mines as he had at the time of exhibitting of the said bill and for 3 yeares before the exhibitting thereof. /s/ Eyre for the Queen[328]

E127/24 f.179r – order no. 16[329]

30 October 1705 – Between the Queen's Attorney General at the relation of Reginald Marriott esquire plaintiff and the honourable Thomas Lord Wharton defendant. By English bill

Upon the mocion of Mr Lechmere of councell with the defendant praying that the said defendant might have time till the first day of next terme to put in his answer to the plaintiffs bill and upon hearing of Mr Turner & Mr Scroope on the same side It is this day ordered by the court that the said relator doe shew cause on this day sevennight why the said defendant shall not have so much time to put in his answer as is desired.

326 The court orders have been transcribed from the entry books of orders (E127). The Latin preambles have been translated.

327 Common minute book, E161/64, f. 98r; Original order, E131/13, Mich 4 Anne, no. 8.

328 Orders were entered in the entry books of orders by the sworn clerk of the party in whose favour the order was made (Bryson 1975, pp. 146f). In this instance the sworn clerk acting for the plaintiffs was Thomas Eyre and for the defendants Thomas Harris (Bryson 1975, p. 192).

329 Common minute book, E161/64, f. 101v; Original order, E131/13, Mich 4 Anne, no. 16.

Easter Term, 5 Anne, 1706

E127/25 f.109r – order no. 261[330]

9 May 1706 – Between the Queen's Attorney General at the relation of Reginald Marriott esquire plaintiff and the honourable Thomas Lord Wharton and others defendants. By English bill

Upon mocion of Mr Dodd of councell with the plaintiff praying liberty to ad[d] the honorable Phillip Wharton esquire a party defendant to the said bill with proper charges which this court doth thinke reasonable and hereby order the same accordingly.

/s/ Eyre for the plaintiff.

Trinity Term, 5 Anne, 1706

E127/25 f.156r – order no. 121[331]

28 May 1706 – Between the Queen's Attorney General at the relation of Reginald Marryott esquire plaintiff and Thomas Lord Wharton Edward Hartley Anthony Wharton and others defendants. By English bill

Whereas by an order of this court made in this cause the 24th day of October last It was ordered by this court, an injunction under the seal of this court should be forthwith awarded against the defendants their servants and agents for restraineing them from disturbing the said relator in sure quiet possession of the mines in question as he had at the time of the exhibiting his said bill or for three years before the exhibiting thereof Now upon the mocion of Sir Edward Northey knight her majesty's Attorney Generall on the behalfe of her majesty and the said relator informing the court that an injuncion have issued under the seale of this court pursuant to the said order directed to the said defendants and every of them their & every of their servants and agents and that James Close senior James Close junior John Sheldon Ralph Metcalfe James Carter James Clarkson Luke Robinson John Hird Nicholas Buckle Simon Harker William Wholley James Hird Stephen Whitell Stephen Hancock Anthony Hird John Williamson Robert {Robertson}Robinson John Burras and James Close the youngest agents and workmen imployed by the defendants or some of them had been duely served with the said injuncion but refused to give obedience to the same and still continued to disturbe the said relator in the possession of the mines in question and that some of them spoke slightly and disrespectfully of the said injuncion as appeared by the affidavits of William Bodinson Thomas Pratt Henry Place Matthew Sisson and Jacob Bayne now read in court her majesty's said Attorney Generall moved that attachments might be awarded against the said James Close senior James Close junior John Sheldon Ralph Metcalfe James Carter James Clarkson Luke Robinson John Hird Nicholas Buckle Simon Harker William Wholley James Heard Stephen Whitell Stephen Hancock Anthony Hird John Williamson, Robert Robinson John Burras and James Close the youngest for disobeying the said injuncion and hearing Mr. Ettrick Mr. Dodd Mr. Browne Mr Phipps and Mr Ward of the same side and Mr Letchmere of councell with the said defendants It is this day ord'red by this

330 Common minute book, E161/65, f. 29r; Original order, E131/13, Easter 5 Anne, no. 261.
331 Common minute book, E161/65, f. 39v; Original order, E131/13, Trinity 5 Anne, no. 121.

court that writts of attachment shall be awarded under the seal of this court against the said James Close senior James Close junior John Sheldon Ralph Metcalfe James Carter James Clarkson Luke Robinson John Hird Nicholas Buckle Simon Harker William Wholley James Hird Stephen Whitell Stephen Hancock Anthony Hird John Williamson Robert Robinson John Burras and James Close the youngest for breach of the said injuncion as desired unless cause be shewd to the contrary on this day fortnight and if the plaintiffs or defendants make use of any new affidavits the same are to be brought into court and filed the Saturday before the end of the terme.

/s/ Eyre for the Queen

E127/25 f.211v – order no. 316[332]

19 June 1706 – Between the Queen's Attorney General at the relation of Reginald Marriott esquire plaintiff and the honourable Thomas Lord Wharton Edward Hartley Anthony Wharton and others defendants. By English bill

Whereas by an order of this court made the 28th day last it was ordered for the reasons therein mencioned that writts of attachment should be awarded under the seale of this court against James Close senior James Close junior John Sheldon Ralph Metcalfe James Carter James Clarkson Luke Robinson John Hird Nicholas Buckle Simon Harker William Wholley James Hird Stephen Whitell Stephen Hancock Anthony Hird John Williamson Robert Robinson John Burrowes and James Close the youngest for breach of the plaintiffs injuncion in this cause unless cause should be shewd to the contrary on that day fortnight and if the plaintiffs or defendants made use of any affidavits the same were to be brought into court and filed & the Satturday before the end of the terme And whereas the said order of the 28th of May by an order of this court made the 11th of this instant June stood over to this day Now upon the mocion of Mr Lechmere of councell with the said James Close senior James Close junior and the rest of the persons aforesaid shewing cause against the said orders and upon reading of the said order of the 28th day of May last and of the affidavits of Nicholas Buckle Richard Close James Clarkson Alice Milner Christopher Hall Leonard Galloway John Sympson John Heard Anthony Heard Anthony Fryer John Sheldon Robert Robertson Francis Whipp and John Milner on the behalfe of the said James Close and the rest of the said persons and upon hearing of Sir Edward Northey knight her majestys Attorney Generall on the behalfe of her majesty and of the said relator and of Mr Serjeant Cheshire Mr Ettrick Mr Dodd Mr Browne and Mr Phipps on the same side and on reading the affidavits of Henry Place William Bodyson Thomas Pratt Matthew Sis[s]on and Jacob Bayne[s] on behalfe of the said plaintiffs and upon hearing of Sir Thomas Powys knight one of her majestys serjeants learned in the law Sir Symon Harcourt knight her majestys Sollicitor Generall Sir Joseph Jekyll knight one of her majestys serjeants learned in the law Sir James Mountague knight one of her majestys councell learned in the law and Mr Serjeant Pratt Mr Turner and Mr Lechmer[e] of councell with the said James Close senior and the rest of the persons above mencioned It is this day ordered by the court that the said orders of the 28th day of May and the 11th June against the said James Close senior and the aforesaid severall persons shall be and are hereby discharged and sett aside.

/s/ Harris for James Close senior et al

332 Common minute book, E161/65, f. 66r; Original order, E131/13, Trinity 5 Anne, no. 316.

Michaelmas Term 1706

E127/25 f.219v – order no. 9[333]

25 October 1706 – Between the Queen's Attorney General at the relation of Reginald Marryott esquire plaintiff and the honourable Thomas Lord Wharton and others defendants. By English bill

Upon the mocion of Mr Ward on the behalfe of the plaintiffs praying the said plaintiffs might be at liberty to reply to the defendant the Lord Wharton's answer without prejudice to the defendants that have not answered which this court doth think reasonable and hereby order accordingly.

/s/ Eyre for the Queen

Easter Term 1707

E127/26 f.100v – order no. 46[334]

2 May 1707 – Between the Queen's Attorney General at the relation of Reginald Marriott esquire plaintiff and the honourable Philip Wharton esquire and others defendants. By English bill

Upon the mocion of Mr Ward of councell with the plaintiff informeing the court that a commission of rebellion had been awarded against the defendant for not answeareing the plaintiffs said bill of complaint and upon reading the returne of the said commission of rebellion It is this day ordered by this court that the sergeant at armes attending the lord high treasurer of England take into custody the said defendant Phillipp Wharton and bring him into this court to answeare his said contempt and it is further order[ed] by this court that all mayors sherriffes constables and all other her majesties ministers and subjects whatsoever shall be aideing and assisting to the said sergeant at armes and his deputy in the execucion of this order

/s/ Eyre for the Queen

Trinity Term 1707

E127/26 f.242r – order no. 224[335]

16 July 1707 – Between the Queen's Attorney General at the relation of Reginald Marriott esquire plaintiff and the noble Thomas Earl Wharton and others defendants. By English Bill

Whereas by an order of this court made in this cause the 24th day of October 1705 It was ordered by the court that an injuncion under the seale of this court should be forthwith awarded against the defendants their sollictors and agents for restraining them from disturbing the said relator in such quiett possession in the mines in question as he had at the time of exhibiteing his said bill and for three years before the exhibiteing thereof Now upon the mocion of Mr Dodd on the behalfe of her majesty and the said relator informeing the court that an injuncion haveing issued under

333 Common minute book, E161/65, f. 85v; Original order, E131/13, Mich 5 Anne, no. 9.

334 Common minute book, E161/65, f. 158r; Original order, E131/14, Easter 6 Anne, no. 46.

335 Common minute book, E161/66, f. 33r; Original order, E131/14, Trinity 6 Anne, no. 224.

the seale of this court pursuant to the said order directed to the said defendants and every of them their and every of their servants and agents and that James Carter John Blaides junior William Whally John Borras James Close Anthony Hird Thomas Miller John Hird Thomas Harrison John Close and Joseph Harkey all of them being servants or workmen under the defendant the Earle Wharton had been duly served with the said injuncion but refused to give obedience to the same and still continued to disturbe the relator in his possession of the mines in question and spoke slightly and disrespectfully of this court as appeared by the affidavits of Edward Smithson esquire and William Bodinson esquire now read in court and upon reading the said injuncion the said Mr Dodd prayed attachments might be awarded against the said James Carter John Blaides junior William Whally John Borras James Close Anthony Hird Thomas Miller John Hird Thomas Harrison John Close and Joseph Harkey for disobeying the said injuncion It is this day ordered by this court that writts of attachment shall be awarded under the seale of this court against the said James Carter John Blaides junior William Whalley John Borras James Close Anthony Hird Thomas Miller John Hird Thomas Harrison John Close and Joseph Harkey as well for the breach of the said injuncion as for the foul and disrespectfull language in the said affidavits mencioned unless cause shall be shewne to the contrary the first day of the next terme and both sides are to file such affidavits as they shall make use of at the said time of shewing cause a week before the first day of next terme

/s/ Eyre for the Relator

Michaelmas Term 1707

E127/26 f.309v – order no. 115[336]

8 November 1707 – Between the most honourable Thomas Lord Wharton now Earl Wharton and the honourable Philip Wharton esquire now commonly called Lord Winchindon an infant by the aforesaid Lord Thomas Wharton his father and guardian plaintiffs and the Queen's Attorney General and Reginald Marriott esquire defendants. By English bill

It is this day ordered by the court that publication of the deposicions taken in this cause shall pass and this cause be putt into the paper of causes to be heard sometime in next terme unless cause be shewne to the contrary on this day sennight

/s/ Harris for the complainants

E127/26 f. 8r – order no. 227[337]

21 November 1707 – Between the Queen's Attorney General at the relation of Reginald Marriott esquire plaintiff and the most honourable Thomas Earl Wharton and the honourable Philip Wharton esquire defendants and between the aforesaid most honourable Thomas Earl Wharton and the honourable Philip Wharton esquire plaintiffs & the aforesaid Reginald Marriott esquire and others defendants. By separate English bills.

Upon the mocion of Mr Ward of councell with the said relator Mr Marriott who came to shew cause against the rule for publicacion passing and the cause being heard the next terme wherein the said Earle Wharton and his sonn the Lord Winchindon

336 Common minute book, E161/66, f. 53; Original order, E131/14, Mich 6 Anne, no. 115.

337 Common minute book, E161/66, f. 57r; Original order, E131/14, Mich 6 Anne, no. 227.

are plaintiffs and upon hearing Mr Ettrick Mr Dodd Mr Phipps and Mr Scroop of the same side and hearing Mr Sergeant Pratt Mr Bridges Mr Turner and Mr Letchmere of councell with the said Earle Wharton who being present in court declared he would not insist upon any priviledge in these causes It is this day ordered by the court by and with the consent of all partyes that the deposicions of witnesses taken in both the said causes shall be forthwith published and that these causes be putt into the paper of causes the last day of causes in the next Hillary terme to be adjourned to the first day of causes after the next terme to be heard together and all the said partyes do likewise consent to appeare gratis at the hearing and it is by the like consent further ordered by the court that the deposicions of witnesses taken in both the said causes [shall be read & made use of in either of the said causes][338] at the said hearing saveing just excepcions and it is further ordered by the court by like consent that the order of this court made the fifteenth day of July last for attachments against James Carter and others therein named shall be and is hereby continued over till the hearing of these causes as aforesaid and all proceedings thereupon are hereby stayed in the meane time And it is further ordered by the like consent that in case the court shall think fitt upon the hearing of the said causes to direct a veiw and a tryall that then such veiw shall be had in Hillary vacacion next and a tryall at barr shall be had in Easter terme after

/s/ Eyre for the Queen /s/ Harris for Earl Wharton & Lord Winchindon

Easter Term 1708

E127/27 f.148r – order no. 173[339]

14 May 1708 – Between the Queen's Attorney General plaintiff and the most honourable Thomas Earl Wharton defendant upon information.

Upon the mocion of Sir Thomas Powis knight her majestyes principall sergeant at law of councell with the said defendant praying that the bill of excepcions tendered in this cause may be allowed and sealed by the barons of this court and hearing of Sir James Mountague knight her majestyes sollicitor generall Mr Sergeant Hooper one of her majestyes sergeants learned in the law Mr Eyre and Mr Letchmere two of her majestys councell learned in the law and of Mr Sergeant Pratt Sir Symon Harcourt knight Mr Bridges Mr Turner and Mr King on the same side and on reading of the said bill of excepcions and hearing of Sir Thomas Parker knight one of her majestyes sergeants learned in the law on the behalfe of her majestyes Attorney Generall It is this day ordered by the court that the said bill of excepcions be left in the hands of the said Earle of Whartons clerke in court *de bene esse* and that coppyes thereof be delivered to the barons of this court without prejudice to either party

/s/ Harris for Earl Wharton

338 By an oversight, the passage enclosed in square brackets has been omitted from the order as enrolled in the entry book of orders and has been restored by reference to the original order.

339 Common minute book, E161/66, f. 119r; Original order, E131/14, Easter 6 Anne, no. 173.

Michaelmas Term 1708

E127/27 f.216v – order no. 22[340]

23 October 1708 – Between the Queen's Attorney General at the relation of Reginald Marriott esquire plaintiff and the most honourable Thomas Earl Wharton and others defendants. By English Bill

Upon the mocion of Mr Letchmere one of her majestys councell learned in the law of councell with the said Earle of Wharton praying liberty at the hearing upon the equity reserved in this cause to move for a new tryall and upon hearing of Mr Sergeant Pratt on the same side and of Mr Ettrick and Mr Dodd of councell with the relator Mr Marriott It is this day ordered by the court that the said Earle of Wharton shall be at liberty to move at the hearing upon the equity preserved in this cause for a new tryall as is desired and it is further ordered by the court by consent that both causes shall stand in the paper of causes to be heard on Thursday sennight and affidavits concerneing a new tryall are to be filed two days before on both sides.

/s/ Harris for Earl Wharton

340 Common minute book E161/66, 157v; Original order, E131/14, Mich 7 Anne, no. 22.

Decrees[341]

Hilary term, 6 Anne, 1708

E126/19 ff. 13v–14r[342]

25 February 1708 – Whereas Sir Edward Northey knight her majesties then Attorney Generall by and at the / relacion of Reginald Marriott esquire plaintiffs did in Michaelmas terme the fourth year of her now majesties / reigne exhibitt their English bill or informacion against the right honourable Thomas Lord Wharton / now Earle of Wharton and the Honourable Philip Wharton esquire now commonly called Lord Winchindon / Sir Solomon Swale and others defendants thereby setting forth That his late majesty King William the / third being seized in fee in right of his crowne of and in all those mines or veines of lead as well / open as to be found within the mannor of Grinton and the territories fields wasts and moores thereof / and particularly of Harkaside Whittaside and Grinton How and the moores and wasts thereof / or thereto belonging did by his lease under the seale of this court dated 18th of August in the / eighth year of his reigne grant the same premisses to George Tushingham gentleman for 31 years / at the yearly rent of twenty shillings and a tenth part of the clear profitts arising out of the said / mines That Sir Solomon Swale having gott into the possession of the said mines and claiming title / thereto the relator in the name of the said Tushingham about eight yeares agoe brought his ejectment / and upon tryall thereof recovered possession and afterwards there was a tryall at barr wherein the / relator had another verdict and had ever since worked the veines at Harkahill and Grinton How / and been in quiet possession thereof and paid the yearly rent and tenth to the crowne amounting / to very great summes of money That Tushingham was only a trustee for the relator and had / since assigned to him That the defendants had lately entred on Harka Moor Harka hill and Grinton / How and laid claime to the mines there and disturbed the crowne and relator and pretend / to be seized of the mannor of Helaugh and that the said Harka moore Harke hill & Grinton / How and all the mines and veines therein are parcell of the said mannor of Helaugh & not of / the mannor of Grinton and pretended to sett out and ride the boundaries of the said mannor of / Helaugh and have taken in a great part of the mannor of Grinton and the moores & wasts / of the same and particularly of Harkaside Whittaside Grinton How and Harkahill or part / thereof and laid claime to the said mines demised by the crowne and forbid the relators workmen / to work further therein and offered to indempnifie them if they would work for the defendants and though / they had not taken in the mouth of the mine then in working by the relator at Grinton How / yet they had {since} sunck a shaft upon the same veine and near adjoyning to the relators shaft / and designe thereby to disturb and disinherit the crowne and relator in their right That the mannor / of Grinton extends from the mannor or lordship of Ellerton on the east and Haverdale Beck / on the west and

341 The decrees have been transcribed from the entry books of decrees (E126).

342 A copy of this decree in the possession of Mr Lawrence Barker has been transcribed and published in the December issue of the *Memoirs of the Northern Cavern & Mine Research Society*, 1967, pp. 26–30; a second transcription was printed along with other documents presented as evidence in a late nineteenth-century Chancery case (Chancery Case 1898; NYCRO, ZYL, p. 187, mic. 2688).

from the river Swale on the north to and beyond the High Carle on the south / adjoyning to the Duke of Boltons mannor there and no part of the mannor of Helaugh does lye / within those boundaries but the crowne and relator are duely intituled to all mines within the same / and that for time imemoriall Harkaside Whittaside Harkahill or moor Whitta Somerlodge / and Grinton How are all within the Grinton mannor and within the territories and feilds of Grinton / and part of the wasts and moores of the same and were formerly part of the possessions of the late / dissolved monastrey of Bridlington in the said county of Yorke That the defendant the Earle of Wharton / under colour of some houses or freehold or coppyhold estate in or near Harkaside pretends his mannor / of Helaugh extends and takes in Harkaside and the wasts thereof and claimes the mines there and / endeavours to conceal and confound the boundaries of the said mannors of Grinton & Helaugh / whereas he nor any of his ancestors ever laid claime to any such boundaries or to have any / seignory or lordshipp within the said mannor of Grinton or the moores or wasts thereof or to any / mines open or not open in or upon the same or in Harkaside Whittaside Harka Whitta or / Grinton How untill two months before the exhibiting of the bill by rideing the boundaries / aforesaid wherefore that the defendants might answer and discover the premisses and that the / boundaries of the mannor of Grinton and Helaugh and the extent of the mines and veines leased / to the relator might be sett out and ascertained and the crown and relator quieted in the possession / and working of the said mines and the testimony of their antient witnesses preserved And the Attorney Generall on the behalfe of the crowne and relator be releived was the scope of the / said bill or informacion To which said bill or informacion the said defendant the Earle of Wharton / being served with a letter and his said sonn with process of this court appeared & answered / and the said Earl of Wharton by his answer said That Philip Lord Wharton his father was in his / lifetime and for sixty yeares before his death seized of & in the mannor of Swaledale and of / Helaugh alias Helaw alias Reyth in Swaledale within the antient boundaries whereof there were / as he beleived and doubted not to prove divers vills townshipps or hamletts with divers moores / wasts and commons thereto belonging and distinguished from each other and in particular the vill / of East Grinton West Grinton and Harkaside and that defendants father quietly held diverse demesne / lands rents services royalties priviledges and profitts in Harkaside & West Grinton and the / moores wasts and commons thereto belonging and that the tennants thereof paid their rents & performed / their suites and services at the courts held for the said mannor and that the coppyhold tennants who held / by coppy of court roll of the mannor of Helaugh in Swaledale or Reyth as also the freehold and / lease hold tennants did all along enjoy right of common or other profitts upon the moores and wasts of / Harkaside and West Grinton separate and distinct from the moores wasts and commons of / East Grinton That the defendants father on or {about} before defendants marryage setled the said premisses on defendant / for life remainder to his first and every other son in tayle male and That the said Philip Wharton / is his only son and intituled to the remainder of the premisses after the defendants death under the said settlement / and the defendants father dying seized about tenn yeares since the defendant did some short time after= / =wards enter upon the premisses and hath ever since enjoyed the same and hath heard that the vill of East Grinton has been sometimes by the owner thereof called a mannor Denyed that / to his knowledge or beleife the townes and places called West Grinton and Harkaside & the / moores and wasts there{of}to belonging and particularly the place where defendant caused a shaft to be sunk or / digged or any part thereof is a parcell of the mannor of Grinton or East Grinton but that the same / were and had been enjoyed by defendant and his ancestors and their tennants distinct from the said pretended / mannor of Grinton or East Grinton Knew not that the premisses in the informacion were part of / the possessions of the monastrey of Bridlington but beleived that monastrey was seized of the vill / of East Grinton and of some particular tenements within the said vill and also of right of common in / Whittaside Harkaside and Grinton How or some of them Knew not that King William the third was / seized of the mines or veines of lead in the informacion but beleives he made some grant of / diverse lead mines to Tushingham and referred thereto when produced beleived there was suites / between relator or his trustee and Sir Solomon Swale and that the relator had some verdict against Sir Solomon / but insisted that he being no party to such suites nor his right in question therein the same was / noe wayes affected thereby beleived after the said tryalls the relator proceeded to work the mines / and veines at Harkaside or Hill and the mine or veine at Grinton How and continued / to worke for some time on Harkahill but then desisted beleived the relator had not quiett / possession thereof nor knew what rents he had answered to the crowne Confessed that there being / divers mines and veines of lead within the moores and wasts of East Grinton and other places to which / the relator pretends title and he having wrought the mines called How Groves in and through / the wasts of Grinton called the How and very near to the utmost extent thereof where the boundaries / thereof divide the same from the wasts of Harkaside and West Grinton unto which defendant is intituled / did in affirmance and ascertaining his right give direcions to disturbe the relator from working / thereon and for rideing the defendants boundaries of Harkaside and West Grinton and afterwards his agents / did by his order sinck one or more shafts pitts or groves near the mouth of the mines working by / the relator at Grinton How and near the relators shafts there but said the same was sunck / within the bounds of the defendants estate That in rideing the boundaries they took in Harka Harkahill / and Harkaside moor but no other of the places in the informacion nor anything except the lands moores / and wasts of Harkaside and West Grinton to which he claimes title Knew not that Harkaside & West Grinton / and the moores and wasts thereof were part of the mannor of Grinton and beleives the boundary paper / contained no more then the mannor of Helaugh or defendants estate denyed the river of Swale was a / boundary either to the mannor of Swaledale or mannor of Helaugh and setts forth the boundaries / and doubted not to prove that the same extend as well to the south as to the north side of the river of Swale / but said he could not sett forth the boundaries of the mannor of Helaugh different and distinct from / the lordship of Swaledale nor what claimes his ancestors made to the mines of West Grinton or / Harkaside or either of them nor how oft or when they have dugg the mines nor had he dugg any since / the estate came to him but hopes the same should not prejudice his right he and his ancestors having / other valuable mines generally open in other parts of Swaledale or Helaugh and continuall enjoyment / of other profitts within the moores and wasts of Harkaside and West Grinton That defendant Philip Wharton / by his answer said he was an infant and was son and heir of the other defendant the Earl of Wharton and as / such had or might have some title or interest in the premisses as tenant in tayle or otherwise under some / settlement made by his father or grandfather and hoped the court would take care to preserve his right and / interest so as he might not be deprived thereof or debarred therefrom by any proceeding in the cause / submitted to the court That afterwards the defendants the Earle of Wharton and his sonn Philip Wharton / exhibited their bill against the relator Marriott and others to the effect of their answer And her majesties / attorney putt in his answer to the said bill and the said relator put in his answer thereto to the effect / of his bill and replicacions were filed in both causes and all parties rejoyned and the said causes / being at issue divers witnesses were examined on both sides in the said causes and their deposicions / duely published and the said causes being in the paper of causes came to be heard this present day in / Serjeants Inne hall in Fleet Street before the right honourable Sir Edward Ward knight lord cheif baron Sir / Thomas Bury knight Mr Baron Price and Mr Baron Smith three other of the barons of this court / whereupon opening of the said bills and answers by councell on both sides and on reading of / diverse records and of diverse deposicions of witnesses on both sides taken in the said causes & hearing / Sir Thomas Parker knight one of her majesties serjeants learned in the law Sir Edward Northey knight / Mr Serjeant Cheshire Mr Serjeant Wynne Mr Conyers Mr Ettrick Mr Dodd Mr Phipps Mr Scroope / and Mr Ward of councell with the said relator and hearing Sir Thomas Powis knight her majesties first / serjeant at law Sir James Montague knight her majesties Sollicitor Generall Sir Joseph Jekill knight Mr / Serjeant Hooper two other of her majesties serjeants learned in the law Mr Robert Eyre one other / of her majesties councell learned in the law Mr Serjeant Pratt Sir Simon Harcourt Mr Bridges / Mr Turner Mr King Mr Lechmere Mr Denton and Mr Bendloes of councell with the defendants / and on debate of the matter **It is** this day ordered by this court that the matters in question / be referred to a tryall at law on a feigned accion or informacion to be brought in this court in the / Queens Remembrancers office as of last terme in the name of Sir Simon Harcourt knight her / majesties late Attorney Generall against the defendant Thomas Earl of Wharton upon these two issues / (vizt) whether the place or places where the shafts lately sunck or wrought by the said Earle / of Wharton in the wasts or moores of Harkaside or West Grinton or any and which of them / be parcell of the mannor of Grinton or the mannor of Helaugh secondly whether the wasts / or moores of Harkaside or West Grinton or any and what part thereof are or is parcell of / the mannor of Grinton or the mannor of Helaugh unto which accion or informacion the said / defendant is to appear and plead so as issue may be joyned and ended as of the said last terme / and so as the same may be tryed at the barr of this court on Fryday the seaventh day of May / next by a speciall jury of the county of Yorke and to that end the sheriffe of the said county of / York is to bring in the freeholders book upon oath and attend the deputy remembrancer of this / court with the same who thereout is to name eight and forty and each side is to strike out / twelve and the remaining four and twenty are to be returned to try the said issues and the / deposicions of such witnesses taken in these causes as are dead or not able to travell are to be / read and made use of at the said tryall And it is further ordered by this court by & with the / consent of all parties that the jury shall in the meantime have a view of the places in question / and that John Hutchinson a surveyor named by the relator and Leonard Brackenbury / another surveyor named by the defendant the Earl of Wharton doe make a mapp of the whole estate / in question with all the known places or markes but without setting forth the claimes of either / party plaintiff or defendant and that the said mapp be delivered to the jury at the time of the view and that / the same view be had upon Thursday the two and twentieth day of Aprill next at tenn of the / clock in the forenoon of the same day by the six first jurymen or by so many of the said six / and as many more of the said jury as shall be present at the said view and they are to be / moderately entertained at the Queens Armes in Richmond at the equall charge of the relator / and the defendant And it is further ordered by the like consent that Sir Hugh Smithson barronet and / Thomas Dickenson gentleman on the part of the relator and Thomas Thomson gentleman and Thomas / Gill gentleman on the part of the Earle of Wharton doe shew the places in question to the jury at the time / of the view but no evidence is then to be given to the jury by either side and these causes are to be / continued in the paper of causes to be further heard after the said tryall be had as aforesaid.

/s/ Eyre for the relator By order of the Queen's Deputy Remembrancer

Michaelmas term, 1708

E126/19 f.74r[343]

4 November 1708 – Between the Attorney General ex rel. Reginald Marriott esquire plaintiff and the honourable Earl Wharton & others defendants and between the aforesaid honourable Thomas Earl Wharton and others plaintiffs and the aforesaid Attorney General and others defendants by separate English bills.

These causes standing in the paper of causes to be heard this day and upon hearing Sir James Montague knight her majesties Attorney Generall on the behalf of her majestie and Sir Edward Northey knight and Mr Ettrick for the relator It is this day ordered by this court That these causes stand over in the paper of causes to be heard this day sennight the first causes.

/s/ Eyre for the Queen By order Henry Stevens[344]

11 November 1708 – Between the Attorney General ex rel. Reginald Marriott esquire plaintiffs and the honourable Earl Wharton & others defendants and between the aforesaid honourable Thomas Earl Wharton and others plaintiffs and the aforesaid Attorney General and others defendants by separate English bills.

These causes standing in the paper of causes to be heard this day and upon hearing Sir James Montague knight her majesties Attorney Generall on the behalfe of her majestie Sir Thomas Parker knight one of her majesties serjeants learned in the law Sir Edward Northey knight and Mr Serjeant Wynn for her majestie and the relator It is this day ordered by this court that these causes stand over in the paper of causes to be heard on Monday next.

/s/ Eyre for the Queen By order Henry Stevens

15 November 1708 – Between the Attorney General ex rel. Reginald Marriott esquire plaintiffs and the honourable Earl Wharton & others defendants and between the aforesaid honourable Thomas Earl Wharton and others plaintiffs and the aforesaid Attorney General and others defendants by separate English bills.

These causes standing in the paper of causes to be heard this day now upon hearing of Sir James Montague knight her majesties Attorney Generall on the behalf of her

343 The following three decretal orders appear together on the same page of the entry book of decrees (E126).

344 Henry Stevens was deputy Queen's Remembrancer on 6 November 1707. He was appointed custodian of the office of Queen's Remembrancer of 23 October 1708 and held that position until 25 June 1709. From June 1709 until July 1713 he again served as deputy Queen's Remembrancer on several occasions (Bryson 1975, p. 189).

majestie Mr Conyer Mr Ettrick Mr Todd Mr Phipps Mr Raymond Mr Latwitch and Mr Ward for her majestie and the relator It is this day ordered by this court that these causes stand over in the paper of causes to be heard this day sennight the first causes.

/s/ Eyre for the Queen By order Henry Stevens

Appendix 1. Calendar of the Case

1705, Oct. 23	E207/111/1	Affidavit of Reginald Marriott.
1705, Oct. 24	E161/64 f.98r	Process re. issue of injunction.
1705, Oct. 24	E127/24 f.178r	Order. Injunction to quiet Marriott in possession for three years.
1705, Oct. 25	E112/95	Bill. EBOR 80. AG *ex rel.* Marriott v. Ld. Wharton.
1705, Oct. 30	E161/64 f.101v	Process. Ld. Wharton asks for time to answer.
1705, Oct. 30	E127/24 f.179r	Order. Ld. Wharton given time to answer.
1705, Nov. 6	E161/64 f.104v	Process. Ld. Wharton to answer the day before next term.
1705, Dec. 8	E161/64 f.131v	Process. AG permitted to amend his Bill.
1705, Dec. 9	E161/64 f.137v	Process. Counsel for Ld. Wharton given till 10 Jan. to file his cross bill.
1706, Easter	E112/915	Bill. EBOR 100. Ld. Wharton v. AG and Marriott.
1706, May 4	E112/915 80	Ld. Wharton's Answer to the AG's Bill + Schedule. Swearing date.
1706, May 9	E161/65 f.29r	Process. AG given leave to add the Hon. Philip Wharton's name to his bill as a defendant.
1706, May 9	E127/25 f.109r	Order. As above.
1706, May 24	E112/915 100	Answer. Marriott's answer to Ld. Wharton's bill. Swearing date.
1706, May 28	E161/65 f.39v	Process. AG v. Ld. Wharton – Affidavits of Matthew Sisson, Wm. Bodinson, Thos. Pratt and Henry Place read. Ld. Wharton v. AG and Marriott – five affidavits read. Attachment of 19 named individuals ordered. Affidavits to be filed.
1706, May 28	E127/25 f.156r	Order. Writs of Attachment of all those who swore against the injunction of 24 Oct. 1705.
1706, June 11	E161/65 f.57r	Process. Order of 28 May to stand over.
1706, June 12	E112/915 100	Commission. For the answer of Toby West, vicar of Grinton.
1706, June 19	E161/65 f.66r	Process. Affidavits of 28 May read. Injunction read. Ld. Wharton's answer read. Order of 28 May for attachment discharged.
1706, June 19	E127/25 f.219v	Order. Orders of 28 May and 11 June for the attachment of James Close and others set aside.
1706, June 19	E112/915 80	Answer. Sir Solomon Swale's answer to the AG's bill. Swearing date.
1706, June 20	E161/65 f.67v	Process. Counsel for Marriott proposes a trial next term.
1706, June 21	E161/65 f.69v	Process.
1706, June 25	E161/65 f.77r	Process. Wharton v. Marriott. Ancient witnesses to be examined. Affidavit of Mathew Smales read.
1706, June 26	E161/65 f.78v	Process. Commissions awarded to both Ld. Wharton and Marriott.
1706, June 26	E161/65 f.79v	Ld. Wharton permitted to amend his bill by adding Philip Wharton's name as a party.
1706, Oct. 25	E161/65 f.85v	AG and Marriott for permission to reply to Ld. Wharton's answer.

1706, Oct. 25	E127/25 f.219v	Order. AG. and Marriott permitted to reply to Ld. Wharton's answer.
1706, Oct. 25	E112/915 80	Replication of AG, Sir Edward Northey.
1706, Nov. 26	E112/915 100	Disclaimer of Toby West, vicar of Grinton. Swearing date.
1706, Dec. 4	E161/65 f.117r	Process. Return of a commission of rebellion read.
1707, May 2	E161/65 f.158r	Process. AG requests a Serjeant at Arms.
1707, May 2	E127/26 f.100v	Order. Commission of rebellion for the arrest of Philip Wharton for contempt in not answering the AG's bill.
1707, May 5	E112/915 80	Commission. For the answers of Philip Wharton.
1710, May 21	E194/6	Return of the Serjeant at Arms. Philip Wharton has not been found.
1707, May 23	E112/915 80	Answer of Philip Wharton. Swearing date.
1707, June 11	E112/915 100	Answer. Marriott's second answer to Ld. Wharton's bill.
1707, June 11	E112/915 100	Answer. Answer and Disclaimer of George Tushingham to Ld. Wharton's bill.
1707, June 14	E161/66 f.4r	Process. AG requests sequestration against Ld. Wharton. Marriott shall not be obliged to answer until Ld. Wharton answers his bill.
1707, June 23	E112/915 80	Replication of the AG, Sir Simon Harcourt.
1707, July 2	E134/6 Anne/Mich 38 & 44	Commissions to take depositions from Marriott's and Ld. Wharton's witnesses.
1707, July 11	E161/66 f.25v	Process. Both commissions to be executed together.
1707, July 16	E161/66 f.33v	Process. Injunction and affidavits of Ed. Smithson and Wm. Bodinson read. Attachment ordered of those who used foul language against the injunction.
1707, July 16	E127/26 f.242r	Order. Affidavits of Smithson and Bodinson read. Affidavits on both sides to be filed for hearing a week before the start of next term.
1707, Oct. 25	E134/6Anne/Mich 38 & 44	Depositions. Date of *liberatur*.
1707, Nov. 8	E127/26 f.309v	Order. Publication of depositions on both sides.
1707, Nov. 18	E161/66 f.53	Process. Counsel for Ld. Wharton proposes that both causes be heard simultaneously.
1707, Nov. 21	E161/66 f.57r	Process. Counsel for the AG and Marriott agree.
1707, Nov. 21	E127/26 f.8r	Order. Publication of depositions. Cause to be heard next Hilary term. Order of 15 July 1707 for the attachment of James Carter and others to be continued over and proceedings to be stayed meanwhile. View of the disputed territory to be had in Hilary vacation and a trial at Bar in the Easter term following.
1707, Nov. 28	E161/66 f.60v	Process. Counsel for Ld. Wharton assents to the last order.
1707/8, Feb. 12	E161/66 f.88v	Process. Both sides granted liberty to prove exhibits.
1707/8, Feb. 25	E126/19 f.13v	Decree. Court orders a trial on a feigned accommodation to be held before a jury of Yorkshire freeholders on 7 May. A map to be made of the disputed territory followed by a view on 22 April.
1707/8, Feb. 28	E161/66 f.102r	Process. Report of the Deputy Remembrancer read.

1708, May 7–10		Trial at the Bar of the Court of Exchequer.
1708, May 14	E127/27 f.148r	Order. Counsel for Ld. Wharton prays that the Bill of Exceptions be allowed. Ordered that the Bill of Exceptions be left in the hands of Ld. Wharton's clerk.
1708, Hilary 6Anne	E159/552	Verdict. Enrolled in the Queen's Remembrancer's Memorandum Roll. Verdict for the AG and Marriott.
1708, June 12	E161/66 f.133r	Process. Counsel for Ld. Wharton pleads that his Bill of Exceptions be allowed.
1708, June 18	E161/66 f.138v	Process. Opinion of the Court sought on the Bill of Exceptions.
1708, June 23	E161/66 f.144r	Process. Court rules that Ld. Wharton's Bill of Exceptions be not received.
1708, July 7	E161/66 f.152v	Process. Ld. Wharton's counsel moves for the verdict to be set aside.
1708, Oct. 23	E161/66 f.157v	Process. Ld. Wharton's counsel move for a new trial.
1708, Oct. 23	E127/27 f.216v	Order. Court permits Ld. Wharton to move for a new trial.
1708, Nov. 4	E126/19 f.74r	Decree. Cause to stand over.
1708, Nov. 11	E126/19 f.74r	Decree. Cause to stand over.
1708, Nov. 15	E126/19 f.74r	Decree. Cause to stand over.

Appendix 2. Lists of Witnesses
with interrogatories and counter-interrogatories answered

1. Witnesses appearing for Marriott – interrogatories answered

Interrogatory number				1	2	3	4	5	6	7	8	9	10	11	12	13	14	15	16	17
Addison, Margarett	Woodale	widdow	79		x															
Addison, Richard	Woodale	yeoman	39		x															
Alderson, Anthony	Cogden	gent	70	x	x	x	x		x	x	x		x	x	x	x				
Appleby, Bryan	Grinton	yeoman	28		x	x				x		x		x	x					
Arnold, George	Ellerton Abbey	gent	45	x							x	x								
Arundale, John	Harkaside		38		x		x						x	x	x					x
Ascough, Francis	Marrick	yeoman	70	x	x	x				x		x		x	x					x
Atkinson, Anthony	Hagg	yeoman	60		x															
Atkinson, Thomas	Fremington	yeoman	92		x															
Atkinson, William	Preston under Scar	miner	55	x	x										x					
Baley, Roger	Dycas	miner	60		x								x	x	x	x				
Bell, Edward	Whitaside	miner	36		x	x	x								x					x
Binks, Christopher	Waitwith		74		x										x					
Binks, William	Marsk	yeoman	82	x	x										x			x		
Blaides, James	Crackpott	yeoman	56		x		x			x			x	x	x					x
Blaides, William	Crackpott	yeoman	57		x									x						
Bland, Edward	Whitaside	miner	52		x	x	x		x	x	x	x	x	x	x					x
Bowes, Matthew	Ellerton	yeoman	50		x							x		x	x					
Bowes, Richard	Grinton	whitesmith	46		x	x				x		x	x	x	x					

Interrogatory number				1	2	3	4	5	6	7	8	9	10	11	12	13	14	15	16	17
Braithwaite, Mathew	Preston	yeoman	58	x	x	x									x					
Breckenfield, Francis	Fremington	miner	40		x	x				x		x		x	x					
Browne, Robert	Inghead	yeoman	86		x							x								
Buckton, Thomas	Oxnip	gent	59	x	x							x			x					
Carter, Ann	Stainton	wife	57		x															
Carter, Henry	Preston	mason	62	x	x										x					
Chambers, Christopher	Grinton	miner	50		x	x	x			x	x	x			x	x				
Chapman, Henry	Carperby	yeoman	77	x	x															
Clarkson, William	Feetham	yeoman	54		x	x	x			x	x	x	x	x	x	x				
Clerkson, John	Whitaside	husbandman	65		x	x	x				x	x		x	x					
Clerkson, John	Leyburn	yeoman	79	x	x															
Close, James	Whitaside	yeoman	60		x					x				x	x					
Close, Ralph	Richmond	butcher	57	x	x										x					
Close, Ralph	Whitaside	miner	80		x		x			x										x
Close, Simon	Red Hurst	yeoman	48		x							x								
Coates, Anthony	Marsk		72		x															
Coates, John	Grinton	miner	49		x	x	x		x	x	x	x	x	x	x					
Coats, Christopher	Aglethorpe	yeoman	53		x															
Cowling, George	Richmond	alderman	56	x	x															
Dawson, Christopher	Marsk		72		x															
Dunn, Alexander	Grinton	miner	50		x					x			x	x	x	x				
Ellerton, Edward	Downham	yeoman	83		x															
Emerson, Arthur	Grinton	miner	40		x	x						x	x	x	x	x				
Emerson, George	Grinton	miner	36		x	x	x			x	x	x	x	x	x	x				x

Interrogatory number				1	2	3	4	5	6	7	8	9	10	11	12	13	14	15	16	17
Etherington, Richard	Fremington		60	x	x	x	x			x		x	x	x	x	x				
Fawcett, Henry	Marrick	yeoman	63		x										x					
Fawcett, John	Marrick	gent	72	x	x	x						x		x	x					
Forster, Henry	East Bolton	yeoman	76	x	x										x					
Fothergill, Christopher	Reeth	miner	56		x	x	x			x	x	x	x	x	x					
Fothergill, James	Grinton	miner	49		x	x	x		x	x		x	x	x	x					x
Freer, John	Ellerton		33		x							x		x						
Fryer, William	Downham	yeoman	40		x	x	x			x	x	x	x	x	x					
Garth, John	Whitaside	yeoman	45		x		x			x			x	x	x					x
Garth, John	Low Spenceley	yeoman	91		x	x	x		x	x		x	x		x	x				
Garth, Richard	Crackpott	yeoman	55		x					x			x	x	x					
Garth, Thomas	Whitaside	yeoman	48		x		x			x			x	x	x					x
Greathead, Francis	Marrick	miner	51		x	x	x			x	x	x		x	x					
Greathead, Philip	Marrick		47	x	x	x				x		x		x	x					
Hall, Francis	Fremington	miner	45	x	x										x					
Hall, George	Fremington	miner	49		x	x	x			x										
Hall, James	Fremington	miner	42		x	x	x			x	x	x		x	x					x
Hall, Nicholas	Booze	gent	66		x	x														
Hall, Thomas	Fremington	miner	34		x									x	x					
Hamond, Richard	Preston under Scar	gent	66	x	x	x									x					
Harland, Bryan	Marsk	yeoman	88		x															
Harland, Roger	Downham	yeoman	65		x															
Haukin, Thomas	Grinton	miner	62		x	x						x	x	x	x	x				
Hawkins, Robert	New Forest	yeoman	40		x							x		x	x					
Haykin, John	Grinton		40		x	x				x	x	x	x		x	x				

Interrogatory number				1	2	3	4	5	6	7	8	9	10	11	12	13	14	15	16	17
Hemsley, Peter	Marrick	yeoman	64	x	x	x				x		x		x	x					
Hetherington, Christopher	Fremington	miner	57	x	x	x	x			x			x	x	x	x				x
Hibert, Alexander	East Bolton	yeoman	79		x										x					
Hillary, Ralph	Marrick	miner	56	x	x	x				x		x		x	x					
Hillary, Thomas	Arkingarthdale		40		x															
Hodgson, John	Fremington	miner	49		x	x	x					x		x	x					x
Hopps, Anthony	Marrick	husbandman	63		x															
Horseman, Mathew	Milfeild	yeoman	50		x	x						x		x	x					
Hudson, Anthony	St John Chapple	miner	42		x	x						x		x	x					
Hudson, John	Marrick	wood wright	66		x							x								
Hudson, Leonard	Fremington	miner	30		x	x	x			x	x	x	x	x	x	x				x
Hudson, Thomas	Marrick	yeoman	76	x	x							x		x	x					
Humphray, William	Wensley	yeoman	78		x															
Hutchinson, John	Skelton	felmonger	63		x	x						x		x						
Hutchinson, Metcalfe	Carperby	yeoman	77	x	x															
Hutchinson, Peter	Kirby Hill	blacksmith	64	x	x															
Jaques, Charles	Arkingarth Dale	miner	59		x									x	x					
Jefferson, Edmund	Catherick	yeoman	73		x							x								
Jefferson, George	Hudswell	yeoman	60		x							x			x					
Jefferson, Reynold	Marrick	smelter	38			x						x		x						
Jeofferson, William	Marrick	miner	65		x							x								
Kay, Christopher	Marrick	yeoman	58	x	x	x				x		x		x	x					
Kendall, John	Grinton	miner	42		x	x	x			x		x	x	x	x	x				x
Kendall, Marke	Grinton	miner	47		x	x						x		x	x					
Lamb, Edmund	Marrick	yeoman	72		x							x								

Interrogatory number				1	2	3	4	5	6	7	8	9	10	11	12	13	14	15	16	17
Lamb, Leonard	Marrick	cooper	68		x							x								
Lamb, Thomas	Marrick		41	x	x	x						x		x	x					
Lee, William	Grinton	miner	38		x	x	x			x	x	x	x	x	x	x				x
Lickley, John	Grinton	miner	73		x	x				x		x		x	x					
Loftas, Christopher	Fremington	miner	45		x									x	x					
Loftus, James	Bucktin in Craven	husbandman	48		x							x		x	x					
Loftus, John	Marrick	yeoman	57	x	x	x						x		x	x					
Maddison, James	Thorpe	yeoman	77		x							x								
Maitchel, Robert	Ridmer	yeoman	63	x	x										x					
Maltus, Robert	East House, Grinton		75	x	x	x	x		x	x	x	x	x	x	x					
Mason, Henry the elder	East Bolton	miner	59	x	x	x				x					x					
Mason, Henry the younger	East Bolton	miner	56	x	x	x									x					
Metcalfe, Brian	Carperby	yeoman	83	x	x															
Metcalfe, John	Grinton	waller	67		x	x	x			x	x	x	x	x	x	x				
Metcalfe, John senior	Carperby	yeoman	63	x	x	x														
Metcalfe, John the younger	Carperby	yeoman	47	x	x															
Miller, George	Harkaside	yeoman	60		x	x	x		x	x	x	x	x	x	x	x				x
Milner, John	Skelton	yeoman	72		x															
Musgrave, Samuell	Skelton		54		x									x						
Musgrave, William	Reeth Moore		65		x							x								
Nelson, Richard	Ellerton	yeoman	51		x							x	x	x	x					
Orton, Ralph	Hellwith	jagger	59		x							x		x	x					
Orton, Stephen	Thorpe	yeoman	46		x	x	x			x		x			x					x
Petty, Absolom	Fremington	yeoman	45		x	x	x			x	x	x	x	x	x	x				x
Place, Henry	Grinton	miner	43		x	x	x		x	x	x	x	x	x	x	x				

Interrogatory number				1	2	3	4	5	6	7	8	9	10	11	12	13	14	15	16	17
Place, James	Brunton upon Swale	yeoman	44		x															
Porter, Thomas	Redhurst		55		x							x								
Pratt, John	Carperby	yeoman	53	x	x															
Pratt, Thomas	Grinton	yeoman	56	x	x	x	x		x	x	x	x	x	x	x	x				x
Pratt, William	Grinton	miner	40		x	x	x		x	x		x	x		x					
Purchase, Benjamin	Midlam	gent	73	x	x	x	x	x		x	x		x	x				x		
Race, William	Fremington	miner	60		x	x	x			x	x	x	x	x	x	x				
Raw, James	Summerlodge	yeoman	54		x								x							
Raw, Richard	Fremington	yeoman	60		x															
Richardson, Robert	Craven Kell	miner	60		x										x					
Robinson, Francis	Fremington	yeoman	43	x	x	x	x		x	x		x	x	x	x	x				x
Robinson, John	Preston under Scar	gent & miner	62	x	x	x	x				x			x						
Robinson, Joseph	Woden	yeoman	36	x	x										x					
Robinson, Luke	Harkaside	mason	49									x								
Robinson, Martha	Harkaside	wife	47		x		x					x	x	x						x
Robinson, Ralph	Carperby	miner	50	x	x															
Robinson, Samuell	Fremington	yeoman	54	x	x	x	x			x		x		x	x					
Scott, Henry	Preston under Scar	yeoman	70	x	x															
Scott, Ralph	Whitaside	husbandman	66		x															
Simpson, John	Carperby	yeoman	83	x	x															
Simpson, John	Grinton	yeoman	68		x	x				x	x	x			x	x				
Snow, Roland	Nappa Scar	gent	56	x	x	x				x		x		x	x					
Spence, John	Midlam	miner	50		x	x						x		x	x					
Spenceley, George	Downham	yeoman	72	x	x										x					
Spenceley, James	Whitaside	rough mason	67		x	x	x		x	x		x	x	x	x	x				x

Interrogatory number				1	2	3	4	5	6	7	8	9	10	11	12	13	14	15	16	17
Spenceley, Leonard	Hallgaites	gent	49		x															
Spencely, William	Whitaside	miner	60		x	x	x		x	x		x	x	x	x					
Staveley, Roger	Wensley		51		x							x		x						
Stewardson, Nicholas	Fremington	miner	58		x	x	x				x	x	x	x	x					x
Stoddart, John	Fremington	miner	56		x	x				x		x		x						
Sympson, Anthony	Whitaside		66		x	x						x	x	x	x					
Sympson, Matthew	Whitaside	miner	38		x	x	x			x		x	x	x	x					x
Taylor, John	Ellerton	jagger	50									x								
Terry, John	York	gent	49		x							x			x					
Theakstone, Bartholomew	Preston under Scar	miner	66	x	x										x					
Thompson, James	Ridmer	yeoman	63	x	x										x					
Thwaites, Parcivall	Redhurst	miner	58		x	x						x		x	x					
Tirry, George	Ridmer	yeoman	76	x	x										x					
Tomlin, Nicholas	Midleham	miner	38		x	x	x			x	x	x	x	x	x	x				x
Waggett, Henry	Marrick	miner	65	x	x	x						x		x	x			x		x
Waggett, Michael	Downholme	clerk	55		x	x				x		x			x					
Waggett, Thomas	Marrick	yeoman	50	x	x	x				x		x		x	x					
Walker, Richard	Marrick	yeoman	52		x							x								
Waller, Robert	Dycas	yeoman	61											x	x	x				
Watson, Mathew	Marrick		53		x					x				x	x					
Watson, Thomas	East Bolton	miner	54	x	x	x														
West, Tobias	Grinton	clerke	58		x	x	x		x	x	x	x	x	x	x	x	x			x
Wheatley, Henry	Preston under Scar	miner	78	x	x										x					
Williamson, John	Feetham	miner	44		x															x
Willis, Charles	Marrick		73	x	x	x				x		x			x					

Interrogatory number				1	2	3	4	5	6	7	8	9	10	11	12	13	14	15	16	17
Willis, John	Marrick	yeoman	42		x															
Wilson, Martin	Carperby	yeoman	60	x	x															
Wilson, Thomas	Gilling	weaver	91	x							x									
Wood, Thomas	Grinton	yeoman	66		x	x	x	x		x		x	x	x	x	x				
Wood, Thomas	Grinton	miner	37		x	x				x		x	x	x	x	x				
Wynn, William	Carlton	yeoman	82		x							x		x	x					
Young, Thomas	Preston under Scar	miner	41	x	x	x									x					
Total				57	169	80	47	2	14	62	29	93	46	87	111	29	1	3	0	29

2. *Witnesses appearing for Marriott – counter-interrogatories answered*

Counter-interrogatory number	1	2	3	4	5	6	7	8	9
Purchase, Benjamin			x						
Alderson, Anthony			x					x	
Robinson, Joseph			x	x					
Maltus, Robert	x								
Haykins, Thomas	x								
Coates, John	x								
Garth, John		x	x						
Close, Ralph		x	x						
Raw, James			x						
West, Tobias	x	x	x	x	x	x	x	x	
Miller, George		x	x		x	x		x	

3. Witnesses appearing for Lord Wharton – interrogatories answered

Interrogatory number				1	2	3	4	5	6	7	8	9	10	11	12	13	14	15	16	17	18	19	20	21	22
Addison, Daniel	Helaugh	yeoman	52				x										x								
Addison, Ralph	Harkaside		65				x	x	x	x	x		x	x		x	x	x							
Alderson, George	Sturfitt Hall	yeoman	64	x	x		x			x							x	x							
Alderson, George	Helaugh	miner	34						x								x								
Alderson, James	Helaugh	yeoman	50				x																		
Alderson, Mathew	Reeth	yeoman	53	x	x		x	x	x	x			x	x		x	x	x							
Alderson, Roger	Reeth		51				x							x		x	x	x							
Alderson, William	Kirkby Ravensworth	miner	33				x																		
Arundale, William	Harkaside	yeoman	32	x	x		x			x	x		x			x		x							
Barningham, Christopher	Arkingarthdale	miner	56		x		x												x						
Blaides, James	Richmond		87				x																		
Blaides, John	West Grinton	yeoman	55	x				x					x												
Blaides, William	East Grinton						x	x	x	x	x		x		x									x	
Bland, Edward	Whitaside	miner	52						x																
Bland, John	Whitaside		65				x	x																	
Boneson, James	Sturtforth Hall	fisherman	70				x																		
Brackenbury, Leonard	Richmond	surveyor of land	56																			x			
Buckle, Margarett	East Grinton	widdow	90														x								
Buckle, Nicholas	Bowz	miner	50		x		x		x	x					x	x	x								
Buckle, Robert	Helaugh		55				x																		
Butterfield, George	Melbeck	yeoman	69				x	x	x																
Carter, Margarett	Whitaside	spinster	60				x										x								
Carter, Thomas	Kerton	taylor	43		x		x																		
Chalder, Francis	Reeth		68		x									x	x	x									
Charder, Margarett	Reeth	wife	64				x		x																

Interrogatory number				1	2	3	4	5	6	7	8	9	10	11	12	13	14	15	16	17	18	19	20	21	22
Chirry, Edward	Fetholme		63				x																		
Chirry, George	Fetholm	smelter	69				x																		
Chirry, Thomas	Fetholme	smelter	57				x																		
Clarkeson, James	Ivelett	yeoman	48	x	x		x	x	x	x	x		x		x			x	x						
Clarkson, Henry	Crackpott	husbandman	58				x											x							
Clarkson, Joseph	Satteron	yeoman	51		x		x																		
Clemetson, Ambrose	Low Raw	smelter	37		x		x																		
Clemitson, Mathew	Fetham	miner	54				x																		
Close, Anthony	Bankhead	yeoman	77				x																		
Close, James	Whitaside	yeoman	59	x	x		x	x	x	x			x			x	x								
Close, James	Whitaside	miner	49		x		x	x	x								x								
Close, John	East Grinton	miner	55				x	x	x	x			x				x								
Close, Miles	Harkaside	yeoman	56	x			x	x	x	x			x												
Close, Richard	Richmond	yeoman	70	x	x	x	x	x	x	x	x		x	x	x	x		x						x	
Colvill, Roger	Fremington		65				x																		
Cowpland, Nicholas	Reeth		45				x	x	x	x						x	x								
Dowson, John	Levell House	miner	66				x	x	x																x
Dunn, William	Grinton		55				x	x																	
Ellott, Ralph	Park Hall	yeoman	64				x										x								
Etherington, Joseph	Hartford	gent	39																				x		
Etherington, Richard	Fremington	miner	60														x								
Fothergill, Anthony	Feetholme	husbandman	58				x																		
Fothergill, Richard	Feetham	husbandman	40				x																		
Fothergill, Thomas	Kirton	miner	40				x																		
Frier, Cornelius	Low Raw	yeoman	39		x		x	x	x	x	x	x	x		x				x						
Fryer, Anthony	Low Row	yeoman	57		x		x	x	x																
Galloway, James	Reeth		65		x		x	x	x	x				x	x	x	x	x							
Galloway, James	Helaugh		34		x		x																		

Interrogatory number				1	2	3	4	5	6	7	8	9	10	11	12	13	14	15	16	17	18	19	20	21	22
Galloway, John	Helaugh		53						x																
Galloway, Leonard	Harkaside	husbandman	50	x				x					x												x
Garth, Mary	Whitaside	widdow	80				x																		
Gibson, Richard	Reeth	husbandman	70	x	x		x		x	x						x	x								
Gorton, James	Low Raw		31				x																		
Hall, Christopher	East Grinton		35				x	x	x																
Harker, Christopher	East Grinton	yeoman	42	x			x	x	x																
Harker, Edmond	Cracky Pott Hall		72		x		x																		
Harker, James	Satteron		78				x																		
Harker, John	Oxnup		73						x															x	
Harrison, Mathew	Reeth	yeoman	59				x																		
Harrison, Thomas	Harkaside	miner	68	x	x	x	x	x	x	x	x		x	x		x	x	x							
Harrison, William	Barugh	miner	56		x		x										x								
Hartley, Marmaduke	Middleton Tyas	gent	51				x			x															
Hird, Anthony	Harkaside	labourer	48				x	x	x	x						x									
Hird, Christopher	Whitaside		50				x	x	x																
Hird, James	Harkaside	yeoman	64	x			x	x	x	x			x			x									
Hodgson, Richard	Kersey Green		72				x		x																
Hutchinson, Arthur	Reeth	yeoman	50	x		x				x						x	x								
Hutchinson, Francis	Reeth	yeoman	54	x	x		x	x	x	x			x	x	x	x	x	x							
Hutchinson, James	Reeth	smith	59	x	x		x	x	x	x			x	x	x	x	x	x							
Hutchinson, Luke	Newsham	yeoman	38				x																		
Hutchinson, Richard	Rucroft	gent	39		x		x																		
Hutchinson, Thomas	Reeth	yeoman	35		x		x	x	x	x	x		x	x		x	x	x							
Hutchinson, William	Reeth		45				x	x	x	x				x		x									
Kirton, Anthony	Oxnop		72		x		x																		
Kirton, William	Helaugh		78		x		x																		
Liddell, James	Heggs	yeoman	44				x																		

Interrogatory number				1	2	3	4	5	6	7	8	9	10	11	12	13	14	15	16	17	18	19	20	21	22
Lonesdale, George	Helaugh	husbandman	47		x		x																		
Lonsdale, Christopher	Helaugh	yeoman	30				x																		
Lonsdale, James	Helaugh	stockinger	55				x										x								
Lonsdale, James	Helaugh	yeoman	23				x																		
Lonsdale, John	Helaugh	yeoman	72				x										x								
Matchell, Henry	Reeth		50	x	x		x	x	x	x			x	x		x	x	x							
Metcalfe, Anthony	Gunnerside	yeoman	60		x		x																		
Metcalfe, Dorothy	Reeth	spinster	64				x										x								
Metcalfe, George	Reeth	miner	68		x		x	x							x	x	x		x						
Metcalfe, Giles	Whitaside	yeoman	60	x	x		x	x	x	x	x		x			x	x								
Metcalfe, James	Thorpe under Stone	miner	54				x																		x
Metcalfe, James	Gunnerside	miner	55		x		x																		
Metcalfe, John	Gunnerside		50				x																		
Milner, Alice	Harkaside	wife	50						x								x								
Milner, James	Calvert House	yeoman	71				x																		
Milner, John	Harkaside	yeoman	70	x	x	x	x	x	x	x	x		x	x	x	x		x	x					x	
Milner, Thomas	Harkaside	yeoman	56	x			x	x	x	x	x		x			x	x	x							
Naylor, James	West Grinton		35	x	x			x	x	x			x					x							
Parkin, James	Pott Ing	husbandman	64				x																		
Peacock, John	Reeth	yeoman	54		x		x																		
Peacock, John	Spring End	husbandman	68				x																		
Peacock, Ralph	Reeth	husbandman	56				x																		
Plewes, William	Richmond	gent	68				x											x							
Pratt, James	Gunnerside	miner	54		x		x																		
Pratt, William	Fetham	husbandman	57				x																		
Raper, Anthony	Whitaside	yeoman	63	x	x	x	x	x	x	x			x	x		x		x						x	
Raper, Christopher	Harkaside	husbandman	70	x	x	x	x	x	x	x				x	x	x		x							
Raw, James	Low Raw	yeoman	58				x																		

Interrogatory number				1	2	3	4	5	6	7	8	9	10	11	12	13	14	15	16	17	18	19	20	21	22
Raw, Leonard	Reeth	yeoman	56				x			x								x							
Robinson, Christopher	Fremington	yeoman	60				x											x							
Robinson, James	Oxnup	yeoman	55		x	x	x			x			x	x			x	x							
Robinson, Robert	Harkaside	mason	69	x	x		x	x	x	x			x			x		x	x					x	
Scott, John	Reeth	husbandman	51		x		x	x	x	x			x	x	x	x		x							
Simpson, Giles	Winterings	husbandman	45				x																		
Simpson, John	Whitaside	husbandman	76	x			x	x																	
Simpson, Ralph	Feetholme	yeoman	79				x																		
Simpson, William	Helaugh	miner	63				x									x	x								
Spenceley, James	Whitaside	mason	67														x								
Spenceley, John	Spring End	yeoman	37				x																		
Spenceley, Thomas	Helaugh	husbandman	50				x																		
Stoddard, Robert	Reeth		62				x								x	x									
Stones, Hugh	Arkingarthdale		50						x																
Storey, George	Ivelett	husbandman	74		x		x	x									x								
Tiplady, James	Smarber		55				x																		
Waggett, Ralph	Reeth	miner	50				x							x				x							
Waller, Robert	Harkaside	yeoman	61					x																	
Waller, William	Helaugh	gent	60	x		x	x			x			x										x		
Wensley, Ralph	West Grinton	yeoman	56	x	x		x	x	x	x			x	x									x		
Wensley, Ralph	Riddings	miner	59				x																		
Wensley, Richard	Featham	yeoman	92				x																		
Wharton, Anthony	Feetham	miner	38				x																		
Wharton, George	Bankheads		40	x	x		x			x	x		x				x								x
Wharton, George	Low Spenceley	yeoman	63				x																		
Wharton, James	Low Raw	husbandman	65				x																		
Wharton, Thomas	Brockabanke	husbandman	68				x																		

Interrogatory number				1	2	3	4	5	6	7	8	9	10	11	12	13	14	15	16	17	18	19	20	21	22
Whipp, Francis	Reeth	husbandman	75		x		x			x				x			x								
Whipp, Isabell	Reeth	wife	80				x										x								
White, George	Gunnerside		58				x		x																
Wilkin, William	Helaugh	blacksmith	48				x										x								
Wilson, Luke	Kirkby Ravensworth	carpenter	55				x																		
Wray, Edward	Fremington	whitesmith	61				x																		
Total				27	45	8	125	43	46	39	12	1	29	20	14	29	39	26	6	0	0	1	1	6	4

4. Witnesses appearing for Lord Wharton – counter-interrogatories answered

Counter -Interrogatory	1	2	3	4	5	6	7
Addison, Ralph	x		x	x			
Alderson, Matthew	x		x	x	x		
Arundale, William		x	x	x		x	
Blaides, James	x						
Blaides, John		x	x	x		x	
Bland, John		x		x	x		
Buckle, Nicholas			x	x	x		
Charder, Francis	x		x	x	x		
Clarkson, Henry	x						
Close, Miles		x	x	x	x	x	
Close, Richard		x	x	x	x	x	
Colvill, Roger			x	x	x		
Fryar, Anthony			x	x			
Gibson, Richard				x			
Harker, Christopher		x	x	x		x	
Harker, John	x						
Harrison, Matthew	x		x	x	x		
Harrison, Thomas			x	x	x	x	
Hird, Anthony	x		x	x			
Hird, James			x	x	x		
Hutchinson, Francis	x		x	x	x		
Hutchinson, James	x		x	x	x		
Kirton, William			x	x	x		
Lonsdaile, John	x		x	x			
Maitchell, Henry	x						
Metcalfe, George		x	x	x	x		
Milner, John			x	x	x	x	
Milner, Thomas			x	x		x	
Naylor, James	x		x	x			
Raper, Anthony			x	x	x	x	
Raper, Christopher	x		x	x	x	x	x
Robinson, James	x		x	x			
Robinson, Robert			x			x	
Waggitt, Ralph			x	x			
Wensley, Ralph	x	x	x	x	x	x	
Whipp, Francis	x		x	x	x	x	

Appendix 3. Biographical Notes on Lawyers appearing for the two sides in the dispute

Lawyers appearing for Reginald Marriott

Chesshyre [Cheshire], Sir John (1662–1738), lawyer. Called to the bar Nov. 1689. Apt. serjeant at law June 1705. Apt. Queen's serjeant Nov. 1711 and re-apt. King's serjeant Jan. 1715. Knighted June 1713. (*DNB*)

Dodd, Sir Samuel (1652–1716), judge. Called to the bar in 1679. Apt. serjeant at law, Oct. 1714. Knighted Oct. 1714 and created chief baron of the Exchequer Nov. 1714. (*DNB*)

Ettrick, [William] (1651–1716), lawyer and politician. Matriculated Trinity College, Oxford, in 1666/7, aged 16. Called to the bar in 1675. MP for Poole, 1685–7 and Christchurch, Hants, 1690–1716. (*Alumni Oxoniensis*; *History of Parliament* 2002)

Northey, Sir Edward (1652–1723), lawyer and politician. Called to the bar May 1674. Knighted June 1702. Attorney General 29 June 1701 – 5 April 1707 and again 14 Oct. 1714 – 13 March 1718. MP for Tiverton 1710–22. (*DNB*; *History of Parliament* 2002)

Parker, Thomas, first Earl of Macclesfield (1667–1732), judge and politician. Called to the bar May 1691. Serjeant at law May 1705 and Queen's serjeant June 1705. Lord Chief Justice, March 1710. Created Baron Parker of Macclesfield, March 1716, and first Earl of Macclesfield, Nov. 1721. Lord Chancellor May 1718 – Jan. 1725. (*DNB*; *History of Parliament* 2002)

Wynne, Sir Richard (c.1652–1742), lawyer. Serjeant at law June 1705. Knighted 1719 (Baker 1984)

The following lawyers are also recorded as having appeared on the same side: Mr Brown, Mr Crawford, Mr Conyers, Mr Johnson, Mr Lutwyche, Mr Phipps, Mr Scroop and Mr Ward.

Lawyers appearing for Lord Wharton

Eyre, Sir Robert (1666–1735), judge and politician. Called to the bar in 1690. Apt. Solicitor General Oct. 1708 – May 1710. Serjeant at law 1710. Knighted May 1710. Justice of the Queen's Bench, May 1710 – Nov. 1723. Lord chief baron of the Exchequer 1723–5. Lord Chief Justice of the Common Pleas 1725. MP for Salisbury 1698 – 6 May 1710. (*DNB*; *History of Parliament* 2002)

Harcourt, Simon, first Viscount Harcourt (1661–1727), lawyer and politician. Called to the bar Nov. 1683. Knighted June 1702. Solicitor General June 1702 – April 1707. Attorney General 25 April 1707 – 12 Feb. 1708 and 16 Sept. 1710 – 16 Oct. 1710. Lord Keeper Oct. 1710 – April 1713. Lord Chancellor April 1713 – Sept. 1714. Created Baron Harcourt of Stanton Harcourt Sept. 1711 and Viscount Harcourt Sept. 1721. MP for Abingdon 1690–1705; Bossiney 1705–1708; Abingdon 1708–1709; Cardigan Boroughs 1710; and Abingdon 1710. (*DNB*; *History of Parliament* 2002)

Hooper, Sir Nicholas (1654–1731), lawyer and politician. Called to the bar 1678. Serjeant at law Nov. 1700. Queen's serjeant June 1702–14. King's serjeant 1714–31. Knighted June 1713. MP for Barnstaple 1695–1715. (*Alumni Cantabrigiensis*; Baker 1984; Sainty 1983; *History of Parliament* 2002)

Jekyll, Sir Joseph (1662–1738), lawyer and politician. Called to the bar 1687. Knighted Dec. 1697. Serjeant at law Nov. 1700. King's serjeant Oct. 1700. Queen's serjeant 1702–14. Prime serjeant Oct. 1714. Master of the rolls July 1717 – Aug. 1738. MP for Eye 1697–1713; Lymington 1713–22; and Reigate 1722–38. (*DNB*; *History of Parliament* 2002)

Lechmere, Sir Nicholas (1675–1727), lawyer and politician. Called to the bar 1698. Solicitor General 1714–15. Attorney General March 1718 – April 1720. Created first Baron Lechmere Sept. 1721. MP for Appleby 1708–10; Cockermouth 1710–17; Tewksbury 1717–21. (*DNB*; *History of Parliament* 2002)

Montague, Sir James (1666–1723), lawyer and politician. Called to the bar 1689. Knighted April 1705. Solicitor General 25 April 1707 – 21 Oct. 1708. Attorney General 21 Oct. 1708 – 19 Oct. 1710. Serjeant at law 1714. Baron of the Exchequer Nov. 1714–22. Chief baron of the Exchequer 1722–3. MP for Tregony 1695–8; Bere Alston 1698–1700; and Carlisle 1705–13. (*DNB*; *History of Parliament* 2002).

Powys, Sir Thomas (1649–1719), judge and politician. Called to the bar 1673. Knighted 1686. Solicitor General 1686–7. Attorney General 1687–8. Serjeant at law June 1702. Prime serjeant 1702–13. Queen's serjeant 1702–13. King's serjeant 1714–19. Justice of the Queen's Bench 1713–14. MP for Ludlow 1701–13. (*DNB*; *History of Parliament* 2002)

Pratt, Sir John (1657–1725), judge and politician. Called to the bar 1682. Serjeant at law Oct. 1700. Knighted Oct. 1714. Justice of the King's Bench 1714–18. Lord Chief Justice 1718–25. Interim Chancellor of the Exchequer Feb.–April 1721. MP for Midhurst 1711–14. (*DNB* ; *History of Parliament* 2002)

The following lawyers are also recorded as having appeared on the same side: Mr Bendloes, Mr Bridges, Mr Denton, Mr Ford, Mr King, and Mr Turner.

Appendix 4. Biographical Notes on those who claimed a right to mine for lead and coal on Grinton Moor, c.1650–1705

Hillary, Robert (d.1623), of Clifford's Inn, London, gentleman. In his will, Roger Hillary bequeathed 'all my freeholde lands which I bought of Mr Simpson with the royalties of the mannor of Grynton' to 'Roger Hillary my kinsman and servant'. Robert Hillary was born in Marrick where he also owned property at the time of his death and his links with both Marrick and Grinton were further acknowledged by bequests of twenty shillings to the poor of each place. Robert Hillary had bought what remained of Grinton manor from the vicar, Henry Simpson, and his son, Geoffrey, in 1622. The precise nature of his kinship with Roger Hillary remains to be determined. At the time of his death, Roger Hillary was married to Anne, daughter of Mr Peeter Havercampe. (Will of Robert Hillary, TNA PROB 11/142)

Hillary, Roger (dates unknown but still living in 1677), gentleman, inherited Grinton manor from Robert Hillary (q.v.) upon the latter's death in 1623. According to the depositions of numerous of Marriott's witnesses, Roger Hillary worked for coal on Grinton moor and for lead on Grinton How. Ann Hillary is named as Roger Hillary's wife in a nineteenth-century copy of an abstract of the title of Caleb Readshaw to the manor of Grinton. (YML, Hailstone MS, Box 5.41)

Hillary, Roger (dates unknown but still living in 1689), is named as the son of Roger Hillary senior (q.v.) in an indenture dated 6 July 1670. (YML, Hailstone MS, Box 5.41)

Hillary, John (dates unknown), was probably the son of Roger Hillary junior (q.v.). In 1705/6, he and his wife, Anne, sold one third of the manor of Grinton to Thomas Bateman and John Budgett as trustees for Reginald Marriott. The deposition of Thomas Pratt, dated August 1707, refers to John Hillary and Roger(?) Hillary as 'both of the citty of Yorke'.

Marriott, Reginald (d.1730), deputy auditor, then auditor, of fee farm rents in the Exchequer, 1689–1716. Under-teller to the receiver-general of excise 1697. In 1697, while employed as an auditor in the Exchequer, Marriott was involved in the fraudulent conversion of Exchequer bills to the value of £7000 but received a royal pardon after confessing his role in the affair. In May 1701 he was cited in articles of impeachment against Lord Somers to whom he and an accomplice had given inside information about undisposed fee farm rents in return for which they received grants to the value of £500. In 1696 he again used his position as auditor of fee farm rents to obtain a lease of the Grinton lead mines in the name of a nominee, George Tushingham. In 1711 Marriott was returned as MP for Weymouth and Melcombe Regis, a position that he occupied until 1715. Reginald Marriott was married to Dorothy, daughter of Thomas Pulleine Esq. of Carlton Hall, near Bedale, with whom he had one son, Hugh Marriott. Reginald Marriott died intestate at his home in Parson's Green, Fulham, in 1730. (*History of Parliament* 2002)

Paulet [Powlett], Charles, first Duke of Bolton (1630/1–1699), politician. Styled Lord St John, he was the son of John Paulet, fifth Marquess of Winchester (1598?–1675). After the death of his first wife, he married Mary, widow of Henry Carey and daughter of Emmanuel Scrope, Earl of Sunderland, who was heiress to large estates in Yorkshire. Charles Paulet represented Winchester in the parliament of 1660 and

Hampshire in the parliament of 1661. On the death of his father in 1675 he succeeded to the title of sixth Marquess of Winchester and took a seat in the House of Lords. A moderate Whig, he was named a privy councillor in 1679, a position he retained to the end of his life. In 1689 he was created first Duke of Bolton. In 1672 Charles Paulet bought the lease of the Grinton mines in the mistaken belief that it was a fee farm rent. However, on discovering that it was an expired lease rent he quickly surrendered it. There is no evidence to indicate that Paulet actually worked any lead mines within the bounds of Grinton manor during the brief period that he held the lease. Paulet died at his residence in Hampshire in February 1699 and was succeeded as second Duke of Bolton by his son, Charles Paulet (*c.*1661–1722). (*DNB*; TNA LRRO 3/85)

Swale, Sir Henry, second baronet (d. 1682, aged 43), of Swale Hall. On his marriage, his father, the first baronet, settled Swale Hall and lands at West Grinton on him by a deed dated December 1663. (Harrison 1879)

Swale, Sir Solomon, first baronet (1610–78), of Swale Hall and South Stainley, Yorkshire. Called to the bar 1635. Created baronet June 1660. Sheriff of Yorkshire 1670–71. As an ardent royalist, Sir Solomon had his lands confiscated during the Civil War though they were later restored to him. In 1649 he purchased Swale Hall and lands at West Grinton from an aged recusant cousin, 'Old' Solomon Swale. MP for Aldborough 1660, 1661–78. Overwhelmed by debt, he died in the King's Bench prison in 1678. (Harrison 1879; *History of Parliament* 1983)

Swale, Sir Solomon, third baronet (d. 1733), of Swale Hall. His fraudulent claim to be lord of a 'manor of West Grinton' involved him in several lawsuits involving the lead mines of Harkaside and West Grinton, all of which he lost. As a result he died in poverty.

Swale, Solomon (*c.*1570–1651+). Often referred to as 'Old' Solomon Swale. A Catholic recusant, in 1649 he sold his lands at West Grinton to his cousin, Sir Solomon Swale, the first baronet, in return for an annuity. Still living in 1651, aged 80.

Wharton, Philip, fourth Baron Wharton (1613–96), politician. Educated at Eton, on coming of age he inherited an estate with an estimated annual income of £8000. Three times married, he had eleven children of whom four sons and four daughters survived. In 1638 he was admitted to Lincoln's Inn. During the Civil War he was prominent on the Parliamentary side, remaining politically active after the Restoration. From the mid-1670s he was one of the leaders of the country party in the House of Lords. As a Presbyterian and anti-Catholic he welcomed the 1688 revolution and was appointed a privy councillor to William and Mary in 1689. Generally regarded as a paternalistic landlord, in the 1680s he took legal action in an attempt to increase the level of fines due from those of his tenants who paid enhanced rents. These tenants were otherwise referred to as the tenants of 'new lands' and were the tenants of that part of Healaugh manor which formally belonged to the Molineux and was bought by Philip Wharton in 1635 from Sir Thomas Vachell of Coley in Berkshire. Although Philip Wharton lost his case against the tenants, as a gesture of goodwill they paid some of his legal costs and agreed that in future such fines should be abandoned in favour of a fixed sum paid over a period of three years. Philip Wharton died in Hampstead in February 1696. He is commonly referred to as the 'old' Lord Wharton by deponents in the Exchequer case of 1705–8. (*DNB*; Fieldhouse and Jennings 1978, chap. 6)

Wharton, Thomas, first Marquess of Wharton, first Marquess of Malmesbury, and first Marquess of Catherlough (1648–1715), politician, was the third but eldest surviving son of Philip, fourth Baron Wharton (q.v.) and his second wife, Jane, daughter of Arthur Goodwin. Most of Thomas (Tom) Wharton's early life was spent on his father's estate at Upper Winchendon in Buckinghamshire. As a dissenter he was educated at home before spending three years in Paris and on the Continent with his younger brother, Goodwin, and his tutor. In September 1673 he married Anne, daughter and heir of Sir Henry Lee. The marriage brought him £8000 in cash and an income of £2000 a year. Anne Wharton died in October 1685 and in 1692 he remarried, to the Hon. Lucy, daughter and sole heir of Adam Loftus, Viscount Lisburne, who brought with her large estates in Ireland. In 1679 he successfully stood for election at Aylesbury as a knight of the shire for Buckinghamshire. On the death of his father, in February 1696, he became the fifth Baron Wharton with a seat in the House of Lords. Thomas Wharton was a leading member of the Whig party, a government minister and one of the leading politicians of his generation. As a nonconformist he opposed the policies of the Catholic King James II. He was made a privy councillor under William and Mary but was dismissed from this and his other posts on the accession of Queen Anne, in March 1702, who installed a Tory government headed by Lord Treasurer Godolphin. Thomas Wharton set up the machinery which led to the succession of a Hanoverian monarch and helped formulate the legislation which brought about the Union with Scotland, in 1706. For this he was rewarded with a peerage, becoming Viscount Winchendon and Earl of Wharton on 23 December 1706. On 4 December 1708 he was appointed Lord Lieutenant of Ireland, a post he occupied until October 1710. Lord Wharton died on 12 April 1715 and was succeeded by his son, Philip Wharton (q.v.). (*DNB*)

Wharton, Philip James, Duke Wharton (1698–1731), politician and spendthrift, was born in December 1698, the only son and heir of Thomas Lord Wharton (q.v.) and his second wife, Lucy (d. 1716). He was educated privately before eloping with Martha Holmes whom he married in March 1715 against his father's wishes. After his father was created Viscount Winchendon, in December 1706, Philip Wharton was styled Lord Winchendon. On his father's death in April 1715 Philip inherited his father's estates which were placed under the control of his mother and several of his father's Whig associates. Philip Wharton then went to France to complete his education. On his return he embarked on a political career but his financial profligacy and scandalous personal behaviour led eventually to his ruin and he died in 1731. (*DNB*)

Appendix 5. Court of Exchequer document classes searched

Documents with material relevant to the case

TNA:

E112	bills and answers
E126	entry books of decrees
E127	entry books of orders
E131	original orders
E134	depositions
E159	Queen's Remembrancer's memorandum roll
E161	common minute books
E194	Deputy Remembrancer's reports and certificates
E207	bille

Other classes of document searched

TNA:

E140	exhibits
E162	minute books
E192	private papers and exhibits
E195	miscellaneous reports
E214	modern deeds
E219	clerk's papers

Appendix 6. Translations of Latin Documents

Both Latin texts translated here are heavily abbreviated and contracted. Suspension marks are also used occasionally, especially in the text of the verdict, which is written in a late court hand. All these factors pose difficulties for the translator, particularly in deciding the tense or voice that was intended by the writer. It is also apparent that the clerk who prepared the final documents made occasional mistakes, both in grammar and orthography, and for this reason it has not always been possible unambiguously to determine the precise meaning of the original text. The translations that follow stick closely to the Latin sentence construction as written even though the result will no doubt strike the modern reader as occasionally awkward. Finally, it is legitimate to speculate as to how far the clerks who wrote out these texts were fluent in Latin at a time when its use as the language of formal record was less all pervasive than had previously been the case.

(a) The 1696 Crown lease to George Tushingham (TNA LR1/207)

William the third by the grace of God king of England Scotland France and Ireland defender of the faith etc To all to whom our present letters shall come Greetings Know you that we for and in consideration of the great costs and expenses to be expended in the future by our well beloved subject George Tushingham gentleman in asserting and defending our right and title to and in the mines and premises hereunder demised As much as for and in consideration of the rents and covenants hereunder mentioned and expressed on the part of the aforesaid George Tushingham his executors and assigns to be paid and performed And by the advice of the well beloved and faithful Sidney Lord Godolphin[1] our councillor Stephen Fox[2] knight Charles Montague[3] esquire our councillor and chancellor and sub-treasurer of our court of exchequer John Smith[4] esquire our councillor and Thomas Littleton[5] knight commissioner of our treasury **We have leased** granted and to farm let and by these presents for our heirs and successors we lease grant and to farm let to the aforesaid George Tushingham All those mines and veins of lead as well opened as to be found and hitherto not opened or wrought growing and being within the territories and fields of Grinton Whitaside and Harkaside[6] in our county of York parcel of the lordship or

1 Sidney, Lord Godolphin, first Earl Godolphin (1645–1712), politician; Lord of the Treasury 1679–84, 1687–90; first Lord of the Treasury 1684–5, 1690–96; Lord High Treasurer 1700–1, 1702–10; Privy Councillor 1680–9, 1690–1712 (*DNB*; *History of Parliament* 1983; Bryson 1975, p. 172).

2 Sir Stephen Fox (1627–1716), politician; Lord of the Treasury 1679–85, 1687–9, 1690–1702 (*DNB*; *History of Parliament* 2002).

3 Charles Montague (1661–1715), politician; Privy Councillor 1694; Commissioner of the Treasury 1692–4; Chancellor of the Exchequer 1694–9; First Lord of the Treasury 1697–9, 1714–15; Auditor of the Exchequer 1699–1714 (*DNB*; *History of Parliament* 2002; Bryson 1975, pp. 174–5).

4 John Smith (*c.* 1655–1723), politician; Privy Councillor 1695; Chancellor of the Exchequer 1699–1701 and 1708–10; Lord of the Treasury 1694–1701; Teller of the Exchequer 1710–12, 1714–23 (*DNB*; *History of Parliament* 2002; Bryson 1975, p. 175).

5 Sir Thomas Littleton (1647–1709), politician and lawyer; Lord of the Treasury 1696–7; Treasurer of the Navy 1699–1710 (*DNB*; *History of Parliament* 2002).

6 The Latin text of the lease is heavily contracted and abbreviated which has been the cause of some difficulty in determining its precise meaning, not only in respect of the present translation, but possibly in the past also. In the enrolled version (TNA LR1/207), the phrase here translated as 'within the

manor of Grinton in the aforesaid county and lately parcel of the possessions of the late monastery of Bridlington in the aforesaid county And all those mines and veins of lead as well opened or to be found and hitherto not opened or wrought growing and being within the territories and fields of Fremington in the said county of York parcel of the possessions of William late Marquis of Northampton[7] and also in any part or parcel in which we our heirs and successors might lawfully or have power to have dig or take lead be it or be they demised or not demised for a term of years or otherwise On condition that the said George Tushingham his executors and assigns from time to time during the term hereunder granted he or they will have covenanted and agreed with the tenants and farmers of the soil and grounds[8] in the said manor of Grinton Whitaside Harkaside and Fremington to them and their assigns to allow and give such a sum of money for the digging of the aforesaid mines existing within the pastures and lands in their tenure in proportion to the harm and detriment they will have suffered by reason of the same digging within their lands and pastures together with all buildings lands commons waters water courses and all other rights jurisdictions profits commodities and emoluments whatsoever in any way belonging or pertaining to the aforesaid premises or to their appurtenances or before this time with them usually demised leased used occupied reputed or enjoyed which premises (inter alia) were so demised to Humphrey Wharton[9] esquire by letters patent of our lord the late king Charles the first under the great seal of England bearing date the first day of May in the fourth year of his reign **To hold** from the Feast of the Annunciation of the Blessed Virgin Mary then last past until the end of the term and for the term of twenty one years from then next following and to be fully completed rendering yearly of and for the aforesaid mines of lead within the lands and fields of Grinton Whitaside and Harkaside aforesaid with their appurtenances twenty shillings and of and for the aforesaid mines of lead within the lands and fields of Fremington aforesaid with their appurtenances twenty shillings at the Feasts of St Michael the Archangel and of the Annunciation of the Blessed Virgin Mary to the receiver of the exchequer or to the hand of the bailiff or receiver of the premises for the time being by equal portions **Also** by these presents for our heirs and successors we have granted

territories and fields of Grinton Whitaside and Harkaside' appears as '*infra territor[ ] et campos de Grinton Whitaside et Harkaside*'. The same contracted form '*territor*[ ]' is likewise employed in the equivalent passage of a draft version of the lease prepared by the Surveyor General, Samuel Travers, in August 1696, which purports to quote Humphrey Wharton's lease of 1623 (TNA E367/3645). However, some, though not all, later instances of what may, or may not, have been intended as the same phrase have been rendered as '*terr*[ ] *et campos*' which is open to an alternative translation as 'the lands and territories' of Grinton Whitaside and Harkaside. Certainly there are a number of undoubted grammatical errors in the Latin lease, both in its draft form and as it was enrolled, which show the Latin text to be occasionally defective, and it is a moot point whether the phrases written as '*terr*[ ] *et campos*' and '*territor*[ ] *et campos*' are strictly synonymous or not. In the present translation the contracted form '*territor*[ ]' had been rendered as 'territories', and '*terr*[ ]' as 'lands', bearing in mind the possibility that by '*terr*[ ] *et campos*' might have been meant '*territor*[ ] *et campos*' throughout, or *vice versa*. This ambiguity could have far reaching consequences if by '*terr*[ ] *et campos*' was meant '*terrae et campos*', perhaps referring only to the *enclosed* fields of Grinton Whitaside and Harkaside as opposed to the unenclosed moors and wastes. Lawyers seeking to challenge the scope of Marriott's lease would be sure to exploit uncertainties of this kind.

7 William Parr (1513–71), first Marquess of Northampton. After his death, his estates, including the manor of Fremington, were escheated to the Crown and then leased by the Crown to Henry, Lord Scrope, and Arthur Phillips, in 1583. Tyler, Spensley and White 1995, 91.

8 Latin '*fundis*'.

9 Humphrey Wharton, of Gillingwood, Esq. (d. 1 Jan. 1635). Humphrey Wharton's lease of the mines of Grinton and Fremington, granted in 1623, was for a period of twenty-one years and was due to expire on 1 May 1649. Thereafter no new Crown lease was granted until August 1696.

to the aforesaid George Tushingham All and singular the arrears and mesne profits of all and singular the premises which from the year AD 1660 hitherto have accrued and fallen due and for and by whatever person or persons received and in the hands of whoever or whomsoever remaining **To have and to hold** all and singular the mines and premises above demised or mentioned to be demised with all their appurtenances to the aforesaid George Tushingham his executors administrators and assigns from the making of these our letters patent until the end of the term and for the term of thirty one years thereafter next following and to be fully completed and ended **Rendering** and paying therefore yearly to us our heirs and our successors from and for the aforesaid mines of lead within the lands and fields of Grinton Whitaside and Harkaside aforesaid with their appurtenances the yearly rent or sum of twenty shillings And of and for the aforesaid mines of lead within the lands and fields of Fremington aforesaid with their appurtenances the annual rent or sum of twenty shillings of lawful English money And also one full tenth part of the clear annual profits from time to time issuing growing falling due and arising of and out of all the mines of lead and the premises above demised accounting for the same yearly upon the oath of George Tushingham himself his executors or assigns before the auditor of the aforesaid county for the time being and paying in cash together with the aforesaid several rents twenty shillings to the hands of our receiver general of the premises for the time being at the Feast of St Michael the Archangel each year during the term aforesaid And moreover rendering and paying to us our heirs and our successors one full tenth part of all the aforesaid mean profits above granted which at some time henceforth during the aforesaid term should be recovered by the aforesaid George Tushingham his executors administrators or assigns to be accounted for upon oath before the auditor aforesaid as is previously stated And to be paid to the hands of the aforesaid receiver general of the county aforesaid within the space of six months next after the recovery and receiving of the same mean profits respectively And the aforesaid George Tushingham for himself his heirs executors administrators and assigns covenants and concedes to and with us our heirs and successors by these presents **That** he the aforesaid George Tushingham his executors administrators and assigns in digging and making a pit and pits in the aforesaid mines for the winning of lead in the same place will make and leave good and sufficient pillars (*columnas*) for the support of the same and will dig and work the aforesaid mines and pits in such manner and form that the aforesaid mines and pits will not fall or become worse by their neglect **And we will** and by these presents we grant on behalf of us our heirs and successors to the aforesaid George Tushingham his executors administrators and assigns that it shall be truly lawful for him and any of them from time to time during the term above granted to them at their pleasure newly to dig and make a pit and pits and trenches and watergates in any place above specified within the territories and fields of Grinton Whitaside Harkaside and Fremington in the aforesaid county for the greater advantage and profit of the aforesaid George Tushingham his executors and assigns at the cost and expense of him George Tushingham his executors or assigns together with good and sufficient wayleave and stayleave for all and all manner of carriages and carts on and over the territories and fields of Grinton Whitaside Harkaside and Fremington aforesaid as often as he or they shall please during the term aforesaid and also that it shall be truly lawful for them and any of them from time to time to seize take and have sufficient wood and timber within the woods of the lordship or manor of Grinton aforesaid by direction and survey of our auditor receiver surveyor steward and bailiff of our heirs and successors in that place for the time being or of any two of the same for the maintaining and supporting of the aforesaid mines and in all things necessary or appertaining to the same mines **Always provided** however that if the aforesaid

George Tushingham his executors administrators or assigns will not account upon oath before our auditor for the time being for the aforesaid premises for all the yearly profits of the aforesaid several premises in the form and manner above mentioned or if it should happen that the several yearly rents of twenty shillings or the aforesaid tenth part of the clear yearly profits of the same premises above mentioned or any part of the same will be in arrears or unpaid in part or in whole for the space of sixty days next after the feasts aforesaid which as is previously stated ought to be paid Or if the aforesaid George Tushingham his executors administrators or assigns will not account upon oath as is previously stated concerning the mean profits above granted and will not pay or cause to be paid to our receiver aforesaid one full tenth part of the same within the space of six months next after the recovery of the same respectively or if these our letters patent be not enrolled before our auditor of the premises aforesaid or his sufficient deputy for the time being within the space of six months next after the date thereof That then and from thenceforth in any such case this our present lease and grant shall be void and null notwithstanding anything in respect thereof in these presents to the contrary In witness of which thing these our letters we have made patent witnessed by the aforesaid well beloved and most faithful commissioners of our Treasury at Westminster on the eighteenth day of August AD 1696

by warrant of the commissioner of the Treasury
Russell[10]

(b) The verdict enrolled in the memorandum roll of the Queen's Remembrancer for Hilary term 1708 (TNA E159/552)

[membrane i *recto*]

Thus far the day to day business of the term of St. Hilary in the sixth year of Queen Anne

Ebor

Concerning a special verdict given at the bar of this court on behalf of the lady the present Queen upon a trial of the several issues joined between the said lady the Queen and Thomas earl Wharton upon an information exhibited by the attorney general of the said lady the Queen against the aforesaid earl touching a mine of lead and shafts [*puteis*] and groves [*fodinis*] in the aforesaid mine dug and wrought by the aforesaid earl in the moors and wastes in Harkaside and West Grinton in the aforesaid county here enrolled

Memorandum that Simon Harcourt knight the present attorney general of the lady the Queen who for the same lady the Queen in this behalf follows being present here in court on the twenty third day of January in this term for the same lady the queen here gives the court to be understood and informed that since on the first of January in the sixth year of the reign of the said lady now the Queen at Richmond in the aforesaid county a certain conference was had and moved[11] between the aforesaid

[10] Endorsed at the foot: Enrolled on the sixteenth day of December 1696 and executed by J.O [and] R.E.

[11] Needless to say, the 'conference' said to have taken place at Richmond on 1 January 1708 was entirely fictional.

present attorney general of the said lady the Queen and the right honourable Thomas earl Wharton concerning the metes bounds and limits between the manor of Grinton and the manor of him Thomas earl Wharton of Helaugh both of and concerning certain mines of lead and certain shafts or groves in and upon that mine lately dug or wrought by the aforesaid earl in the moors or wastes of Harkaside and West Grinton And upon that conference the same attorney general of the lady the queen kept affirming that the place and places in which and where the aforesaid shafts or groves were lately dug or wrought by the aforesaid earl Wharton in the moors or wastes of Harkaside and West Grinton were and are and each and all of them was and is parcel of the aforesaid manor of Grinton and not parcel of the aforesaid manor of Helaugh which affirmation indeed the aforesaid earl up till then wholly denied alleging that the place and places in which and where the aforesaid shafts or groves were lately dug and wrought by him the earl Wharton in the moors or wastes of Harkaside and West Grinton was and are and each of them were and is parcel of the aforesaid manor of Helaugh and not parcel of the aforesaid manor of Grinton And since the aforesaid earl Thomas then and there (namely) on the same first day of January in the sixth year abovesaid at Richmond aforesaid in the county aforesaid in consideration of twenty shillings paid to him then and there by him the attorney general of the lady the Queen took upon himself and to the same attorney general of the lady the Queen faithfully promised to pay to the same attorney general of the lady the Queen five pounds if the aforesaid places in which the aforesaid shafts and groves were lately dug or wrought by the aforesaid earl Wharton in the moors or wastes of Harkaside and West Grinton were and are parcel of the aforesaid manor of Grinton and if any such place or places and not all of them were parcel of the aforesaid manor of Grinton then to pay him five pounds for every such place so existing within that manor And the aforesaid attorney general of the lady the Queen in fact says that all the aforesaid places in which the aforesaid groves or shafts were lately dug or wrought by the aforesaid earl Wharton in the moors or wastes of Harkaside and West Grinton were and are and each of them was and is parcel of the aforesaid manor of Grinton And when also on the same first day of January in the aforesaid sixth year aforesaid at Richmond aforesaid a certain other conference had been held and moved between the aforesaid attorney general of the lady the Queen and the aforesaid Thomas earl Wharton concerning the metes bounds and limits between the aforesaid manor of Helaugh and the aforesaid manor of Grinton and of and concerning the moors or wastes of West Grinton and Harkaside And upon that conference the aforesaid attorney general of the lady the Queen kept affirming that the aforesaid moors or wastes of West Grinton and Harkaside and every part thereof were and are parcel of the aforesaid manor of Grinton and not parcel of the aforesaid manor of Helaugh which same affirmation the aforesaid earl then and there wholly denied alleging that those moors or wastes and every part of them were and are parcel of the aforesaid manor of Helaugh and not parcel of the aforesaid manor of Grinton And upon this the same earl then

[membrane i *verso*]

and there (namely) on the same first of January in the sixth year aforesaid at Richmond aforesaid in the county aforesaid in consideration of another twenty shillings then and there paid to him by the aforesaid attorney general of the lady the Queen took upon himself and faithfully promised to the same Attorney General of our lady the Queen to pay to him another five pounds if the aforesaid moors or wastes of West Grinton and Harkaside of any part of them were or was parcel of the manor of Grinton And the aforesaid attorney general of the lady the Queen in fact says that those moors or wastes were and are and every part of them was and is parcel of the aforesaid manor of Grinton aforesaid However the earl not caring for his several promises and assumptions aforesaid made in the form aforesaid but deviously and fraudulently intending callously and craftily to deceive and defraud the aforesaid attorney general in that behalf has not yet paid or in any manner satisfied the same attorney general the aforesaid several sums of money or any of money in that behalf Although for this the same earl afterwards (namely) on the same first day of January in the aforesaid sixth year abovesaid and often afterwards at Richmond aforesaid in the aforesaid county was requested by the same attorney general but he himself has thus far altogether refused to pay and still refuses whereupon the same attorney general for the same lady the Queen prays the advice of the court here in the premises and due legal process against the aforesaid earl in this behalf to proceed for the responding in the premises etc **Whereupon** it was agreed that an order be issued to the aforesaid Thomas earl Wharton by writ of the said lady the Queen under the seal of this Exchequer of being here etc to answer in the premises And here under a penalty of one hundred pounds which etc If not etc And the aforesaid same earl Thomas in the form aforesaid Thus etc On the third of February on which day the aforesaid Thomas earl Wharton by his attorney Thomas Harris comes and says that the aforesaid attorney general of the lady the Queen ought not to have or maintain his information aforesaid against him the earl because he says that it is good and true that he himself took upon himself in manner and form just as the aforesaid attorney general has alleged above but as to his promises and assumptions aforesaid first mentioned above the same earl says that the aforesaid places in which the aforesaid shafts or groves were lately dug or wrought by him the earl Wharton in the moors or wastes of Harkaside and West Grinton are and every part of them is parcel of the aforesaid manor of Helaugh and not parcel of the aforesaid manor of Grinton just as the attorney general of the lady the Queen has alleged above And about this matter he puts himself on the country And the aforesaid attorney general of the lady the Queen likewise etc And as to his promise and assumption aforesaid secondly mentioned the same earl says that the aforesaid moors or wastes of West Grinton and Harkaside are and every part of them is parcel of the aforesaid manor of Helaugh and not parcel of the aforesaid manor of Grinton as the aforesaid attorney general of the lady the queen has alleged above And about this he puts himself upon the country And the aforesaid attorney general of the lady the Queen similarly etc Therefore let there be an inquisition in that behalf ordered And the aforesaid sheriff of the county of York [was ordered] not to omit etc And to *venire facias*[12] here on the twelfth of February twelve free and lawful men of the body of his county aforesaid each of whom etc by whom etc And who [are] not etc to recognise[13] in the premises And the same day is given here to the aforesaid Thomas earl Wharton on which day the same Thomas earl Wharton should come here as before And the sheriff (vizt) Henry Iveson Esq returned here the aforesaid writ together with the list of names of jurors attached to the same writ which are on the file of writs executed on behalf of the said lady the Queen and others of the term of St Hilary in this year on behalf of him the Remembrancer in York And the jurors did not come Therefore the aforesaid sheriff [ordered] that he distrain upon jurors themselves for their lands etc So etc on

[12] A writ of *venire facias* requires the recipient to attend court (literally to 'come here').
[13] 'Recognise' in the legal meaning of to warrant.

the seventh of the month of May[14] And the same day was given here to the aforesaid earl Wharton On which day the same Thomas earl Wharton comes here as before And the sheriff (vizt) the aforesaid Henry Iveson returns here the aforesaid writ and orders that the aforesaid jurors be distrained for their lands etc Wherefore the issues etc just as appears in the lists sewn to the same writ which

[membrane ii *recto*]

are in the file of writs executed on behalf of the said lady the Queen for the Easter term of the seventh year of the reign of the said lady the Queen on behalf of this Remembrancer of York And the aforesaid jurors having been called came on the same day just as appears in the same list And thereupon public proclamation was made here in court for the said Queen just as is the custom that if anyone should wish to inform the serjeant at law of the said lady the Queen or the attorney general of the said lady the Queen or the aforesaid jurors concerning the premises or to challenge the jury he should come and be heard And Sir Thomas Parker knight serjeant at law of the said lady the Queen offered to do this himself for which there is a process to take the aforesaid inquisition by the jury now here appearing who to the truth of and in the premises declared elected tried and sworn as to the aforesaid promise and assumption in the aforesaid information first mentioned the aforesaid jury upon their aforesaid oaths say that all the aforesaid places in which the aforesaid shafts or groves were dug or wrought by the aforesaid Thomas earl Wharton in the moors or wastes of Harkaside and West Grinton were and every part of them was and is parcel of the aforesaid manor of Grinton and not parcel of the aforesaid manor of Helaugh And as to the aforesaid promise and assumption in the aforesaid information secondly mentioned the aforesaid jurors on their aforesaid oath furthermore say that the aforesaid moors or wastes of West Grinton and Harkaside were and are and every part of them was and is parcel of the aforesaid manor of Grinton and not parcel of the aforesaid manor of Helaugh

14 Friday 7 May 1708 was in fact the first day of the trial.

Index of Personal Names

Because of the frequency with which the names of the principal mineral lords occur in the texts it would serve no useful purpose to include them in the index. They have therefore been omitted from the index. Profiles of the mineral lords appear in Appendix 4.

Page references followed by (D) indicate depositions made by witnesses. The names of witnesses who gave evidence at the trial in London are marked with an asterisk (*).